How to Excel in Mathematics

How to Excel in Mathematics

Ashok Kumar Aggarwal

GenNext Publication

Published by

GenNext Publication
5, Ansari Road
Daryaganj, New Delhi-110002
Phone: 011-47034999, 9811692060
E-mail: gennextpublication@hotmail.com

Distribution Network
gyanbooks.com
India, USA, Canada, UK, Australia, France

ISBN: 978-93-5663-953-9 (PB)
"An Imprint of Kalpaz Publication"
First Published, 2023

Printed at: Gyan Press, Delhi.

HOW TO EXCEL IN MATHEMATICS
Author: ASHOK KUMAR AGGARWAL

ACKNOWLEDGEMENTS

I am an Engineering graduate of 1972 batch and a retired Central Civil Service officer. I had realized early during my school days that Mathematics is a subject that requires least effort and memorizing if different topics are understood at concept level and in turn the concepts are applied on one's own to solve different problems. This approach helped me in solving many unknown problems in the examinations. Sometimes, solution to certain problems will not occur even after considerable effort. But to my surprise, in most cases, the solution will become evident well after I gave up, sometimes after a good night rest. My conclusion was that the mind keeps working on such problems in the background and a mind at rest is a must to give your best.This understanding of mine got further enhanced as I progressed towards pre-engineering and engineering classes. I found that my approach to mathematics helped me in doing well in Physics as well. I also realized that this approach enhances your logical skills.

As our elder daughter, Kamna, reached secondary school level in late 1990s, she developed fascination for Biology and was very clear that she wanted to study biology but did not want to pursue MBBS course to be a Doctor as she did not like dissection. Since we did not have any knowledge of the field, we approached her school (Springdales, Pusa Road) counsellor, a senior teacher, who advised us that a new area of Bio-chemistry is coming up and Kamna should take Mathematics apart from usual subjects of Physics, Chemistry and Biology for non-medical Group. This combination was to be quite demanding but she was fine with it as she was able to pursue Biology. Kids going for private tuitions was quite common in those days. I wanted Kamna to adopt my approach to grasp fundaments and apply those on her own to Mathematics and Physics and offered her help if she gets stuck anywhere in these two subjects. She accepted the idea and decided not to go for any tuition saying that if she ever needs any help for Chemistry she would get it from her friend's mother who was a Chemistry teacher. This approach worked wonders and she secured in high 90s each in her CBSE examination for Physics, Mathematics and Biology. She went on to graduate in Bio-Chemistry and post graduate in Molecular Biology from Delhi University and did exceptionally well. She further secured fellowship to pursue PhD programme in Biological Sciences at a leading American University in her first attempt where she discovered a new gene as part of her PhD work and named it "Rudra". She is now a well-established scientist in the US. I believe her approach to the study helped her in a long way. I am grateful to her school and the counsellor teacher who guided her so appropriately at that crucial juncture of her study.

Our younger daughter, Aparna, also studied in the same school as her elder sister. As she reached secondary school level in early 2000s, she was quite clear that she wanted to study History. Once again since we were not clear what to do, we approached the same school counsellor teacher, who advised us that for pursuing History, she has to essentially study English and Geography. Apart from these three subjects, she can take Mathematics and Economics and though the school did not have such a combination for Humanities students, the schooladministration were willing to create it for her or any other student interested in that combination. This combination was to be quite demanding but Aparna was fine with any other subjects as long as she was able to pursue History. Only one more student went for that combination. The school adjusted their class time-table such that these two students studied English, History and Geography with their Humanities class, Economics with Commerce class and Mathematics with Non-Medical Group class. Hats off to such a progressive school.I once again offered to help her in solving any Mathematics problem she gets stuck. She too did not go for any tuitions and did exceedingly well. She secured 99% in Mathematics in CBSE Class XII Board examination which was a school record for a Humanities students for years to come. She went on to pursue graduation in History (Hons) at a reputed Delhi college. Thereafter, on advice of one of her college teachers and her own career choice, she gave up the idea of pursuing

History beyond graduation and started working towards getting into a reputed Management School for MBA. She did exceptionally well in CAT and a few other Management entrance examinations by securing over 99 percentile ranking. Despite doing exceptionally well in Mathematics at her school level, she had started believing that she was not very keen on numbers. Hence she decided to go for MBA in Marketing at a reputed Management School in Mumbai. She is a successful Management professional and has gradually moved towards financial sector and is currently running her own practice as Investment Consultant based in Mumbai.

I am thankful to both my daughters for believing in my approach to study in general and Mathematics in particular and proved it by achieving great heights in studies as well as in their respective professions.

I stopped working in the year 2012. I wanted to guide a few students voluntarily in my approach to Mathematics. To this offer, two students from my neighborhood, who could not afford private tuitions, came forward. Their names are Afzal Malik and Bhaumik Singh. They were both studying in 6th grade in a nearby Public School. I started giving them an hour for 5 days a week except when I was not in Delhi, which at times was quite long. I realized that the school teachers were virtually feeding them solutions to most of the problems and there was hardly any emphasis on testing students own question solving skills. There was also very little exposure to unknown problems. I gradually started exposing them to unknown and difficult problems and guided them to apply their mind and keep trying till they are able to reach a solution. Even if they failed, they had to share with me the efforts made by them. I continued to guide them till senior secondary level and they gradually realized the fruitfulness of making serious attempt to solve questions using their own skills. Ultimately both of them have been able to get into Engineering Colleges in year 2021 with very affordable fee. During about 7 years of interaction with them, idea of putting on paper my understanding of how one should learn Mathematics, came to my mind and germinated in starting to write this book. Majority of the book was ready during this period. I am thankful to Afzal Malik and Bhaumik Singh for motivating me to write this book.

I realized that the course content for Senior Secondary level Mathematics had changed a lot since even our daughters' student days. Hence I also started learning many new concepts and sought a lot of information/concepts from the web. I am thankful to many web portals like https://en.wikipedia.org , www.youtube.com, https://brilliant.org, https://mathworld.wolfram.com, https://www.cut-the-knot.org, https://amsi.org.au, https://www.handakafunda.com, https://cjquines.com, https://www.nycmathteam.org, http://jwilson.coe.uga.edu, https://www.studiestoday.com, https://www.cuemath.com, https://byjus.com, https://math.stackexchange.com, https://www.toppr.com, https://phys.libretexts.org, https://www.vedantu.com, http://www.blazartheory.com, https://www.brainkart.com, https://www.doubtnut.com, https://www.shaalaa.com, https://www.youtube.com/c/nehaagrawalmathematicallyinclined and https://mathemerize.com that I consulted off and on to learn certain new concepts and clearing many doubts. These portals are doing excellent work in offering this freely available study material.

Last but not the least, once draft of the book was ready, I earnestly requested my friend Prof. (Retired) of IIT Delhi, Hari Mohan Gupta ji to review the draft book. Even though he was quite busy, he found time to diligently go through the draft and made many valuable suggestions and comments. At the same time he appreciated contents of the book. But for his review, I may not have had the confidence and courage to get this book published. I am grateful to Prof. Hari Mohan Gupta ji for his extra ordinary help.

I am also thankful to Shri Rajender Vashisht, who painstakingly translated contents of my book from word format to printable format.

Last but not the least, I am so grateful to the love and understanding of my wife, Nirdosh Aggarwal, who never complained about the time spent by me in writing and reviewing contents of the book during last few years. At times I was busy for hours on certain days.

Ashok Kumar Aggarwal

PREFACE

This book is meant for school studentsof secondary and senior secondary level who are not able to do well in the subject of Mathematics despite putting in significant effort.

Mathematics builds a base that helps in advanced subjects like Engineering, Statistics, Economics, Commerce, Accounting, Molecular Biology and many more.

We will see as we go along that if we understand numbers and mathematical concepts rather than try to memorize everything, mathematics is fun. It is a subject that taxes your mind or memory the least, while yielding very good returns.

This book in no way is a replacement for text books in Mathematics. The student must go through the text book thoroughly with a view to understand different mathematical concepts. This book will help him/her grasp and recall those concepts more easily. The book will also cover many unorthodox ideas that will help develop students' mathematical and logical skills.

So, let us walk together and enjoy our journey called Mathematics. I assure every school student that excelling in Mathematics is not difficult. And this is based on my personal experience both during my study days as well as while guiding my children who did exceptionally well not only in Mathematics but also other subjects without any coaching.

The book contains Practice problems with Answers at end of every chapter. It is expected that the students will try to solve Practice Problems given at the end of different chapters on their own. In case they are not able to solve some of the problems, it is ideal to discuss with their friends or the class teacher and arrive at the solution.

However, in case the student wants to check the solution after making a serious attempt on his/her own, a separate book containing solutions to all Practice problems will be available soon for help. Enjoy solving these Practice Problems as well as many other unknown problems of different topics.

I would love to receive any feedback on my book from the students. Feedback can be sent at ashok4363@gmail.com.

March 2023 **Ashok Kumar Aggarwal**

CONTENTS

1

NUMBER SYSTEM

1.1 Numbers

Decimal number system is the most popular among all number systems. As we all know, numbers go from **0-9** in the decimal system and as we go beyond 9, the number in the **unit position** (extreme right) goes to 0 and the next position to the left of the unit position assumes value 1and this position is called **Tens** position. Thus the number after 9 is 10. Number 10 is called a two digit number and is the smallest 2-digit number.

This process continues till we reach the highest 2-digit number which is ninety nine (99). As we go one unit further, we have to move one more position to the left, which is termed as **Hundreds** position. Thus number after 99 is '100' which implies 1 in Hundreds position and 0 in Tens and Unit position. '100' is the smallest 3-digit number and '999' is the largest 3-digit number. With '1' more than '999', we go to '1000'- the smallest 4 digit number. The numbering system goes on and on like this.

Numbers can be positive or negative. A number line is depicted below:

-5 -4 -3 -2 -1 0 1 2 3 4 5

-- ---------------------------

Middle number is '0'. So numbers larger than 0 are positive (also written as +ve) and numbers smaller than 0 are negative (also written as -ve). Negative integers start with -1 and go on as -2, -3, -4 and so on. Thus -1 is larger than -2, -2 is larger than -3 and so on. This is a very important concept and must be understood very well.

1.1.1 Odd and Even numbers

If an integer can be split into two equal integer numbers, such numbers are called **Even numbers**. Examples are 2,4,6,8 etc. All numbers between two even numbers are **Odd numbers** and such numbers cannot be divided into two equal integers. Examples of odd numbers are 1,3,5.7 etc.

Note: Zero is an EVEN number.

1.1.2 Factorial

Factorial of an integer is defined as product of all integers starting with 1 up to the given integer. It is written as number followed by! For example $3! = 3\times2\times1 = 6$ and $5! = 5\times4\times3\times2\times1 = 120$.

1.1.3 Prime numbers:

Prime numbers are numbers that can be divided only by 1 or itself. Examples are 1,2,3,5,7,11 etc.

An important property of integer numbers is that they can always be expressed as a product of two or more prime numbers. Examples are: $6 = 2\times3$, $27 = 3\times3\times3$, $56 = 2\times2\times2\times7$ ***etc.***

1.1.4 Absolute value of numbers:

Absolute value of a number is the value of the number ignoring sign of the number. Thus absolute value of positive numbers is same as the number itself while absolute value of negative numbers is negative value of the negative number. We will talk about it later on as well.

1.2 Number system

In India, two number naming systems are popular for decimal numbers. These are Arabic system and International system. Both these systems are listed below for clarity.

International System	Numeric	Arabic System
One	1	One
Ten	10	Ten
Hundred	100	Hundred
Thousand	1000	Thousand

International System	Numeric	Arabic System
Ten Thousand	10000	Ten thousand
Hundred thousand	100,000	One Lakh
One Million	1,000,000	Ten lakhs
Ten Million	10,000,000	One Crore
Hundred Million	100,000,000	Ten Crore
Thousand Million	1,000,000,000	One Arab
Ten thousand Million	10,000,000,000	Ten Arab
Hundred thousand Million	100,000,000,000	One Kharab
One Billion	1000,000,000,000	Ten Kharab

Very large numbers are not easy to define (Examples- numbers larger than multiple billions or a number as ratio of two numbers - one very very large and the other very very small fraction with value close to zero - that cannot be counted or listed). Such numbers are referred to as **Infinity** and denoted by the *symbol* ∞. Any number divided by 0 is also referred to as infinity. ***A young boy once explained to me the term Infinity in a very interesting way. He said, take 1 and keep putting zeros after that till you get tired. That is the number Infinity. I found it quite apt.***

1.3 Managing Numbers

Everyone has to deal with numbers in Arithmetic. Speed and accuracy in dealing with numbers is a very basic need for doing well in Mathematics.

One needs to perform basic operations of addition, subtraction, multiplication and division on numbers. In order to improve speed of arithmetical operations, let us put decimal numbers in two categories- numbers 1 to 5 as they are and numbers 6-9 as 4 less than 10, 3 less than 10, 2 less than 10 and 1 less than 10 respectively. This process limits your mental count only up to 5 which is far simpler than counting up to 10. Let us look at a few examples:

1.3.1 Addition

If you need to add 5 and 8, 8 being 2 less than 10, it will be easier to count 2 less than 5, that is 3 and add 10 to make it 13. In conventional way, you will count 8 on your fingers and then 5 more to make it 13. Similarly, if you want to add 9 and 8, just subtract 1 from 8 and add 10, to make it 17. Practice this method and have fun in adding two single digit numbers. The same process can be thought of for adding two 2-digit numbers if the numbers are close to 100.

Single digit addition process described above will also help in speeding up conventional method of adding multi digit numbers.

To speed up and perfect your mental addition skill, attempt adding digits of Registration number of moving vehicles that you see in front of you when you are travelling. You will find that gradually your speed and accuracy of addition will improve.

Adding a positive and negative number amount to subtraction of the absolute value of the negative number from the positive number.

1.3.2 Subtraction

Similar process can be thought of for subtraction. For example if you need to subtract 7 from 12, 7 being 3 less that 10, add 3 to 12 and subtract 10 to arrive at 5. Similarly if you need to subtract 9 from 16, simply add 1 to 16 and subtract 10, to arrive at 7.

Try practicing this process for multi digit numbers and you will see significant improvement in your speed of subtraction.

Subtraction can also be thought about as addition of negative value of the number to be subtracted as mentioned above. This concept will be handy in Algebra.

1.3.3 Multiplication and Division

Multiplication is basically repeated addition. Thus multiplication is a simpler process of adding a number to itself multiple times. If we need to add a number say 6, four times to itself, we get the required value by multiplying 6×4. For example, $6+6+6+6 = 24$ and $6\times4 = 24$.

Division is the reverse of multiplication. Division is a process to split a number into multiple equal numbers depending upon the divisor.

In order to quickly multiply and divide numbers, it is essential to practice number Tables. This is one aspect on which I will lay extra stress. Every school student must practice and memorize number Tables 2-20, with each table up to 20. This implies that the student should be able to answer instantly products like 7×9, 4×15, 18×19, 14×17 etc. in a split second.

This task may sound very difficult but if a student decides to do it sincerely, he/she can do it in a couple of months. And believe me; having done it, you will reap its fruits all your life.

In the beginning, you can treat numbers like 7, 8, 9 etc., as 3 less than 10, 2 less than 10 and 1 less than 10 respectively. This would mean when you practice Table of 9, you keep adding 10 and counting backwards by 1. For example, $3\times9 = 27$; $4\times9 = 1$ less than $37 = 36$; $5\times9 = 46-1= 45$; $6\times9 = 55-1 = 54$ and so on. This process will be quite helpful for Tables like 17, 18, 19 etc. However, ultimately the tables have to be learnt by rote.

Another trick every student must learn is to assess the approximate answer of an operation to make sure that the actual answer seems correct. For example, if you multiply 19×13, he/she should be able to see that the answer will be less that 260 (20×13) by 13. Thus if answer is less than 250 but more than 240 (correct answer being 247) it will be correct. Another hint is that when you multiply 9×3, the product must end with 7 in the unit place, once again confirming that 247 is correct. Every student can devise his/her own methods of approximation.

Tables are even more important for process of Division. If you know your Tables well, you will waste very little time in optimizing quotient at every step of division.

1.4 Divisibility Rules

In many situations we need to know whether a large number is divisible by a specific prime factor. Actual division to know the answer could be a laborious process. Divisibility Rules have been established to quickly assess whether the given large number is divisible by a given prime factor. These Rules are quite handy and are listed below:

(Courtsey Dr. Stuart Savory, Reference document at www.savory.de/maths1.htm)

Divisibility factor	Method
2	The number should be even
3	Sum of all digits should be divisible by 3
4	Last two digits should be divisible by 4
5	Last digit should be 0 or 5
6	The number should be even and sum of all digits should be divisible by 3
7	Subtract 2 times the unit digit from the truncated number ignoring the unit digit. If the resulting number is divisible by 7, the original number is also divisible by 7. To be clearer about this process, we can say that if the given number is abcd. Let (abc - 2d) = xyz. If xyz is divisible by 7 then, abcd is also divisible by 7.

Divisibility factor	Method
	The process described above can be applied over and over again till it is easy to test. The process described above can be applied over and over again and tested. For example : to check divisibility of 51632, we go: 5163–2×2 = 5159; 515–2×9 = 497, that can be easily seen to be divisible by 7. Hence 51632 is divisible by 7.
8	Last three digits should be divisible by 8
9	Sum of all digits should be divisible by 9
11	The process is quite similar to divisibility test for 7 except that here we subtract unit digit (and not 2 times the unit digit) from the truncated number ignoring the unit digit. If the resulting number is divisible by 11, the original number is also divisible by 11. This process can also be applied over and over again.
13	Add 4 times the unit digit to the truncated number ignoring the unit digit. If the resulting number is divisible by 13, the original number is also divisible by 13. This process can also be applied over and over again.
17	Subtract 5 times the unit digit from the truncated number ignoring the unit digit. If the resulting number is divisible by 17, the original number is also divisible by 17. This process can be applied over and over again.
19	Add 2 times the unit digit to the truncated number ignoring the unit digit. If the resulting number is divisible by 19, the original number is also divisible by 19. This process can be applied over and over again.
23	Add 7 times the unit digit to the truncated number ignoring the unit digit. If the resulting number is divisible by 23, the original number is also divisible by 23. This process can be applied over and over again
29	Add 3 times the unit digit to the truncated number ignoring the unit digit. If the resulting number is divisible by 29, the original number is also divisible by 29. This process can be applied over and over again.
31	Subtract 3 times the unit digit from the truncated number ignoring the unit digit. If the resulting number is divisible by 31, the original number is also divisible by 31. This process can be applied over and over again.
37	Subtract 11 times the unit digit from the truncated number ignoring the unit digit. If the resulting number is divisible by 37, the original number is also divisible by 37. This process can be applied over and over again
41	Subtract 4 times the unit digit from the truncated number ignoring the unit digit. If the resulting number is divisible by 41, the original number is also divisible by 41. The process described above can be applied over and over again

Divisibility factor	Method
43	Add 13 times the unit digit to the truncated number ignoring the unit digit. If the resulting number is divisible by 43, the original number is also divisible by 43. The process described above can be applied over and over again
47	Subtract 14 times the unit digit from the truncated number ignoring the unit digit. If the resulting number is divisible by 47, the original number is also divisible by 47. The process described above can be applied over and over again

It is generally adequate to know Divisibility Rules upto 9. Rules for prime numbers 11 and beyond are for more curious readers.

How about multiplying and dividing positive and negative numbers. It is obvious that a positive (+ve) number multiplied by a +ve number results in a positive number. However it is important to remember that a +ve number multiplied by a negative (–ve) number **results in a negative** number but multiplication of two -ve numbers **results in a positive number**.Same is true for Division process. A +ve number divided by a +ve number results in a +ve number, a +ve number divided by a –ve number or a –ve number divided by a +ve number result in a –ve number while a –ve number divided by a –ve number results in a +ve number.

This concept can be a little tricky to remember for some of the students. An easy way to recall this concept is as below:

Assume that the +ve sign is your Friend and –ve sign is your Enemy. Now a Friend's Friend is a Friend (+ve×/÷ +ve = +ve), a Friend's Enemy is an Enemy (+ve ×/÷ –ve = –ve), and an Enemy's Enemy is (generally) a Friend (–ve ×/÷–ve = +ve).

Another important concept to remember is that (any number) multiplied by '0' = 0

1.5 Complex operations on numbers

Numbers can be expressed having multiple operations of multiplication, division, addition or subtractions. We make use of brackets to separate out certain operations. For example: 3 multiplied by sum of 5 and 2 can be expressed as 3× (5+2). Or 15 divided by sum of 3 and 2 as 15 ÷ (3+2). In these cases we need to assess value of numbers inside the bracket first before carrying out other operations.

We also have an operation ***'of'*** which is like multiplication. For example $\frac{1}{2}$ of 10 is 5, $\frac{3}{4}$ of 16 is 12.

For complex arithmetical operations we use a sequence of operations (from left to right) that is denoted by an easy to remember acronym called BODMAS. BODMAS sequence is as below:

B- Bracket

O- Of / Exponent (power of numbers)

D- Division

M- Multiplication

A Addition

S- Subtraction

This process will be clear from the following examples:

Example 1:

Solve: $3\{(-6)+20\}+(-7+9)(2)-12$

$=3\{-6+20\}+(2)(2)-12$ .. (We solve smaller brackets first)

$=3\{14\}+4-12$

$=42+4-12=34$

Example 2:

Solve: $\frac{1}{7} of\, 441 \div 18+14$

$=\frac{441}{7}\div 18+14$ .... (solved ***Of*** first)

$=\frac{63}{18}+14$ ....(solved Division next)

$=\frac{7}{2}+14=\frac{35}{2}$

1.6 Fractions

Fractions are used to denote numbers between whole numbers. You should have a clear understanding of a fraction. For example, if you have a fraction $\frac{3}{4}$, it means you have 3 out of 4 parts that make a whole. 3 being less than 4, this fraction is less than 1.Such fractions are called proper fractions. In this fraction, 3 is called numerator and 4 is called denominator. Thus in proper fraction, numerator is less than (<) the denominator. This will be clear from the following figure.

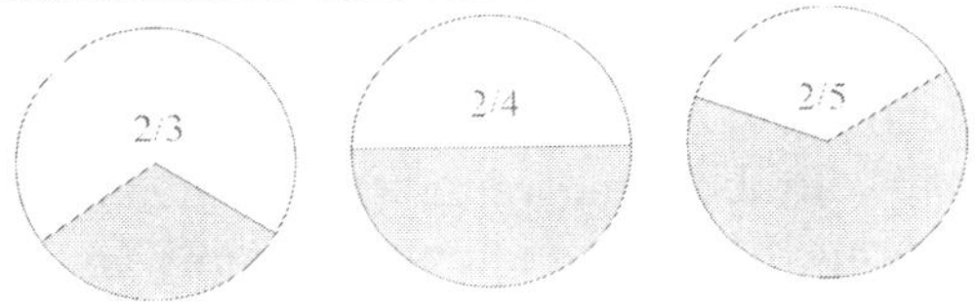

Denominator can also be understood to mean size of the parts of a whole. If denominator is say 3, size of parts is larger than if the denominator is say 5. Thus 2/3 is larger than 2/5.

In case, numerator is larger than the denominator but not divisible by the denominator, such fractions are called improper fractions as these are a mix of whole number and a proper fraction. For example, if the fraction is 5/4, you have 5 parts while 4 parts make a whole. Thus this fraction is more than 1. 5/4 can also be written as 1¼ (Illustration below).

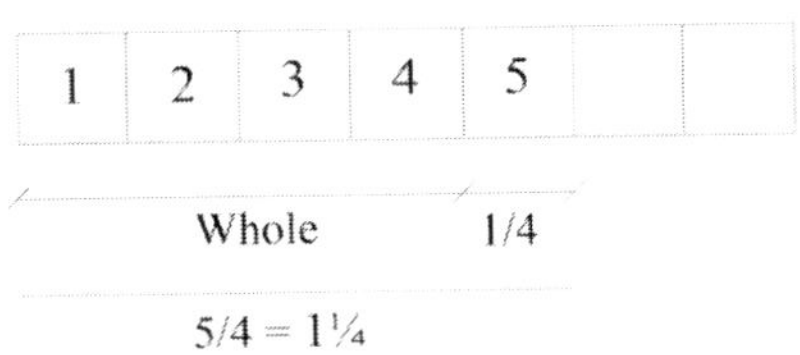

1.6.1. Rational and irrational numbers

If a number can be expressed as ratio of two integers, it is called a rational number. The word comes from "ratio". All other numbers are called irrational. A famous irrational number is π that is used for calculating area or perimeter of circles and other circular shapes. Often value of π is taken as $\frac{22}{7}$ for simplicity. This ratio equals 3.1428571428571….which is only an approximate value. Its real value accurate to over 30 decimal places is3.1415926535897932384626433832795...So don't get misled into believing that π equals $\frac{22}{7}$ and is rational. Another important irrational number is Euler's number 'e' that we will discuss later in this chapter.

1.6.2 Addition/ Subtraction of fractions

We can add and subtract two fractions easily if they have a common denominator. The resulting fraction is simply the sum/difference of the numerator with denominator being same. This is generally not the case. For adding/subtracting two fractions, with different denominators, their denominators need to be brought to the same value. This is achieved by finding Least Common Multiple (LCM) of denominators and suitably multiplying denominators and numerators so as to bring denominators (size of the parts) to the same value. Once the denominator is same in the two fractions, their numerators are simply added or subtracted to achieve the result.

For example, to add $\frac{1}{2}$ and $\frac{1}{4}$, we find LCM of denominators (2 and 4) to be 4. Now by multiplying numerator and denominator by 2, $\frac{1}{2}$ changes over to $\frac{2}{4}$ and $\frac{1}{4}$ remains unchanged. Sum of these two fractions is simply $\frac{1+2}{4}$ i.e. $\frac{3}{4}$. Difference of $\frac{1}{2}$ and $\frac{1}{4}$ will similarly be $\frac{(2-1)}{4}$ i.e. $\frac{1}{4}$. Same principle applies to all other fractions.

1.6.3 Multiplication/ Division of fractions

Multiplication of fractions is achieved by simply multiplying numerator and denominator and cancelling out any common factors in resultant numerator and denominator. For example multiplication of $\frac{1}{2}$ and $\frac{1}{4}$, results in value $\frac{1}{8}$.

Division being reverse of multiplication, the numerator and denominator of the divisor fraction are reversed and multiplied to the numerator and denominator of the dividend to get the result in the same way as multiplying two fractions. Thus $\frac{1}{2} \div \frac{1}{4}$ means $\frac{1}{2} \times \frac{4}{1} = 2$

From the above discussion, it is evident that if we divide a number by a number more than 1, the result is smaller than the number being divided. Conversely, if we divide a number by a number smaller that '1' (a proper fraction), the result is larger than the given number. For example, $3 \div \frac{1}{3} = 9$. Now for a given dividend, if divisor goes smaller and smaller, the resultant number keeps growing, so much so that if divisor approaches value close to '0', the result is extremely large known as ***Infinity*****. (Any number) ÷0, is another definition of Infinity.**

1.6.4 Fractions expressed in decimal form:

Fractions are more conveniently expressed in decimal form. In a proper fraction, since numerator is smaller than the denominator, if we divide the numerator by the denominator, we will get quotient as zero and remainder same as the numerator. Now we add a decimal point after quotient and add a 0 after the remainder resulting in a number ten times the remainder. This is now dividedby the denominator to get a quotient and the remainder. This process is continued till we get zero as remainder or we get to the desired number of places after the decimal point.

For example, $\frac{1}{2} = 0.5$, $\frac{1}{4} = 0.25$, $\frac{1}{25} = 0.04$ and $\frac{1}{3} = 0.3333....$ up to desired number of places

1.7 Operations with Decimals

1.7.1 Adding and Subtracting Decimals

Adding and subtracting decimals is easy if you're comfortable with carrying and borrowing. Just line up the decimal points, and (if necessary) add zeros to the end of one number, so they have the same number of decimal places. Then add (or subtract) as you usually would do for numerals.

In your answer, the decimal point should go exactly below where it is in the numbers being added (or subtracted).

Let us consider an example for clarity. Let us say we need to add 13.492 and 7.8. Since the first number has 3 places to the right of the decimal point, we need to add 2 zeros to the right of second number to make it 7.800. We then add as usual, as depicted below:

$$\begin{array}{r} 13.492 \\ +7.800 \\ \hline 21.292 \end{array}$$

Similar process is followed for subtraction.

1.7.2 Multiplying and Dividing with Decimals

1.7.2.1 Multiplying Decimals

To multiply decimals, first just multiply the numbers as if they were whole numbers without caring for the location of decimal points.

Then count the total number of places to the right of the decimal point in BOTH numbers you're multiplying. Let's call this number n. In your answer, start from the right and move n places to the left, and put a decimal point. Here is an example:

We need to multiply 1.2 by 0.9. As explained above, we will multiply the two numbers by ignoring the decimal point i.e we multiply 12 by 9 arriving at the value 108. Now if we count total number of places to the right of the decimal point together for the two numbers, its 2. Hence we place a decimal point two places from the right to arrive at 1.08 as the answer. So multiplying two numbers with decimal point is quite straight forward.

1.7.2.2 Dividing with Decimals

Dividing with decimals is a bit more difficult. Recall that in the division problem involving whole numbers, we need to assess value of the quotient and multiply the divisor by the quotient such that the remainder is less than the divisor. This process is continued till we exhaust the number in the dividend.

Process for dividing with decimals is exactly same except for the following:

If divisor is not a whole number, move the decimal place of the divisor ***n*** places to the right to make it a whole number. Then move the decimal place in the dividend the same number of places to the right (adding some extra zeros if necessary).

Divide as usual. If the divisor doesn›t go exactly, add a decimal point in the quotient and zeros to the right of the dividend and keep dividing until we get zero as remainder, or until a repeating pattern shows up or we reach the required number of places beyond the decimal in the quotient.

Check your answer against your estimate to see if it›s reasonable.

Let us look at an example to divide 2.7 by 0.36.

As discussed above, we move decimal point two places to the right in the divisor to make it 36 and also move decimal by two places in the dividend adding zero as necessary to arrive at 270. Dividing these whole numbers as usual and adding decimal point as necessary we get answer as 7.5

1.8 Exponents (or raised to the power) of a number

When we multiply a number repeatedly by itself, it is easy to represent it by a notation called exponent. For example $3\times3\times3$ can be represented as 3^3 and $1000 = 10\times10\times10$ can be represented as 10^3. Large numbers can be conveniently represented in exponential form. For example, 1 Lakh = 10^5, 1Million = 10^6 and 1 Billion = 10^{12}

Now let us quickly learn operations on exponents. As an example, $3\times3\times3\times3\times3 = (3\times3\times3) \times (3\times3)$ which can be represented as $3^5=3^3\times3^2$, meaning thereby that if two numbers with same base are multiplied,their exponents add. Another example $16\times4 = (4^2) \times (4^1) = 4^{2+1} = 4^3 = 64$

Similarly, if two numbers with the same base are divided, the exponents get subtracted. For example $125\div5 = (5^3)\div(5^1)=5^{3-1}= 5^2= 25$

If a number expressed in exponent form is raised to another power (squared/ cubed etc.), exponent of the number gets multiplied by power to which it is raised. For example $9^3 = (3^2)^3 = 3^{2x3}= 3^6= 729$. Similarly, root of a number expressed in exponent form results in exponent value getting divided by the root value, ex. exponent will get divided by 2 for square root, by 3 for cube root and so on. For example $\sqrt{81} = (81)^{1/2} = (3^4)^{1/2} =(3)^{4\times1/2}=(3)^2 = 9$.

On the other hand if two numbers with same exponents are multiplied, their bases get multiplied. For example $8 \text{ x } 27 = (2^3) \text{ x } (3^3) = (2\text{x}3)^3= 6^3= 216$. Same principle is true for division. For example $15^3\div3^3= (15/3)^3=5^3=125$

1.8.1 Special cases of exponent

1.8.1.1 Exponent as a fraction

We have already learnt that root of a number expressed in exponent form results in the exponent getting divided by the root value. Hence if we assume that number stand alone, it means that number with exponent 1 and square root of the number would mean (the number)$^{1/2}$, cube root as (the number)$^{1/3}$ and so on. At the moment think about value of repeated square root of a number. We will discuss it in some time for an important concept.

1.8.1.2 Exponent as a negative integer

We have learnt that in a division process, exponent of a dividend gets reduced by exponent of the divisor. Example, $(5^3)\div(5^1)=5^{3-1}= 5^2 = 25$. Now let us find out value of 5/125 in exponent form. It means

$5^1 \div 5^3 = 5^{-2}$.This will hold true whenever dividend is smaller than the divisor.Thus we can see that a number with negative value of exponent always means it is a fraction less than 1, so much so that ***as negative value of exponent keeps increasing, the fraction tends to the value zero.*** We can say that (any number)$^{-\infty} = 0$

1.8.1.3 Special case of exponent value '0'

It is a little tricky to understand exponent value of zero as we cannot imagine a number being multiplied by itself zero number of times. So we take advantage of concept of fraction exponents.

As we have seen earlier, a number with exponent ½ means square root of that number. Square root of the resultant number means (the number)$^{1/4}$. Repeated square root of resultant number means the original number with exponent value of $\frac{1}{8}\frac{1}{16}\frac{1}{32}\frac{1}{64}$ etc. If we keep going in this way, we will see that the exponent will keep diminishing, approaching a value '0' and the number on which repeated square root is performed approaches the value 1.

Two examples are shown below for clarity:

Exponent		**Original no.**	**Resul-tant no.**	**Original no.**	**Resultant no.**
Fraction	**Decimal**	**x = 23645**		**y = 156785**	
1/2	**0.5**	$x^{1/2}$	**153.77**	$y^{1/2}$	**395.96**
1/4	**0.25**	$x^{1/4}$	**12.4**	$y^{1/4}$	**19.898**
1/8	**0.125**	$x^{1/8}$	**3.521**	$y^{1/8}$	**4.46**
1/16	**0.0625**	$x^{1/16}$	**1.876**	$y^{1/16}$	**2.11**
1/32	**0.03125**	$x^{1/32}$	**1.37**	$y^{1/32}$	**1.45**
1/64	**0.015625**	$x^{1/64}$	**1.17**	$y^{1/64}$	**1.205**
1/128	**0.0078125**	$x^{1/128}$	**1.08**	$y^{1/128}$	**1.0978**
1/256	**0.00390625**	$x^{1/256}$	**1.04**	$y^{1/256}$	**1.0477**

Thus you can see that even a 5 or 6 digit number reduces to a value close to 1 as we repeatedly square root the number. Thus we can generalize that (any number)$^{1/\infty}$ = (any number)^0=1. As you grow, you will find frequent use of exponent in representing large quantities and scientific calculations. A little bit of practice will help in firming up the concepts of exponents.

All Rules of exponents are summarized in the table below. Here the base can be any positive number and exponent can be any number- positive, negative, fraction or even 0. Same rules apply for negative base numbers except that the sign of resultant number is determined by the value of exponent as explained above.

Multiplication of two numbers with same base and different exponents	$a^b \times a^c = a^{b+c}$
Division of two numbers with same base and different exponents	$a^b \div a^c = a^{b-c}$
Exponent of a number with exponent	$(a^b)^c = a^{bc}$
Multiplication of two numbers with different bases and same exponent	$a^b \times c^b = (ac)^b$
Division of two numbers with different bases and same exponent	$a^b \div c^b = (a/c)^b$
Exponent value is negative	$a^{-b} = 1/a^b$
Exponent value is 0	$a^0 = 1$

Practice will help in firming up the concepts of exponents.

1.9. Logarithm - another way of crunching large numbers and carrying out mathematical operation on complex numbers

We have seen technique of representing large numbers conveniently using exponent. Another way is to use logarithm for mathematical operations on large and complex numbers. Logarithm is written as 'log' in short and logarithm of number a to the base b is written and defined as below:

If $\log_b a = c$, it implies that $a = b^c$.

A popular base for logarithm is 10. Thus $\log_{10} 1 = 0$ because $10^0 = 1$. Similarly $\log_{10} 10 = 1$ as $10^1 = 10$. There are logarithm tables for value of $\log_{10}$ of different numbers. In olden days, when calculators were not very common, log tables were used extensively for multiplication or division of large (numerals/decimal) numbers. A log table is shown at end of the chapter for curious readers. It's good to understand the process of multiplying/dividing using log tables.

Let us consider log of product of two numbers a and b. If $\log_{10} a = p$ and $\log_{10} b = q$; by definition $10^p = a$ and $10^q = b$.

$$\therefore ab = 10^p \times 10^q = 10^{p+q} \quad ...(1)$$

If $\log_{10} ab = c$; $ab = 10^c$ ...(2)

Equating 1 and 2, $c = p+q \Rightarrow \log_{10} ab = \log_{10} a + \log_{10} b$

We get back the product ab by finding antilog of c

Let us use the log table to multiply 2.34 by 43.5

From the log table we can see that log of number 234 can be seen in the row starting 23 and column starting with 4.

So log 234 = 3692 yielding log value of 2.34 = 0.3692 (decimal value as 2.34 < 10)

Similarly, log value of 435 = 6385 yielding log of 43.5 = 1.6385 (number being >1<100)

Hence log of product of 2.34 and 43.5 = 0.3692+1.6385 = 2.0077

Now seeing antilog of 0077 from the logtable, we get value 1018.

Hence 2.34 × 43.5 =101.8 (decimal after 3 digits as log value is 2+)

It can be verified that this result is quite close to the actual multiplication value of 101.79.

We thus see that problem of multiplication changes to that of addition, making it much easier.

Similarly we can use the formula $\log_{10} \frac{a}{b} = \log_{10} a - \log_{10} b$ *for division using log tables. Another formula relating to logarithm that can be proved easily by log tables is* $\log_{10} a^b = b \times \log_{10} a$

Another base used for logarithm is constant 'e' which will be defined below in sec 1.10

1.10. Mathematical constant 'e'

The number *e* known as Euler's number, after the Swiss mathematician Leonhard Euler, is a famous irrational number, and is one of the most important numbers in mathematics. The first few decimal digits of 'e' are **2.718281828459045**...... The decimal value can be easily memorized as 7 followed by 1828 twice and followed by angles of isosceles right triangle 459045 (courtesy: mathsisfun.com)

The number e can be calculated as the sum of the infinite series

$$\sum_{n=0}^{\infty}\frac{1}{n!}=\frac{1}{1}+\frac{1}{1}+\frac{1}{1\times 2}+\frac{1}{1\times 2\times 3}+\frac{1}{1\times 2\times 3\times 4}\ldots\ldots$$

We will study much more of Euler's number in Chapter 8 on Limits, Continuity and Derivatives.

LOGARITHMS

	0	1	2	3	4	5	6	7	8	9	Mean Difference								
											1	2	3	4	5	6	7	8	9
10	0000	0043	0086	0128	0170	0212	0253	0294	0334	0374	4	8	12	17	21	25	29	33	37
11	0414	0453	0492	0531	0569	0607	0645	0682	0719	0755	4	8	11	15	19	23	26	30	34
12	0792	0828	0864	0899	0934	0969	1004	1038	1072	1106	3	7	10	14	17	21	24	28	31
13	1139	1173	1206	1239	1271	1303	1335	1367	1399	1430	3	6	10	13	16	19	23	26	29
14	1461	1492	1523	1553	1584	1614	1644	1673	1703	1732	3	6	9	12	15	18	21	24	27
15	1761	1790	1818	1847	1875	1903	1931	1959	1987	2014	3	6	8	11	14	17	20	22	25
16	2041	2068	2095	2122	2148	2175	2201	2227	2253	2279	3	5	8	11	13	16	18	21	24
17	2304	2330	2355	2380	2405	2430	2455	2480	2504	2529	2	5	7	10	12	15	17	20	22
18	2553	2577	2601	2625	2648	2672	2695	2718	2742	2765	2	5	7	9	12	14	16	19	21
19	2788	2810	2833	2856	2878	2900	2923	2945	2967	2989	2	4	7	9	11	13	16	18	20
20	3010	3032	3054	3075	3096	3118	3139	3160	3181	3201	2	4	6	8	11	13	15	17	19
21	3222	3243	3263	3284	3304	3324	3345	3365	3385	3404	2	4	6	8	10	12	14	16	18
22	3424	3444	3464	3483	3502	3522	3541	3560	3579	3598	2	4	6	8	10	12	14	15	17
23	3617	3636	3655	3674	3692	3711	3729	3747	3766	3784	2	4	6	7	9	11	13	15	17
24	3802	3820	3838	3856	3874	3892	3909	3927	3945	3962	2	4	5	7	9	11	12	14	16
25	3979	3997	4014	4031	4048	4065	4082	4099	4116	4133	2	3	5	7	9	10	12	14	15
26	4150	4166	4183	4200	4216	4232	4249	4265	4281	4298	2	3	5	7	8	10	11	13	15
27	4314	4330	4346	4362	4378	4393	4409	4425	4440	4456	2	3	5	6	8	9	11	13	14
28	4472	4487	4502	4518	4533	4548	4564	4579	4594	4609	2	3	5	6	8	9	11	12	14
29	4624	4639	4654	4669	4683	4698	4713	4728	4742	4757	1	3	4	6	7	9	10	12	13
30	4771	4786	4800	4814	4829	4843	4857	4871	4886	4900	1	3	4	6	7	9	10	11	13
31	4914	4928	4942	4955	4969	4983	4997	5011	5024	5038	1	3	4	6	7	8	10	11	12
32	5051	5065	5079	5092	5105	5119	5132	5145	5159	5172	1	3	4	5	7	8	9	11	12
33	5185	5198	5211	5224	5237	5250	5263	5276	5289	5302	1	3	4	5	6	8	9	10	12
34	5315	5328	5340	5353	5366	5378	5391	5403	5416	5428	1	3	4	5	6	8	9	10	11
35	5441	5453	5465	5478	5490	5502	5514	5527	5539	5551	1	2	4	5	6	7	9	10	11
36	5563	5575	5587	5566	5611	5623	5635	5647	5658	5670	1	2	4	5	6	7	8	10	11
37	5682	5694	5705	5717	5729	5740	5752	5763	5775	5786	1	2	3	5	6	7	8	9	10
38	5798	5809	5821	5832	5843	5855	5866	5877	5888	5899	1	2	3	5	6	7	8	9	10
39	5911	5922	5933	5944	5955	5966	5977	5988	5999	6010	1	2	3	4	5	7	8	9	10
40	6021	6031	6042	6053	6064	6075	6085	6096	6107	6117	1	2	3	4	5	6	8	9	10
41	6128	6138	6149	6160	6170	6180	6191	6201	6212	6222	1	2	3	4	5	6	7	8	9
42	6232	6243	6253	6263	6274	6284	6294	6304	6314	6325	1	2	3	4	5	6	7	8	9
43	6335	6345	6355	6365	6375	6385	6395	6405	6415	6425	1	2	3	4	5	6	7	8	9
44	6435	6345	6454	6464	6474	6484	6493	6503	6513	6522	1	2	3	4	5	6	7	8	9
45	6532	6542	6551	6561	6571	6580	6590	6599	6609	6618	1	2	3	4	5	6	7	8	9
46	6628	6637	6646	6656	6665	6675	6684	6693	6702	6712	1	2	3	4	5	6	7	7	8
47	6721	6730	6739	6749	6758	6767	6776	6785	6794	6803	1	2	3	4	5	5	6	7	8
48	6812	6821	6830	6839	6848	6857	6866	6875	6884	6893	1	2	3	4	4	5	6	7	8
49	6902	6911	6920	6928	6937	6946	6955	6964	6972	6981	1	2	3	4	4	5	6	7	8
50	6990	6998	7007	7016	7024	7033	7042	7050	7059	7067	1	2	3	3	4	5	6	7	8
51	7076	7084	7093	7101	7110	7118	7126	7135	7143	7152	1	2	3	3	4	5	6	7	8
52	7160	7168	7177	7185	7193	7202	7210	7218	7226	7235	1	2	2	3	4	5	6	7	7
53	7243	7251	7259	7267	7275	7284	7292	7300	7308	7316	1	2	2	3	4	5	6	6	7
54	7324	7332	7340	7348	7356	7364	7372	7380	7388	7396	1	2	2	3	4	5	6	6	7
	0	1	2	3	4	5	6	7	8	9	1	2	3	4	5	6	7	8	9

LOGARITHMS

	0	1	2	3	4	5	6	7	8	9	Mean Difference								
											1	2	3	4	5	6	7	8	9
55	7404	7412	7419	7427	7435	7443	7451	7459	7466	7474	1	2	2	3	4	5	5	6	7
56	7482	7490	7497	7505	7513	7520	7528	7536	7543	7551	1	2	2	3	4	5	5	6	7
57	7559	7566	7574	7582	7589	7597	7604	7612	7619	7627	1	2	2	3	4	5	5	6	7
58	7634	7642	7649	7657	7664	7672	7679	7686	7694	7701	1	1	2	3	4	4	5	6	7
59	7709	7716	7723	7731	7738	7745	7752	7760	7767	7774	1	1	2	3	4	4	5	6	7
60	7782	7789	7769	7803	7810	7818	7825	7832	7839	7846	1	1	2	3	4	4	5	6	6
61	7853	7860	7868	7875	7882	7889	7896	7903	7910	7917	1	1	2	3	4	4	5	6	6
62	7924	7931	7938	7945	7952	7959	7966	7973	7980	7987	1	1	2	3	3	4	5	6	6
63	7993	8000	8007	8014	8021	8028	8035	8041	8048	8055	1	1	2	3	3	4	5	5	6
64	8062	8069	8075	8082	8089	8096	8102	8109	8116	8122	1	1	2	3	3	4	5	5	6
65	8129	8136	8142	8149	8156	8162	8169	8176	8182	8189	1	1	2	3	3	4	5	5	6
66	8195	8202	8209	8215	8222	8228	8235	8241	8248	8254	1	1	2	3	3	4	5	6	6
67	8261	8267	8274	8280	8287	8293	8299	8306	8312	8319	1	1	2	3	3	4	5	5	6
68	8325	8331	8338	8344	8351	8357	8363	8370	8376	8382	1	1	2	3	3	4	4	5	6
69	8388	8395	8401	8407	8414	8420	8426	8432	8439	8445	1	1	2	2	3	4	4	5	6
70	8451	8457	8463	8470	8476	8482	8488	8494	8500	8506	1	1	2	2	3	4	4	5	6
71	8513	8519	8525	8531	8537	8543	8549	8555	8561	8567	1	1	2	2	3	4	4	5	5
72	8573	8579	8585	8591	8597	8603	8609	8615	8621	8627	1	1	2	2	3	4	4	5	5
73	8633	8639	8645	8651	8657	8663	8669	8675	8681	8686	1	1	2	2	3	4	4	5	5
74	8692	8698	8704	8710	8716	8722	8727	8733	8739	8745	1	1	2	2	3	4	4	5	5
75	8751	8756	8762	8768	8774	8779	8785	8791	8797	8802	1	1	2	2	3	3	4	5	5
76	8808	8814	8820	8825	8831	8837	8842	8848	8854	8859	1	1	2	2	3	3	4	5	5
77	8865	8871	8876	8882	8887	8893	8899	8904	8910	8915	1	1	2	2	3	3	4	4	5
78	8921	8927	8932	8938	8943	8949	8954	8960	8965	8971	1	1	2	2	3	3	4	4	5
79	8976	8982	8987	8993	8998	9004	9009	9015	9020	9025	1	1	2	2	3	3	4	4	5
80	9031	9036	9042	9047	9053	9058	9063	9069	9074	9079	1	1	2	2	3	3	4	4	5
81	9085	9090	9096	9101	9106	9112	9117	9122	9128	9133	1	1	2	2	3	3	4	4	5
82	9138	9143	9149	9154	9159	9165	9170	9175	9180	9186	1	1	2	2	3	3	4	4	5
83	9191	9196	9201	9206	9212	9217	9222	9227	9232	9238	1	1	2	2	3	3	4	4	5
84	9243	9248	9253	9258	9263	9269	9274	9279	9284	9289	1	1	2	2	3	3	4	4	5
85	9294	9299	9304	9309	9315	9320	9325	9330	9335	9340	1	1	2	2	3	3	4	4	5
86	9345	9350	9355	9360	9365	9370	9375	9380	9385	9390	1	1	2	2	3	3	4	4	5
87	9395	9400	9405	9410	9415	9420	9425	9430	9435	9440	0	1	1	2	2	3	3	4	4
88	9445	9450	9455	9460	9465	9469	9474	9479	9484	9489	0	1	1	2	2	3	3	4	4
89	9494	9499	9504	9509	9513	9518	9523	9528	9533	9538	0	1	1	2	2	3	3	4	4
90	9542	9547	9552	9557	9562	9566	9571	9576	9581	9586	0	1	1	2	2	3	3	4	4
91	9590	9595	9600	9605	9609	9614	9619	9624	9628	9633	0	1	1	2	2	3	3	4	4
92	9638	9643	9647	9652	9657	9661	9666	9671	9675	9680	0	1	1	2	2	3	3	4	4
93	9685	9689	9694	9699	9703	9708	9713	9717	9722	9727	0	1	1	2	2	3	3	4	4
94	9731	9736	9741	9745	9750	9754	9759	9763	9768	9773	0	1	1	2	2	3	3	4	4
95	9777	9782	9786	9791	9795	9800	9805	9809	9814	9818	0	1	1	2	2	3	3	4	4
96	9823	9827	9832	9836	9841	9845	9850	9854	9859	9863	0	1	1	2	2	3	3	4	4
97	9868	9872	9877	9881	9886	9890	9894	9899	9903	9908	0	1	1	2	2	3	3	4	4
98	9912	9917	9921	9926	9930	9934	9939	9843	9948	9952	0	1	1	2	2	3	3	4	4
99	9956	9961	9965	9969	9974	9978	9983	9987	9991	9996	0	1	1	2	2	3	3	3	4
	0	1	2	3	4	5	6	7	8	9	1	2	3	4	5	6	7	8	9

2

ALGEBRA

2. Introduction

Algebra is primarily the arithmetic involving unknown numbers. The unknown numbers can be "variables" (that can have different values satisfying the algebraic expression) or constants (non variables). Unknown numbers are denoted by alphabets like a,b,c,.....x,y,z etc. and these alphabets can have any number value – integer, fraction, positive, negative or in certain situations even zero. The symbols are connected by arithmetical operations BODMAS to form an Algebric Expression. No sign in front of a symbol implies that the symbol is to be treated as positive and no arithmetical operator between two symbols or between a numeral and a symbol implies multiplication. For example 3a implies 3 times **a** and ab implies a × b. In case there is no constant affixed in front of a variable, the constant is assumed to be 1

A –ve sign will always be placed before the symbol to treat it as a negative symbol. We will learn more about variables and constants as we go along.

2.1 Algebraic expressions

In case algebraic symbols are related as an expression with arithmetical operators on different symbols separated by an equality sign, such an expression is called an equation. We have a way of expressing equality. If a number $a + b$ has the same value as sum of two other numbers 'c' and 'd', we write this expression as

$a + b = c + d$. For example, to denote that a variable "x" has only one value of 7 irrespective of any conditions, the equation denoting it will be $x = 7$. If 2 variables x and y add up to 7, then this conditionwill be expressed as $x + y = 7$

The algebraic equation can comprise of a single term or multiple terms. An equation with a single term is called **Monomial**, with two terms a **Binomial** and with three terms as Trinomial and multiple terms as **Polynomial**. Thus xy is Monomial, $x + y$ is binomial and xy+z+yz is Trinomial etc.

2.2 Arithmetical Operations in Algebra

2.2.1 Addition/subtraction

In case of addition of 2 expressions, similar terms are added separately. For example, sum of 4x+3y and & x+8y which is $(4x + 3y) + (x + 8y) = (4x + x) + (3y + 8y) = 5x + 11y$

Now let us start with an expression $a + b = c + d$. Two sides of the equation are called Left hand Side (LHS) and Right Hand Side (RHS) respectively. To keep the equation unchanged, if any action is to be done, it should be done on both the sides.

Now consider we subtract number 'c' on the LHS. To maintain equality, we will need to subtract the same number on the RHS. Thus the equation will change to

$a + b - c = c + d - c$

or $a + b - c = d$ (since $c - c = 0$)

This effectively means that if we move 'c' from RHS to LHS, its symbol goes from ' + ' to '–' and vice versa. Thus

If $a+b = c-d$,

Moving d from RHS to LHS will imply that

$a + b + d = c$

Subtracting an algebraic number from another algebraic number implies adding a negative version of the number to be subtracted. Thus

$a - b = a + (-b)$

In case the number to be subtracted has multiple algebraic components, sign of each algebraic component will be reversed when moving from one side to the other, viz. + ve to –ve and vice versa.

For example:

If $(a+b)=(c+d)$ it transforms to

$$(a+b)-(c+d)=0 \Rightarrow a+b-c-d=0$$

Here both c and d turn negative as they are moved from RHS to LHS.

2.2.2 Multiplication and Division

An algebraic symbol may be multiplied or divided by another algebraic symbol or by itself. Thus **a** multiplied by **b** is written as ab or **a** multiplied by **a** implies a^2.

Similarly (a + b) divided by c is written as $\frac{a+b}{c}$ and **a** divided by (b + c) is $\frac{a}{b+c}$

We know that when we have a number multiplying another number expressed within the brackets like 3× (4+5), we are supposed to solve the expression within bracket before multiplying it by 3 (BODMAS principle). Alternatively, we can expand the bracket by multiplying each item within the bracket by the multiplier outside the bracket.

The same principle applies in Algebra. However since the expression within the bracket in most cases will be of unknown numbers, the latter principle applies generally.

For example if there is an expression a(b + c), if value of $b + c$ is not known, we multiply both b and c individually by a to arrive at the result as $a(b + c) = ab + ac$.

Same principle is true in case there is a common element in different parts of an expression. If we need to separate out the common element, it can be brought outside the bracket by dividing each part of the expression by the common element. For example, abc + bcd + bce = bc(a + d + e). It is also important to remember that we represent $a \times a$ as a^2, called **'a' square**. And $a \times a \times a$ as a^3, called **'a' cube**.

Now let us see how we transpose multiples or divisors in an equation.

If $\frac{a}{b}$ = c, and if we multiply both sides by 'b', it would imply $\frac{a}{b} \times b = c \times b$.

Now $\frac{a}{b} \times b = a$. Hence a = cb.

Meaning thereby that moving an algebraic quantity being multiplied or divided on LHS will end up changing to division or multiplication respectively when moved from LHS to RHS and vice versa. Thus if $\frac{a}{b} = \frac{c}{d}$, by cross multiplication, it changes to $ad = bc$

ab = c implies $a = \frac{c}{b}$, xy = ab implies $x = \frac{ab}{y}$ and a + b = cd implies $\frac{a+b}{c} = d$

Note carefully that when c moves from RHS to LHS, it ends up dividing the entire LHS iesum of a + b by c **NOT** either a or b. As already mentioned, principles of BODMAS apply as much to algebra as to numbers.

It is very important to remember that if we divide an algebraic expression by an algebraic symbol or expression, the divisor algebraic symbol/ expression in no case should be able to assume a value '0'. Otherwise the result will end up in a fallacy, as any number divided by 0 is undefined and results in infinity which is not a specific number but rather a concept defined in Chapter 1.

2.2.3 Categories of Equations:

2.2.3.1 Linear equations:

A linear equation is an algebraic equation in which each term is either the product of a constant and (the first power of) a variable or a constant. Linear equations can have one or more variables. Examples are a + 3c = 4 or x + y = 9 etc. A generic linear equation in 2 variables is ax + by = c where x and y are variables and a, b, c are constants.

In case there are two variables, the given linear equation will hold good for any number of pair of values of the two variables. We will learn later that if we represent such pair of values on X–Y coordinates, the points will form a straight line.We will learn much more on linear equations in Chapter 7 on Co–ordinate Geometry.

2.2.3.2 Quadratic equations

A **quadratic equation** is a second–order polynomialequation in a single variable. Standard from of a Quadratic equation is $ax^2 + bx + c = 0$. Such an equation has two possible solutions ie. two possible values of x that satisfy the equation.

2.2.3.3 Simultaneous equations:

Many real life problems can be expressed in terms of simultaneous equations having multiple variables. Such problems are solved by eliminating variables one by one. Let us look at an example with two simultaneous linear equations in two variables:

$2x + 3y = 18$ ......(1) and

$3x + 4y = 25$.....(2) .we need to find value of x and y.

There are two commonly used approaches to solve such a problem of simultaneous equations.

One obvious method is to find value of x in terms of y from first equation and substitute it in the other equation to find value of y. Thereafter we substitute value of y in either of the equations and get value of x.

In the above example, from first equation, we get $x = \frac{18-3y}{2}$

Substituting in the second equation, we get

$$3 \times \frac{18-3y}{2} + 4y = 25$$

$$\frac{54-9y}{2} + 4y = 25$$

$$27 - \frac{9}{2}y + 4y = 25$$

$$-\frac{1}{2}y = 25 - 27$$

$$y = (-2) \div (-\frac{1}{2}) = (-2) \times (-\frac{2}{1}) = 4$$

Substituting in equation 1,

$$2x + 3 \times 4 - 18$$

$$2x = 18 - 12$$

$$x = \frac{6}{2} = 3$$

Thus solution to the given equations is $x = 3$ and $y = 4$

The other method is by elimination process. Here we need to eliminate x or y to get an equation involving only the other variable. We multiple both equations suitably such that we are able to eliminate one of the variables by subtracting one equation from the other. In this approach, let us multiply first equation by 3 and the second equation by 2 to arrive at the following equations:

$$6x + 9y = 54$$

$$6x + 8y = 50$$

Subtracting second equation from the first we again get $y = 4$. By substituting value of y in either of the equations, we will get $x = 3$. The same result can be achieved by multiplying equation (1) by 4 and equation (2) by 3 and subtracting.

This way we get $8x + 12y = 72$ and $9x + 12y = 75$ leading to $x = 3$

By substituting value of x in either equation, we get $y = 4$.

We can verify the answer by substituting values in either of the equations.

$$2(3) + 3(4) = 18$$

$$3(3) + 4(4) = 25$$

This quick check proves very handy even while solving equations in examinations.

We will see that the second method in this case is much simpler. There is no hard and fast rule in favor of one approach or the other. We need to judge which of the two methods will yield quicker result.

If an equation has both numerator and denominator, we follow process of cross– multiplication to simplify the equation. However, it needs to be ensured that denominator is common to all terms in the numerator. Let us take an example

$$\frac{a}{3} - 2b + 8 = \frac{3b}{4} \qquad ...(1)$$

$$2a + 3b = 30 \qquad ...(2)$$

We take LCM on LHS of equation 1 yielding

$$\frac{a - 6b + 24}{3} = \frac{3b}{4},$$

By cross multiplication,

$4a - 24b + 96 = 9b$

$4a - 33b = -96,$ ...(3)

Multiplication of equation..(2) by 2 gives

$4a + 6b = 60$ ...(4)

Subtraction equation (3) from equation (4),

$39b = 156 \Rightarrow b = 4$

Substituting in equation 2; $2a = 30 - 12 \Rightarrow a = 9$

This is just an illustration of a problem involving two variables. Algebraic principles can be used to solve simultaneous equations involving more than 2 variables. ***An important principle to remember is that we need as many simultaneous equations as the number of variables to solve the system of equations. For example to solve for three variables, we need three simultaneous equations.***

An Important Note: *It is not essential that the two linear equations will always yield a solution. The following are the two exceptions:*

Case 1: *Ratios of index of x, y and the constant are same in the two equations. In that case, we will have infinite number of solutions.*

Case 2: *Ratios of index of x, y is same but ratio of the constants is different in the two equations. In that case, there will be no solution.*

In all other cases, there will be a definite solution of the equations. We will learn more about these concepts in co–ordinate geometry.

2.2.4 Multiplication of two Linear Equations:

For simplicity, let us see what happens when we multiply (a + b) by itself.

$$\begin{array}{lll} a + b & & \\ a + b & & \\ \hline a^2 + ab & & \\ \quad\;\; ab & + b^2 & \\ \hline a^2 + 2ab & + b^2 & \end{array}$$

We start multiplying the upper expression by each component of the lower expression Left to Right one by one and reach the above result.

Thus $(a+b)^2 = a^2 + 2ab + b^2$ ***(An important equation to remember)***

We can also derive the formula for $(a+b)^3$ as below:

$$(a+b)^3 = (a+b)\times(a+b)^2 = (a+b)(a^2+2ab+b^2)$$

$$= a^3 + 2a^2b + ab^2 + a^2b + 2ab^2 + b^3$$

$$= a^3 + 3a^2b + 3ab^2 + b^3 = a^3 + 3ab(a+b) + b^3$$

Division process is reverse of multiplication $\frac{a^3}{a} = a^2$ and $\frac{a^2b}{a} = ab$

Now let us practice division process by dividing $a^3 + b^3 + c^3 - 3abc$ by $(a + b + c)$

$a+b+c\ )\ a^3 + b^3 + c^3 - 3abc\ (\ a^2 + b^2 + c^2 - ab - bc - ca$

$a^3 \qquad + a^2b + a^2c$

$- \qquad\quad - \quad -$

$b^3 + c^3 - 3abc - a^2b - a^2c$

$b^3 \qquad + ab^2 + b^2c$

$- \qquad\quad - \quad -$

$c^3 - 3abc - a^2b - a^2c - ab^2 - b^2c$

$c^3 \qquad + ac^2 + bc^2$

$- \qquad\quad - \quad -$

$-3abc - a^2b - a^2c - ab^2 - b^2c - ac^2 - bc^2$

$-abc - a^2b \qquad - ab^2$

$+ \qquad + \qquad +$

$-2abc - a^2c - b^2c - ac^2 - bc^2$

$-abc \qquad - b^2c \qquad - bc^2$

$-abc - a^2c \qquad - ac^2$

$-abc - ac^2 \qquad - a^2c$

Thus $a^3 + b^3 + c^3 - 3abc = (a + b + c)(a^2 + b^2 + c^2 - ab - bc - ca)$

The students should practice division to master the process.

A few important algebraic equations to remember are as follows:

$(a + b)^2 = a^2 + b^2 + 2ab$ (1)

$(a + b)^3 = a^3 + b^3 + 3ab(a + b)$ (2)

$(a + b)^4 = a^4 + b^4 + 2ab(2a^2 + 2b^2 + 3ab)$ (3)

$(a + b)\times(a - b) = (a^2 - b^2)$ (4)

There is no need to remember the following similar equations as these can be arrived at by substituting b by –b in equations (1) to (3) respectively

$(a-b)^2 = a^2 - 2ab + b^2$ (5)

$(a-b)^3 = a^3 - b^3 - 3ab(a-b)$ (6)

$(a-b)^4 = a^4 + b^4 - 2ab(2a^2 + 2b^2 - 3ab$ (7)

The same principle holds good whenever sign of a variable needs to be changed to arrive at a new equation.

2.2.5 Solving Quadratic equations:

Let us look at an equation $(ax + p) \times (x + q) = 0$. Let us try to multiply these two linear equations:

$ax + p$

$x + q$

$ax^2 + xp + axq + q^2$

By taking x common in 2nd and 3rd term we can write the result as:

$ax^2 + (p + aq)x + pq = 0$

The above process implies that the quadratic equation $ax^2 + (p + aq) x + pq = 0$ has two factors $(ax + p)$ and $(x + q)$. Thus the equation is satisfied by two values of x;

$x = -\frac{p}{a}$ and $x = -q$ since in either case the value of the expression on LHS becomes zero.

Thus $-\frac{p}{a}$ and $-$ q are called solutions of the above equation that can also be written as $x^2 + \left(\frac{p}{a} + q\right) x + \frac{p}{a} \times q = 0$. These values p and q are also called roots of the equation.

If we need to find roots of a quadratic equation $ax^2 + (p + aq) x + pq = 0$, first attempt should be to see if we can find a pair of numbers that add up to the coefficient of x (be it positive or negative) and multiply up to the product of coefficient of x^2 and the constant value again keeping in view sign of this product. In case the above product is negative, it implies that the two numbers must be of opposite polarity and there sum (will actually be the difference of their absolute values) can be either positive or negative depending upon value of the two numbers. However if the constant happens to be positive, it implies that the two numbers must be either positive or negative. This process is also known as factorization of the equation. Let us look at a few examples:

$x^2 + 5x + 6 = (x + 2)(x + 3)$ (the pair of numbers being 2 and 3)

$x^2 – 5x + 6 = (x – 2)(x – 3)$ (the pair of numbers being –2 and –3)

$x^2 + 4x – 117 = (x + 13)(x – 9)$ (the pair of numbers being 13 and –9)

$2x^2 – 13x + 20 = (2x – 5)(x – 4)$ (the pair of numbers being –5 and –8)

$6x^2 – 13x – 5 = (3x + 1)(2x – 5)$ (the pair of numbers being 2 and –15)

Solutions of such quadratic equations are found by equating each factor to 0, thereby giving the solution in terms of values of x. In the first case above, solution of the equation is x = –2 and x = –3. Similarly in the last case, solution is x = –1/3 and x = 5/2.

However all quadratic equations will not have suitable number pairs to form factors. In such situations we need to follow the solution offered for Generic quadratic equation as below.

2.2.5.1 Generic Quadratic Equation

A quadratic equation in generic form is written as $ax^2 + bx + c = 0$. This can also be written as $x^2 + x\left(\frac{b}{a}\right) + \frac{c}{a} = 0$

If we compare it with the equation $x^2 + x\left(\frac{p}{a} + q\right) + \frac{p}{a} \times q = 0$ where $-\frac{p}{a}$ and $-q$ are roots of the equation, we can see that if coefficient of x^2 is 1, coefficient of x equals negative of the sum of the roots and the constant is the product of the roots. This gives us a very important property of the quadratic equations viz–

Sum of the roots of a Quadratic Equation = negative of the coefficient of x (irrespective whether the coefficient is positive or negative) ensuring that coefficient of x^2 is 1. **And**

Product of the roots of a Quadratic Equation = value of constant ensuring again that coefficient of x^2 is 1

In generic terms we can say that for a Quadratic Equation $ax^2 + bx + c = 0$

Sum of the roots = $-\frac{b}{a}$ and Product of the roots = $\frac{c}{a}$

General formula for the Roots of the above equation is

$$\frac{-b \pm \sqrt{b^2 - 4ac}}{2a}$$

This will yield two roots as

$$\frac{-b + \sqrt{b^2 - 4ac}}{2a} \text{ and } \frac{-b - \sqrt{b^2 - 4ac}}{2a}$$

Readers may try adding and multiplying these two roots and will find that these will satisfy the sum and product values given above. ***Sometimes wemay find value of $b^2 - 4ac$ in the above formulae to be negative. this is a fallacy as no real number (positive or negative) has a negative square, or we can say negative number does not have a square root and hence such a quadratic equation does not have a solution.***

One must practice solving algebraic equations with good understanding of concepts outlined above and he/she will find it quite easy to master Algebra.

Mastering Algebra is quite important as it is often used while solving Word problems, Mensuration, Arithmetical series and in higher engineering mathematics.

It is always a good practice to verify your answer by substituting the answer values in the given algebraic equations for checking validity of the answer.

2.2.6 Higher order polynomials (Binomial Theorem)

We noticed in Section 2.2.4 that finding cube of a binomial expression $(a+b)$ required repeated (3 times) multiplication. If we need to multiply higher order binomial expression, we realize that the manual process of multiplication gets much more complicated.

We have a simpler solution to the problem by using ***Binomial Theorem*** that will be described in this section. Let us recall the following binomial equationsof different indices/exponents and observe them carefully:

$(a + b)^0 = 1 \ (a + b \neq 0)$

$(a + b)^1 = a + b$

$(a + b)^2 = a^2 + 2ab + b^2$

$(a + b)^3 = a^3 + 3a^2b + 3ab^2 + b^3$

One more equation is added here that the reader can verify manually.

$(a + b)^4 = a^4 + 4a^3b + 6a^2b^2 + 4ab^3 + b^4$

We notice that

1. Number of terms of the expansion (on the Right Hand Side) are 1 more than the index value of the expression.
2. Each term of the expansion is product of the two components of the binomial (a and b here)
3. Index of the first component'a' starts with index of the binomial and keeps reducing by 1 till it goes to 0 and index of the 2nd component 'b' starts with 0 and increases by 1 in succeeding terms till it reaches index of the binomial.
4. In each term sum of indices of the two components (a and b) equals index of the expression (a + b).
5. Coefficients of different terms of expansion shows a pattern starting with 1, reaching a peak and reducing again back to 1.

We need to find an easy way of calculating coefficient of different terms. For this the reader should refer to Chapter 11, sections 11.2.1 and 11.2.3 for definition of Factorial and formula for combinations.

Coefficient of different terms for expression with index n starts with ${}^nC_0$ and keeps increasing to ${}^nC_1, {}^nC_2, \ldots\ldots {}^nC_n$ ..**These coefficients are sometimes referred to as C_0, C_1, C_2,......................C_n.**

On the above basis, we can write the expansion of $(a + b)^n$ as

$$(a+b)^n = {}^nC_0 a^n b^0 + {}^nC_1 a^{n-1} b^1 + {}^nC_2 a^{n-2} b^2 \ldots\ldots\ldots + {}^nC_{n-1} a^1 b^{n-1} + {}^nC_n a^0 b^n$$

$$\text{where } {}^nC_r = \frac{n!}{(n-r)! \times r!} = \frac{n \times (n-1) \times (n-2) \times \ldots\ldots (n-r+1)}{r \times (r-1) \times (r-2) \times \ldots\ldots 2 \times 1}$$

$$\text{As an example, } {}^nC_3 = \frac{n!}{(n-3)! \times 3!} = \frac{n \times (n-1) \times (n-2)}{3 \times 2 \times 1}$$

To practice Binomial theorem, let us write expansion for $(a + b)^5$

$$(a+b)^5 = {}^5C_0a^5b^0 + {}^5C_1a^4b^1 + {}^5C_2a^3b^2 + {}^5C_3a^2b^3 + {}^5C_4a^1b^4 + {}^5C_5a^0b^5$$

Calculate values of ${}^5C_0, {}^5C_1, {}^5C_2, {}^5C_3, {}^5C_4$ & ${}^5C_5$ which will yield values of 1, 5, 10, 10, 5, 1 respectively.

Thus $(a+b)^5 = a^5 + 5a^4b + 10a^3b^2 + 10a^2b^3 + 5ab^4 + b^5$

2.2.6.1 Special cases of Binomial theorem

Case 1:

$$(1+x)^n = {}^nC_0x^0 + {}^nC_1x^1 + {}^nC_2x^2 \,\ldots\ldots\ldots + {}^nC_{n-1}x^{n-1} + {}^nC_nx^n$$

This special case yields two very important outcomes. Substituting value of x = 1, we get

$$2^n = {}^nC_0 + {}^nC_1 + {}^nC_2 \ldots\ldots\ldots + {}^nC_{n-1} + {}^nC_n \ldots..(1)$$

And if we substitute x = −1, we get

$0 = {}^nC_0 - {}^nC_1 + {}^nC_2 - {}^nC_3 \ldots\ldots\ldots - {}^nC_{n-1} + {}^nC_n \ldots\ldots(2)$. Here it will be noticed that all odd powers of x (alternate terms) yield negative coefficients.

Adding equations (1) and (2), we will get

$2^n = 2\,{}^nC_0 + 2\,{}^nC_2 + 2\,{}^nC_4 \ldots\ldots\ldots + 2\,{}^nC_n$ (assuming n to be even), Or

$2^{n-1} = {}^nC_0 + {}^nC_2 + {}^nC_4 \ldots\ldots\ldots + {}^nC_n$

Similarly, subtracting (2) from (1) will yield,

$2^{n-1} = {}^nC_1 + {}^nC_3 + {}^nC_5 \ldots\ldots\ldots + {}^nC_{n-1}$

$$\text{Thus } {}^nC_0 + {}^nC_2 + {}^nC_4 \ldots\ldots\ldots + {}^nC_n = {}^nC_1 + {}^nC_3 + {}^nC_5 \ldots\ldots\ldots + {}^nC_{n-1} = 2^{n-1}$$

Case 2:

$$\left(1+\frac{1}{n}\right)^n = {}^nC_0\left(\frac{1}{n}\right)^0 + {}^nC_1\left(\frac{1}{n}\right)^1 + {}^nC_2\left(\frac{1}{n}\right)^2 \ldots\ldots\ldots + {}^nC_{n-1}\left(\frac{1}{n}\right)^{n-1} + {}^nC_n\left(\frac{1}{n}\right)^n$$

$$= 1 + n\left(\frac{1}{n}\right) + \frac{n(n-1)}{2}\left(\frac{1}{n^2}\right) + \frac{n(n-1)(n-2)}{2\times 3}\left(\frac{1}{n^3}\right) \ldots\ldots\ldots + n\left(\frac{1}{n^{n-1}}\right) + \left(\frac{1}{n^n}\right)$$

If value of n is very large approaching infinity, values of *(n–1), (n–2), (n–3).... etc. are close to n*and values of terms with exponent n or close to n like $\left(\frac{1}{n^{n-1}}\right), \left(\frac{1}{n^n}\right)$ etc. will be very very small tending to zero. Hence for very large n,

$$\left(1+\frac{1}{n}\right)^n = 1+1+\frac{\sim n^2}{1\times 2}\left(\frac{1}{n^2}\right)+\frac{\sim n^3}{1\times 2\times 3}\left(\frac{1}{n^3}\right)..........$$

As 0! = 1, we can re–write the above equation for very large n as

$$\left(1+\frac{1}{n}\right)^n = \frac{1}{0!}+\frac{1}{1!}+\frac{1}{2!}+\frac{1}{3!}..........$$

The expression on Right Hand Side can be expressed as $\sum_{n=0}^{\infty}\frac{1}{n!}$ which is nothing but the **Euler's number (e)** with value $= 2.71828....$ Relevance of Euler's number will be seen when we discuss limits and logarithms later on in Chapter 8.

Case 3:

$${}^nC_r + {}^nC_{r-1} = {}^{n+1}C_r$$

Proof:

Expanding left hand side we find

$${}^nC_r + {}^nC_{r-1} = \frac{n!}{(n-r)!\times r!}+\frac{n!}{(n-r+1)!\times(r-1)!}$$

$$= \frac{n!}{(n-r)!\times r!}\left(1+\frac{r}{n-r+1}\right) = \frac{n!}{(n-r)!\times r!}\left(\frac{n-r+1+r}{n-r+1}\right)$$

$$= \frac{n!}{(n-r)!\times r!}\left(\frac{n+1}{(n+1)-r}\right) = \frac{(n+1)!}{((n+1)-r)!\times r!} = {}^{n+1}C_r$$

Case 4:

$${}^nC_r = \left({}^{n-1}C_{r-1} + {}^{n-2}C_{r-1} + {}^{n-3}C_{r-1} + ... + {}^rC_{r-1} + {}^{r-1}C_{r-1}\right)$$

Proof:

In this case, proof can be understood by a different process of finding number of possible combinations. Let us assume an example of finding number of ways of choosing 3 out of 6 given colours say

1. Blue
2. Green
3. Yellow
4. Orange
5. Red
6. Pink

The straight away answer to the given question is ${}^{6}C_{3}$

Alternate way of finding the solution is to consider possible ways of selecting 3 out of 6 colours as below:

If we select 4th colour (Orange), there is only 1 way of selecting both the colours below it that can be defined as ${}^{2}C_{2}$. Now if we select colour 3 (yellow), we can select 2 out of 3 colours below yellow by ${}^{3}C_{2}$ ways. Similarly by selecting colour 2 (Green) and colour 1 (Blue) we will have ${}^{4}C_{2}$ and ${}^{5}C_{2}$ ways respectively.

Thus the total number of combinations are ${}^{2}C_{2} + {}^{3}C_{2} + {}^{4}C_{2} + {}^{5}C_{2}$

Thus ${}^{6}C_{3} = {}^{5}C_{2} + {}^{4}C_{2} + {}^{3}C_{2} + {}^{2}C_{2}$

This relationship can be expressed in general terms as

$${}^{n}C_{r} = \left({}^{n-1}C_{r-1} + {}^{n-2}C_{r-1} + {}^{n-3}C_{r-1} + \ldots\ldots\ldots\ldots + {}^{r}C_{r-1} + {}^{r-1}C_{r-1}\right)$$

The above formula derived in case of colours can be easily verified as below:

$${}^{6}C_{3} = \frac{6\times5\times4}{3\times2\times1} = 20 \text{ and}$$

$${}^{5}C_{2} = \frac{5\times4}{2\times1} = 10,\ {}^{4}C_{2} = \frac{4\times3}{2\times1} = 6^{3}.\ {}^{3}C_{2} = \frac{3\times2}{2\times1} = 3,\ {}^{2}C_{2} = 1$$

This is a very important formula again. The students can try and verify the formula by assuming any other values of n and r.

2.2.7 Trinomial expansion

Trinomial expansion is of use to easily solve certain problems. So let us look at expansion of $(a + b + c)^4$. A conventional approach could be to split trinomial into a binomial and use Binomial theorem as below:

$$(a+b+c)^4 = \{a+(b+c)\}^4$$

$$= {}^4C_0a^4 + {}^4C_1a^3(b+c) + {}^4C_2a^2(b+c)^2 + {}^4C_3a(b+c)^3 + {}^4C_4(b+c)^4$$

$$= a^4 + 4a^3b + 4a^3c + 6a^2(b^2 + 2bc + c^2) + 4a(b^3 + 3bc(b+c) + c^3)$$

$$+ (b^4 + 4b^3c + 6b^2c^2 + 4bc^3 + c^4)$$

$$= a^4 + 4a^3b + 4a^3c + 6a^2b^2 + 12a^2bc + 6a^2c^2 + 4ab^3 + 12ab^2c +$$

$$+ 12abc^2 + 4ac^3 + b^4 + 4b^3c + 6b^2c^2 + 4bc^3 + c^4$$

We can notice there is a pattern here. Sum total of exponents of all variables in each term is 4 and first exponent of **a** goes from 4 to 0, then exponent of **b** goes from 0 to 4 as in case of binomial expansion.

The expansion can be rewritten as

$$= a^4 + 4a^3b + 4a^3c + 6a^2b^2 + 6a^2c^2 + 12a^2bc + 12ab^2c + 12abc^2 +$$

$$+ 6b^2c^2 + 4ab^3 + 4ac^3 + 4b^3c + 4bc^3 + b^4 + c^4$$

It can be noticed that coefficients of each term also follow a pattern like 1, 4, 6, 12, 6, 4, 1

A Trinomial expansion formula has been developed to solve tedious trinomial expansions with any positive integer exponent value n as below:

For trinomial expansion of

$$(a + b + c)^n = \sum_{\substack{i,j,k \\ i+j+k=n}} \binom{n}{i,j,k} \times a^i b^j c^k,$$

Where $\binom{n}{i,j,k} = \dfrac{n!}{i! \times j! \times k!}$ and i, j and kacquire all possible sequential values

In the given example above, n = 4, giving different values of i, j, k as below:

Using these values, we can straight away write the expansion of given trinomial as

i	j	k
4	0	0
3	1	0
3	0	1
2	2	0
2	1	1
2	0	2
1	3	0
1	2	1
1	1	2
1	0	3
0	4	0
0	3	1
0	2	2
0	1	3
0	0	4

Using these values, we can straight away write the expansion of given trinomial as:

$$(a+b+c)^4 = \frac{4!}{4!\times 0!\times 0!}a^4 + \frac{4!}{3!\times 1!\times 0!}a^3b + \frac{4!}{3!\times 0!\times 1!}$$

$$a^3c + \frac{4!}{2!\times 2!\times 0!}a^2b^2 + \frac{4!}{2!\times 1!\times 1!}a^2bc + \frac{4!}{2!\times 0!\times 2!}a^2c^2 + \frac{4!}{1!\times 3!\times 0!}ab^3$$

$$+\frac{4!}{1!\times 2!\times 1!}ab^2c + \frac{4!}{1!\times 1!\times 2!}abc^2 + \frac{4!}{1!\times 0!\times 3!}ac^3 + \frac{4!}{0!\times 4!\times 0!}$$

$$b^4 + \frac{4!}{0!\times 3!\times 1!}b^3c + \frac{4!}{0!\times 2!\times 2!}b^2c^2 + \frac{4!}{0!\times 1!\times 3!}bc^3 + \frac{4!}{0!\times 0!\times 4!}c^4 \Rightarrow$$

$$(a+b+c)^4 = a^4 + 4a^3b + 4a^3c + 6a^2b^2 + 12a^2bc + 6a^2c^2 + 4ab^3$$

$$+12ab^2c + 12abc^2 + 4ac^3 + b^4 + 4b^3c + 6b^2c^2 + 4bc^3 + c^4$$

This is a bit tedious formula but it is good to know because you may come across use of this formula in competitive exams.

2.3 Solved examples

2.3.1 Linear equations

1. Find the value of k for which the following pair of equations has no solution:

   $x - ky = 2$ and $3x + 6y = -5$

Solution:

We know that for no solution, the lines representing the linear equations need to be parallel requiring *Ratios of coefficient of x and y of the two equations should be same but ratio of the constants should be different.*

Ratio of coefficients of x is $\frac{1}{3}$ while ratio of constants is $\frac{-2}{5}$. As these ratios are different, the lines will be parallel if ratio of indices of y $= \frac{1}{3}$

Hence $\frac{1}{3} = \frac{-k}{6} \Rightarrow -k = \frac{6}{3} \therefore k = -2$

2. 3 video cassettes and 2 audio cassettes cost Rs 425. 3 audio cassettes and 2 video cassette cost Rs 350. What are the prices of an audio cassette and a video cassette.

Solution:

Let price of an audio cassettes be x and that of a video cassette be y.

As per given parameters, $2x + 3y = 425$ ..(1) and $3x + 2y = 350$ ..(2)

Multiplying Equation (1) by 3 and equation (2) by 2, we get

$6x + 9y = 1275$ ....(3) and $6x + 4y = 700$ ....(4)

Subtraction equation (4) from equation (3), we get

$5y = 575$ or $y = 115$

Substituting value of y in equation (1) we get

$2x = 425 - 345$ or $x = \frac{80}{2} = 40$

Thus price of audio cassette is Rs 40 and that of a video cassette is 115.

This result can be easily verified by substituting values of x and y in equation (2).

3. A boat goes 30 kilometers (kms) upstream and 44 kms downstream in 10 hours. In 13 hours it can go 40 kms upstream and 55 kms downstream. Determine the speed of the stream and that of the boat on still water.

Solution:

Let speed of the boat in still water be x and speed of the stream be y.

Then speed of the boat upstream $= x - y$ and

speed of the boat downstream $= x + y$

As per given conditions, time for the first set of data

$10 = \frac{30}{x-y} + \frac{44}{x+y}$ ....(1) and as per second set of data

$13 = \frac{40}{x-y} + \frac{55}{x+y}$ ....(2)

Multiplying eq (1) by 4 and eq (2) by 3 and subtracting eq (1) from eq (2) we get

$$(39-40) = \frac{165-176}{x+y}$$

Or $-1 = \frac{-11}{x+y}$ *or* $\boldsymbol{x + y = 11}$

Substituting value of x + y in equation (1) we get

$10 = \frac{30}{x-y} + \frac{44}{11}$ or $6 = \frac{30}{x-y}$ leading us to $\boldsymbol{x - y = 5}$

By adding and subtracting values of *(x – y) and (x + y), we get*

x = 8 and y = 3

∴Speed of the boat in still water =8 kmph and speed of the stream =3 kmph

The above result can be easily verified by substituting values in Eq (2)

4. A fraction becomes $\frac{9}{11}$, if 2 is added to both the numerator and the denominator. If 3 is added to both the numerator and the denominator, it becomes $\frac{5}{6}$. Find the fraction.

Solution:

Let the fraction be $\frac{x}{y}$

As per given conditions, $\frac{x+2}{y+2}=\frac{9}{11}$ ....(1) and $\frac{x+3}{y+3}=\frac{5}{6}$ ....(2)

Cross multiplying both the equations we get,

$11x - 9y + 4 = 0$ ......(3) and $6x - 5y + 3 = 0$....... (4)

Multiplying eq (3) by 6 and eq (4) by 11 and subtracting, x terms cancel out and we get

$-54y + 55y + 24 - 33 = 0 \Rightarrow y = 9$

Substituting in equation (3),

$11x - 81 + 4 = 0 \Rightarrow x = 7$

Thus the fraction is $\frac{7}{9}$

5. The age of two friends Anil and Brij differs by 3 years. Anil's father Dharam is twice as old as Anil and Brij is twice as old as his sister Kiran. The ages of Kiran and Dharam differ by 30 years. Find the ages of Anil and Brij.

Solution:

Let the present age of Anil be x and present age of Brij be y.

Let the present age of Dharam be 'a' and present age of Kiran be 'b'.

As per given conditions,

$x - y = 3..(1),\ a = 2x..(2),\ y = 2b...(3)$ *and* $a - b = 30$.........(4)

From eq (2) $x=\frac{a}{2}$ .....(5)

Subtracting eq (3) from eq (5) and comparing with eq (1),

$\frac{a}{2}-2b=3$ or $a-4b=6$ ......(6)

Subtracting eq (6) from eq (4) gives us

$3b = 24$ *or* $b = 8$ *and* $a = 38$

Substituting value of a in eq (2), we get $x = 19$ and

Substituting value of xin eq (1) yields $y = 16$

So Age of Anil is 19 years and that of Brij is 16 years

2.3.2 Quadratic equations

6. Solve by factorization

$4x^2 - 4a^2x + (a^4 - b^4) = 0$

Solution:

$4x^2 - 4a^2x + a^4 = b^4$

$4x^2 - 2a^2x - 2a^2x + a^4 = b^4$

$2x(2x - a^2) - a^2(2x - a^2) = b^4$

Or $(2x - a^2)^2 = b^4$

And $(2x - a^2)^2 - b^4 = 0$, leading to

$$(2x - a^2 + b^2)(2x - a^2 - b^2) = 0 \Rightarrow x = \frac{a^2 - b^2}{2} \;\&\; x = \frac{a^2 + b^2}{2}$$

7. By the method of completion of squares show that the equation $4x^2 + 3x + 5 = 0$ has no real roots.

Solution:

The first term is square of $2x$. The middle term being $3x$, to complete the square the second term should be $\frac{3x}{2x \times 2} = \frac{3}{4}$

Thus $4x^2 + 3x + 5 = \left(2x + \frac{3}{4}\right)^2 + 5 - \left(\frac{3}{4}\right)^2 = 0$

Or $\left(2x + \frac{3}{4}\right)^2 = -\left(5 - \left(\frac{3}{4}\right)^2\right) = -\frac{71}{16}$

Since a negative number has no real roots, hence the given equation has no real roots

8. A dealer sells a toy for Rs.24 and gains as much percent as the cost price of the toy. Find the cost price of the toy.

Solution:

Let cost price be x. As per given condition, Sale price = cost price + gain i.e. $x + \frac{x}{100} \times x = x\left(1 + \frac{x}{100}\right) = 24$

Or $x + \frac{x^2}{100} - 24 = 0$ leading to

$x^2 + 100x - 2400 = 0$

Factorising $x^2 + 120x - 20x - 2400 = 0$ or

$(x + 120)(x - 20) = 0$ giving us value of $x = 20$ or $x = -120$

However, since price cannot be negative, the only option is x = 20.

Thus cost price of the toy is Rs 20.

9. A *lotus* with its stem is 2m above the water in a pond. Due to wind the lotus slides on the side and stem completely submerges in water with its top visible at a distance of 10m from the original position. Find the depth of water in the pond

Solution:

The given situation is depicted below. Here depth of the pond is assumed as x meters. So length of the lotus stem = *x + 2 metres*

Now after the stem slide, it makes a right angled triangle with diagonal as (x + 2), base as 10 metre and depth as x.

Here AD is an arc with centre at C and radius AD

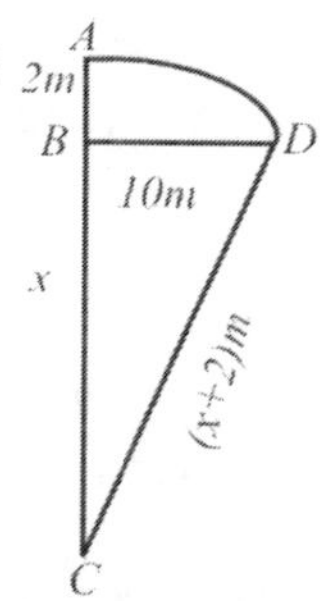

$\therefore 10^2 + x^2 = (x+2)^2$, leading to

$(x+2)^2 - x^2 = 100$

Or $4x + 4 = 100$

$x = 24$

Thus depth of the pond = 24 metres

10. A teacher on attempting to arrange the students for mass drill in the form of a solid square found that 24 students were left over. When he increased the size of the square by one student he found he was short of 25 students. Find the number of students.

Solution:

The problem under consideration is akin to forming a solid square with dots. Suppose the square with x students on each side leaves out 24 students and total number of students are y, the given condition implies that $x^2 + 24 = y$

Now when the next square has students on each row of the square, the number of students are short by 25.

Hence $y=(x+1)^2-25 \Rightarrow x^2+24=(x+1)^2-25$

Or $x^2+24=x^2+2x+1-25=x^2+2x-24 \Rightarrow 2x=48 \Rightarrow x=24$

Total number of students $y = 24^2+24=600$

2.3.3 Binomial theorem

11. Expand the following by Binomial theorem

$\left(x^2+\frac{2}{x}\right)^4, x \neq 0$

Solution:

$$\left(x^2+\frac{2}{x}\right)^4 = {}^4C_0(x^2)^4\left(\frac{2}{x}\right)^0 + {}^4C_1(x^2)^3\left(\frac{2}{x}\right)^1 + {}^4C_2(x^2)^2\left(\frac{2}{x}\right)^2 + {}^4C_3(x^2)^1\left(\frac{2}{x}\right)^3 + {}^4C_4(x^2)^0\left(\frac{2}{x}\right)^4$$

$$= x^8+4x^6\left(\frac{2}{x}\right)+6x^4\left(\frac{4}{x^2}\right)+4x^2\left(\frac{8}{x^3}\right)+\left(\frac{16}{x^4}\right)$$

$$= x^8+8x^5+24x^2+32x^{-1}+16x^{-4}$$

12. First 3 terms in the expansion $(1+x)^n$ are 1, 10 & 40. Find the expansion.

Solution:

$$(1+x)^n = {}^nC_0x^0 + {}^nC_1x^1 + {}^nC_2x^2 \ldots\ldots\ldots + {}^nC_{n-1}x^{n-1} + {}^nC_nx^n$$

The first term is obviously 1 as ${}^nC_0$ as well as x^0 are 1.

As per given conditions, the second term nx = 10 .....(1) and

the third term $\frac{n(n-1)}{2}x^2 = 40$ or $\frac{nx(n-1)x}{2} = 40$

Substituting value of nx, we get $\frac{n-1}{2}x = \frac{40}{10} = 4$ ....(2)

Dividing equation (2) by equation (1),

$\frac{n-1}{2n} = \frac{4}{10}$ or *10n – 10 = 8n* leading to $n = \frac{10}{2} = 5$

Substituting value of n in eq (1)

$5x = 10$ or $x = 2$

Thus the given expansion is (1 + 2)⁵

13. Using binomial theorem, prove that $(3^{2n+2} - 8n - 9)$ is divisible by 64, where n is a positive integer.

Solution:

$$3^{2n+2} = 3^{2(n+1)} = 9^{n+1} = 9 \times 9^n$$

$$= 9(1+8)^n = 9\left(1 + {}^nC_1 \times 8 + {}^nC_2 \times 8^2 + {}^nC_3 \times 8^3 + ...\right)$$

$$= 9 + 72\,{}^nC_1 + 9\left({}^nC_2 \times 8^2 + {}^nC_3 \times 8^3 + ...\right) \Rightarrow$$

$$(3^{2n+2} - 8n - 9) = 9 + 8n + 64\,{}^nC_1 + 9\left({}^nC_2 \times 8^2 + {}^nC_3 \times 8^3 + ...\right) - 8n - 9$$

$$= 64\,{}^nC_1 + 9\left({}^nC_2 \times 8^2 + {}^nC_3 \times 8^3 + ...\right)$$

$$= 64\left\{{}^nC_1 + 9\left({}^nC_2 + 8 \times {}^nC_3 + ...\right)\right\}$$

The above expression is clearly divisible by 64 as the terms within the brackets is a whole number for all values of n. For n = 1, value of the expression $(3^{2n+2} - 8n - 9)$ ***= 64, for n = 2, its 64 × 11 and so on. Hence*** $(3^{2n+2} - 8n - 9)$ ***is divisible by 64 for all n.***

14. If x^p occurs in the expansion of $\left(x^2 + \frac{1}{x}\right)^{2n}$, prove that its coefficient is $\dfrac{2n!}{\left[\frac{4n-p}{3}\right]!\left[\frac{1}{3}(2n+p)\right]!}$

Solution:

$$\left(x^2 + \frac{1}{x}\right)^{2n} = \left(\frac{x^3+1}{x}\right)^{2n} = \frac{(x^3+1)^{2n}}{x^{2n}}$$

(r + 1)th term in expansion of $(x^3 + 1)^{2n}$ is

$${}^{2n}C_r \times x^{3\times(2n-r)} = \frac{2n}{(2n-r)! \times r!} \times x^{3\times(2n-r)}$$

Exponent of x in the (r + 1)th term of $\frac{(x^3+1)^{2n}}{x^{2n}}$ will therefore be (6n–3r–2n) = 4n–3r

As per given conditions, p = 4n–3r or 3r = 4n–p or $r = \frac{4n-p}{3}$

Substituting value of r in coefficient of (r + 1)th term, we get

$$= \frac{2n!}{(2n-r)!r!} = \frac{2n!}{\left(2n - \frac{4n-p}{3}\right)!\left(\frac{4n-p}{3}\right)!} = \frac{2n!}{\left(\frac{2n+p}{3}\right)!\left(\frac{4n-p}{3}\right)!}$$

Hence proved.

15. Find the value of r, if coefficient of (2r + 4)th& (r–2)th term in the expansion (1 + x)^{18} are equal

Solution:

Coefficient of (r + 1)th term in expansion of $(1+x)^{18}$ is $\frac{18!}{(18-r)!\times r!}$

Accordingly, coefficient of (2r+4)th term = $\frac{18!}{(18-(2r+3))!\times(2r+3)!}$

And coefficient of (r–2)th term = $\frac{18!}{(18-(r-3))!\times(r-3)!}$

$$\therefore \frac{18!}{(18-(2r+3))!\times(2r+3)!} = \frac{18!}{(18-(r-3))!\times(r-3)!}$$

Here $(2r+3)$ is the larger term in expression on Left hand side and $18-(r-3)$ is the larger term in expression on the Right hand side. Equating the two, we get $(2r+3) = 18-(r-3)$

Or $2r+3+r-3=18$ leading to $\boldsymbol{r = 6}$

16. In the binomial of (a + b)^m, $m \geq 5$ and sum of 5th & 6th terms is zero, find $\frac{a}{b}$

Solution:

5th term of (a + b)^m = ${}^mC_4 \times a^{m-4} \times b^4$

And 6th term of (a + b)^m = ${}^mC_5 \times a^{m-5} \times b^5$

Sum of 5th & 6th terms = ${}^mC_5 \times a^{m-5} \times b^5 + {}^mC_4 \times a^{m-4} \times b^4$ ....(1)

We know that $^mC_4 = \times\frac{m(m-1)(m-2)(m-3)}{4\times3\times2\times1}$ and

$$^mC_5 = \times\frac{m(m-1)(m-2)(m-3)(m-4)}{5\times4\times3\times2\times1}$$

$$\therefore\ ^mC_5 = {^mC_4}\times\frac{m-4}{5}$$

Substituting in eq (1),

$$\text{Sum of }5^{th}\ \&\ 6^{th}\text{ terms} = {^mC_4}\times\left(\frac{m-4}{5}\times a^{m-5}\times b^5 + a\times a^{m-5}\times\frac{b^5}{b}\right)$$

$$= {^mC_4}\times a^{m-5}\times b^5\times\left(\frac{m-4}{5}+\frac{a}{b}\right)=0$$

Since $^mC_4\times a^{m-5}\times b^5$ is $\neq 0$

$$\therefore\left(\frac{m-4}{5}+\frac{a}{b}\right)=0$$

Or $\frac{a}{b} = -\frac{m-4}{5} = \frac{4-m}{5}$

17. Find coefficient of term independent of x in expansion of

$$(1+x+2x^3)\left(\frac{3}{2}x^2-\frac{1}{3x}\right)^9$$

Solution:

Binomial expansion of the second term will yield different exponents of x. We will need coefficients of terms independent of x, x^{-1} and x^{-3} to arrive at the solution.

General coefficient of power of x will be

$$^9C_r\left(\left(\frac{3}{2}x^2\right)^{(9-r)}\times\left(-\frac{1}{3x}\right)^r\right) = {^9C_r}\left(\left(\frac{3}{2}\right)^{(9-r)}\times\left(-\frac{1}{3}\right)^r x^{2(9-r)-r}\right)$$

For term independent of x, $2(9-r)-r=0$ leading to

18 – 3r = 0 **or r = 6**

For coefficient of x^{-1}, $18-3r=-1$ leads to value of r as a fraction which is not feasible. Hence there will be no x^{-1} term.

For coefficient of x^{-3}, $18-3r=-3$ leading to **r = 7**

Hence, coefficient of term independent of

$$x = {}^9C_6\left(\left(\frac{3}{2}\right)^3 \times \left(-\frac{1}{3}\right)^6\right)$$

$$= \frac{9\times 8\times 7}{3\times 2}\left(\left(\frac{1}{2\times 3}\right)^3\right) = \frac{3\times 4\times 7}{6\times 6\times 6} = \frac{7}{18}$$

And coefficient of term

$$x^{-3} = {}^9C_7\left(\left(\frac{3}{2}\right)^2 \times \left(-\frac{1}{3}\right)^7\right)$$

$$= -\frac{9\times 8}{2}\left(\frac{1}{2^2\times 3^5}\right) = -\frac{1}{27}$$

Multiplying the above coefficients with 1 and 2 (coefficient of x^3) respectively we get, overall coefficient of term independent of

$$x = \frac{7}{18} - 2\times\frac{1}{27} = \frac{21-4}{54} = \frac{17}{54}$$

18. If N is a prime number that divides

$${}^{39}P_{19} + {}^{38}P_{19} + {}^{37}P_{19} + \ldots\ldots {}^{20}P_{19}, \text{ find N}$$

Solution:

Reference section 11.2.2 and 11.2.3, ${}^nP_r = \dfrac{n!}{(n-r)!}$ and ${}^nC_r = \dfrac{n!}{(n-r)!\times r!}$ leading to ${}^nP_r = r!\times {}^nC_r$

Using the above equation, the problem reduces to

$${}^{39}P_{19} + {}^{38}P_{19} + {}^{37}P_{19} + \ldots\ldots {}^{20}P_{19} =$$

$$19!\times\left({}^{39}C_{19} + {}^{38}C_{19} + \ldots\ldots.. + {}^{20}C_{19}\right)$$

Recall the following formula:

$${}^nC_r = \left({}^{n-1}C_{r-1} + {}^{n-2}C_{r-1} + {}^{n-3}C_{r-1} + \ldots\ldots\ldots\ldots + {}^rC_{r-1} + {}^{r-1}C_{r-1}\right)$$

For n = 40, and r = 20 in the above equation, we get

${}^{40}C_{20} = \left({}^{39}C_{19} + {}^{38}C_{19} + {}^{37}C_{19} + \ldots\ldots\ldots\ldots + {}^{20}C_{19} + {}^{19}C_{19}\right)$, leading to

$$\left({}^{39}C_{19} + {}^{38}C_{19} + \ldots\ldots.. + {}^{20}C_{19}\right) = {}^{40}C_{20} - 1$$

Or $${}^{39}P_{19} + {}^{38}P_{19} + {}^{37}P_{19} + \ldots\ldots {}^{20}P_{19} = 19! \times \left({}^{40}C_{20} - 1\right)$$

$$= 19! \times \left(\frac{40 \times 39 \times 38 \times \ldots\ldots.. \times 21}{20!} - 1\right)$$

$$= 19! \times \left(\frac{(41-1)(41-2)(41-3)\ldots\ldots..(41-20)}{20!} - 1\right)$$

This will yield a sum of different exponents from 20 to 1 of 41 with different coefficients + product of all numbers from 1 to 20 as below where **a** depicts a suitable integer value

$$= 19! \times \left(\frac{41 \times a + 1 \times 2 \times 3 \times \ldots\ldots\ldots \times 20}{20!} - 1\right)$$

$$= 19! \times \left(\frac{41 \times a + 20!}{20!} - 1\right)$$

$$= 19! \times \left(\frac{41 \times a}{20!} + 1 - 1\right) = \frac{41 \times a}{20!}$$, a product of 41

Thus N = 41 is the prime number that divides the given sum.

19. If $\frac{2}{9!} + \frac{2}{3! \times 7!} + \frac{1}{5! \times 5!} = \frac{a^9}{b!}$, find a, b.

Solution:

We know from section 2.6.6.1 that

$${}^{n}C_1 + {}^{n}C_3 + {}^{n}C_5 \ldots\ldots\ldots. + {}^{n}C_{n-1} = 2^{n-1}$$

Substituting n = 10, we get

$$2^9 = {}^{10}C_1 + {}^{10}C_3 + {}^{10}C_5 + {}^{10}C_7 + {}^{10}C_9$$

Since ${}^{n}C_r = {}^{n}C_{n-r}$, we find that ${}^{10}C_1 = {}^{10}C_9$ and ${}^{10}C_3 = {}^{10}C_7$, leading to

$$2^9 = 2 \times \left({}^{10}C_1 + {}^{10}C_3\right) + {}^{10}C_5 = 2 \times \left(\frac{10!}{9!} + \frac{10!}{7! \times 3!}\right) + \frac{10!}{5! \times 5!}$$

Dividing both sides by 10!, we get

$$\frac{2^9}{10!} = \frac{2}{9!} + \frac{2}{3! \times 7!} + \frac{1}{5! \times 5!}$$

Thus **a = 2 and b = 10**

20. If a_1, a_2, a_3, a_4 are coefficients of any 4 consecutive terms in the expansion of (1 + x)[n], then find $\frac{a_1}{a_1+a_2}+\frac{a_3}{a_3+a_4}$

Let $a_1 = {}^nC_{r-1}$, then $a_2 = {}^nC_r$, $a_3 = {}^nC_{r+1}$ and $a_4 = {}^nC_{r+2}$

From section 2.2.6.1 case 3, we know that ${}^nC_r + {}^nC_{r-1} = {}^{n+1}C_r$

Thus $a_2 + a_1 = {}^{n+1}C_r$ and $a_4 + a_3 = {}^{n+1}C_{r+2}$, leading to

$$\frac{a_1}{a_1+a_2} = \frac{{}^nC_{r-1}}{{}^{n+1}C_r} = \frac{r(n+1-r)}{(n+1)(n-r+1)} = \frac{r}{n+1} \text{ and}$$

$$\frac{a_3}{a_3+a_4} = \frac{{}^nC_{r+1}}{{}^{n+1}C_{r+2}} = \frac{(r+2)(n-1-r)}{(n+1)(n-r-1)} = \frac{r+2}{n+1} \text{ leading to}$$

$$\frac{a_1}{a_1+a_2}+\frac{a_3}{a_3+a_4} = \frac{r}{n+1} + \frac{r+2}{n+1} = \frac{2(r+1)}{n+1}$$

Similarly we can find that $\frac{a_2}{a_2+a_3} = \frac{r+1}{n+1}$

Thus we can also say that $\frac{a_1}{a_1+a_2}+\frac{a_3}{a_3+a_4} = \frac{2a_2}{a_2+a_3}$

21. If sum of coefficients of 1st, 2nd, 3rd terms of $\left(x^2+\frac{1}{x}\right)^m$, where m is a natural number, is 46, find coefficient of the term independent of x.

Solution:

Coefficients of the first 3 terms are ${}^mC_0$, ${}^mC_1$ and ${}^mC_2$ respectively. These translate to *1, m,* $\frac{m(m-1)}{2}$.

As per given condition, $1+m+\frac{m(m-1)}{2}=46$ or

$2+2m+m^2-m=92$, leading to

$m^2+m-90=0$ or

$(m+10)(m-9)=0$, giving us $m=9$

Let us assume that (r + 1)[th] term is independent of x. Thus

${}^9C_r \times x^{2(9-r)} \times x^{-r} \Rightarrow 18-2r-r=0$ or $\boldsymbol{r = 6}$

Thus coefficient of the term independent of

$$x = {}^9C_6 = {}^9C_3 = \frac{9\times8\times7}{3\times2\times1} = 3\times4\times7 = 84$$

2.4 Practice problems

1. Solve the following pair of linear equations:

$$\sqrt{2}x+\sqrt{3}y=0 \text{ and } \sqrt{3}x-\sqrt{8}y=0$$

2. The taxi charges in a city consist of a fixed charge together with the charge for the distance covered. For a distance of 10 km, the charge paid is Rs. 105 and for a journey of 15 km, the charge paid is Rs. 155. What are the fixed charges and the charges per km? How much does a person have to pay for travelling a distance of 25 km?

3. Five years hence, the age of Jacob will be three times that of his son. Five year ago, Jacob's age was seven times that of his son. What are their present ages?

4. Solve the following pairs of equations:

$$\frac{1}{3x+y}+\frac{1}{3x-y}=\frac{3}{4} \text{ and } \frac{1}{2(3x+y)}-\frac{1}{2(3x-y)}=-\frac{1}{8}$$

5. A train covered a certain distance at a uniform speed. If the train would have been 10 km/h faster, it would have taken 2 hours less than the scheduled time. And, if the train were slower by 10 km/h, it would have taken 3 hours more than the scheduled time. Find the distance covered by the train.

6. The sum of areas of two squares is $468m^2$. If the difference of their perimeters is 24m, find the sides of the two squares.

7. The hypotenuse of a right triangle is 20m. If the difference between the length of the other sides is 4m. Find the sides.

8. The positive value of k for which $x^2 + kx + 64 = 0$ .....(1) & $x^2 - 8x + k = 0$ ...(2) will have real roots.

9. A motor boat whose speed in still water is 18 km/h takes 1 hour more to go 24 kms upstream than to return downstream to the same spot. Find the speed of the stream.

10. A squirrel and a crow are at the top of a pole of height 6m, whose base was at a distance of 10m from a point A on the ground. The squirrel descends the pole and went straight to the point A. The crow flew vertically up to a height x metres and

then flew in a straight line to the point A, the distance travelled by each being the same. Find the value of x.

11. Find coefficient of x^4 in the expansion of $(1 + x)^n(1-x)^n$

12. In the expression $(1+x)^n$, three successive coefficients are 462, 330 and 165 respectively, find the value of n and r.

13. Find the value of r, if coefficient of $(2r + 4)^{th}$ & $(r-2)^{th}$ term in the expansion $(1 + x)^{18}$ are equal

14. Show that $(101)^{50} > (100)^{50} + (99)^{50}$

15. Find the value of k so that the term independent of x in $\left(\sqrt{x}+\frac{k}{x^2}\right)^{10}$ is 405.

16. In the expansion of $\frac{1+x}{1!}+\frac{(1+x)^2}{2!}+\frac{(1+x)^3}{3!}+$............ find coefficient of x^n

17. In the expansion of $(1 + 3x + 2x^2)^6$, find coefficient of x^{11}

18. Which will be the first negative term in the expansion of $(1 + x)^{\frac{27}{5}}$, given that $0 < x < 1$?

19. If $(1+x)^n = C_0 + C_1x + C_2x^2 + ...C_nx^n$, find the value of $\frac{(C_0+C_1)(C_1+C_2)(C_2+C_3)............(C_{n-1}+C_n)}{C_1C_2C_3................C_n}$ where $C_r = {}^nC_r$

20. If $s_n = \sum_{r=0}^{n} \frac{1}{{}^nC_r}$ and $t_n = \sum_{r=0}^{n} \frac{r}{{}^nC_r}$, find the value of $\frac{t_n}{s_n}$

Answers to Practice Problems:

1. x = 0 and y = 0
2. –255
3. Jacob 40 year and his son 10 years
4. x = 1 and y = 1
5. 600 kms
6. x = 18 and y = 12
7. 12 and 16 metres
8. 16
9. 6 km/h
10. $x = \frac{30}{11}m$
11. $\frac{n(n-1)}{2}$
12. n = 11, r = 6
13. t ***r* = 6**
15. k = ± 3
16. $x^n = \frac{e}{n!}$
17. $x^{11} = 576$
18. 8th term
19. $\frac{(n+1)^n}{n \times (n-1)(n-2)............2 \times 1} = \frac{(n+1)^n}{n!}$
20. $\frac{t_n}{s_n} = \frac{n}{2}$

3

GEOMETRY

3. Introduction:

Geometry is a branch of Mathematics concerned with questions of shape, size, relative position of figures, and the properties of space. You will encounter properties of different shapes and practical knowledge concerning lengths, areas and volumes.

We will deal here with important properties of different shapes and attempt to understand logic behind those properties so as to have easy recall.

3.1 Lines and Angles:

3.1.1 Lines:

A line segment is the shortest distance between 2 points. A ray has one end point and straight line has neither a starting point nor an end point. When two lines meet each other they form a set of angle(s)

Perpendicular bisector of a line is a line segment that divides it into two equal parts and forms a right angle to the given line. An important property of a perpendicular bisector is that

All points on the perpendicular bisector of a line segment are equidistant from the end points of the line.

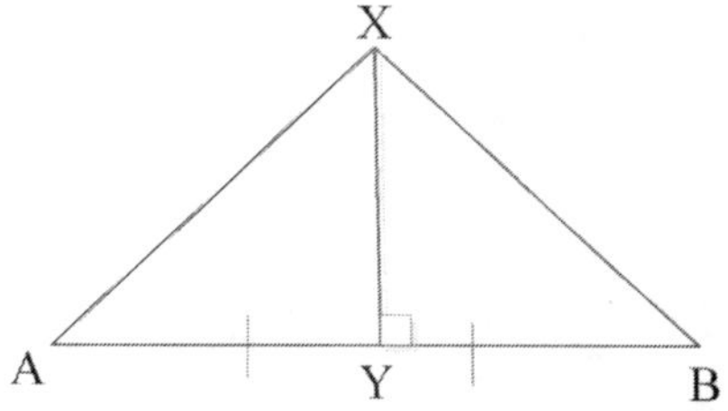

Figure 3.1

For illustration, XY is perpendicular bisector of line segment AB in the figure 3.1 above. And all points on XY are equidistant from A and B.

The converse is also true. That is if points on a line segment are equidistant from end point of a line segment, the point lies on perpendicular bisector of the line. This can be correlated to properties of an isosceles triangle as we go forward.

It is not intended to give proof for all properties being listed for geometrical figures in this chapter. The students can refer to their text books for proof.

3.1.1.1 Perpendicular lines:

When there is a right angle between two lines, the lines (OA and OB in the figure 3.2 below) are said to be perpendicular to each other.

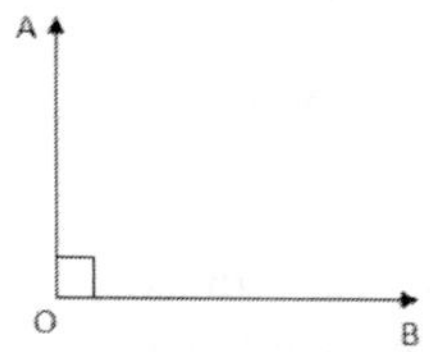

Figure 3.2

3.1.1.2 Parallel lines:

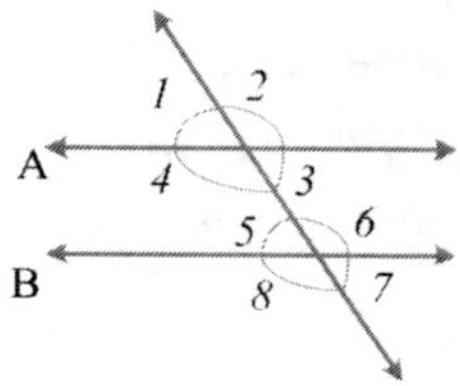

Figure 3.3

Here, A and B are two parallel lines, intersected by a line p. The line p is called a transversal. As seen in the figure above, when a transversal intersects two lines, 8 angles are formed.Let us consider the details in a tabular form for easy reference.

Types of Angles	Angles
Interior Angles	$\angle 3$, $\angle 4$, $\angle 5$, $\angle 6$
Exterior Angles	$\angle 1$, $\angle 2$, $\angle 7$, $\angle 8$
Vertically opposite Angles	($\angle 1$, $\angle 3$), ($\angle 2$, $\angle 4$), ($\angle 5$, $\angle 7$), ($\angle 6$, $\angle 8$)
Corresponding Angles	($\angle 1$, $\angle 5$), ($\angle 2$, $\angle 6$), ($\angle 3$, $\angle 7$), ($\angle 4$, $\angle 8$)
Interior Alternate Angles	($\angle 3$, $\angle 5$), ($\angle 4$, $\angle 6$)
Exterior Alternate Angles	($\angle 1$, $\angle 7$), ($\angle 2$, $\angle 8$)
Interior Angles on the same side of transversal	($\angle 3$, $\angle 6$), ($\angle 4$, $\angle 5$)

When a transversal intersects two parallel lines,
The corresponding angles are equal.
The vertically opposite angles are equal.
The alternate interior angles are equal.
The alternate exterior angles are equal.
The pair of interior angles on the same side of the transversal is supplementary.
We can also say that the lines are parallel if we can verify at least one of the afore mentioned conditions.

Note: Parallel lines when extended don't meet. However non parallel lines when extended will meet and intersect each other at some point.

3.1.2 Angles

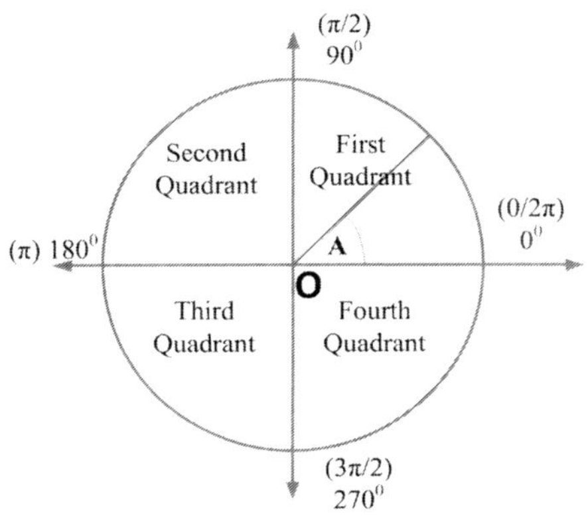

Figure 3.4: Measures of Angles

Angles are measured in degrees or radians. Conventionally, positive value of angle is measured in counter clockwise direction in a circle. Measure is negative if we go clockwise. Diagram 3.4 below shows a circle with diameters crossing each other vertically.

The angle formed at the centre all around the circle measures 360 degree or 2π radians by definition. Unit π is used extensively in measurement of circumference/area of circles. An angle A is drawn in first quadrant (quarter of the circle) of the circle. Measure of angle in first quadrant varies from 0^0 to 90^0, in 2nd quadrant from 90^0 to 180^0 and so on. In terms of radians, measure of angle in first quadrant varies from 0 to $\frac{\pi}{2}$, in 2nd quadrant from $\frac{\pi}{2}$ to π and so on.

Angles are divided in four basic categories:

- **Acute angle** is less than 90^0 ($\pi/2$)
- **Obtuse angle** is more than 90^0 and less than 180^0 (π)
- **Right angle** is of 90^0
- **Straight angle** is of 180^0

3.1.2.1 Angle Bisectors

> Here CP bisects the angle APB. If a point lies on the angle bisector, it is equidistant from the arms of the angle. That is CA=CB where $CA \perp PA$ and $CB \perp PB$
>
> Converse is also true, viz if a point in interior of the angle is equidistant from the arms of the angle, it lies on the bisector of the angle.

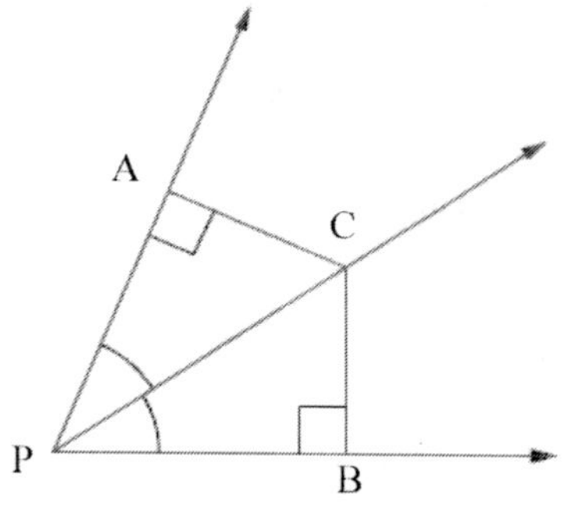

Figure 3.5

3.1.2.2 Pairs of angles

3.1.2.2.1 Adjacent angles

The angles that have a common arm and a common vertex are called adjacent angles. In the figure above, ∠CPA and ∠CPB are adjacent angles. Their common arm is PC and common vertex is 'O'. ∠APC + ∠CPB = ∠APB.

3.1.2.2.2 Supplementary angles

If the sum of two angles is 180° then the angles are called supplementary angles.The pair of adjacent angles whose sum is a straight angle is also called a linear pair.

In the figure 3.6 below angles AOC and COB are supplementary

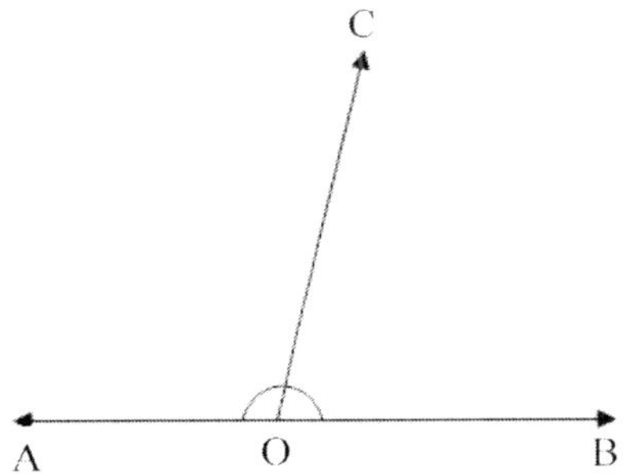

Figure 3.6

3.1.2.2.2 Complementary angles

Sum of the Complementary angles is 90^0. In figure 3.7 below angles COA and AOB are complementary. i.e. ∠COA + ∠AOB = 90°

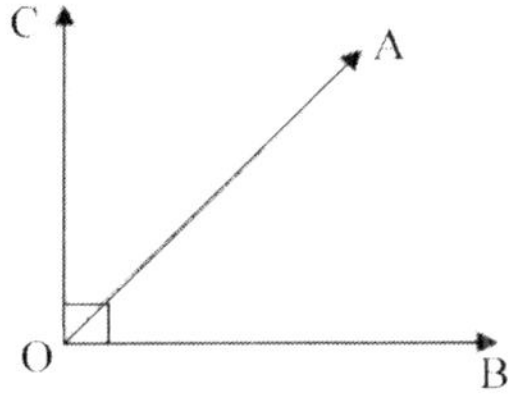

Figure 3.7

3.1.2.2.3 Vertically opposite angles

When two lines intersect, the angles formed opposite to each other at the point of intersection (vertex) are called vertically opposite angles. Vertically opposite angles are equal.

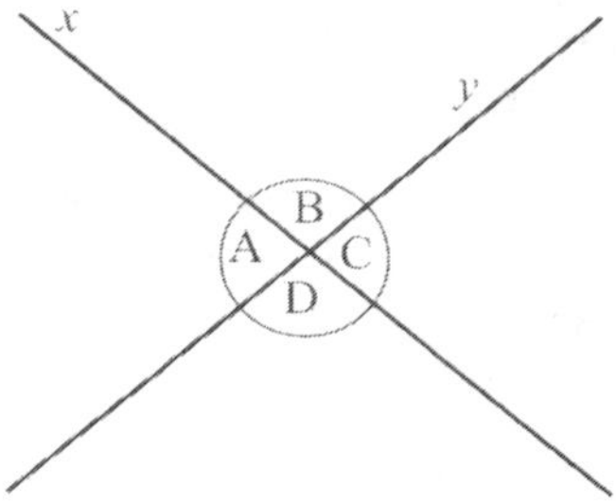

Figure 3.8

In the figure 3.8 above, x and y are two intersecting lines.

∠A and ∠C make one pair of vertically opposite angles and ∠B and ∠D make another pair of vertically opposite angles

3.2 Triangles

Triangles are the most important part of Geometry as properties of triangles find frequent use in solving problems not only for triangular shapes but also for other geometrical shapes like quadrilaterals, pentagons, circles and other polygons.

> *It is very important to understand various properties of different type of triangles as well as to develop an eye to recognize those properties in complex shapes and figures to quickly solve problems.*

3.2.1 Different type of Triangles

There are two broad category of triangles- Acute and Obtuse.

There are four types of triangles - Equilateral, Isosceles, Scalene and Right Angled Triangle. Different types of triangles are depicted in figure 3.9 below.

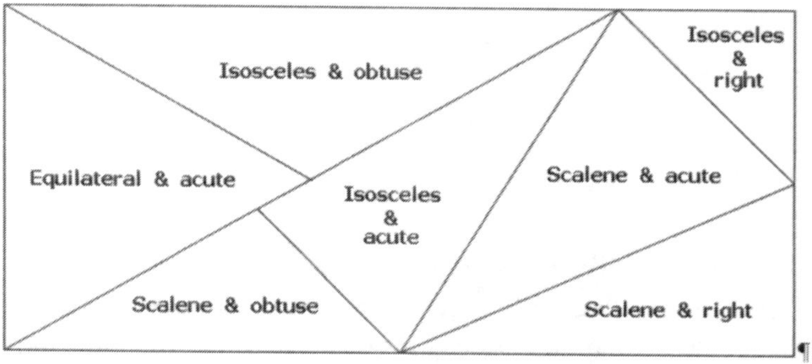

Figure 3.9 (Courtsey: *First Steps in Geometry* by G. A. Wentworth and G. A. Hill)

The basic elements of any triangle are its sides and angles. Traditionally, angles are often denoted by the capital letter of the vertex of the triangle. For example, in triangle ABC, the angles could be denoted as Angle A, Angle B and Angle C respectively. Similarly the side opposite to an angle is denoted by the same small case letter. So side opposite Angle Ais denoted as a, side opposite Angle Bis denoted as b and so on.

Triangles are classified depending on relative sizes of their elements.As regard their sides, triangles are classified as

1. ***Scalene*** (all sides are different)
2. ***Isosceles*** (two sides are equal)
3. ***Equilateral*** (all three sides are equal)

And as regard their angles, triangles are classified as

- ***Acute*** (all angles are acute)
- ***Right*** (one angle is right)
- ***Obtuse*** (one angle is obtuse)
- ***Equiangular*** (all angles are equal)

Common properties of any type of triangle:

1. Sum of its interior angles is 180 degrees. This is also known as ***Angle Sum Property***
2. ***An Exterior angle*** formed by extending a side of a triangle ***equals sum of the other two angles of the triangle,*** leaving the angle adjacent to the exterior angle.
3. Sum of any two sides is greater than the third side. This also implies that any side is lesser than sum of the other two sides and greater than difference of the other two sides. So we can say that $(a+b)>c>(a-b)$
4. Length of a side is proportional to the angle opposite to the side. It thus implies that sides opposite to equal angles are equal. And side opposite to the larger angle is larger.
5. See section 3.2.5 for another interesting property relating to heights of the triangle from the three vertices

3.2.2 Congruence of Triangles:

Two triangles are said to be congruent if their size and shape are exactly the same. Congruent triangles will match exactly if they are placed one above the other. There are 3 tests/Rules of congruence that should be understood very clearly.

SSS Rule:

If all 3 sides of a triangle are equal to all 3 sides of another triangle, the triangles will be congruent and their corresponding angles will be equal.

SAS Rule:

If two sides of a triangle and the included angle are equal to the corresponding sides and angle in another triangle, the triangles will be congruent and the third side and other two angles will also be same.

AAS Rule:

If any two angles and a side of a triangle are equal to the corresponding angles and side in another triangle, the triangles will be congruent and the other two sides will also be equal. Please note that if two angles are equal, it implies all 3 angles are equal (Angle Sum property. ASA Rule, therefore, effectively means that all angles are equal and one of the sides is equal.

3.2.3 Similar Triangles:

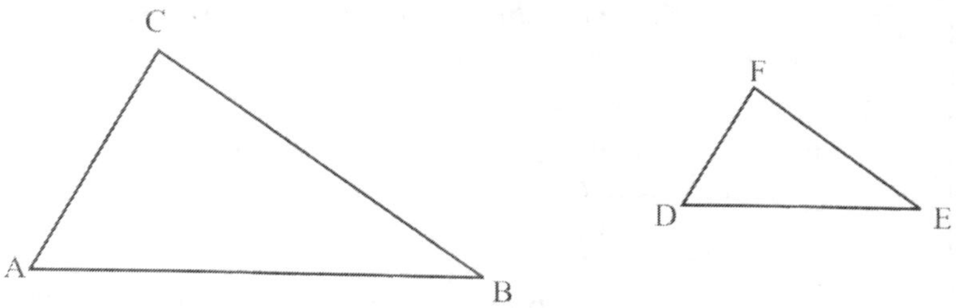

Figure 3.10

Reference figure 3.10

Two triangles ABC and DEF are said to be Similar if

(i) All 3 angles in triangle ABCare equal to all 3 angles in another triangle (DEF) but none of the sides are equal.

(ii) Two sides are proportional and the included angle is equal

(iii) All three sides of one triangle are proportional to the corresponding sides in the other triangle.

In each of the above cases, all three sides of the similar triangles are directly proportional to each other by a common factor. So in the above figure $\frac{AB}{DE} = \frac{BC}{EF} = \frac{CA}{FD} = k$

3.2.4 Specific properties of different types of Triangles:

3.2.4.1 Equilateral Triangle:

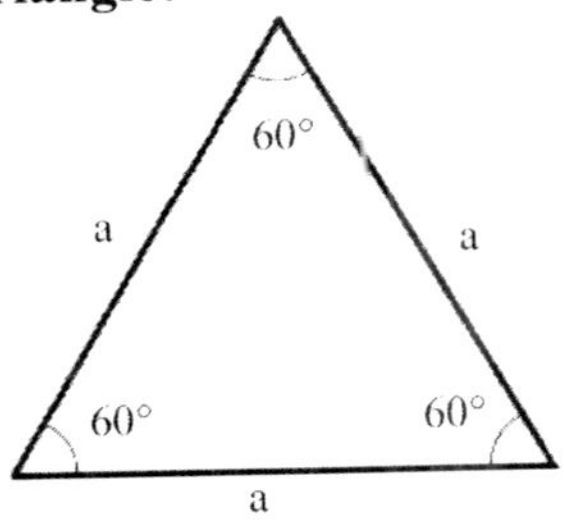

Figure 3.11

> 1. *Each angle is equal and is of 60 degree (Sum of the angles being 180 degree).*
> 2. *Each side is equal, being opposite to equal angles.*
> 3. *A perpendicular drawn from the vertex to the base (known as altitude) bifurcates the base. It also bisects the vertex angle. Clearly it is perpendicular bisector of the base and passes through the vertex.*
> 4. *All altitudes are of equal length.*
> 5. *Height from any vertex to the opposite side equals sum of the distances from any interior point to its sides (For proof for this property refer to section 3.2.4.4*

3.2.4.2 Isosceles Triangle:

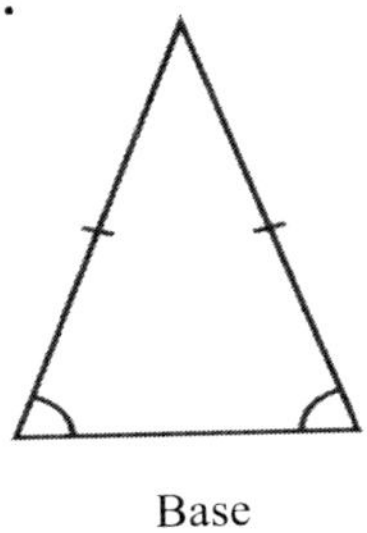

Figure 3.12

1. *Two sides are equal*
2. *Angles opposite to equal sides are equal.*
3. *A perpendicular, drawn from point of intersection of equal sides to the third side (generally base of the triangle), bifurcates the third side. It also bisects the vertex angle. Conversely, if a line is drawn from point of intersection of equal sides to the centre of the third side, it will meet the third side at right angle and hence will be perpendicular bisector of the base.*

3.2.4.3 Right Angled Triangle

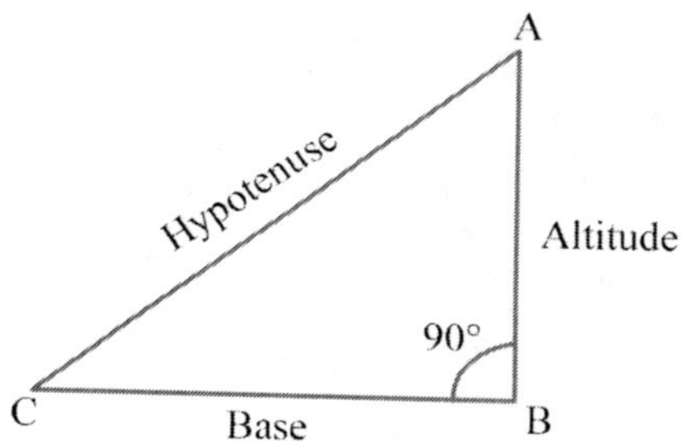

Figure 3.13

1. *One angle is 90^0 and hence sum of other two angles is also 90 degrees.*
2. *Two arms of the Right angle are generally named Base and Altitude of the triangle. The third side (opposite to the Right angle is called Hypotenuse*
3. *Sides are related by Pythagoras theorem as $(Hypotenuse)^2 = Base^2 + Altitude^2$*
4. *Numbers that satisfy Pythagoras theorem are called Pythagorean triplets*

Hint: ***It is good to remember Pythagorean triplets like (3,4,5), (5,12,13), (7,24,25) and (8,15,17) for quick calculations. Triplets with a common multiple to the above triplets also qualify. Example (3,4,5), (4.2,5.6,7), (6,8,10) etc.***

Note: Refer to Chapter 4 on Number series for more insight on Pythagorean triplets.

Important information

We will deal with area and other measurements of different geometrical figures in Chapter 6 on Mensuration. However, in order to understand proof for Pythagoras theorem below, it is important to know that area of a triangle = 1/2× base× height of the triangle from apex to the base.

An interesting proof of Pythagoras theorem:

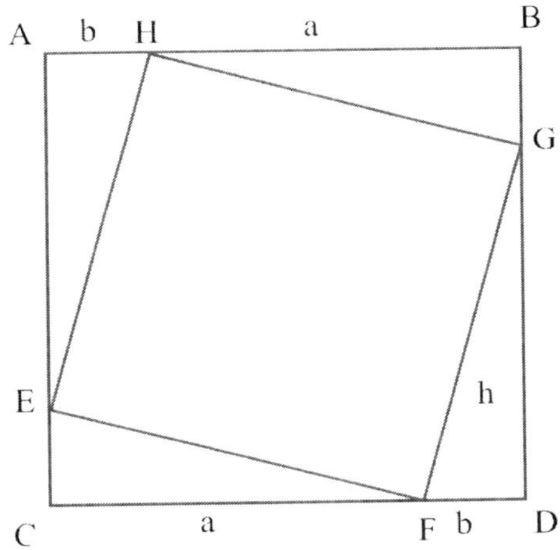

Figure 3.14

In figure 3.14 above, square ABCD has length L. Each side is divided into two parts of length a and b as shown. These points are marked as E, F, G and H. and joined.

Notice that we have 4 right angles congruent triangles, CEF, DFG, GHB and HEA (as two sides in each triangle are equal to their corresponding sides in other triangles). Area of each of these 4 triangles is $\frac{1}{2}ab$.

Take triangle CEF. Sum of angles CEF and angle EFC is 90^0. Since triangles are congruent, Sum of angles AEH and angle CEF is also 90^0. Thus angle HEF is 90^0, implying thereby that EFGH is a square. Let side length of this square be ***h***.

Thus outer square of area L^2 comprises of a smaller square of area h^2 and 4 right angled triangles of area $\frac{1}{2}ab$ each.

Thus, $L^2 = h^2 + 4 \times \frac{1}{2}ab = h^2 + 2ab$

Also $L^2 = (a+b)^2 = a^2+b^2+2ab$

Equating the above two equations,

$h^2 + 2ab = a^2+b^2+2ab$

or $h^2 = a^2+b^2$ - Hence proved.

3.2.4.4 Scalene Triangle:

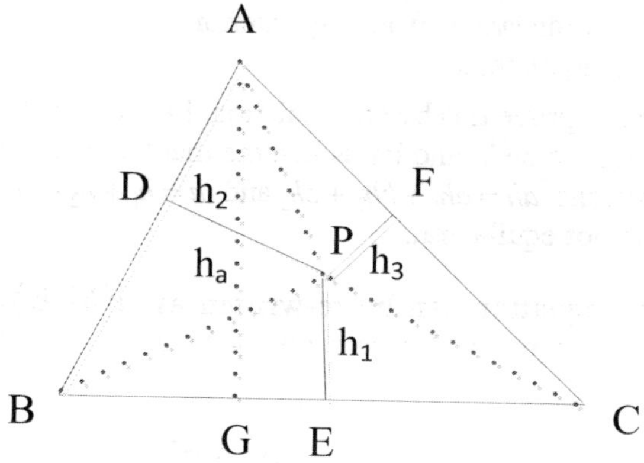

Figure 3.15

A Scalene triangle has all three angles and all three sides unequal. All common properties of triangles listed above are applicable to scalene triangles.

An interesting property is that if h_a(AG) is height from vertex A to the base BC of length a of any scalene triangle ABC and h_1, h_2 and h_3 be the distances from any interior point P to the sides having lengths a, b and c of this triangle, then $ah_a = ah_1 + bh_2 + ch_3$

Proof:

Reference figure 3.15.

We can see that the triangle ABC is split into 3 triangles BCP and CAP and ABP each having base length **a**(BC), **b**(CA) and **c** (AB) and heights h_1(PE), h_2(PF) and h_3(PD) respectively.

If height of the triangle is h_a (AF) from base a, we get

$$\frac{1}{2}ah_a = \frac{1}{2}ah_1 + \frac{1}{2}bh_2 + \frac{1}{2}ch_3 \Rightarrow ah_a = ah_1 + bh_2 + ch_3$$

The above equation can also be rewritten as $h_a = h_1 + \frac{b}{a}h_2 + \frac{c}{a}h_3$

And if the triangle is equilateral, a = b = c and $h_a = h_b = h_c = h$ leading to $h = h_1 + h_2 + h_3$. This is known as ***Viviani's theorem.***

Converse of this property is also true i.e. if height of a triangle equals to sum of distances from any interior point to the three sides, the triangle is Equilateral.

Let us try to prove this by contradiction. Let height of the triangle on side of length **a** be h, and let us assume that the triangle satisfying the two equations $ah = ah_1 + bh_2 + ch_3$ and $h = h_1 + h_2 + h_3$ where h is a constant, is not equilateral.

The first equation can be re-written as $a(h - h_1) = bh_2 + ch_3$ and the second equation by multiplying both sides by a, gives us $a(h - h_1) = ah_2 + ah_3$

The above equations imply that $ah_2 + ah_3 = bh_2 + ch_3$ ...... (1)

Since h_2 and h_3 are two distinct positive numbers, eq (1) above implies that both **b** and **c** must be equal to **a** in order to satisfy this equation.

Hence a = b = c confirming that the triangle has to be equilateral.

Here is a tricky problem related to Viviani's theorem.

Problem:

Point P is chosen inside an equilateral triangle ABC and D, E and F are the feet of the perpendiculars from P to AC, BC and AB respectively. It is given PD: PE: PF:: 1 : 2 : 4. Find the ratio of the areas of quadrilateral CDPE and triangle ABC.

Solution:

Refer to the figure 3.16 below depicting the given conditions.

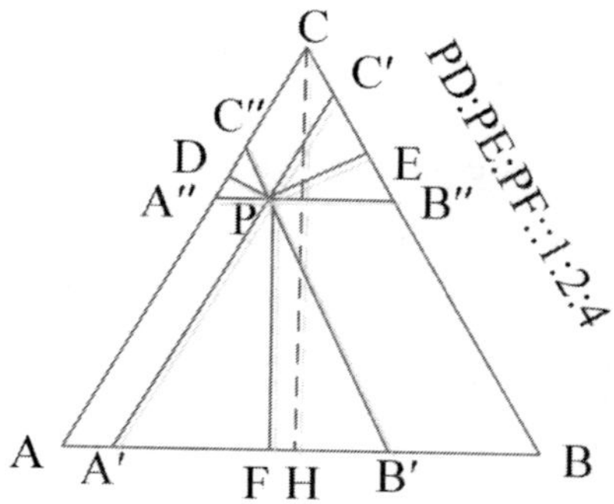

Figure 3.16

Lines A′C′, A″B″ and B′C″ are drawn parallel to the sides AC, AB and BC respectively. Let us say length of PD, PE and PFis p, 2p and 4p respectively. Hence height of the equilateral triangle ABC equals 7p (as per Viviani's theorem). This leads to the conclusion that height of $\Delta CA"B"$= CH - PF = 7p - 4p = 3p.

In order to find area of the quadrilateral CDPE, we can take area of $\Delta CA"B"$ and subtract area of $\Delta s\, A"PD \,\&\, B"PE$. We can easily visualise that $\Delta s\, A"PC"\&\, B"PC'$ are equilateral (recall properties of parallel lines). Hence area of $\Delta s\, A"PD \,\&\, B"PE$ are respectively half of area of $\Delta s\, A"PD \,\&\, B"PE$.

Now let us recall that area of an equilateral triangle is proportional to square of its side length or its height. Since in this case we know proportion of heights, we can see that ratio of areas of $\Delta CA"B"$: ΔABC :: 9:49.

Similarly ratio of areas of $\Delta s\, A"PD : \Delta ABC \,\&\, \Delta B"PE : \Delta ABC$ are 1:49 and 4:49.

Going by the above arguments ratio of area of quadrilateral CDPE to area of ΔABC will be $\frac{9}{49}-\frac{1}{2}\left(\frac{1}{49}+\frac{4}{49}\right)=\frac{9}{49}-\frac{5}{2\times 49}=\frac{13}{98}$

3.2.5 Perimeter and area of a Triangle:

Perimeter equals sum of three sides of a triangle and is logically 3 times side of an Equilateral triangle and 2 times equal sides of Isosceles triangle plus its base.

Area of a triangle is arrived at by multiplying base and altitude divided by 2. Thus we need to calculate altitude of the triangle if not known for finding its area. This also gives us the information that for fixed area, altitude of a triangle is inversely proportional to length of its base. This leads us to another interesting property in continuation to ***Common properties of a triangle*** listed in section 3.2.1.

We saw there that if a, b, c are the three sides of the triangle, $(a+b)>c>(a-b)$. Now if h_1 is the altitude from vertex A to the side **a**, h_2 is the altitude from vertex B to the side **b** and h_3 is the altitude from vertex C to the side **c**, we get $a \propto \frac{1}{h_1}, b \propto \frac{1}{h_2} \;\&\; c \propto \frac{1}{h_3}$.

Substituting in the above inequality, we get

$$\left(\frac{1}{h_1}+\frac{1}{h_2}\right)>\frac{1}{h_3}>\left(\frac{1}{h_1}-\frac{1}{h_2}\right) \Rightarrow \left(\frac{h_1+h_2}{h_1 h_2}\right)>\frac{1}{h_3}>\left(\frac{h_1-h_2}{h_1 h_2}\right).$$

Taking the inverse we get

$\left(\frac{h_1 h_2}{h_1+h_2}\right)<h_3<\left(\frac{h_1 h_2}{h_1-h_2}\right)$. *This is again a very important inequality for solving related questions. For an example refer to Q6 of practice sums in section 3.4*

Common formulae for area of different type of triangles are as below:

Equilateral triangle with side length **a** $= \frac{\sqrt{3}}{4}a^2$.

Isosceles triangle with equal sides **a** and base as **b** $= \frac{1}{4}b \times \sqrt{4a^2-b^2}$.

Right angled triangle with base **b** and height **h** $= \frac{1}{2} \times bh$

Scalene Triangle with sides **a, b, c** $= \sqrt{s(s-a)(s-b)(s-c)}$ where s is the semi-perimeter of the triangle, that is $s=\frac{a+b+c}{2}$. This is Heron's formula.

Proof for Heron's formula:

Reference figure 3.17 below. BD is perpendicular to AC dividing it in two parts of length x and (b – x). Height of the triangle BD = h leading to Area of the triangle = $\frac{1}{2}bh$......(1)

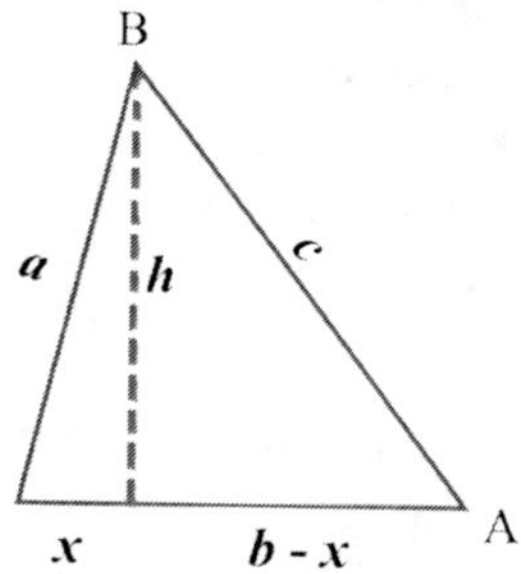

Figure 3.17

Using Pythagoras theorem, in ΔBCD,

$$x^2 + h^2 = a^2 \Rightarrow x^2 = a^2 - h^2 ....(2) \Rightarrow x = \sqrt{a^2 - h^2} ....(3)$$

Similarly, in ΔABD, $(b-x)^2 + h^2 = c^2 \Rightarrow b^2 + x^2 - 2bx = c^2 - h^2$

Substituting values of x and x^2 from equations (2) and (3),

$$b^2 + a^2 - h^2 - 2b \times \sqrt{a^2 - h^2} = c^2 - h^2 \Rightarrow$$

$$b^2 + a^2 - c^2 = 2b \times \sqrt{a^2 - h^2} ....(4)$$

Squaring both sides of equation 4,

$$\left(b^2 + a^2 - c^2\right)^2 = 4b^2 \times \left(a^2 - h^2\right) \Rightarrow h^2 = \frac{4a^2b^2 - \left(b^2 + a^2 - c^2\right)^2}{4b^2}$$

$$= \frac{(2ab)^2 - \left(b^2 + a^2 - c^2\right)^2}{4b^2} \Rightarrow$$

$$h^2 = \frac{\left(2ab + b^2 + a^2 - c^2\right)\left(2ab - b^2 - a^2 + c^2\right)}{4b^2}$$

$$= \frac{\left[(a+b)^2 - c^2\right]\left(c^2 - (a-b)^2\right)}{4b^2} \Rightarrow$$

$$h^2 = \frac{(a+b+c)(a+b-c)(c+a-b)(c-a+b)}{4b^2} \Rightarrow 2bh$$

$$= \sqrt{(a+b+c)(a+b-c)(c+a-b)(c-a+b)}$$

Now area of $\Delta ABC = \frac{1}{2}bh$ leading to Area of

$$\Delta ABC = \frac{\sqrt{(a+b+c)(a+b-c)(c+a-b)(c-a+b)}}{4} =$$

$$\sqrt{\frac{(a+b+c)}{2}\frac{(a+b-c)}{2}\frac{(c+a-b)}{2}\frac{(c-a+b)}{2}}$$

Now with semi-perimeter $s = \frac{a+b+c}{2}$, the above expression can be re-written as

Area of $\Delta ABC = \sqrt{s(s-c)(s-b)(s-a)}$, Hence proved.

Refer to Chapter 6 Mensuration for Algebraic proof for Area of Equilateral, isosceles and scalene triangles.

3.2.6 Advance properties of Triangles

We now learn about advanced properties of triangles. Proof for different properties are dealt in ***Chapter 6 on Mensuration.***

3.2.6.1 A line joining any vertex to mid-point of the opposite side is called its Median.

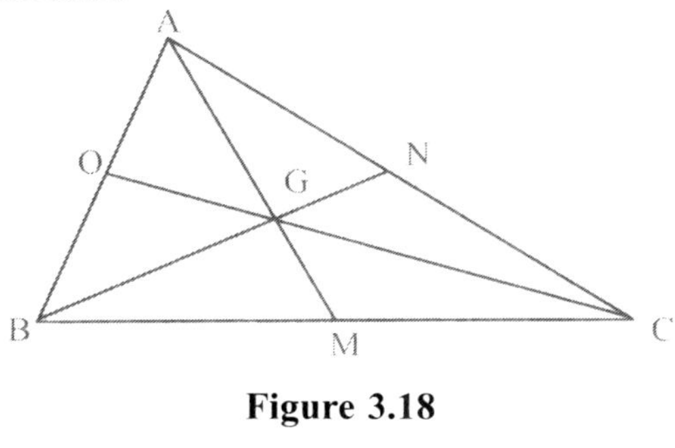

Figure 3.18

Each Median divides the triangle into two parts of equal area. *This can be visualized easily since the two parts have same base length (being half of the original side) and same height* ***All three Medians***

meet at a point called Centroid (G in the diagram above), which is also the centre of mass of the triangle.

> ***A few important properties that are derived from the Medians are:***
>
> a. ***Line segments joining mid points of two sides viz ON, OM and MN are parallel to the third side viz BC, AC and AB respectively and are half of their size. Thus ON= ½BC and so on. This is also known as Midpoint Theorem.***
> b. ***Converse is also true. That is the line drawn through the mid-point of one side of a triangle and parallel to another side bisects the third side.***
> c. ***The 3 medians divide the triangle into 6 triangles each of equal area.***
> d. ***Distance from each vertex to the Centroid is 2/3 of the length of that Median***

3.2.6.2 Perpendicular bisectors of a triangle meet at a point called Circumcentre of the triangle.

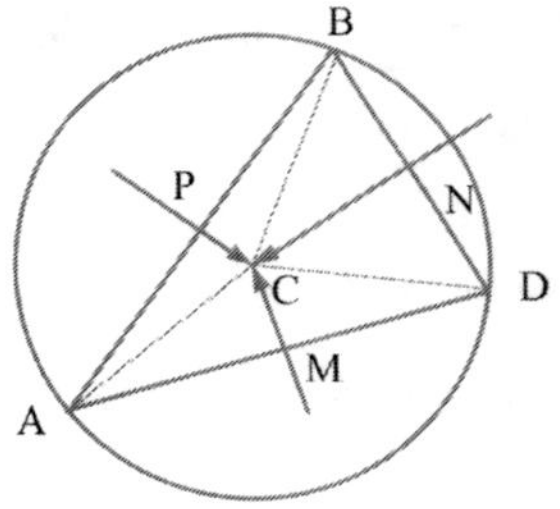

Figure 3.19

In the figure 3.19 above, PC and NC are perpendicular bisectors of sides AB and BD meeting at C. Similarly perpendicular bisector of side AD will also meet at point C. Remember that all points on a perpendicular bisector of a line segment are equidistant from ends of the line. Thus Point C is equidistant from the three vertices of the triangle leading us to AC=BC=DC and C being ***Circumcenre*** *of the triangle. If we draw a circle with Circumcentre as centre and distance to a vertex as radius, it touches all three vertices of the triangle and is called* ***Circumcircle.***

> ***A few of important corollaries to remember:***
>
> a. ***In case of Acute triangles, circumcentre always lies within the triangle***
>
> b. ***In case of Obtuse triangles, circumcentre always lies outside the triangle.***
>
> c. ***In case of Right angled triangle, circumcenter lies on the hypotenuse and at its centre. i.e. hypotenuse is diameter of the circumcircle.***

Refer to section 6.1.3 in Chapter 6 where the following formula for Circumradius (R) of a triangle with side lengths a, b and c is derived

$$R = \frac{abc}{4 \times \Delta(ABC)}$$

3.2.6.3 The three line segments drawn perpendicular from each vertex to the line opposite to it (i.e. the altitudes) meet at a point.

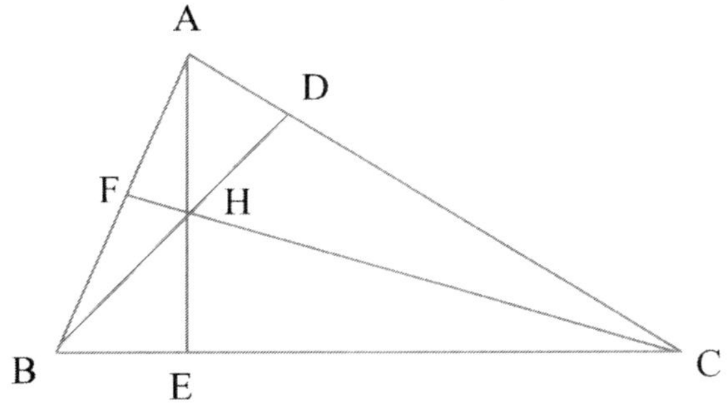

Figure 3.20

Point of concurrency (point H in the figure 3.20 above) is called. ***Orthocenter*** *of the Triangle. It may or may not lie within the triangle. In case of an Acute triangle it always lies within the triangle.*

There is no direct formula for Orthocenter of a triangle. One needs to use knowledge of coordinate geometry to find out Orthocenter.

An Important property:

Orthocenter, Circumcenter and the Centroid of any triangle are always collinear ie. they lie on a straight line.

Distance between Orthocenter and Centroid = 2 times distance between Centroid and Circumcenter.

Refer to Chapter 7 Section 7.2.9 on Co-ordinate Geometry for proof of the above relationship.

3.2.6.4 Angle bisector theorem

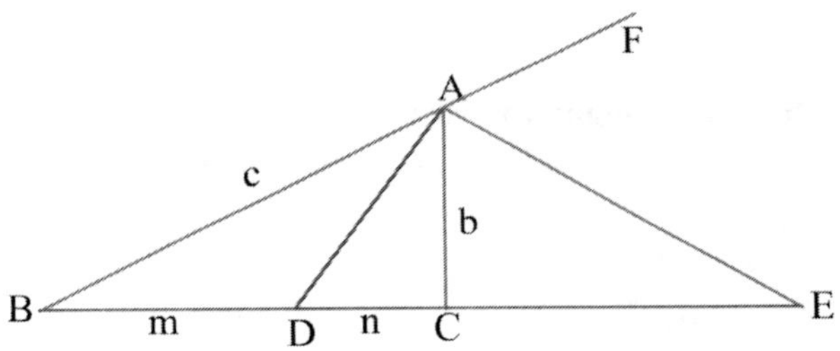

Figure 3.21

The angle bisector of an angle in a triangle divides the opposite side internally into the ratio of the sides containing the angle. It means that, as shown in figure 3-21 above, if there be a triangle ABC and AD be the angle bisector of angle A of this triangle and AD divides the side BC in m:n ratio then,

$\frac{AC}{AB}=\frac{b}{c}=\frac{CD}{BD}=\frac{n}{m}$. This theorem holds for both interior as well as exterior angles.

AE is angle bisector of exterior angle A, leading to $\frac{AC}{AB}=\frac{b}{c}=\frac{CE}{BE}=\frac{n}{m}$

The three angle bisectors of a triangle meet at a point called InCenter of the Triangle.

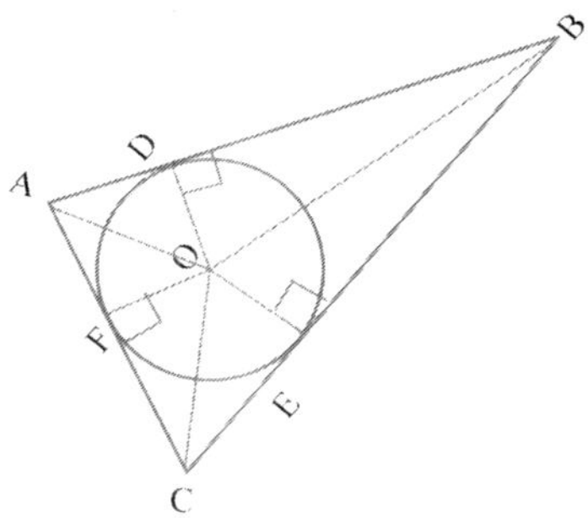

Figure 3.22

We can find radius of Incircle, also called inradius by refering to section 3.2.4.4 where we found relation between height of a triangle and distaces from an interior point to the three sides by the formula

$$ah_a = ah_1 + bh_2 + ch_3$$

Now we can see that ah_a = 2× Area of the triangle and if the interior point under consideration is centre of the incircle, then $h_1 = h_2 = h_3 = r(radius\ of\ the\ incircle)$ leading us to 2× Area of the triangle = $r(a+b+c)$ = Perimeter of the triangle

We thus find that inradius (r) is given by the formula

$$r = \frac{2 \times \Delta(ABC)}{Perimeter} \quad ...(1)$$

From the above expression, we can also see that

Area of triangle $\triangle ABC = r \times s \text{ where } s = semi-Perimeter$

Refer to section 6.1.4 in Chapter 6 where another formula as below is derived for the inradius.

$r = \sqrt{\dfrac{xyz}{x+y+z}}$, where x, y and z are the distance from the vertices to the points where the circle touches the three sides of the triangle.

We have seen in section 3.2.6.2 that formula for circumradius (R) is

$$R = \frac{abc}{4 \times \Delta(ABC)} \quad ..(2)$$

From (1) and (2) we can deduce that ***the radius r of in-circle and R of circumcircle are related as*** $r \times R = \dfrac{abc}{2 \times perimeter}$

> ***Another Important property:***
> ***In case of an Equilateral triangle, Incentre, Circumcentre, Orthocentre as well as Centroid meet at a common point.***

3.2.7 Unorthodox theorems related to triangles

3.2.7.1 Pompeiu's theorem

Let P be a point inside an equilateral triangle ABC. Then there exists a triangle with side lengths PA, PB, and PC.

Proof:

Reference figure 3.23 below. ABC is the equilateral triangle. To prove the theorem, we draw line AR at an angle of 60^0 to the line segment AP with AR = AP and join R to B. Since AR = AP, $\angle APR = \angle ARP$ and since $\angle RAP = 60^0$, $\angle APR \,\&\, \angle ARP$ are both 60^0 leading to the conclusion that ΔRAP is equilateral and PR = AP, we can see that in ΔRBP two sides are same as PA and PB.

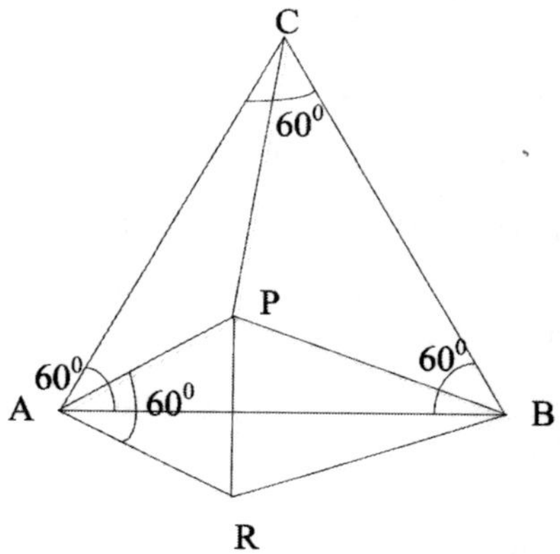

Figure 3.23

So, to prove the theorem, we need to prove that BR = CP so that ΔRBP is proved to be the triangle formed by sides of length PA, PB, and PC.

This can be done if we are able to prove that ΔAPC and ΔARB are congruent. We can see that $\angle CAP = \angle RAB$, each being $= 60^0 - \angle PAB$. Now since AC =AB and AP =AR,

By SAS Rule, ΔAPC and ΔARB are congruent.

Hence it is proved that PA, PB, and PC form a triangle.

3.2.7.2 Ceva's Theorem

Given a triangle *ABC*, let the lines *AO*, *BO* and *CO* be drawn from the vertices to a common point *O* (not on one of the sides of *ABC*), to meet opposite sides at *D*, *E* and *F* respectively. (The segments *AD, BE,* and *CF* are known as cevians.(Median and the angle bisector are special forms of cevian) Then, using lengths of segments,

$$\frac{AF}{FB} \times \frac{BD}{DC} \times \frac{CE}{EA} = 1$$

In other words, the length *XY* is taken to be positive or negative according to whether *X* is to the left or right of *Y* in some fixed orientation of the line. For example, *AF*/*FB* is defined as having positive value when *F* is between *A* and *B* and negative otherwise.

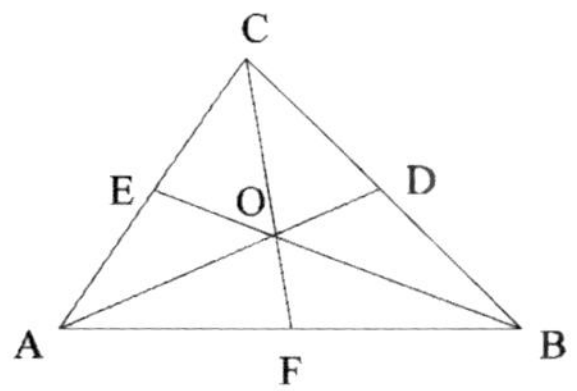

Figure 3.24

We can see that height of ΔBOD & ΔCOD is same. And area of two triangle shaving same height is proportional to their respective base length. Hence area of ΔBOD $(|\Delta BOD|)$ & ΔCOD are related as $\frac{|\Delta BOD|}{|\Delta COD|} = \frac{BD}{DC}$. Similarly, $\frac{|\Delta BAD|}{|\Delta CAD|} = \frac{BD}{DC}$. The above equations lead to

$$|\Delta BOD| \times DC = |\Delta COD| \times BD \;\; \& \;\; |\Delta BAD| \times DC = |\Delta CAD| \times BD$$

Subtracting like terms we get,

$$BD\times\left(|\Delta CAD|-|\Delta COD|\right)=DC\times\left(|\Delta BAD|-|\Delta BOD|\right)\Rightarrow\frac{|\Delta ABO|}{|\Delta CAO|}=\frac{BD}{DC}$$ ..... (1)

Similarly, $\frac{|\Delta CAO|}{|\Delta BCO|}=\frac{AF}{FB}$ ..... (2) and $\frac{|\Delta BCO|}{|\Delta ABO|}=\frac{CE}{EA}$ ...... (3)

Multiplying eqs.. (1), (2) and (3), we get

$$\frac{BD}{DC}\times\frac{AF}{FB}\times\frac{CE}{EA}=\frac{|\Delta ABO|}{|\Delta CAO|}\times\frac{|\Delta CAO|}{|\Delta BCO|}\times\frac{|\Delta BCO|}{|\Delta ABO|}=1$$ ... ***Hence proved***

3.2.7.3 Stewart's theorem

Stewart's theorem gives a relationship between lengths of sides of a triangle and a line segment that intersects a vertex and its opposite side.

Let ***a,b*** and ***c*** be the lengths of sides of a triangle. Let ***d*** be the length of the cevian to the side of length ***a*** dividing it in two segments of length ***m & n*** with m adjacent to c and n adjacent to b, then $b^2m+c^2n=a\left(d^2+mn\right)$

Proof:

Refer to figure 3.25. Here we can see that $m+n=a$ .... (1)

A perpendicular is drawn from the vertex A to meet BC at P. We can prove Stewart's theorem by using Pythagoras theorem as below:

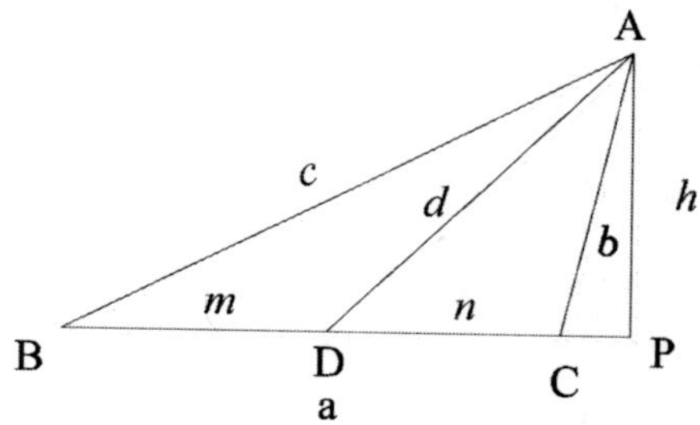

Let $DP=x$

Figure 3.25

In ΔADP, $d^2=h^2+x^2$ .... (2)

Similarly, in $\Delta ABP\ \&\ \Delta ACP$, we have

$c^2 = h^2 + (x+m)^2$ ....(3)

and $b^2 = h^2 + (x-n)^2$ ....(4)

Multiplying (3) by n and (4) by m we get,

$c^2 n = h^2 n + n \times (x+m)^2 \Rightarrow c^2 n = h^2 n + nx^2 + nm^2 + 2mnx$ ......(5)

$b^2 m = h^2 m + m \times (x-n)^2 \Rightarrow b^2 m = h^2 m + mx^2 + mn^2 - 2n$......(6)

Adding (5) and (6), we get

$b^2 m + c^2 n = h^2 (m+n) + x^2 (m+n) + mn(m+n)$

Since, $m + n = a$, we get $b^2 m + c^2 n = a\left(h^2 + x^2 + mn\right) = a\left(d^2 + mn\right)$ ...Hence proved.

3.2.7.4 Apollonius's Theorem:

Apollonius's Theorem states that the sum of the square of any two sides of a triangle equals twice the sum of the square of half the third side and the median bisecting the third side.

If ABC be any triangle as shown in figure 3.26 below where AM is the median of the triangle bisecting BC then, $AB^2 + AC^2 = 2\left(AM^2 + BM^2\right)$

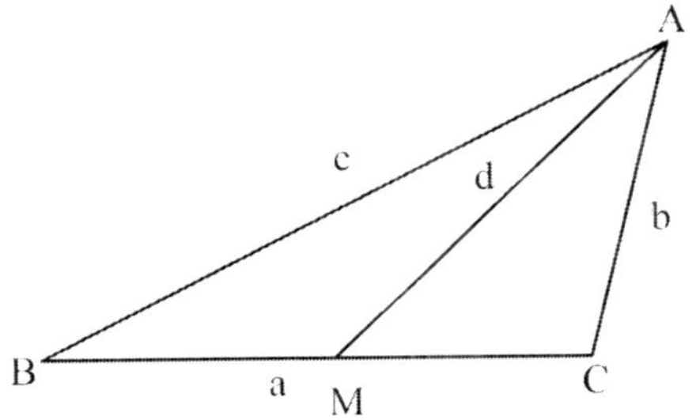

Figure 3.26

Proof:

If we substitute $m = n = \frac{a}{2}$ in the Steward's theorem, we get

$$\frac{a}{2}\times\left(b^2+c^2\right)=a\left(d^2+\left(\frac{a}{2}\right)^2\right)\Rightarrow b^2+c^2=2\left(d^2+\left(\frac{a}{2}\right)^2\right)$$

With reference to figure 3.26, the above expression translates to $AC^2+AB^2=2\left(AM^2+BM^2\right)$........ Hence proved.

3.2.7.5 Mass Point Geometry

Now we will deal with a very interesting concept. You would have observed a weighing balance and a seesaw. They work on the principle that there is a balancing point in the middle and balance is achieved when weight on both sides is equal as depicted below:

Now if we shift the balance point off centre, it is found that weight required to balance the weight on the longer side has to be more than the other weight. Ratio of the two weights is exactly same as the ratio of the two unbalance lengths.

Thus we can say that as shown in the diagram if weights are w_1 and w_2 at a distance of d_1 and d_2(d_2>d_1) respectively from the balance point, we have $w_1d_1=w_2d_2\Rightarrow\frac{w_1}{w_2}=\frac{d_2}{d_1}$ and the effective weight at the balance point is w_1+w_2.

This concept is also useful as a geometry problem solving technique for triangles and intersecting cevians. All problems that can be solved by this concept can also be solved by using properties of similar triangles and area ratio etc. but many problems can be solved more easily by Mass Point Geometry concept.

Let us look at a couple of examples for clarity of the concept.

Example 1:

In triangle ABC, E is on AC such that CE=3AE and F is onAB so that BF= 5AF. If BE and CF intersect at O and line AO intersects BC at D, compute $\frac{OB}{OE}$ & $\frac{OD}{OA}$.

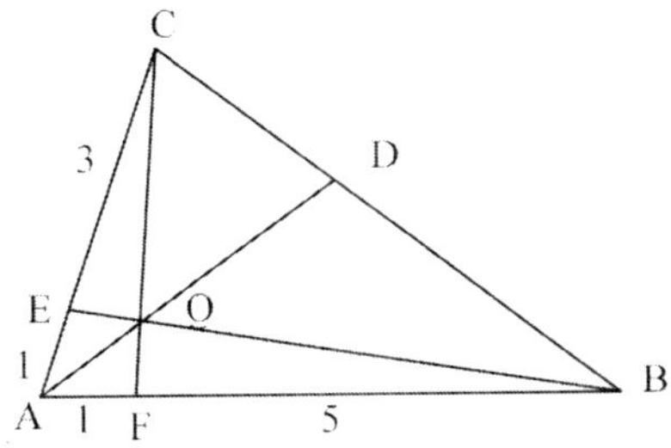

Diagram for example 1

As per Mass Point Geometry concept, with reference to side AC, mass at points A, C and E will be 3, 1 and 4 respectively. Similarly, mass at points A, B and F will be 5, 1 and 6 respectively.

It is convenient to have integer value of mass at all points. Since point A has two possible values of mass (3 and 5), we need to have mass at A that is LCM of the two masses i.e. 15. Using, mass at points A as 15, we get mass at points B, C, E and F as 3, 5, 20 and 18 respectively.

This also implies that mass at point D will be 3 +5 = 8 and at the intersection point O will be 15+8, 5 + 18 or 3 + 20 = 23.

Now we can easily calculate $\frac{OB}{OE} = \frac{20}{3}$ & $\frac{OD}{OA} = \frac{15}{8}$

Example 2:

In triangle ABC, points D, E and F are on sides BC, CA and AB respectively such that AE = AF = CD = 2, BD = CE= 3 and BF = 5. If DE and CF intersect at O, compute $\frac{OD}{OE}$ & $\frac{OC}{OF}$.

Reference diagram below:

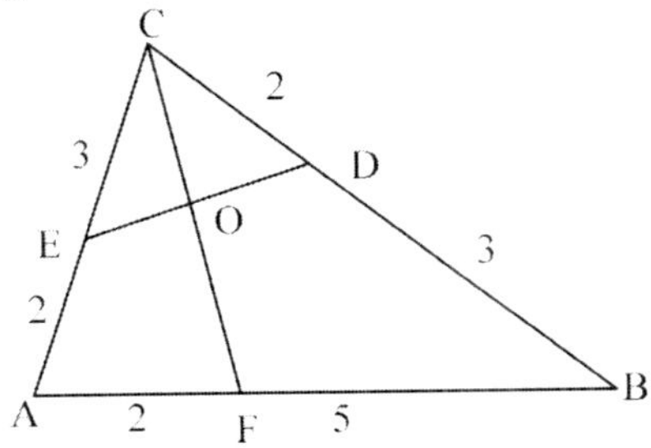

Diagram for example 2

Here we have a transversal line (line that does not touch any of the vertices) DE. In such a case point C will have split mass values.

If we assign a mass value of 15 to point A (based on the same logic of LCM used in example 1), B will have mass point of 6 and Point C will have split mass values of 9 from B side and 10 from A side. Mass point values for D, E and F will be 15, 25 and 21 respectively. Mass point at O will be 15 + 25 = 40 and for C it is sum of the two parts as 10 + 9 = 19.

Hence $\frac{OD}{OE} = \frac{25}{15} = \frac{5}{3}$ and $\frac{OC}{OF} = \frac{21}{19}$

Example 3:

In triangle ABC, points D and E are on sides BC and CA respectively and points F & G are on side AB with G between F and B. BE intersects CF at O_1 and BE intersects DG at O_2 If FG = 1, AE = AF = DB = DC = 2 and BG = CE = 3. Compute $\frac{O_1O_2}{BE}$.

Reference diagram below depicting the given conditions

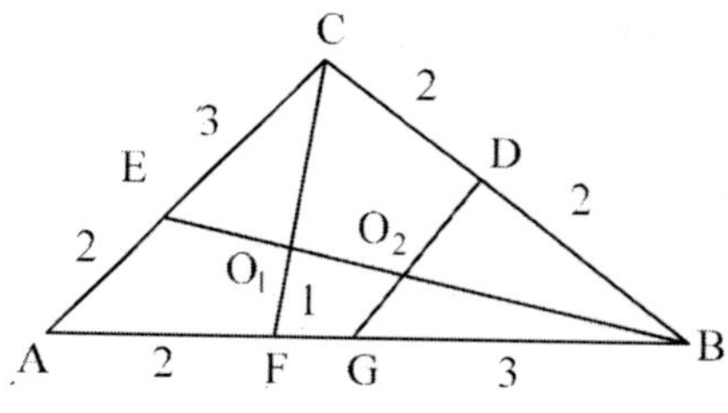

Diagram for example 3

Here we will need to consider ΔABC in two different ways.

In the first case, we ignore segment DG. In this case A will have two mass points 4 and 3. Considering it as 12 (LCM of the two masses), mass points at B, C, E, F and O_1 will be 6, 8, 20, 18 and 26 respectively. This will give us $\frac{EO_1}{BO_1} = \frac{6}{20} = \frac{3}{10}$ & $\frac{EO_1}{BE} = \frac{6}{26} = \frac{3}{13}$

In the second case, we ignore segment CF. Here again as in example 2, DG is a transversal. If A is assigned a mass point of 3, C will have mass point 2 and E will have mass point of 5. B will have two mass points 3 towards AB and 2 towards BC. Hence mass point at B will be 2 + 3 = 5 and at D, G and O_2 will be 4, 6 and 5 + 5 = 10 respectively.

Therefore, $\frac{EO_2}{BO_2}=\frac{5}{5}=1$ & $\frac{EO_2}{BE}=\frac{5}{10}=\frac{1}{2}$

So we find that

$$\frac{O_1O_2}{BE}=\frac{EO_2-EO_1}{BE}=\frac{EO_2}{BE}-\frac{EO_1}{BE}\Rightarrow\frac{O_1O_2}{BE}=\frac{1}{2}-\frac{3}{13}=\frac{7}{26}$$

3.2.7.6 Menelaus' theorem:

Menelaus› theorem relates ratios obtained by a line cutting the sides of a triangle. It states that if a line intersects sides or extended side(s) of a ΔABC at points D, E and F (as shown in figure 3.27 below), the following relation holds $\frac{AD}{DB}\times\frac{BE}{EC}\times\frac{CF}{FA}=1$

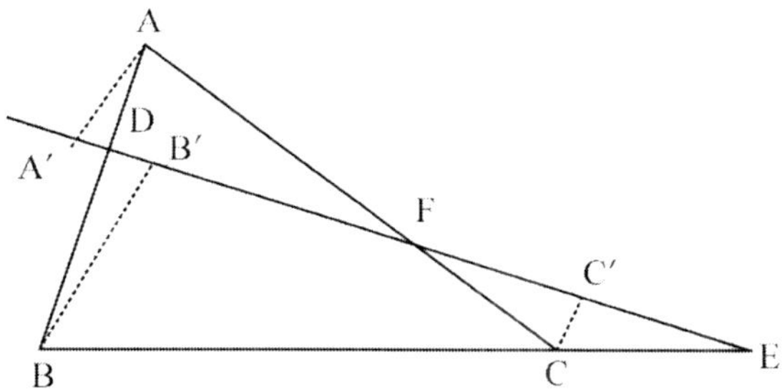

Figure 3.27

Proof:

Line DFE is a transverse line meeting sides AB, extended BC and CA at D, E and F respectively. Perpendiculars are drawn from points A, B and C on the transverse line DFE meeting it at A′, B′ and C′ respectively.

We can notice that the three pairs of right angled triangles ADA′ & BDB′, AFA′ & CFC′ and BEB′& CEC′ are similar. This gives us

$\frac{AD}{DB}=\frac{AA'}{BB'}, \frac{BE}{EC}=\frac{BB'}{CC'}$ & $\frac{CF}{FA}=\frac{CC'}{AA'}$

Multiplying LHS terms with RHS terms we get

$$\frac{AD}{DB}\times\frac{BE}{EC}\times\frac{CF}{FA}=\frac{AA'}{BB'}\times\frac{BB'}{CC'}\times\frac{CC'}{AA'}=1\text{..........Hence proved}$$

The converse of the theorem (i.e. three points on a triangle are collinear if and only if they satisfy criteria proved above) is also true and is extremely powerful

3.2.7.7 Routh's theorm:

Routh's theorem determines the ratio of areas between a given triangle and a triangle formed by the pairwise intersections of three cevians. Refer to the figure 3.28 below:

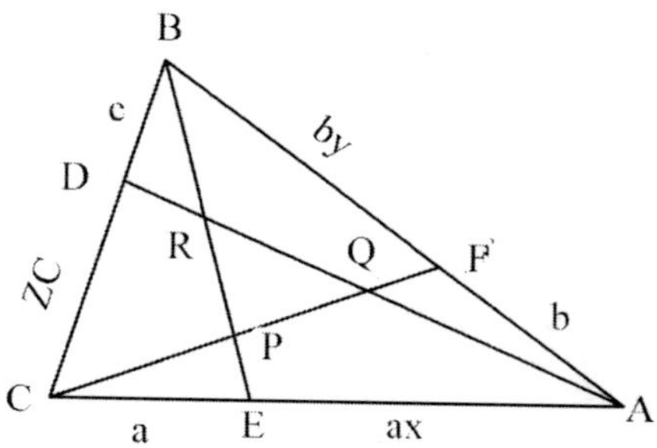

Figure 3.28

The theorem states that if in triangle ABC, points D, E and F lie on segments BC, CA and AB respectively, such that $\frac{CD}{BD}=z$, $\frac{AE}{CE}=x$ and $\frac{BF}{AF}=y$, then area of

the triangle formed by the cevians AD, BE and CF = $\frac{(xyz-1)^2}{(xy+y+1)(yz+z+1)(zx+x+1)}\times S_{\Delta ABC}$

Proof:

Using **Menelaus' theorem** on ΔBCE and transverse line AQD, we get $\frac{BD}{DC}\times\frac{CA}{AE}\times\frac{ER}{RB}=1\Rightarrow\frac{ER}{RB}=\frac{DC}{BD}\times\frac{AE}{CA}=\frac{ax}{a(x+1)}\times\frac{cz}{c}=\frac{zx}{x+1}$

Adding 1 on both sides, we get

$$\frac{ER+RB}{RB}=\frac{zx}{x+1}+1 \Rightarrow \frac{EB}{RB}=\frac{zx+x+1}{x+1}$$

If two triangles have shared base and a common vertex like ΔABR and ΔABE having shared base BRE and common vertex A, their areas are in the ratio of their respective base length (BR and BE respectively) as they have a common altitude.

Hence areas of $\frac{\Delta ABR}{\Delta ABE}=\frac{BR}{BE}$. Similarly, areas of

$$\frac{\Delta ABE}{\Delta ABC}=\frac{AE}{AC}=\frac{ax}{a(x+1)}=\frac{x}{x+1}$$

Now we can see that areas of

$$\frac{\Delta ABR}{\Delta ABE}\times\frac{\Delta ABE}{\Delta ABC}=\frac{BR}{BE}\times\frac{AE}{AC} \Rightarrow \frac{\Delta ABR}{\Delta ABC}=\frac{x+1}{zx+x+1}\times\frac{x}{x+1} \Rightarrow$$

Area of $\Delta ABR=\frac{x}{zx+x+1}\times area\ of\ \Delta ABC$

Similarly Area of $\Delta BCE=\frac{y}{xy+y+1}\times area\ of\ \Delta ABC$ and

Area of $\Delta ACF=\frac{z}{yz+z+1}\times area\ of\ \Delta ABC$

We can now see that

$$Area\ of\ \Delta PQR = Area\ of\ \Delta ABC - Area\ of\ \Delta ABR - Area\ of\ \Delta BCE - Area\ of\ \Delta CFA$$

Substituting value of areas of different triangles, we get

$$Area\ of\ \Delta PQR = Area\ of\ \Delta ABC \times \left(1-\frac{x}{zx+x+1}-\frac{y}{xy+y+1}-\frac{z}{yz+z+1}\right) \Rightarrow$$

$$\frac{Area\ of\ \Delta PQR}{Area\ of\ \Delta ABC}=\left(1-\frac{x}{zx+x+1}-\frac{y}{xy+y+1}-\frac{z}{yz+z+1}\right)$$

We can simplify the expression on the Right Hand side further as below, though the simplification is a little tricky.

Eliminating the numerator in three terms on extreme right we get,

$$1-\frac{x}{zx+x+1}-\frac{y}{xy+y+1}-\frac{z}{yz+z+1}=1-\frac{1}{z+1+\frac{1}{x}}-\frac{1}{x+1+\frac{1}{y}}-\frac{1}{y+1+\frac{1}{z}} \quad \text{........(1)}$$

Let $z+\frac{1}{x}=a$, $x+\frac{1}{y}=b$ and $y+\frac{1}{z}=c$, leading to

$$(a+b+c)=(x+y+z)+\left(\frac{1}{x}+\frac{1}{y}+\frac{1}{z}\right) \quad \text{.... (2)}$$

And by simple multiplication,

$$abc=\left(z+\frac{1}{x}\right)\times\left(x+\frac{1}{y}\right)\times\left(y+\frac{1}{z}\right)=.$$

$$xyz+(x+y+z)+\left(\frac{1}{x}+\frac{1}{y}+\frac{1}{z}\right)+\frac{1}{xyz} \quad \text{...... (3)}$$

From equations (2) and (3), we get

$$abc-(a+b+c)=xyz+\frac{1}{xyz} \quad \text{............ (4)}$$

Substituting values of $z+\frac{1}{x}$, $x+\frac{1}{y}$, $y+\frac{1}{z}$ in equation (1) above, we get

$$\frac{Area\ of\ \Delta PQR}{Area\ of\ \Delta ABC}=\left(1-\frac{1}{a+1}-\frac{1}{b+1}-\frac{1}{c+1}\right)$$

.

$$=\frac{(a+1)(b+1)(c+1)-(b+1)(c+1)-(a+1)(c+1)-(a+1)(b+1)}{(a+1)(b+1)(c+1)}$$

$$=\frac{(b+1)(c+1)(a+1-1)-(a+1)\{(b+1)+(c+1)\}}{(a+1)(b+1)(c+1)}$$

$$=\frac{a(bc+b+c+1)-a\times(b+c+2)-(b+c+2)}{(a+1)(b+1)(c+1)}=\frac{a(bc-1)-(b+c+2)}{(a+1)(b+1)(c+1)}$$

$$= \frac{abc - (a+b+c) - 2}{(a+1)(b+1)(c+1)}.$$

By substituting the values of a, b and c assumed above and value of $abc - (a+b+c)$ from equation (4) above, we get

$$\frac{Area\ of\ \Delta PQR}{Area\ of\ \Delta ABC} = \frac{xyz + \frac{1}{xyz} - 2}{\left(z+1+\frac{1}{x}\right)\left(x+1+\frac{1}{y}\right)\left(y+1+\frac{1}{z}\right)}$$

$$= \frac{\frac{(xyz)^2 + 1 - 2xyz}{xyz}}{\frac{(zx+x+1)(xy+y+1)(yz+z+1)}{xyz}}$$

leading to $\frac{Area\ of\ \Delta PQR}{Area\ of\ \Delta ABC} = \frac{(xyz-1)^2}{(1+x+xz)(1+y+xy)(1+z+yz)}$

This is a very interesting result.

Special cases:

If $x = y = z = say\ p$, the above expression changes to

$$\frac{Area\ of\ \Delta PQR}{Area\ of\ \Delta ABC} = \frac{(p^3-1)^2}{(p^2+p+1)^3} = \frac{(p-1)^2(p^2+p+1)^2}{(p^2+p+1)^3} = \frac{(p-1)^2}{p^2+p+1}$$

If p =1, the three cevians turn into medians and the above ratio goes to zero leading to area of triangle PQR to be zero. Thus this is another proof that medians are concurrent.

There is another particular case with x = y = z = 2. This is popularly known as the one-seventh area triangle as in this case area of triangle PQR is one-seventh of area of triangle ABC.

An alternative proof of Routh's theorem is as below:

If in ΔABC as shown below, three cevians are drawn from vertices A, B and C meeting their opposite sides at D, E and F respectively dividing respective opposite sides in the ratio CA as ***a : ax,*** AB as ***b : by*** and BC as ***c : cz.***

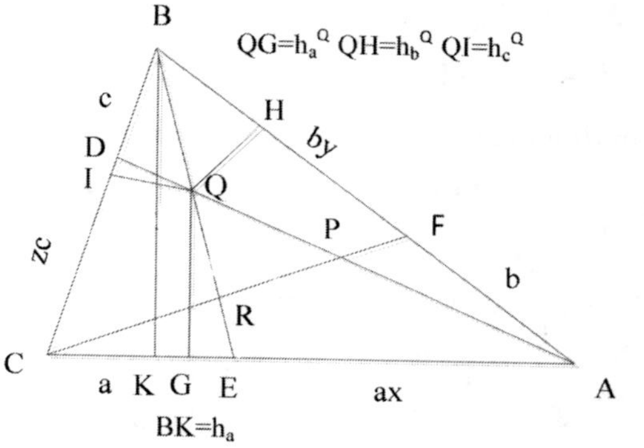

Figure 3.29

Let height of the triangle from three vertices be denoted respectively as h_a, h_b and h_c and let heights from Q to the three sides are denoted as QG = h_a^Q, QH = h_b^Q and QI = h_c^Q. Similarly, height from P and R to the base AC are denoted as h_a^P and h_a^R respectively. In order to find area of the ΔPQR, let us first find distances of points P, Q and R mentioned above from the three sides.

Let $S_{\Delta ABC}$ stand for area of the ΔABC.

We can see that

$$S_{\Delta ABC} = S_{\Delta ABQ} + S_{\Delta BCQ} + S_{\Delta ACQ}$$

$$S_{\Delta ABD} = S_{\Delta ABQ} + S_{\Delta BQD} \text{ and}$$

$$S_{\Delta ABE} = S_{\Delta ABQ} + S_{\Delta AQE}$$

The above equalities lead us respectively to the following equations:

$$a(1+x)h_a = b(1+y)h_b^Q + a(1+x)h_a^Q + c(1+z)h_c^Q \quad \ldots(1)$$

$$ch_c = b(1+y)h_b^Q + ch_c^Q \quad \ldots(2)$$

$$axh_a = b(1+y)h_b^Q + axh_a^Q \quad \ldots(3)$$

We also know that

$$a(1+x)h_a = b(1+y)h_b = c(1+z)h_c = 2 \times S_{\Delta ABC} \quad \ldots(4)$$

Subtracting eq (14) from eq (12),

$$by\left(h_b - h_b{}^P\right) = c\left(1+z\right)h_c{}^P \Rightarrow b\left(h_b - h_b{}^P\right) = \frac{c\left(1+z\right)}{y}h_c{}^P \quad \dots (16)$$

From eq. (15) and eq. (16),

$$\frac{c\left(1+z\right)}{y}h_c{}^P = cz\left(h_c - h_c{}^P\right) \Rightarrow h_c{}^P\left(\frac{\left(1+z\right)}{y} + z\right) = zh_c \quad \dots. (17)$$

As per eq. (5), $h_c = \dfrac{2S_{\Delta ABC}}{c\left(1+z\right)}$

Substituting above value of h_c in eq. (17),

$$h_c{}^P\left(\frac{\left(1+z\right)}{y} + z\right) = z \times \frac{2S_{\Delta ABC}}{c\left(1+z\right)} \Rightarrow h_c{}^P\left(\frac{\left(1+z+zy\right)}{y}\right)$$

$$= \frac{z}{c\left(1+z\right)} \times 2S_{\Delta ABC}$$

Hence $h_c{}^P = \dfrac{yz}{c\left(1+z\right)\left(1+z+zy\right)} \times 2S_{\Delta ABC}$ ..... (18)

From equation 16, we get

$$bh_b = bh_b{}^P + \frac{c\left(1+z\right)}{y}h_c{}^P$$

Substituting value of $h_c{}^P$ from eq. (18) above,

$$bh_b = bh_b{}^P + \frac{c\left(1+z\right)}{y} \times \frac{yz}{c\left(1+z\right)\left(1+z+zy\right)} \times 2S_{\Delta ABC} \Rightarrow$$

$$\frac{2S_{\Delta ABC}}{1+y} = bh_b{}^P + \frac{z}{\left(1+z-zy\right)} \times 2S_{\Delta ABC} \Rightarrow bh_b{}^P =$$

$$2S_{\Delta ABC}\left(\frac{1}{1+y} - \frac{z}{\left(1+z+zy\right)}\right) \Rightarrow$$

$$h_b{}^P = \frac{1+z+zy-z-zy}{b\left(1+y\right)\left(1+z+zy\right)} \times 2S_{\Delta ABC} = \frac{1}{b\left(1+y\right)\left(1+z+zy\right)} \times 2S_{\Delta ABC} \quad \dots.(19)$$

Substituting value of h_b^P from eq. (19) above and h_b from eq. (5) in equation 14, we get

$$a(1+x)h_a^P = bh_b - b\frac{1}{b(1+y)(1+z+zy)} \times 2S_{\Delta ABC} =$$

$$\frac{b \times 2S_{\Delta ABC}}{b(1+y)} - b\frac{1}{b(1+y)(1+z+zy)} \times 2S_{\Delta ABC} \Rightarrow$$

$$h_a^P = \frac{1}{a(1+x)} \times \left(\frac{1}{(1+y)} - \frac{1}{(1+y)(1+z+zy)}\right) \times 2S_{\Delta ABC} =$$

$$= \frac{1}{a(1+x)} \times \left(\frac{(1+z+zy)-(1+y)}{(1+y)(1+z+zy)}\right) \times 2S_{\Delta ABC} \Rightarrow$$

$$h_a^P = \frac{1}{a(1+x)} \times \left(\frac{z(1+y)}{(1+y)(1+z+zy)}\right) \times 2S_{\Delta ABC}$$

$$= \frac{z}{a(1+x)(1+z+zy)} \times 2S_{\Delta ABC} \ldots. (20)$$

An interesting observation:

We have seen from eq. (11) that

$$h_b^Q = 2S_{\Delta ABC} \times \left(\frac{x}{b(1+y)(1+x+xz)}\right)$$

By cyclic permutation, we can replace **b** by the previous/next letter **a or c**, **P** similarly by **R or Q, y** by **x or z, x** by **z or y** and **z** by **y or x** respectively to get

$$h_a^P = \frac{z}{a(1+x)(1+z+zy)} \times 2S_{\Delta ABC}$$

This is precisely the result we derived in eq. (20) above. On the same basis, value of h_b^P derived in eq. (19) will yield the value of h_a^R by cyclic permutation using preceding values as below:

$$h_a^R = 2S_{\Delta ABC} \times \left(\frac{1}{a(1+x)(1+y+xy)}\right)$$

Same way we can derive distances of the other two points from all sides. All results are tabulated below for convenience:

$h_a^Q = 2S_{\Delta ABC} \times \dfrac{xz}{a(1+x)(1+x+xz)}$	$h_b^Q = 2S_{\Delta ABC} \times \dfrac{x}{b(1+y)(1+x+xz)}$	$h_c^Q = 2S_{\Delta ABC} \dfrac{1}{c(1+z)(1+x+xz)}$
$h_c^P = 2S_{\Delta ABC} \times \dfrac{yz}{c(1+z)(1+z+yz)}$	$h_a^P = 2S_{\Delta ABC} \times \dfrac{z}{a(1+x)(1+z+yz)}$	$h_b^P = 2S_{\Delta ABC} \times \dfrac{1}{b(1+y)(1+z+yz)}$
$h_b^R = 2S_{\Delta ABC} \times \dfrac{xy}{b(1+y)(1+y+xy)}$	$h_c^R = 2S_{\Delta ABC} \times \dfrac{y}{c(1+z)(1+y+xy)}$	$h_a^R = 2S_{\Delta ABC} \times \dfrac{1}{a(1+x)(1+y+xy)}$

We can see from figure 3.29 that $2 \times S_{\Delta PQR} = 2 \times \left(S_{\Delta AQE} - S_{\Delta APC} + S_{\Delta CRE}\right)$ leading to $2 \times S_{\Delta PQR} = axh_a^Q - a(1+x)h_a^P + ah_a^R$

Substituting values of different parameters derived above, we get

$$2 \times S_{\Delta PQR} = 2 \times S_{\Delta ABC}\left(\frac{ax \times xz}{a(1+x)(1+x+xz)} - \frac{a(1+x)\times z}{a(1+x)(1+z+zy)} + \frac{a}{a(1+x)(1+y+xy)}\right)$$

$$\text{Or } \frac{S_{\Delta PQR}}{S_{\Delta ABC}} = \left(\frac{x}{1+x} \times \frac{xz}{(1+x+xz)} - \frac{z}{(1+z+zy)} + \frac{1}{(1+x)(1+y+xy)}\right)$$

This formula is different from the formula derived earlier and reproduced below but with the same result. Readers can try it out with any values of x, y and z and will find the result same with either of the formulae. $\dfrac{Area\ of\ \Delta PQR}{Area\ of\ \Delta ABC} = \left(1 - \dfrac{x}{zx+x+1} - \dfrac{y}{xy+y+1} - \dfrac{z}{yz+z+1}\right)$

If we try with x = y = z = 2 with this alternate formula, we find

$$\frac{S_{\Delta PQR}}{S_{\Delta ABC}} = \left(\frac{2}{3} \times \frac{4}{7} - \frac{2}{7} + \frac{1}{3 \times 7}\right) = \frac{8}{21} + \frac{1}{21} - \frac{2}{7} = \frac{9-6}{21} = \frac{1}{7}$$

So the result is same. ***This alternate formula is useful in solving some of the problems related to Routh's theorem. See problem 20 in the Practice problems as an example.***

3.3 Quadrilaterals:

A Quadrilateral is a four sided figure. A general Quadrilateral may have all unequal and non-parallel sides and unequal angles. A quadrilateral is said to be convex if all angles are less than 180^0 and concave if an angle is more than 180^0.

There are special types of Quadrilaterals as below:

3.3.1 Kite:

A **Kite** is a quadrilateral with two distinct pairs of adjacent sides that are congruent.

3.3.2 Parallelogram:

A **parallelogram** has two pairs of equal and parallel sides.

In case a Kite has all 4 equal sides, its sides turn parallel. In case Kite or Parallelogram has all 4 equal sides, it is known as a **Rhombus.**

Special parallelogram with right angles is a **Rectangle** and a special rectangle with equal sides is a **Square**. A **Square** is also a special case of the rhombus.

3.3.3 Trapezium

A Trapezium has two parallel and two non-parallel sides. All 4 sides are unequal in a general Trapezium. However if the non-parallel sides are equal, it is called **Isosceles Trapezium**.

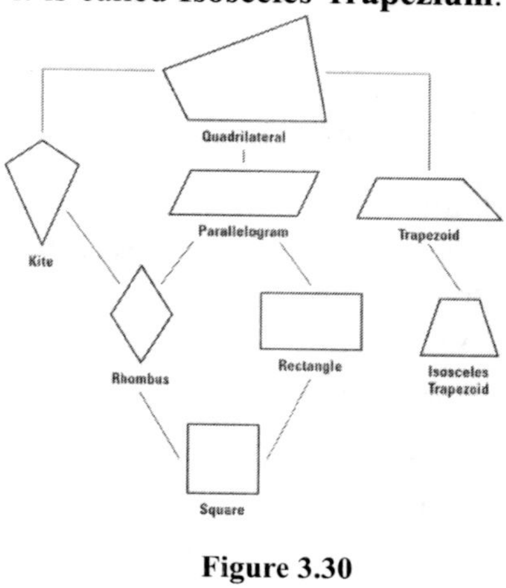

Figure 3.30

Hierarchy of Quadrilaterals is depicted in the above quadrilateral family tree (Fig 3-30).

Shapes of different types of special quadrilaterals are also shown in figure 3.31 below.

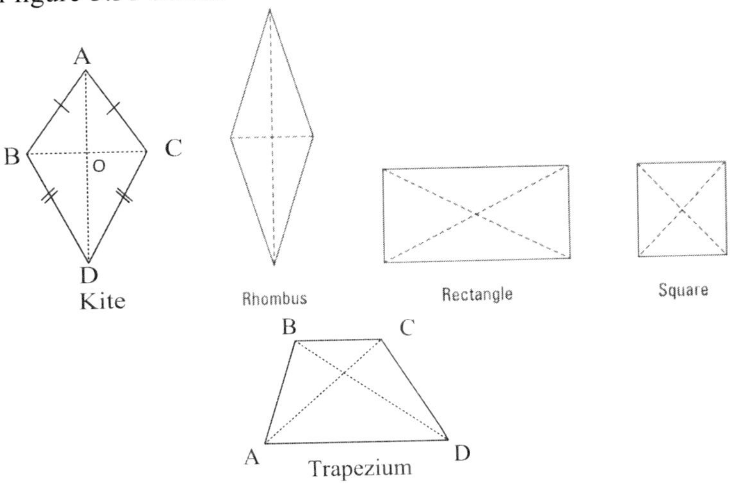

Figure 3.31

3.3.4 Properties of different Quadrilaterals:

General Quadrilateral:

3. *Sum of interior angles of any Quadrilateral is always 360^0.*
4. *Perimeter= Sum of the 4 sides*
5. *Area has many different formulae that will be discussed in Chapter 6 Mensuration. An easiest formula to recall knowing length of the 4 sides* $= \sqrt{(s-a)(s-b)(s-c)(s-d)}$ *where s is semi-perimeter and a, b, c and d are lengths of the four sides. It is very interesting to note that this general formula for area can be used to calculate area of all special quadrilaterals using their specific properties. For example, in case of a square a = b = c = d (say x). Semi-perimeter s will be equal to 2x. Substituting these values in the above formula yields area of the square* $= \sqrt{x \times x \times x \times x} = \sqrt{x^4} = x^2$
6. *Diagonals for a convex quadrilateral intersect inside the quadrilateral. However, diagonals for a concave quadrilateral may or may not intersect inside the quadrilateral.*
7. *Product of the area of opposite set of two triangles each, formed by the two diagonals, is equal. Proof for this property is discussed below:.*

Let ABCD be a general quadrilateral depicted below:

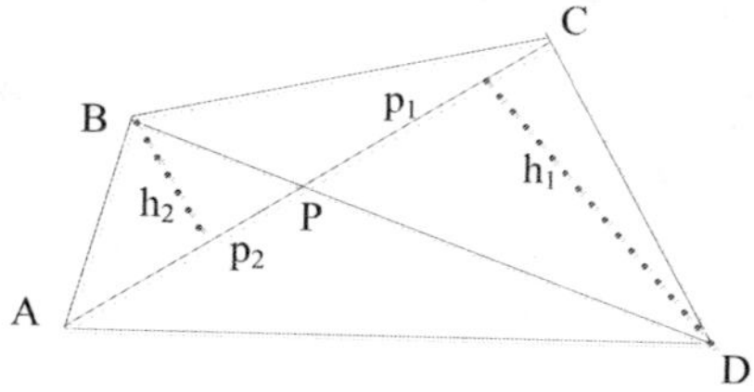

Figure 3.32

We can see that PC = p_1, PA = p_2 leading to area of the opposite triangles APB and CPD as $\frac{1}{2}p_2h_2$ & $\frac{1}{2}p_1h_1$ leading to their product as $\frac{1}{4}p_1p_2h_1h_2$.

Similarly area of the opposite triangles BPC and APD as $\frac{1}{2}p_1h_2$ & $\frac{1}{2}p_2h_1$ leading to their product as $\frac{1}{4}p_1p_2h_1h_2$.

Hence it can be seen that ***product of the area of opposite set of two triangles each, formed by the two diagonals in any quadrilateral, is equal.***

Kite:

1. *One pair of opposite angles is congruent.*
2. *Diagonals meet at right angle and the main diagonal (AD) bisects the cross diagonal BC*
3. *Area= ½ × diagonal$_1$ × diagonal$_2$*

Parallelogram/Rhombus:

1. *Parallel sides are equal.*
2. *Opposite angles are equal*
3. *Consecutive angles are supplementary*
4. *The diagonals bisect each other.*
5. *Area= Base × Height*
6. *Perimeter= Sum of the 4 sides*

Apart from all other properties above, Rhombus, being a special case of kite as well as parallelogram, has all 4 sides equal, diagonals are perpendicular bisectors of each other and Area= ½ × diagonal$_1$ × diagonal$_2$

Rectangle:

1. *Rectangle being a special case of a Parallelogram has all properties of a parallelogram. In addition,*
2. *Each angle is a right angle.*
3. *The two diagonals are equal.*

   Each diagonal= $\sqrt{(Side_1)^2 + (Side_2)^2}$
4. *Area=* $Side_1 \times Side_2$
5. *Perimeter=* $2 \times (side_1 + side_2)$

Square:

1. *All properties of rectangle apply.*
2. *In addition, diagonals of a square are perpendicular bisectors of each other.*
3. *Each diagonal =* $\sqrt{2} \times side$
4. *Area=* $side^2 = \frac{1}{2} \times diagonal^2$
5. *Perimeter=* $4 \times side$

3.3.5 Unorthodox theorems related to quadrilaterals

3.3.5.1 Carpet theorem

If two carpets of equal area overlap, then, the overlap aside, their remaining parts have equal areas. The carpets could of course be of different shapes.Another possibility is that suppose two carpets are expected to fully cover some area, if there is overlap, area left to be covered equals area of the overlap.

This leads to some interestingresults. Let us see a couple of examples:

Example 1:

In figure 3.33.1 below, triangular carpets CMD and CBD have same area (as they have same base and height). Part CND overlaps. Hence area of the triangles DNM and BNC is equal, though it is not very evident from shape of the triangles. Similarly, sum of the area of triangles ADM and CBM = area of triangle ABD.

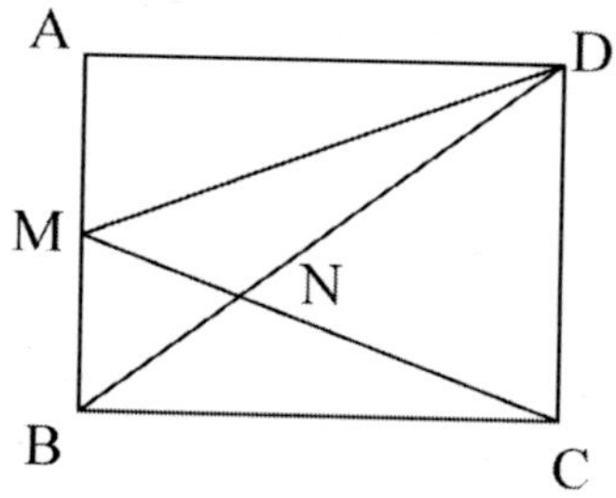

Figure 3.33.1

Example 2:

In the following example (fig 3-33-2), ABCD is a rectangle. Areas of triangles AMP & NRC and quadrilateral BMQN are given. We need to find area of quadrilateral DPQR.

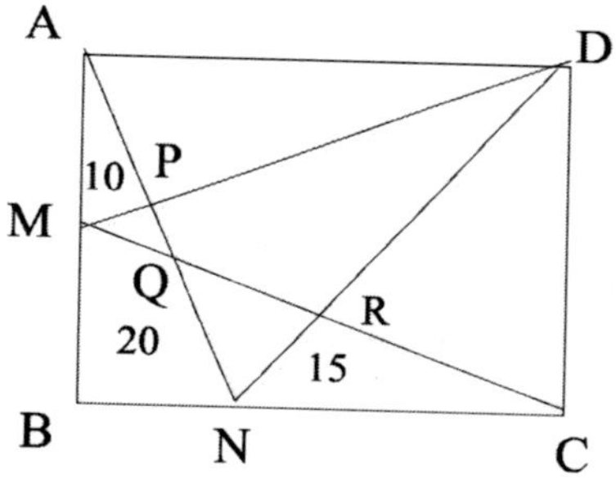

Figure 3.33.2

We can see that two triangular carpets CMD and DNA are of equal area and each has area equal to half the area of the rectangle. Therefore, the carpet area is just adequate to cover the entire rectangular space. Since the carpets overlap, areas of triangle AMP & NRC and quadrilateral BMQN are left to be uncovered due to the overlap which is quadrilateral DPQR.

Hence area of the quadrilateral DPQR = sum of areas of triangle AMP & NRC and quadrilateral BMQN, leading to the conclusion that area of the quadrilateral DPQR = 10+20+15 = 45 units.

3.3.5.2 British Flag theorem:

Given a rectangle ABCD and a point P inside it as shown below, then sum of square of the distances from point P to the opposite corners of the rectangle are equal i.e.

$$AP^2 + CP^2 = BP^2 + DP^2$$

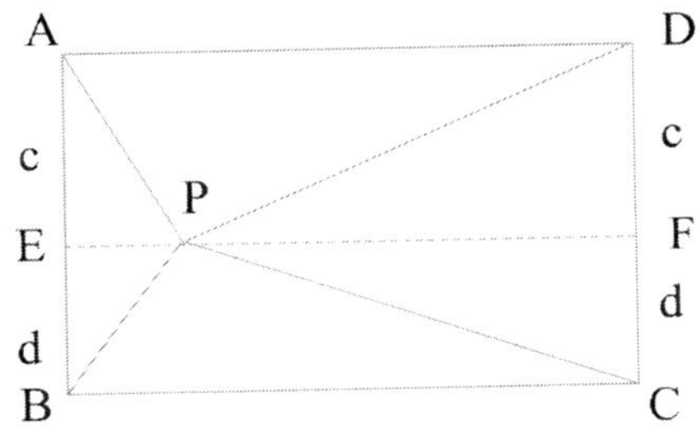

Figure 3.34

Proof:

PE and PF are drawn perpendicular to AB and CD respectively. ABCD being a rectangle, PEF will be a straight line parallel to AD/ BC leading to AE = DF = say c and BE = CF = say d.

Using Pythagoras theorem, $AP^2 = c^2 + PE^2$ ...(1)

$CP^2 = d^2 + PF^2$ ... (2)

and $BP^2 = d^2 + PE^2$ .....(3), $DP^2 = c^2 + PF^2$ ... (4)

Adding eqs. (1), (2) and (3), (4) we get

$AP^2 + CP^2 = c^2 + d^2 + PE^2 + PF^2$. ....(5)

and $BP^2 + DP^2 = d^2 + c^2 + PE^2 + PF^2$ ....(6)

Compairing equations (5) and (6), we can see that

$AP^2 + CP^2 = BP^2 + DP^2$ .... Hence proved.

An interesting observation**: If P coincides with any of the vertices, say A, value of AP will go to zero. By substituting P by A in the above equality, we get* $AC^2 = AB^2 + AD^2$, ***which is nothing but the relationship of diagonal and sides of a rectangle.

Other Polygons:

You may sometimes need to deal with regular polygons having more than 4 sides. A regular polygon has all equal sides and all equal angles. Recall that an equilateral triangle (regular triangle) has each angle of 60^0 and a square (regular quadrilateral) has each angle of 90^0. This leads to a common formula for internal angles of a regular polygon as below:

Internal angle of a regular polygon of n sides= $\frac{2\pi(n-2)}{n}$ and Perimeter = $n\times$ side length

We will explore formulae for area of regular polygons in chapter 6 on Mensuration.

3.4 Circles:

Circle is defined as locus (path) of a point equidistant from another point. Parts of a circle are as below:

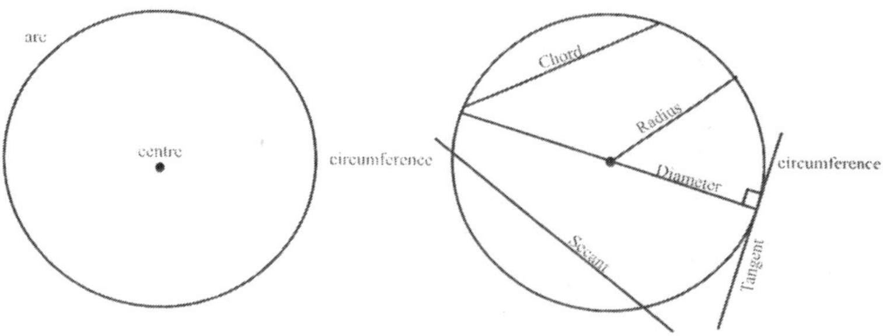

Figure 3.35

Central point from which each point is equidistant is called **Centre** of the Circle.The distance from centre to the circle is termed as **Radius**. A part of the circle is termed as **Arc.** Entire length of the circle is called **Circumference.**

If a line segment is drawn from end to end of the circle passing through the centre, it is called **Diameter** and equals2 x Radius. A line segment touching two points in the circle but not passing through the centre is called a **Chord**.A line that intersects at two points on the circle is called a **Secant** of the circle. A chord can also be understood as the interval of a secant that lies between the points at which it intersects the curve

And a line segment touching the circle is called **Tangent.** If a line is drawn from centre to the point of contact of the tangent, this line segment is perpendicular to the tangent.

Half circle is called Semi circle. Total angle subtended at the centre of the circle is 360^0.Accordingly,angle subtended by Semicircle is 180^0 and by Quarter circle is 90^0

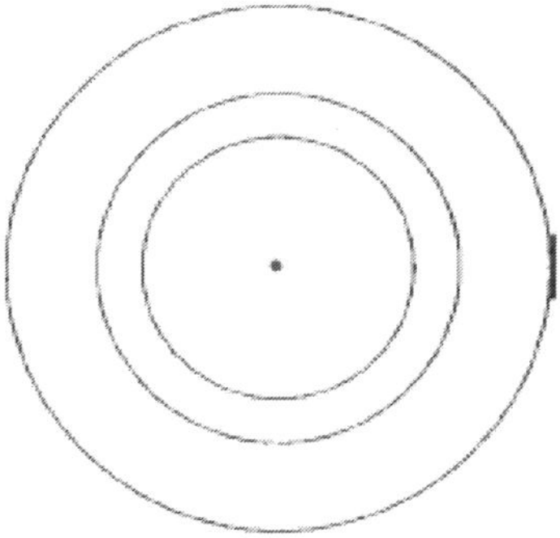

Figure 3.36: Concentric circles

Concentric circles are circles that have common centre. **Equal circles** have equal radius.

Properties of Circles:

3.4.1 Arcs:

Arc is part of a circle's circumference

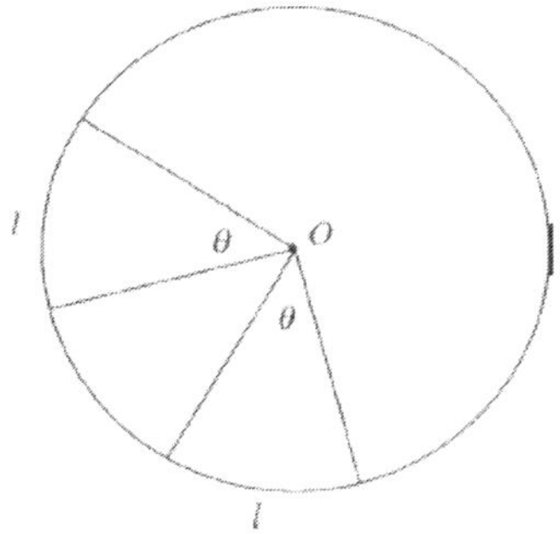

Figure 3.37.1

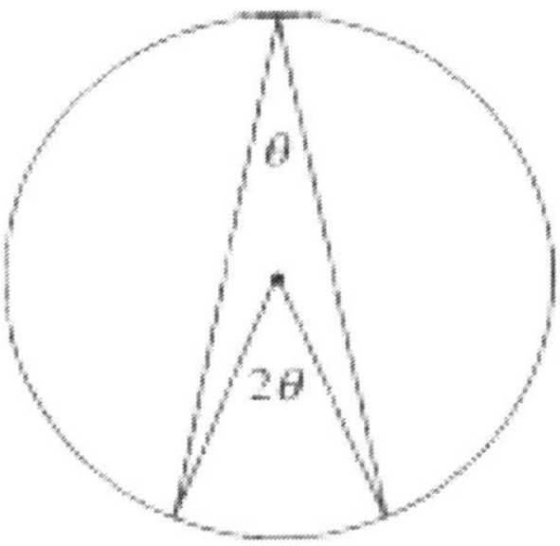

Figure 3.37.2

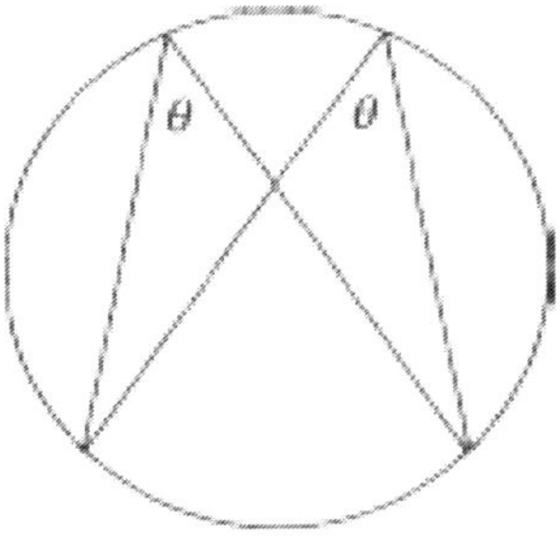

Figure 3-37.3

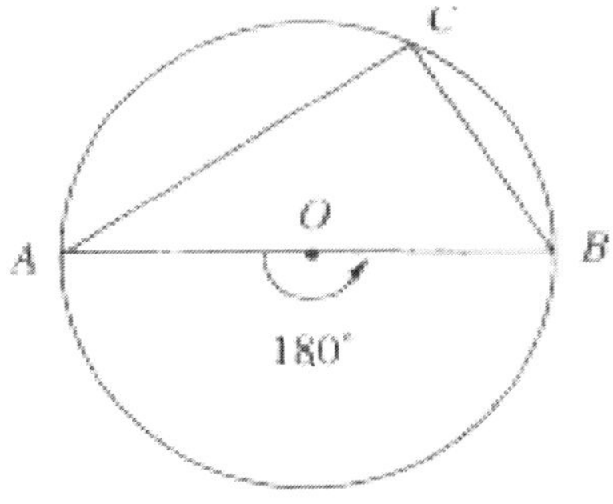

Figure 3-37.4

Properties of Arcs:

a. Equal arcs subtend equal angle at the centre (Figure 3.37.1).
b. Converse, that is if two arcs subtend same angle at the centre, the arcs are equal, is also true.
c. Angle subtended by an arc at the centre of the circle is twice the angle subtended by it at the circumference. It follows that angle formed in a semicircle equals 90^0 (Figures 3.37.2 and 3.37.4)
d. Angles formed by an arc in the same segment of the circle are equal (Figure 3.37.3)

An interesting property of a triangle drawn in a semi circle

Recall that angle subtended in a semi circle equals 90^0. Thus, any triangle formed in a semi circle with diameter as its base, will be a Right Angled triangle. Reference figure 3.38 below. In such a triangle ABD, if a perpendicular line is drawn from apex D on to the base (diameter) AB, meeting the base at D, it can be seen that there are three right angled triangles ACD, BCD and ADB here.

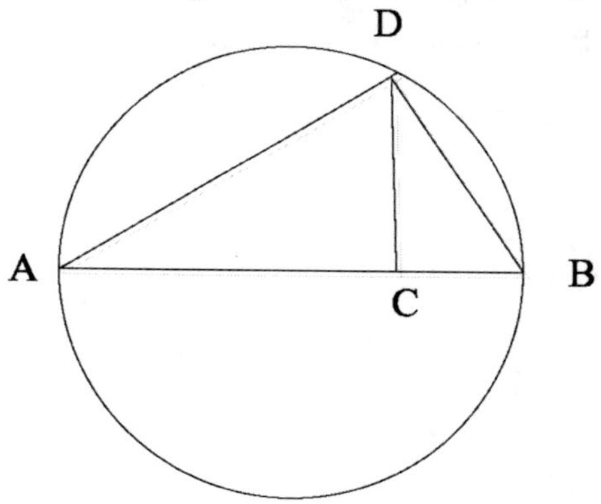

Figure 3.38

By the Pythagoras theorem,

$AC^2 + CD^2 = AD^2$ ... (1)

$BC^2 + CD^2 = BD^2$ ... (2)

and $AD^2 + BD^2 = AB^2$ ... (3)

Now AC + BC = AB

Squaring both sides

$AC^2 + BC^2 + 2 \times BC \times AC = AB^2$ ...(4)

Adding equations (1) and (2) above,

$AC^2 + BC^2 + 2\ CD^2 = AD^2 + BD^2$ ....(5)

Comparing equations (3), (4) and (5), it can be easily seen that

2 x AC x BC = 2 CD^2, or **AC x BC= CD^2**

This implies that measure of line segment CD is square root of the product of measure of line segments AC and BC. Thus we can find square root of a number geometrically by the above property as below:

Split the number into any two factors. Draw a line AB of length equal to sum of the two factors and mark the point C equivalent to the point of contact of the two factor lengthson this line. Now draw a circle with line segment AB as diameter. Draw a perpendicular from point C. Length of the perpendicular upto the point where it meets the circle (CD in the above diagram) is Square Root of the given numbers.

3.4.2 Chords:

Chord is a line segment joining two points on a circle

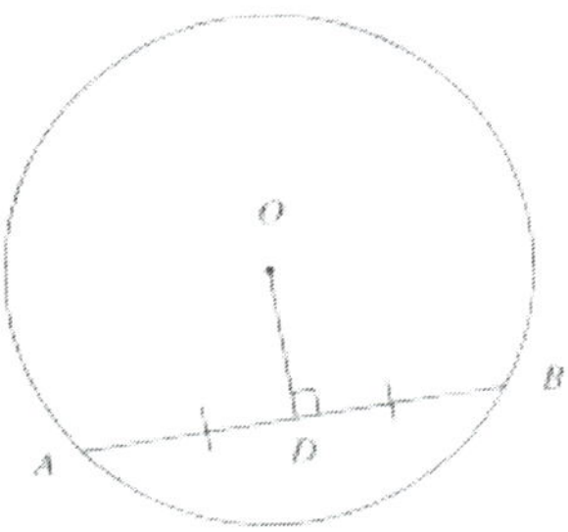

Figure 3.39.1

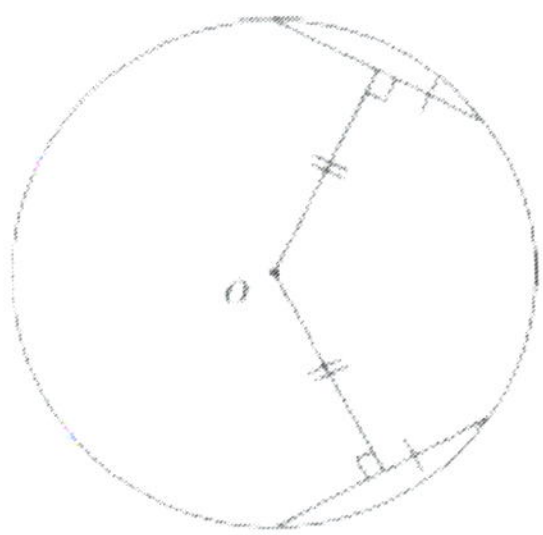

Figure 3.39.2

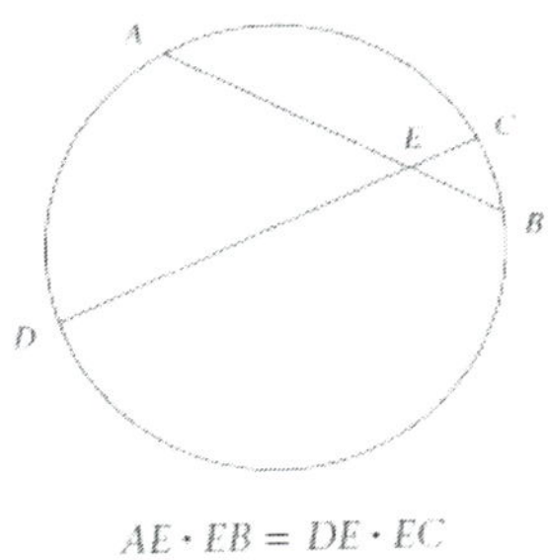

Figure 3.39.3

Properties of Chords:

> a. *A perpendicular line from centre to the chord, bisects the chord. Alternatively, a line from centre of a circle to the mid point of a chord is perpendicular to the chord. (Figure 3.39.1)*
>
> b. *Equal chords are equidistant from the centre. Alternatively, chords that are equidistant from the centre are equal (Figure 3.39.2).*
>
> c. *Products of intercepts of intercepting chords are equal(Figures 3.39.3)*

3.4.3 Concyclic points:

Concyclic points are points that lie on the **circumference** of a circle. In 3-40-1 below, points A, B, C, D, E are Concyclic

> Any 3 ***non-collinear*** points (points not lying on a straight line) are ***concyclic***. They lie on a unique circle, with centre at the point of intersection of the perpendicular bisectors of the *lines* joining these points. Recall the discussion on Circumcenter of a triangle.See figure 3.40.2.
>
> ***Four or more*** non-collinear points ***may not*** necessarily ***lie on a circle.***

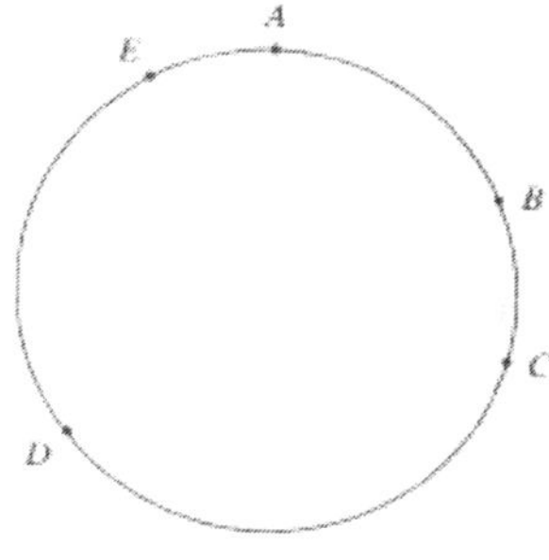

Figure 3.40.1

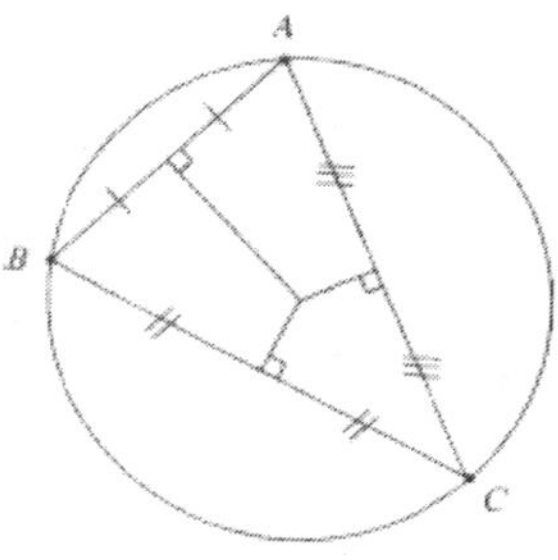

Figure 3.40.2

Properties of Concyclic points:

3.4.3.1 Cyclic Quadrilateral

*A **Cyclic quadrilateral** is a figure whose **4 vertices** are **concyclic points.** In the figure 3-41-1below, ABCD is a **Cyclic Quadrilateral***

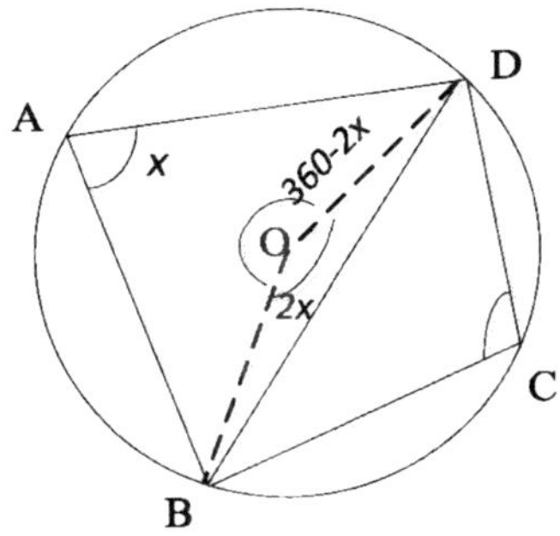

Figure 3.41.1

Properties of Cyclic Quadrilateral

3.4.3.1.1 The opposite angles in a cyclic quadrilateral are supplementary ie. their sum is 180^0 (Figure 3.41.1)

Proof:

Chord BD subtends an angle of x degree at the major arc (point A) and angle 2x at the centre O (as per basic property of the circle). This leads to angle (360-2x) degree on other side of the centre giving angle of (180-x) at the minor arc, point C. Similar is the case for the other two vertices B and D. Hence the property is proved,

Converse of this property i.e. ***if the opposite angles of a quadrilateral are supplementary, then the quadrilateral is cyclic*** is also true.

3.4.3.1.2 The exterior angle at a vertex of a cyclic quadrilateral is equal to the **interior opposite angle.**

This is evident as interior angle opposite to angle x ***of the quadrilateral is*** (180-x) yielding value of the exterior angle as

$180-(180-x)=x$

3.4.3.1.3 The product of the diagonals of a cyclic quadrilateral is equal to the sum of the product of its two pairs of opposite sides. This is known as Ptolemy Theorem

Proof

Reference figure 3.41.2 below.

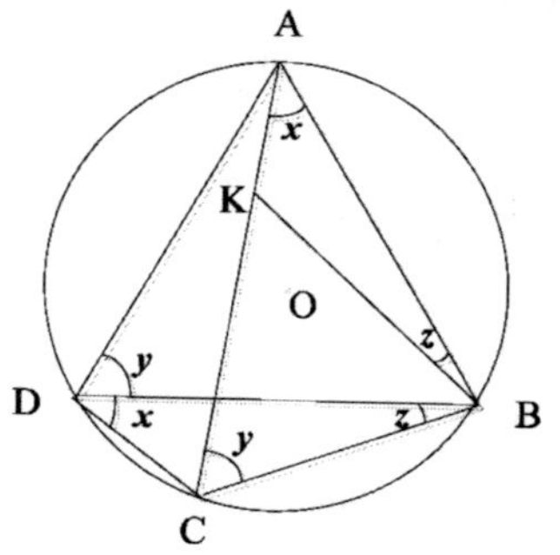

Figure 3.41.2

ABCD is a cyclic quadrilateral with AC and BD as two diagonals. We need to prove $AC \times BD = AB \times CD + AD \times BC$

We will use properties of similar triangles to achieve the above result. Let us draw line BK such that $\angle ABK = \angle DBC$. Let us denote this angle as $\angle z$. We can see that $\angle BDC = \angle BAC = x$ (being angles formed by the same chord BC). Similarly $\angle BCA = \angle BDA = y$ (being angles formed by the same chord AB).

We can also see that $\angle ABD = \angle KBC = \angle z + \angle KBD$

Now with the above information, we find that pair of triangles ΔABK & ΔDBC and ΔABD & ΔKBC are similar leading us to

$$\frac{AB}{DB} = \frac{AK}{DC} \text{ and } \frac{BC}{BD} = \frac{CK}{DA}$$

By cross multiplication, we get

$AB \times DC = AK \times DB.....(1)$ & $BC \times DA = CK \times BD....(2)$

Adding eq. (1) and (2),

$$AB \times CD + BC \times AD = BD\left(AK + KC\right) = BD \times AC$$

Hence proved

***3.4.3.1.4 A special case of this property is in the form of a theorem known as Van* Schooten's theorem** stated below:

If ABC is an equilateral triangle and P is a point on minor arc of its circumcircle as shown in the figure 3.41.3 below, then PA = PB + PC

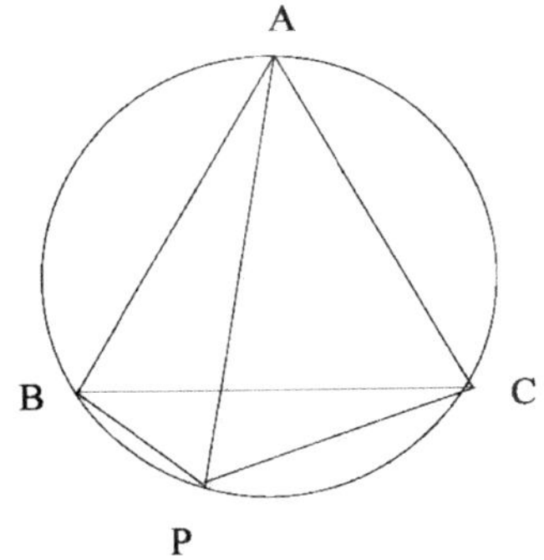

Figure 3.41.3

Proof:

Here we can see that ABPC is a cyclic quadrilateral. Using property (c) above, we get $PA \times BC = AC \times PB + AB \times PC$ .... (1)

Since ΔABC is equilateral, $AB = BC = AC$, say = a.

Hence equation (1) above leads to

$PA \times a = a(PB + PC) \Rightarrow PA = PB + PC$. Hence proved.

3.4.3.1.5 An interesting corollary of Van Schooten's theorem:

Let us mark a point D on PA such that PD = PB. Thus AD = PA - PD = PA - PB PC.

Now we join points BD and extend to point E on the circle and also join AE as shown in figure 3.41.4 below.

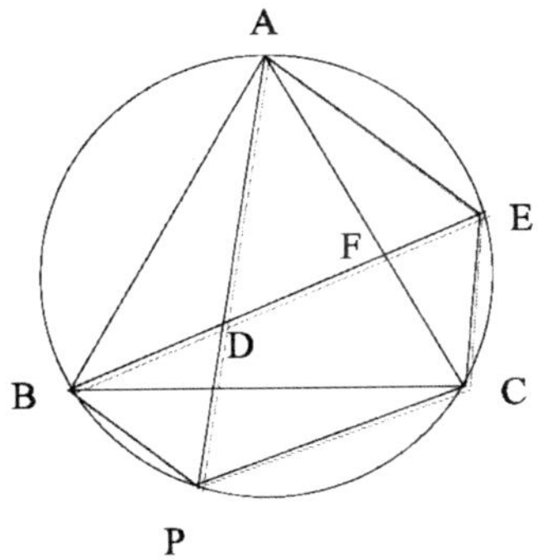

Figure 3.41.4

With the above arrangement we will see that

(a) ΔBDP is equilateral

(b) ΔAED is equilateral

(c) PDEC is a parallelogram

Proof:

(a) ΔABC being equilateral, we can easily see in quadrilateral ABPC that $\angle BPC = 180^0 - \angle BAC = 120^0$. Similarly in quadrilateral AECD, $\angle AEC = 120^0$ $\angle APC = \angle ABC = 60^0$ (Angles formed by the same chord AC on the circle) leading to $\angle BPD = 120^0 - 60^0 = 60^0$

Since PD = PB, it means $\angle PBD = \angle BDP$. And since, it implies that each of the three angles of ΔBDP are of 60^0 and hence ΔBDP is equilateral

(b) $\angle BEC = \angle BAC = 60^0$ (Angles formed by the same chord AC on the circle). $\angle ADE = \angle BDP = 60^0$ (opposite angles) and $\angle AED = \angle AEC - \angle BEC = 60^0$. With two angles of ΔAED $\angle ADE$ & $\angle DEA = 60^0$ each, ΔAED is equilateral leading to DA = ED

(c) In quadrilateral PDEC, ED = AD = PC.

Since $\angle APC = \angle ADE = 60^0$, being corresponding angles for lines DE and PC, lines DE and PC are parallel. $\angle ADE = 60^0$ alsoimplies that $\angle PDE = 120^0$. We have also seen that $\angle APC = \angle BEC = 60^0$ Hence $\angle PCE = 120^0$ leading to the conclusion that quadrilateral PDEC is a parallelogram, thereby CE = PD.

As a corollary, if ΔBDP & ΔADE are equilateral, BDE is a straight line.

3.4.3.1.6 The perpendicular bisectors of the four sides of the inscribed quadrilateral intersect at the centre O of the circle.

Reference figure 3.41.2 above. We can consider the quadrilateral ABCD to consist of two triangles ADC and ABC and the circle in which the quadrilateral is inscribed will be circumcircle of these two triangles. From properties of a triangle we know that perpendicular bisectors of the sides of a triangle meet at centre of the circumcircle. Hence we can easily see that perpendicular bisectors of all the 4 sides of the quadrilateral AB, BC, CD and AD (being sides of the two triangles ADC and ABC will meet at O, centre of the circle. .....***Hence proved.***

3.4.3.1.7 The area of a quadrilateral inscribed in a circle with side lengths a, b, c and d is given by the formula:

$$\text{Area}=\sqrt{s(s-a)\,(s-b)\,(s-c)\,(s-d)} \text{ where } s=\frac{a+b+c+d}{2}$$

This is known as Brahmagupta's formula.

Proof:

There are multiple ways of proving the above formula. We will deal here with the proof based on Heron's formula.

Let ABCD be a cyclic quadrilateral with side lengths a, b, c and d as shown in figure 3.41.5below. Sides BC and AD are extended to meet at X with length DX =x and CX = y

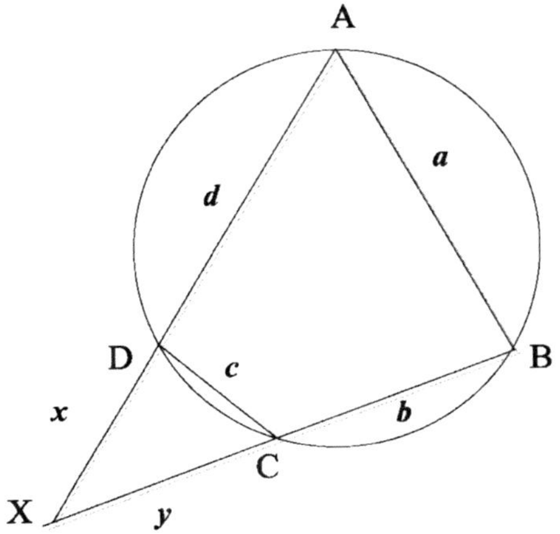

Figure 3.41.5

We can see that $\angle DCX = \angle BAX$ as exterior angle of a vertex of cyclic quadrilateral equals interior opposite angle (property (b) above).

Hence ΔBAX & ΔDCX are similar with angle X being common.

Let ratio of the corresponding sides of the above similar triangles be Z. This gives us $\frac{AB}{CD}=\frac{BX}{DX}=\frac{AX}{CX}=Z \Rightarrow \frac{a}{c}=\frac{b+y}{x}=\frac{x+d}{y}=Z$

We can now see that Area of the quadrilateral ABCD equals Area of $\Delta BAX - \Delta DCX$

Let S_1, S_2, S_3 and S_4 be the 4 factors in Heron's formula for area of ΔDCX. Since ratio of the corresponding sides of the two triangles = Z, corresponding factors for ΔBAX will be ZS_1, ZS_2, ZS_3 and ZS_4 respectively. This leads us to

Area of the quadrilateral

$$\begin{aligned}\text{ABCD} &= \sqrt{(ZS_1)(ZS_2)(ZS_3)(ZS_4)}-\sqrt{(S_1)(S_2)(S_3)(S_4)}\\ &= Z^2\times\sqrt{(S_1)(S_2)(S_3)(S_4)}-\sqrt{(S_1)(S_2)(S_3)(S_4)}\\ &=\left(Z^2-1\right)\times\sqrt{(S_1)(S_2)(S_3)(S_4)}\\ &=\sqrt{S_1(Z-1)S_2(Z-1)S_3(Z+1)S_4(Z+1)}\\ &=\sqrt{(S_1Z-S_1)(S_2Z-S_2)(S_3Z+S_3)(S_4Z+S_4)} \quad \text{....(1)}\end{aligned}$$

Semi-perimeter of the quadrilateral $s=\frac{a+b+c+d}{2}$

From the above figure we have

$$S_1Z=\frac{a+b+y+x+d}{2} \;\&\; S_1=\frac{y+x+c}{2} \text{ leading to}$$

$$S_1Z-S_1=\frac{a+b+y+x+d}{2}-\frac{y+x+c}{2}=\frac{a+b+d-c}{2}=(s-c)$$

Similarly,

$$S_2Z-S_2=\frac{b+y+x+d-a}{2}-\frac{y+x-c}{2}=\frac{b+c+d-a}{2}=(s-a)$$

and $S_3Z+S_3=\frac{a+x+d-(b+y)}{2}+\frac{y+c-x}{2}=\frac{a+d-b+c}{2}=(s-b)$

$$S_4Z + S_4 = \frac{a+b+y-(x+d)}{2} + \frac{c+x-y}{2} = \frac{a+b-d+c}{2} = (s-d)$$

Substituting the above values in equation (1) above,

Area of the quadrilateral = $\sqrt{(s-a)(s-b)(s-c)(s-d)}$ .....***Hence proved.***

This formula also yields back Heron's formula with value of d = 0

Alternate form of Brahmgupta formula:

We can also express Brahmgupta formula for area A of the cyclic quadrilateral in terms of length of its 4 sides as below:

$$A = \sqrt{(s-a)(s-b)(s-c)(s-d)} =$$

$$\sqrt{\frac{(b+c+d-a)(a-b+c+d)(a+b-c+d)(a+b+c-d)}{16}}$$

Rearranging different terms, we get

$$A = \sqrt{\frac{[(c+d)-(a-b)][(c+d)+(a-b)][(a+b)-(c-d)][(a+b)+(c-d)]}{16}} \Rightarrow$$

$$A = \sqrt{\frac{\left[(c+d)^2-(a-b)^2\right]\left[(a+b)^2-(c-d)^2\right]}{16}}$$

$$= \sqrt{\frac{\left[(c+d)^2(a+b)^2-\left(a^2-b^2\right)^2-\left(c^2-d^2\right)^2+(a-b)^2(c-d)^2\right]}{16}}$$

$$= \sqrt{\frac{\left[\left(c^2+d^2+2cd\right)\left(a^2+b^2+2ab\right)+\left(a^2+b^2-2ab\right)\left(c^2+d^2-2cd\right)-\left(a^4+b^4-2a^2b^2\right)-\left(c^4+d^4-2c^2d^2\right)\right]}{16}}$$

$$= \sqrt{\frac{\left[2\left(c^2+d^2\right)\left(a^2+b^2\right)+8abcd-\left(a^4+b^4+c^4+d^4\right)+2a^2b^2+2c^2d^2\right]}{16}}$$

$$= \sqrt{\frac{\left[2\left(a^2c^2+a^2d^2+b^2c^2+b^2d^2+a^2b^2+c^2d^2\right)+8abcd-\left(a^4+b^4+c^4+d^4\right)\right]}{16}}$$

The first term can be re-arranged as

$$2\left(a^2c^2+a^2d^2+b^2c^2+b^2d^2+a^2b^2+c^2d^2\right)=$$

$$\left(a^2+b^2+c^2+d^2\right)^2-\left(a^4+b^4+c^4+d^4\right)\Rightarrow$$

$$A=\sqrt{\frac{\left[\left(a^2+b^2+c^2+d^2\right)^2-\left(a^4+b^4+c^4+d^4\right)+8abcd-\left(a^4+b^4+c^4+d^4\right)\right]}{16}}\Rightarrow$$

$$A=\frac{1}{4}\times\sqrt{\left[\left(a^2+b^2+c^2+d^2\right)^2+8abcd-2\left(a^4+b^4+c^4+d^4\right)\right]}$$

3.4.3.2 Properties of Tangent to a circle

a) The tangent to a circle is perpendicular to the radius drawn from the point of contact. The perpendicular distance is the shortest distance — any other distance would be greater than the radius. (Figure 3.42.1)

b) The converse is also true: The line perpendicular to the radius at the point where it meets the circle is a tangent to the circle at that point.

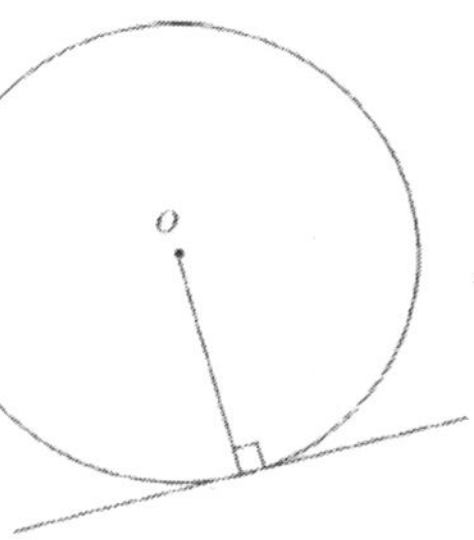

Figure 3.42.1

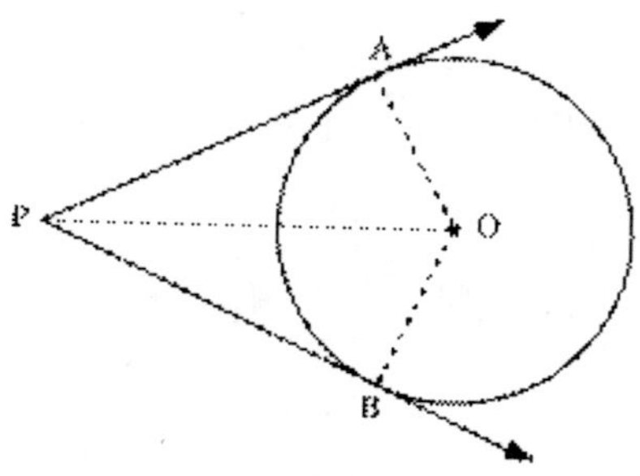

Figure 3.42.2

c) ***Tangents PA and PB to a circle from an exterior point are equal. (Figure 3.42.2).*** *Here O is centre of the circles. Consider ΔPAO and ΔPBO. These are congruent right angled triangles as $AO=BO$ (radius), $\angle PAO$ and $\angle PBO = 90^0$ and PO is common. Hence $PA=PB$.*

d) *When two circles touch externally or internally, they have a common tangent at point of their contact. (Figure 3.43.1). If we now draw a radius line from point of their contact to centres of the two circles, both the radius lines will be perpendicular to the common tangent. Hence these radius lines will either overlap (in case circles are touching internally) or form a straight line (in case circles are touching externally).* ***Hence, the line through the centres of the two circles passes through their point of contact***

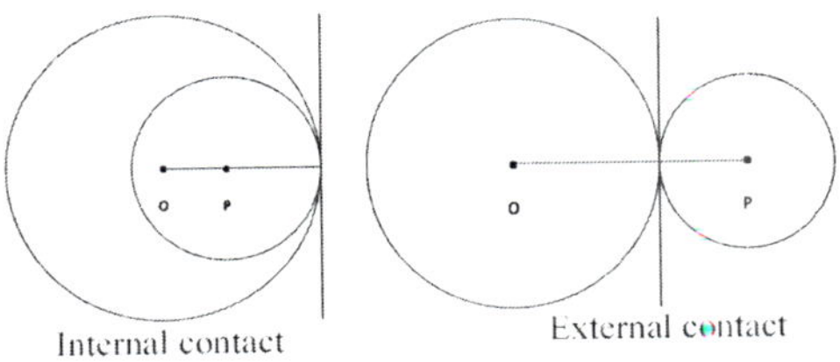

Figure 3.43.1

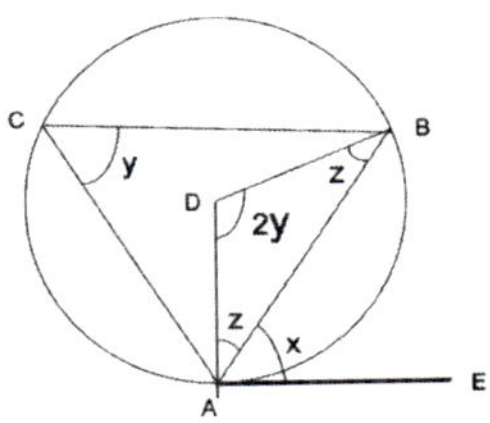

Figure 3.43.2

e) The angle between a tangent and a chord through the point of contact is equal to the angle in the alternate segment. (Figure 3.43.2 above). This can be seen easily. Let us assume arc AB subtends angle y at C. Then $\angle ADB=2y$, being the angle subtended by the same arc AB at centre of the circle. $\angle DAB=\angle DBA=z$ (ΔADB being isosceles.) Hence $2y+2z=180^0$ or $y+z=90^0$. Now $x+z$ is also 90^0 as the radius AD is perpendicular to the tangent AE. So $\angle x=\angle y$. Hence proved.

f) The square of the length of the tangent from an external point is equal to the product of the intercepts of the secant passing through this point. This is a very useful theorem and is known as Tangent Secant theorem. It implies that as in figure 3.44.1 below, $PQ^2 = QR\times QS$ where PQ is a tangent to the circle.

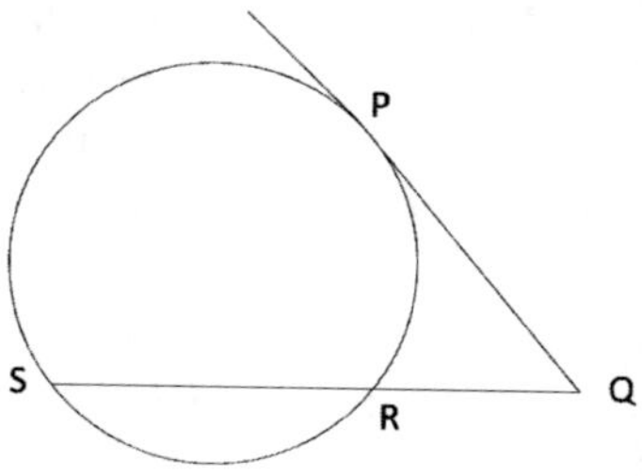

Figure 3.44.1

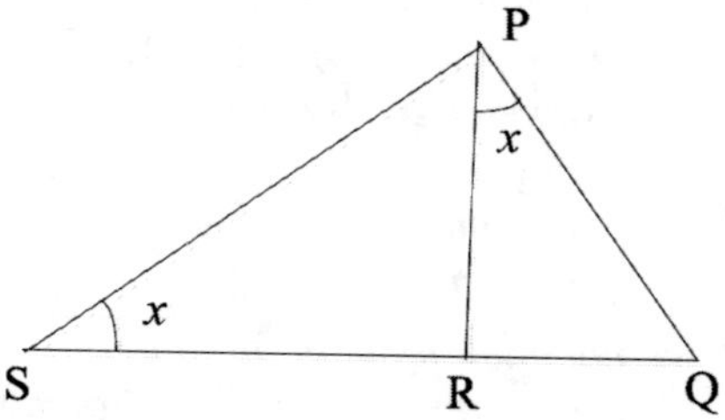

Figure 3.44.2

Proof:

Connect P to R and S as in Figure 3-36-2.

Consider ΔRPQ and ΔPSQ

$\angle RPQ = \angle PSQ = \angle x$ (Property at (e) above)

$\angle$ PQR = $\angle$ *PQS (common angle)*

Hence ΔRPQ and ΔPSQ are similar.

Therefore, $\frac{QS}{PQ} = \frac{PQ}{QR}$

Or, PQ^2=QR x QS.

3.5 Solved examples:

2. Given the rectangle as shown, find the measures of angle 1 and angle 2:

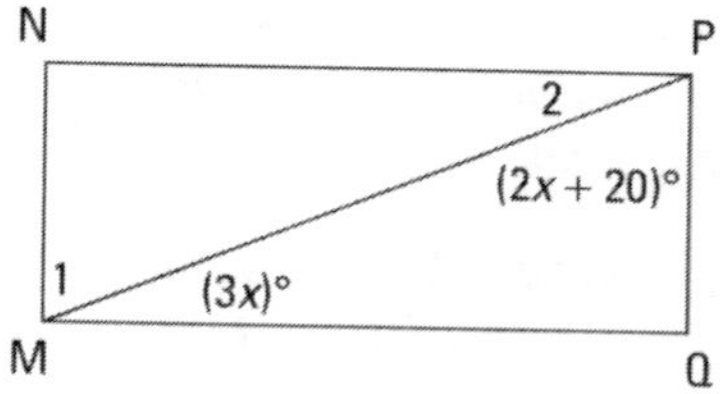

Figure 3.45

MNPQ is a rectangle, so angle $Q = 90°$. Thus, sum of the remaining angles of triangle MQP is also 90^0.

Hence, 3x + 2x + 20= 90or 5x= 70 giving x=14

Now angle M being 90^0,

Angle 1= 90 - 3 x 14 = 48^0 and

Angle 2= 90 - 48 = 42^0

2. Find the perimeter of rhombus *RHOM below:*

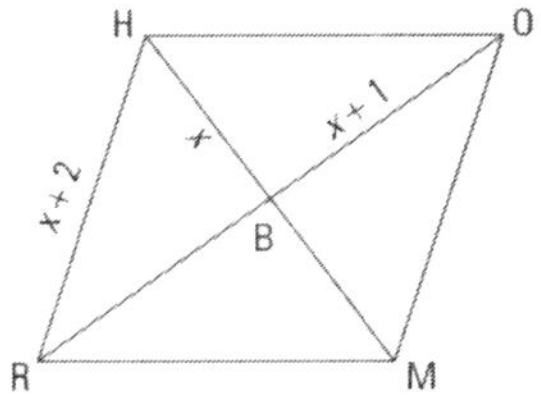

Figure 3.46

All the sides of a rhombus being equal, $HO = x + 2$.

And because the diagonals of a rhombus are perpendicular, triangle *HBO* is a right Δ

By the Pythagoras Theorem:

$x^2 + (x+1)^2 = (x+2)^2$

$x^2 + (x^2 + 2x + 1) = x^2 + 4x + 4$

$x^2 - 2x - 3 = 0$ or $(x - 3)(x + 1) = 0$

$x = 3$ or $x = -1$

Rejecting $x = -1$ because length cannot be negative, we get $x=3$.

Hence perimeter = 4×(x+2) = 20

3.6 Practice problems:

1. In ΔABC, AB = 13 cm, BC = 14 cm and AC = 15 cm. AD is an altitude and line segment EF is perpendicular to BC. Also EF divides ΔABC in two equal areas. What is the area of the trapezium ADFE?
2. ABCD is a quadrilateral with AB = 5cm, CD = 7cm, BC = 17cm and AD = 25cm. Also $\angle ABC + \angle BCD = 270^0$, find the area of the quadrilateral ABCD. (Mass Point Geometry)
3. ABCD is a rectangle with AB = 16 cm and AD = 12 cm. AEFG is a square of side 4 cm. Find area of ΔBOC.
4. In the ΔABC given below, $\frac{AP}{PB} = \frac{1}{2}$ and $\frac{CQ}{AQ} = \frac{1}{4}$. AD is median of the ΔABC. Find the ratio $\frac{AR}{RD}$?

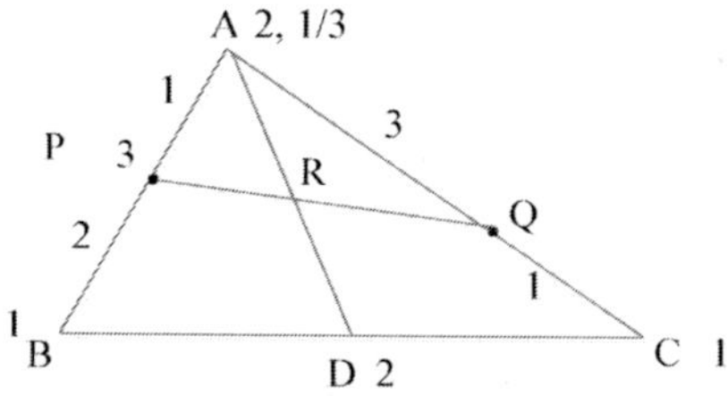

5. ABC is a triangle with sides 3, 4 and 5. Let P be a point inside ΔABC and let x, y, z be the length of perpendiculars drawn from P to the sides AB, BC and AC respectively. What can be the maximum value of the product xyz.
6. In a triangle, two altitudes are 4cm and 12cm respectively and all the altitudes are integers. What will be the largest possible length of the third altitude.
7. ABCD is a quadrilateral. The diagonals of ABCD intersect at point P. Area of the triangles APB and CPD are 12 and 27 respectively. If area of the triangles APD and BPC are equal, then find area of the triangle APD.
8. A circle is drawn with diameter as height of an equilateral triangle. As shown in the figure, the circle cuts side AB and AC at P and Q respectively. Find ratio of AP:PB

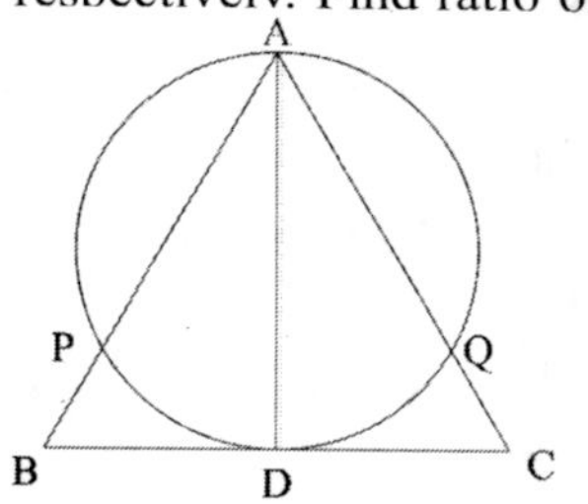

9. In the figure given below ABC is a triangle with AB =9 and AC = 13. E is the mid point of BC, AD bisects angle BAC and $\angle ADB = 90^0$, find length of DE.

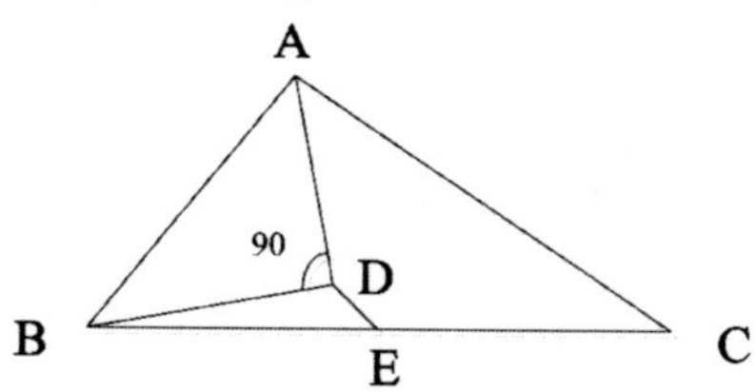

10. Side of the square ABCD is 3 cms. AF = DG = CH = BE = 1 cm. Find area of the square IJKL.

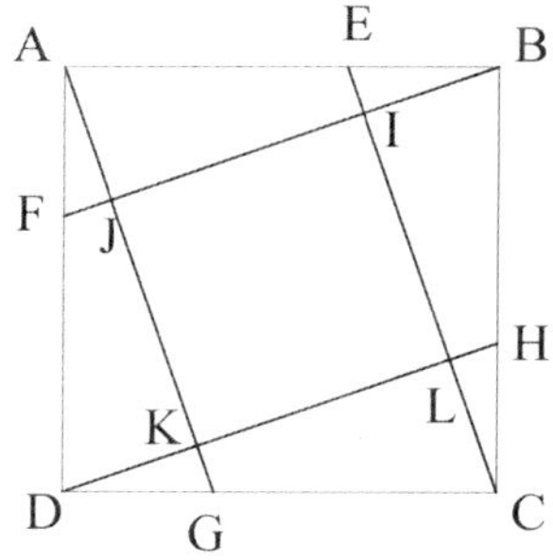

11. ABCD is a square (figure below) with side 10 cm. P is mid-point of CD and Q is mid-point of BC. BP and DQ intersect at O. Find area of the quadrilateral POQC

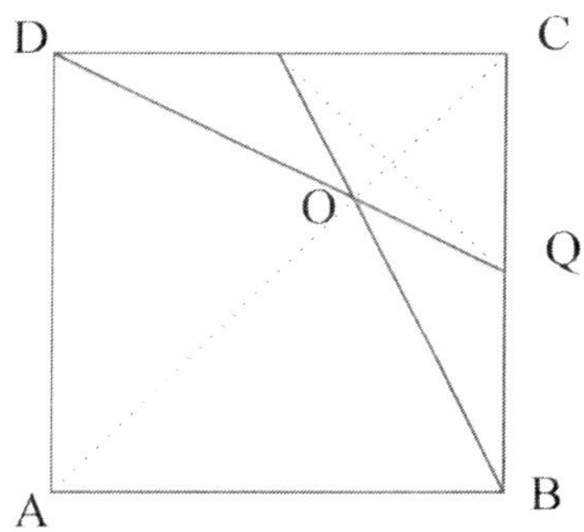

12. In an equilateral triangle, 3 coins of radii 1 unit each are kept such that they touch each other and also the sides of the triangle. Find area of the triangle.

13. In triangle ABC, point D and E are on BC and AC respectively. If AD and BE intersect at point F such that AF:FD = 4:1 and BD:DC = 2:1 then find ratio AE:EC and BF:BE (Mass point Geometry)

14. In a quadrilateral ABCD, M, N, O, P are the mid points of sides AB, BC, CD, DA as shown in the diagram below. Lines joining the vertices to the mid-point of the side opposite to that vertex, these lines meet at points R, S, T, U. Find ratio of the area of quadrilateral RSTU and sum of the areas of 4 triangles formed at the four vertices.

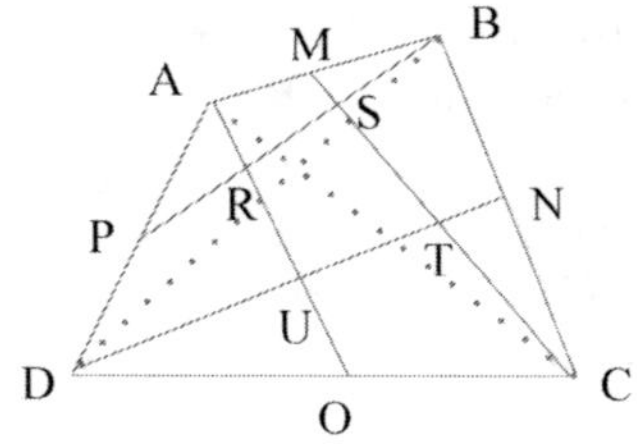

Figure

15 If in ΔABC shown below, AC = 6, AB = 8, BC = 12 and BD = 8. Find length GD where G is centroid of the triangle.

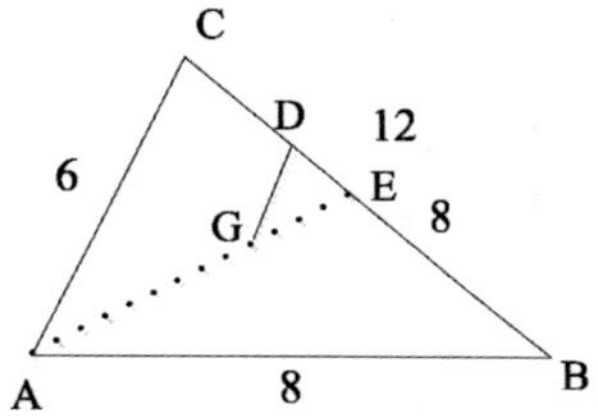

16. Two of the sides of a triangle have length 4 and 9. The median drawn to the third side of this triangle has a length of 6. Compute the length of the third side of this triangle.

17. In DADC, DB is drawn such that DB bisects ÐADC. If AB = 3, AD = 6 , and CD = 8, compute DB.

18. In the triangle given below, if AD bisects angle BAC, prove that $d^2 = bc - mn$

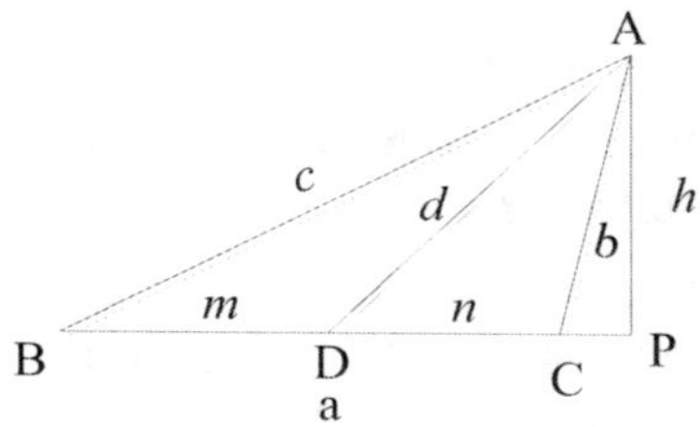

19. Consider a right triangle ABC with legs of length a and b and hypotenuse of length c. Two cevians are drawn from C, intersecting the hypotenuse such that the cevians trisect the hypotenuse. Let p and q be the lengths of the cevians and x be the length of one of the segments the hypotenuse is divided into. Show that $p^2 + q^2 = 5x^2$.

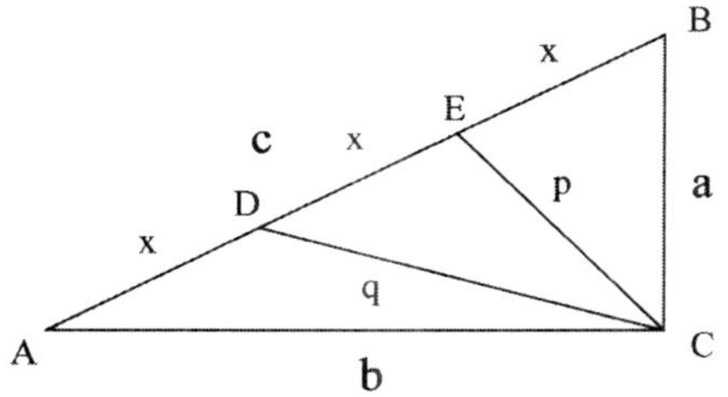

20. If in ΔABC as shown below, three cevians are drawn from vertices A, B and C meeting their opposite sides at D, E and F respectively dividing respective opposite sides in the ratio CA a:ax, AB b:by and BC c:cz. Find

(a) Ratio of all figures formed by sides of the triangle and the 3 cevians to $S_{\Delta ABC}$

(b) Ratios as above for special cases with x = y = z =2.

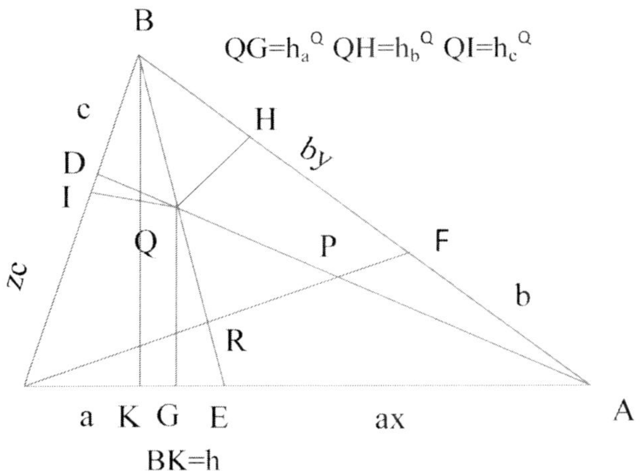

Answers to Practice Problems:

1. 12 cm^2
2. 90 cm^2
3. 54
4. $\frac{6}{7}$
5. $\frac{15}{16}$
6. 6
7. 18
8. 3:1
9. 2
10. $\frac{40}{9}$ cm^2
11. $\frac{50}{3}$ cm^2
12. $6+4\sqrt{3}$
13. 8:3 and 11:15
14. 1
15. 2
16. $\sqrt{50}$
17. 6

20. (a) $\frac{S_{\Delta CRE}}{S_{\Delta ABC}}=\frac{1}{(1+x)(1+y+xy)}$,

$$\frac{S_{\Delta BQD}}{S_{\Delta ABC}}=\frac{1}{(1+y)(1+z+yz)},\ \frac{S_{\Delta APF}}{S_{\Delta ABC}}=\frac{1}{(1+z)(1+x+xz)}.$$

$$\frac{S_{\Delta PQR}}{S_{\Delta ABC}}=\left(\frac{x}{1+x}\times\frac{xz}{(1+x+xz)}-\frac{z}{(1+z+zy)}+\frac{1}{(1+x)(1+y+xy)}\right),$$

$$\frac{Area\ of\ quadrilateral\ CDQR}{S_{\Delta ABC}}=$$

$$\left(\frac{y}{(1+y+xy)}-\frac{1}{(1+z)(1+x+xz)}\right),$$

$$\frac{Area\ of\ quadrilateral\ BFPQ}{S_{\Delta ABC}}=\frac{x}{(1+x+xz)}-\frac{1}{(1+y)(1+z+yz)}$$

and $$\frac{Area\ of\ quadrilateral\ AERP}{S_{\Delta ABC}}=$$

$$\frac{z}{(1+z+yz)}-\frac{1}{(1+x)(1+x+xy)},$$

(b) $\frac{1}{21},\frac{1}{21},\frac{1}{21},\frac{1}{7},\frac{5}{21},\frac{5}{21},\frac{5}{21}$

4

NUMBER SERIES

4. Introduction:

A **Number series** is a form of numbers in certain sequence. The series could be of a fixed number of terms (finite series) or of infinite terms (infinite sequence).

4.1 Arithmetic progression

An **Arithmetic Progression** (AP) is a sequence of numbers such that the difference between the consecutive terms is constant. For instance, the sequence 5, 8, 11, 14, 17 ... is an arithmetic progression with common difference of 3.

The series may have finite terms or infinite terms. If common difference is positive, then the terms will grow towards positive infinity and if common difference is negative, then the terms will grow towards negative infinity.

If the initial term of an arithmetic progression is '**a**' and the common difference of successive members is'**d**', then:

n^{th} term of the sequence is given by $a_n = a+(n-1)d$

and in general if m^{th} term is known, $a_n = a_m+(m-n)d$

The sum of a finite arithmetic progression with initial term a, common difference d and number of terms n is given by:

$$\text{Sum} = \frac{(2a+(n-1)d)}{2} \times n = \frac{First\ term + Last\ term}{2} \times number\ of\ terms$$

$\therefore$ *For n terms with a_1 as first term and a_n as the last term,*

$$Sum = \frac{(a_1 + a_n)}{2} \times n \text{ ......(1)}$$

Arithmetic Mean (AM) of 2 numbers a and b = (a+b)/2

Similarly, Mean of an AP of n terms = (Sum of the AP)/n.......(2)

Hence from equations (1) and (2) above, ***Arithmetic Mean (AM)*** *of n terms with a_1 as first term and a_n as the last term* $= \frac{(a_1 + a_n)}{2}$

4.1.1 Harmomic Progression, a variant of Arithmetic Progression

A **Harmonic Progression (HP)** is a sequence of real numbers that are reciprocals of numbers in an AP. So if the initial term of an arithmetic progression is '**a**' and the common difference of successive numbers is '**d**', then

It is an infinite sequence of the form $\frac{1}{a}, \frac{1}{a+d}, \frac{1}{a+2d}, ...$ where

'a' is non zero and **'−a/d'** is not a natural number ***or*** a finite sequence of the form $\frac{1}{a}, \frac{1}{a+d}, \frac{1}{a+2d}, ... \frac{1}{a+kd}$ where **'a'** is non zero, **'k'** is a natural number and **'−a/d'** is not a natural number or is greater than **'k'**

A general say n^{th} *term of the above HP is* $\frac{1}{a+(n-1)d}$

Mean of an HP is Inverse of the Mean of Inverse of terms of the HP (that are actually in AP). For the above given HP, ***Mean of n-terms*** *can be found by (using formula for sum of an AP of n terms) as*

$$\frac{1}{\dfrac{a+(a+d)+(a+2d)+......+(a+(n-1)d)}{n}}$$

$$=\frac{n}{\dfrac{2a+(n-1)d}{2}\times n}=\frac{2}{2a+(n-1)d}$$

Another property is that the sequence is a harmonic progression when each term is the harmonic mean of the neighboring terms.

It will be noticed that it is not easy to find sum of an HP. An approximation formula for

Sum of HP of n terms is $\frac{1}{d}\times\ln\left(\frac{2a+(2n-1)d}{2a-d}\right)$ ***where*** $2a\neq d$

and $d\neq 0$. ***When n is quite large compared to d, this formula gives a good approximation.***

Derivation of this formula requires knowledge of integration that will be discussed later in Chapter 9.

Harmonic Progression finds its applications in Geometry, Engineering, Music etc.

4.1.2 A few interesting observations:

(a) Sum of n terms of natural numbers $=\frac{n+1}{2}\times n=\frac{n(n+1)}{2}$

(b) It can be seen that Odd and even numbers are in AP. Let us closely look at Odd numbers starting with 1. Here first term 'a' is 1 and common difference 'd' is 2.

nth term of the series will be $1+(n\text{-}1)\times 2=2n\text{-}1$

Sum of first n terms will work out as below by substituting values of first and the nth term:

Sum of first n odd numbers $=\left(\frac{1+2n-1}{2}\right)\times n=n\times n=n^2$

It will thus be noticed from the above that ***sum of n odd numbers starting with 1 equals n²***. This can also be verified for fun. For example, sum of first 4 odd numbers is 1+3+5+7 = 16 = 4^2. Similarly sum of first 5 odd numbers equals sum of 4 odd numbers (16) + (5th odd number) 9 = 25 = 5^2

Now if you observe carefully, 5th odd number (9) is a perfect square (3^2), sum of first 4 as well as first 6 odd numbers are also perfect squares, leading to the conclusion $4^2 + 3^2 = 5^2$. This is nothing but the Pythagorean triplet (4, 3, 5).

We can generalize that when sum of certain first odd numbers is an odd number and a perfect square say 'n', sum of first $\frac{n-1}{2}$ and $\frac{n+1}{2}$ odd terms yield sum as $\left(\frac{n-1}{2}\right)^2$ and $\left(\frac{n+1}{2}\right)^2$ respectively.

Thus $\frac{n-1}{2}$, $\sqrt{n}$, $\frac{n+1}{2}$ form a Pythagorean triplet. Thus taking n as 25, the Pythgorean triplet is 12, 5, 13.

It further follows that for every odd number say m, there is a Pythagorean triplet with that number in the middle and two numbers as $\frac{m^2 \pm 1}{2}$ on either side as above. Let us look at an example. Let the odd number be 7. The numbers on either side will be $\frac{7^2 - 1}{2} = 24$ and $\frac{7^2 + 1}{2} = 25$ and the Pythagorean triplet is (24, 7, 25). Other similar triplets are (40, 9, 41) and (60, 11,61) and so on for every odd number.

4.2 Geometic progression:

A **Geometric Progression**, also known as a **Geometric Sequence**, is sequence of numbers where each term after the first is found by multiplying the previous one by a fixed, non-zero number called the *common ratio*. For example, the sequence 2, 6, 18, 54, ... is a geometric progression with common ratio 3. Similarly 10, 5, 2.5, 1.25, ...is a geometric sequence with common ratio $\frac{1}{2}$.

The general form of a geometric sequence is **a, ar, ar², ar³ etc.** where $r \neq 0$ is the common ratio and ***a*** is a scalar quantity equal to the sequence's start value.

Generally, to check whether a given sequence is geometric, one simply checks whether successive entries in the sequence all have the same ratio.

Properties of Geometric progression:

> The *n*th term of a geometric sequence with initial value ***a*** and common ratio ***r*** is given by $\mathbf{a_n = ar^{n-1}}$ **for every integer** n ≥ **1.**

It may be recalled that power to which a number is raised is referred to as exponent for the number. If base number and common ratio are more than 1, subsequent values in the sequence keep growing as exponent number gets raised, Such growth is known as exponential growth, Value of the common ratio of a geometric sequence may be negative, resulting in an alternating sequence, with numbers switching from positive to negative and back.

For instance, 1, −3, 9, −27, 81, −243, ...is a geometric sequence with common ratio −3.

The behavior of a geometric sequence depends on the value of the common ratio. If the common ratio is:

- Greater than 1, there will be exponential growth towards positive or negative infinity (depending on the sign of the initial term).
- 1, the progression is a **constant** sequence.
- Between −1 and 1 but not zero, there will be exponential decay towards zero.
- −1, the progression is an alternating sequence
- Less than −1, there is exponential growth towards **(unsigned) infinity,** due to the alternating sign.

Geometric sequences (with common ratio not equal to −1, 1 or 0) show exponential growth or exponential decay, as opposed to the linear growth (or decline) of an arithmetic progression such as 4, 15, 26, 37, 48, … (with common *difference* 11).

> An interesting result of the definition of a geometric progression is that for any value of the common ratio, any three consecutive terms *a*, *b* and *c* will satisfy the following equation:
>
> $\mathbf{b^2 = ac}$ where *b* is also known as the ***Geometric Mean*** **(GM)** between *a* and *c*.

Sum of a Geometric progression series

If a is the first term, n is the number of terms, and r is the common ratio

for r > 1, Sum of the series $= \dfrac{a(r^n - 1)}{r-1}$

and for r < 1, sum of the series $= \dfrac{a(1-r^n)}{1-r}$.

If n is large, say > 5, **r^n** will reach value close to 0. Hence sum of such a GP is $\dfrac{a}{1-r}$ These formulae work for any real numbers a and r (except $r = 1$), which results in sum $= \dfrac{0}{0}$

For r=1, the GP is a constant sequence and **sum = na**

If one were to begin the sum not from first term **a** (ar^0), but from a different term, say ar^m, then $\sum_{k=m}^{n} ar^k = \dfrac{a(r^m - r^n)}{1-r}$

4.3 Relationship between AM, HM and GM of two numbers:

Property I: $AM >= GM >= HM$

If a and b are the two numbers,

AM= $\dfrac{(a+b)}{2}$ and GM= $\sqrt{ab} \Rightarrow$

$$\text{AM} - \text{GM} = \frac{(a+b)}{2} - \sqrt{ab} = \frac{1}{2}\left(\sqrt{a} - \sqrt{b}\right)^2$$

Similarly we can see that if we have 3 numbers in GP like a, ar, ar^2, their

$$\text{AM} = \frac{a + ar + ar^2}{3} = \frac{a(1+r+r^2)}{3}$$ and

$$\text{GM} = \sqrt[3]{a \times ar \times ar^2} = \sqrt[3]{a^3 r^3} = ar \Rightarrow$$

$$\text{AM} - \text{GM} = \frac{a(1+r+r^2)}{3} - ar = \frac{a}{3}\left(1+r+r^2-3r\right) = \frac{a}{3}(1-r)^2$$

Since square of a number cannot be negative, it follows that AM – GM is positive.

As a special case for r =1, $AM - GM = 0 \Rightarrow AM = GM$

Hence AM >= GM.

$$HM = \frac{2}{\left(\frac{1}{a}+\frac{1}{b}\right)} = \frac{ab}{\frac{a+b}{2}} = \frac{(GM)^2}{AM} \Rightarrow (GM)^2 = AM \times HM$$

Since AM > = GM, the above equality implies that HM < = GM

Thus **AM > = GM > = HM**

Readers can try and verify this property for any set of numbers.

> **Property II:** *If A be the Arithmetic Mean and G be the Geometric Mean between two positive numbers m and n, then the quadratic equation whose roots are* **m, n** *is* $\boldsymbol{x^2 - 2Ax + G^2 = 0}$.

Proof:

Here A= $\frac{1}{2}(m+n)$ and G = $\sqrt{mn} \Rightarrow (m+n) = 2A$ and $mn = G^2$

Remember generic form of a quadratic equation $ax^2 + bx + c$. This can also be written as or $x^2 + \frac{b}{a} + \frac{c}{a} = 0$

Sum of the roots of this equation is $-\frac{b}{a}$ and product of the roots is $\frac{c}{a}$

Hence the equation can be rewritten as x^2 – (Sum of the roots) x+ (products of the roots) = 0

Substituting Sum and Product of roots defined above in terms of A and G, **the equation is $\boldsymbol{x^2 - 2Ax + G^2 = 0}$**

> **Property III:** *If A be the Arithmetic Mean and G be the Geometric Mean between two positive numbers, then the numbers are* $A \pm \sqrt{A^2 - G^2}$.

Reference equation in Property 2 viz $\boldsymbol{x^2 - 2Ax + G^2 = 0}$

Roots of this equation viz numbers **n** and **m** in terms of AM (A) and GM (G) respectively of the numbers $\frac{2A \pm \sqrt{4A^2 - 4G^2}}{2} = \mathbf{A} \pm \sqrt{A^2 - G^2}$

Property IV: *If ratio of the Arithmetic Mean of two numbers m and n to their Geometric Mean is p:q, then, m : n =* $(p + \sqrt{p^2 + q^2}) : (p - \sqrt{p^2 + q^2})$.

Proof:

From Property III, if we divide the equation for numbers m and n by G, we get

$$\frac{m}{G} = \frac{A}{G} + \sqrt{\frac{A^2}{G^2} - 1} \quad \text{and} \quad \frac{n}{G} = \frac{A}{G} - \sqrt{\frac{A^2}{G^2} - 1}$$

Substituting value of $\frac{A}{G} = \frac{p}{q}$, **we get**

$$\frac{m}{n} = \frac{\frac{p}{q} + \sqrt{\frac{p^2}{q^2} - 1}}{\frac{p}{q} - \sqrt{\frac{p^2}{q^2} - 1}} = \frac{p + \sqrt{p^2 - q^2}}{p - \sqrt{p^2 - q^2}}$$

4.4 Sum of squares of first n natural numbers:

We can find ***sum of first n natural numbers*** by using principles of AP series with first number a = 1 and common difference d = 1. The sum is given as $\frac{n+1}{2} \times n = \frac{n(n+1)}{2}$

We shall deal with sum of squares of first n natural numbers now. We can find this sum using the equation given below:

$$n^3 - (n-1)^3 = n^3 - (n^3 - 3n(n-1) - 1) = 3n^2 - 3n + 1$$

Substituting values of n starting with 1, we get

$$1^3 - 0^3 = 3 \times 1^2 - 3 \times 1 + 1$$

$$2^3 - 1^3 = 3 \times 2^2 - 3 \times 2 + 1$$

$$3^3 - 2^3 = 3 \times 3^2 - 3 \times 3 + 1 \ldots.$$

....

....

$$(n-1)^3 - (n-2)^3 = 3(n-1)^2 - 3(n-1) + 1$$

$n^3-(n-1)^3=3n^2-3n+1$

Adding both sides we find that on the left hand side all items cancel out except n^3-0^3 and the right hand side = $3\times(1^2+2^2+3^2+....+n^2)-3(1+2+3+....+n)+1\,(ntime))$

If we denote Sum of squares of the first n natural numbers as S_2, we will get

$n^3-0^3=3\times S_2$ - $3\times$ *(Sum of first n natural numbers)* + n

Substituting value of Sum of first n natural numbers and transposing terms, we get

$$3S_2 = n^3+3\times\left(\frac{n(n+1)}{2}\right)-n$$

$$Or\ S_2 = \frac{n^3}{3}+n\left(\frac{n+1}{2}-\frac{1}{3}\right)=\frac{n}{6}\times\left(2n^2+3n+3-2\right)$$

$$= \frac{n}{6}\times\left(2n^2+3n+1\right)=\frac{n}{6}\times\left(2n+1)(n+1\right)=\frac{n\left(2n+1)(n+1\right)}{6}$$

Readers are encouraged to verify the above formula by substituting a few values of n.

By using the same method, and starting equation as $n^4-(n-1)^4=n^4-(n^4-4n^3+6n^2-4n+1)$, we can find sum of cubes of first n natural numbers (S_3), which is given by the formula

$S_3=\left(\frac{n(n+1)}{2}\right)^2$, which incidentally is the square of sum of first n natural numbers.

4.5 Solved examples:

1. What is the sum of all 3 digit numbers that leave a remainder '2' when divided by 3?

Solution

First such number is 101 and subsequent numbers will be at a gap 3 each. Hence the series is

101, 104, 107, ...998

In this AP, a = 101 and d = 3. To find number of terms 'n', we can use formula for the nth term i.e. $a+(n-1)d = 998$ yielding

$(n-1)\times 3 = 998-101 = 897$

Or $n = \frac{897}{3}+1 = 299+1 = 300$

Sum of the series

$= \frac{a+n^{th}term}{2}\times n = \frac{101+998}{2}\times 300 = 1099\times 150 = 164850$

2. The sum of third and ninth term of an A.P is 8. Find the sum of the first 11 terms of the progression.

Solution:

3rd term = a+2d and 9th term = a+8d

Sum of 3rd and 8th term = 2a + 10d = 8 ...(1)

Now sum of first 11 terms = $\frac{a+a+10d}{2}\times 11 = \frac{2a+10d}{2}\times 11$

Substituting value of 2a + 10d from (1) above, sum of first 11 terms $= \frac{8}{2}\times 11 = 44$

3. How many 2-digit positive integers are divisible by 4 or 9?

First two digit number divisible by 4 is 12 and that divisible by 9 is 18

Solution:

The numbers divisible by 4 are 12, 16, 20 ....96 (common difference d =4)

And numbers divisible by 9 are 18, 27, 36 ....99 (common difference d =9)

Number of terms of an AP are $\frac{n^{th}term - 1^{st}term}{d}+1$

Hence numbers divisible by 4 = $\frac{96-12}{4}+1 = 22$ and

Nnumbers divisible by 9 = $\frac{99-18}{9}+1 = 10$

Now 2 numbers 36 and 72 are divisible by both 4 and 9, Hence

Total numbers divisible by 4 OR 9 = 22 + 10 – 2 = 30

4. If a rubber ball consistently bounces back $\frac{2}{3}$ of the height from which it is dropped, what fraction of its original height will the ball bounce after being dropped and bounced four times without being stopped?

Solution:

Height of the bounce of the ball form a GP with initial height say h and common ration r = $\frac{2}{3}$. The series will be

$h, \frac{2}{3}h, \left(\frac{2}{3}\right)^2 h, ...$

Height of the ball after n bounces is clearly $\left(\frac{2}{3}\right)^n$

$\therefore$ For 4 bounces, height after 4th bounce will be

$$\left(\frac{2}{3}\right)^4 h = \frac{2^4}{3^4} h = \frac{16}{81} h$$

5. If log 2, log $(2^x - 1)$ and log $(2^x + 3)$ are in A.P, then find value of x

Solution:

Since log 2, log $(2^x - 1)$ and log $(2^x + 3)$ are in A.P,

$$2 \times log\ (2^x - 1) = \log 2 + log\ (2^x + 3)$$

Or $log\ (2^x - 1)^2 = log\ \{2 \times (2^x + 3)\}$

$$\therefore (2^x - 1)^2 = 2 \times (2^x + 3)$$

Or $2^{2x} - 2 \times 2^x + 1 = 2 \times 2^x + 6$

$$2^{2x} - 4 \times 2^x - 5 = 0$$

Factorizing $(2^x - 5) \times (2^x + 1) = 0$

Thus $2^x = 5$ or $2^x = -1$

Value of 2^x cannot be negative. Hence $2^x = 5$ or $\boldsymbol{x = log_2 5}$

6. If first term of an AP is **a** and sum of first p terms is zero, show that sum of the next q terms = $\frac{-a(p+q)q}{p-1}$

Solution

Let common difference be d with a being the first term,

Sum of first p terms = $\dfrac{a + p^{th}\ term}{2} \times p$

Since this sum is 0, pth term must be –a.

Hence leading us to $a+(p-1)d=-a \Rightarrow$ or $d=\dfrac{-2a}{p-1}$.

First term of the series of next q terms will be $-a+d$ and last term will be $(-a+d\ )+(q-1)d=-a+qd$

∴ Sum of the next q terms

$$=\frac{-a+d+(-a+qd)}{2}\times q=\frac{-2a+(q+1)d}{2}\times q$$

Substituting value of d, we get Sum of next q terms

$$=\frac{-2a+(q+1)\times\dfrac{-2a}{p-1}}{2}\times q=\frac{-2aq(p-1+q+1)}{2(p-1)}=\frac{-aq(p+q)}{p-1}$$

.....Hence proved

Alternate solution:

Since sum of first p terms is zero, sum of first (p+q) terms = sum of q terms after p terms. Hence Sum of next q terms = $\dfrac{a+a+(p+q-1)d}{2}\times(p+q)$.

Substituting value of d (devrived above), we get Sum of next

$$\text{q terms} = \frac{2a+(p+q-1)\times\dfrac{-2a}{p-1}}{2}\times(p+q)$$

$$=\frac{2a\{(p-1)-(p+q-1)\}}{2(p-1)}\times(p+q)$$

$$=\frac{a\times(p+q)}{p-1}\times -q=\frac{-a\times(p+q)}{p-1}\times q$$ ....Hence proved

7. A side of an equilateral triangle is 20 cm long. A second equilateral is inscribed in it by joining mid points of sides of the first triangle. The process is continued, Find the perimeter of the 6th inscribed triangle.

Solution:

As per properties of triangles, line joining mid points of two sides is parallel to the third side and half its length. Thus sides of the first inscribed triangle will be half of those of the given triangle i.e. 10 cms.

We thus notice that each succeeding triangle will have side lengths half of the previous triangle. In other words side lengths will form a GP series with first term 'a'= 20 and common ratio 'r'= $\frac{1}{2}$.

Now, n[th] term of a GP = ar^{n-1}.

Side length of 6th triangle will, therefore, be $20\times\left(\frac{1}{2}\right)^5 = \frac{20}{32} = \frac{5}{8}$ cms,

Perimeter of this triangle = $\frac{5}{8}\times 3 = 1\frac{7}{8}$ cms.

8. Find sum of the series $\left(3^3-2^3\right)+\left(5^3-4^3\right)+\left(7^3-6^3\right)+$.....to (i) n terms (ii) 10 terms

Solution:

Each term in the given series is of the form $\left[\left(a_i+1\right)^3-a_i^{\ 3}\right]$ where a_i takes values of 2, 4, 6,.........2(n-1)

$$\left[\left(a_i+1\right)^3-a_i^{\ 3}\right]=(a_i^{\ 3}+3a_i(a_i+1)+1)-a_i^{\ 3}= 3a_i^{\ 2}+3a_i+1$$

$$\therefore \text{ Sum of first n terms } = 3\times\sum_{i=1}^{n} a_i^{\ 2}+3\times\sum_{i=1}^{n} a_i+\sum_{i=1}^{n} 1$$

Since a_i is a series of consecutive even numbers starting with 2, if we take 2 as common factor, the series $\frac{a_i}{2}$ will be consecutive natural numbers starting with 1.

Thus

$$3\times\sum_{i=1}^{n} a_i^{\ 2}+3\times\sum_{i=1}^{n} a_i+\sum_{i=1}^{n} 1=3\times 4\times\sum_{i=1}^{n}(\frac{a_i}{2})^2+3\times 2\times\sum_{i=1}^{n}\frac{a_i}{2}+\sum_{i=1}^{n} 1$$

Hence

$$3\times 4\times\sum_{i=1}^{n}(\frac{a_i}{2})^2+3\times 2\times\sum_{i=1}^{n}\frac{a_i}{2}+\sum_{i=1}^{n} 1 = 12\times\sum_{i=1}^{n}(n_i)^2+6\times\sum_{i=1}^{n} n_i+\sum_{i=1}^{n} 1$$

We know that

$$\sum_{i=1}^{n}(n_i)^2 = \frac{n(n+1)(2n+1)}{6}, \sum_{i=1}^{n} n_i = \frac{n(n+1)}{2} \text{ and } \sum_{i=1}^{n} 1 = n$$

Substituting, Sum of first n terms

$$= 12\times\frac{n(n+1)(2n+1)}{6} + 6\times\frac{n(n+1)}{2} + n$$

$$= 2n(n+1)(2n+1) + 3n(n+1) + n$$

$$= n(n+1)[2(2n+1)+3] + n$$

$$= n(n+1)[4n+5] + n = n[(n+1)(4n+5)+1]$$

Accordingly sum of first 10 terms will be

$$= 10[11\times 45+1] = 10\times(495+1) = 4960$$

9. Find rth term of an AP sum of whose n term is $2n+3n^2$

Solution:

Let first term be 'a' and common difference'd'.

rth term will be $a+(r-1)d$

Sum of first n terms of an AP with first term a and common difference d is $\frac{a+a+(n-1)d}{2}\times n = 2n+3n^2$ (Given)

Simplifying, $2a+(n-1)d = 4+6n$

Or $(2a-d)+nd = 4+6n$ leading us to

$d=6$ and $2a-d = 4$ *or* $a = 5$

Thus the rth term will be $5+(r-1)\times 6$ or $6r-1$

10. If A is the arithmetic mean and G_1 and G_2 be two geometric means between any two numbers, then prove that $2A = \frac{G_1^2}{G_2} + \frac{G_2^2}{G_1}$

Solution:

Let the numbers be a and b. In the given scenario, the four numbers a, G_1, G_2 and b should be in GP.

Let common ration of such a GP be r.

$\therefore G_1 = ar,\ G_2 = ar^2$ and $b = ar^3$

Then Arithmetic Mean A = $\frac{a+ar^3}{2}$; and

$$\frac{G_1^{\ 2}}{G_2}+\frac{G_2^{\ 2}}{G_1}=\frac{(ar)^2}{ar^2}+\frac{(ar^2)^2}{ar}=a+ar^3=2A$$ .. Hence proved.

11. If 5th term of an HP is 10 and 10th term is 18, find the 15th term

Solution:

Since inverse of the terms of an HP are in AP, 5th term of the AP is $\frac{1}{10}$ and 10th term is $\frac{1}{18}$. Since gap between the terms is 5, common difference

$$d=\frac{1}{5}\left(\frac{1}{18}-\frac{1}{10}\right)=\frac{1}{5}\left(\frac{5-9}{90}\right)=\frac{-4}{450}$$

Thus 15th term of the AP, being 5 terms beyond 10th term =

$$\frac{1}{18}+\left(\frac{-4}{450}\right)\times 5=\frac{1}{18}-\left(\frac{4}{90}\right)=\frac{1}{90}$$

Hence the 15th term of the HP is 90

12. Which number should be added to the numbers 13, 15, 19 so that the resulting numbers are consecutive terms of an HP

Solution:

We need to find a number say n to be added to the denominator of $\frac{1}{13},\frac{1}{15},\frac{1}{19}$ such that difference between the resulting numbers is same.

So $\frac{1}{13+n}-\frac{1}{15+n}=\frac{1}{15+n}-\frac{1}{19+n}\Rightarrow$

$$\frac{1}{13+n}+\frac{1}{19+n}=\frac{2}{15+n}\Rightarrow\frac{1}{13-n}+\frac{1}{19+n}=\frac{2}{15+n}\Rightarrow$$

Or $2(13+n)(19+n)=(15+n)(32+2n)\Rightarrow$

$$(13+n)(19+n)=(15+n)(16+n)\Rightarrow$$

$$247+32n+n^2=240+31n+n^2\Rightarrow n=-7$$

13. If pth term of an HP is q and qth term be p, then find it's rth term

Solution:

The question implies that pth term of an AP is $\frac{1}{q}$ and qth term of that AP is $\frac{1}{p}$

Hence if **a** is first term of that AP and **d** is the common difference, p[th] and q[th] terms are

$a+(p-1)d$ & $a+(q-1)d$ respectively.

$\therefore a+(p-1)d=\frac{1}{q}$......(1) and $a+(q-1)d=\frac{1}{p}$........(2)

Hence $aq+(pq-q)d=ap+(pq-p)d \Rightarrow a(q-p)=(q-p)d$

Or $a=d$

This implies that both equations (1) and (2) yield $apq=1 \Rightarrow a=d=\frac{1}{pq}$

Hence the rth term of the AP is $a+(r-1)d=a+(r-1)a=ar$ or $\frac{r}{pq}$

Consequently r[th] term of the HP is $\frac{pq}{r}$

14. If $a_1, a_2, a_3, ...a_n$ are in HP. Find the value of $a_1a_2+a_2a_3+.....a_{n-1}a_n$

Solution:

From the given data, $\frac{1}{a_1}, \frac{1}{a_2}, \frac{1}{a_3} .... \frac{1}{a_n}$ are in AP.

Hence $\frac{1}{a_2}-\frac{1}{a_1}=d \Rightarrow \frac{a_1-a_2}{a_1a_2}=d \Rightarrow \frac{a_1-a_2}{d}=a_1a_2$

Similarly, $a_2a_3=\frac{a_2-a_3}{d}$ .... $a_{n-1}a_n=\frac{a_{n-1}-a_n}{d}$

Adding all terms, $a_1a_2+a_2a_3+.....a_{n-1}a_n=$

$$\frac{(a_1-a_2)+(a_2-a_3)+(a_3-a_4)+.........(a_{n-1}-a_n)}{d}$$

Since alternate terms cancel each other,

$$a_1a_2+a_2a_3+.....a_{n-1}a_n=\frac{(a_1-a_n)}{d}$$

In the given AP,

$$\frac{1}{a_n}-\frac{1}{a_1}=(n-1)d \Rightarrow \frac{a_1-a_n}{a_1a_n}=(n-1)d \Rightarrow \frac{(a_1-a_n)}{d}=(n-1)a_1a_n$$

Hence $a_1a_2+a_2a_3+.....a_{n-1}a_n=(n-1)a_1a_n$

15. If a, b, c, d are in HP. Find the value of ab + bc + cd

Solution:

From the given data, $\frac{1}{a},\frac{1}{b},\frac{1}{c},\frac{1}{d}$ are in AP.

Hence if d′ is the common difference of this AP,

$$\frac{1}{b}-\frac{1}{a}=d' \Rightarrow \frac{a-b}{ab}=d' \Rightarrow ab=\frac{a-b}{d'}$$

Similarly, $bc=\frac{b-c}{d'}$ and $cd=\frac{c-d}{d'} \Rightarrow$

$$ab+bc+cd=\frac{a-b}{d'}+\frac{b-c}{d'}+\frac{c-d}{d'}=\frac{a-d}{d'}$$

Now $\frac{1}{d}-\frac{1}{a}=3d' \Rightarrow \frac{a-d}{ad}=3d' \Rightarrow \frac{a-d}{d'}=3ad$

Hence $ab+bc+cd=3ad$

4.6 Practice problems:

1. The sum of 3 numbers in AP is 21 and product of the first and 3rd numbers is 45. What are the three number?
2. Given A = 2^{65} and B = $(2^{64}+2^{63}+2^{62}+...+2^0)$, which of the two numbers is larger?
3. x, 17, 3x - y^2 - 2 and 3x + y^2 - 30 are four consecutive terms of an increasing arithmetic sequence. The sum of the four numbers is divisible by what single digit number?
4. A piece of equipment cost a certain factory Rs. 600,000. If it depreciates in value, 15% the first year, 13.5 % the next year, 12% the third year, and so on, what will be its value at the end of 10 years, all percentages applying to the original cost?
5. A man accepts a position with initial salary of Rs 5200 per month with a provisio that he will get an increment of Rs 320 every month thereafter. Find

(a) His salary for the tenth month

(b) His total earnings in the first year.

6. If pth and qth terms of a GP are q and p respectively. Show that its (p+q)th term is $\left(\frac{q^p}{p^q}\right)^{\frac{1}{p-q}}$

7. We know that sum of interior angles of a triangle is 180°. Show that sum of interior angles of polygons with 3, 4, 5, 6,… sides form an AP. Find sum of interior angles of a 21 sided polygon.

8. If $a_1, a_2, a_3, \ldots\ldots, a_n$ are in AP such that $a_i > 0$ for all i, show that

$$\frac{1}{\sqrt{a_1}+\sqrt{a_2}}+\frac{1}{\sqrt{a_2}+\sqrt{a_3}}+\frac{1}{\sqrt{a_3}+\sqrt{a_4}}+\ldots\ldots\frac{1}{\sqrt{a_{n-1}}+\sqrt{a_n}}=\frac{n-1}{\sqrt{a_1}+\sqrt{a_n}}$$

9. If the sum of p terms of an AP is ‘q’ and sum of q terms is ‘p’, show that sum of (p + q) terms is – (p + q). Also find sum of first (p - q) terms q (p > q)

10. If pth, qth and rth terms of an AP and GP are both x, y and z respectively, show that $x^{y-z} \times y^{z-x} \times z^{x-y} = 1$

11. If 7th term of an HP is 8 and 8th term is 7, find the 15th term

12. If H is the harmonic mean between p and q, find value of $\frac{H}{p}+\frac{H}{q}$

13. Find Harmonic Mean between roots of the equation $x^2 - 10x + 11 = 0$

14. If $\frac{a^{n+1}+b^{n+1}}{a^n+b^n}$ be the Harmonic Mean between a and b, find value of n.

15. If a, b, c be in HP, prove that $a^2 + c^2 > 2b^2$

Answers to Practice Problems

1. The three numbers are 5, 7, 9 (for d = 2) or 9, 7, 5 (for d = –2)
2. A and B are equal to the nearest unit digit.
3. Sum of the 4 numbers is divisible by two single digit numbers 2 and 4.
4. Depreciated value at the end of 10 years = 105000
5. His salary for the tenth month = 8080 and total earnings in the first year = 77520.
7. 3420°
9. Sum of first (p – q) terms $= \left[\dfrac{(p+2q)(p-q)}{p}\right]$
11. $\dfrac{56}{15}$
12. $\dfrac{2}{pq}$
13. $\dfrac{1}{5}$
14. n = 1

5

TRIGONOMETRY

Trigonometry is a branch of mathematics that studies relationships involving lengths and angles of triangles.

Trigonometry is normally associated with right-angled triangles. The applicability to non-right-angle triangles exists, but, since any non-right-angle triangle can be bisected to create two right-angle triangles, most problems can be reduced to calculations on right-angle triangles. Thus the majority of applications relate to right-angled triangles. Pythagoras theorem is basis for the most fundamental trigonometric identities.

Tables exist for values of different trigonometric functions. We can find ratio of base to hypotenuse, height to hypotenuse or height to base for a right angled triangle with any acute angles of the triangle. This helps us in solving many practical problems like assessing height of a hillock or area of a triangular field knowing base and one of its base angles etc.

5.1 Overview

If one angle of a triangle is 90 degrees, as in the figure below, and one of the other angles is known, the third angle can thereby be easily calculated as the two acute angles add up to 90 degrees. Once the angles are known, the ratios of the sides are determined, regardless of the overall size of the triangle. If one acute angle and length of one of the sides is known, the other two can be determined. These ratios are given by the following trigonometric functions of the known angle A, where a, b and c refer to the lengths of the sides in the accompanying figure 5.1:

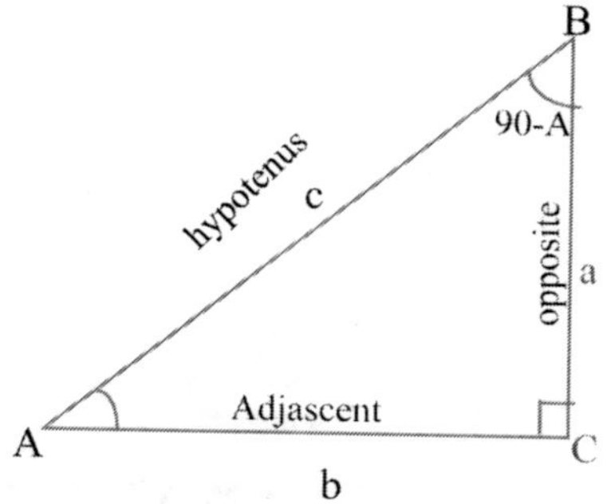

Figure 5.1

As we know, the **hypotenuse** is the side opposite to the 90 degree angle in a right triangle; it is the longest side of the triangle and one of the two arms of angle A (Side c in the above figure). The **adjacent** leg is the other arm of angle A. This is also often referred to as Base of the triangle (Side b in the above figure).

The **opposite** side is the side that is opposite to angle A and is often referred to as perpendicular or altitude of the triangle (Side ain the above figure). Angle B equals 90-A and is also known as complement of angle A.

Trignometric functions are defined as below:

- **Sine** function (sin) is defined as the ratio of the side opposite the angle to the hypotenuse.

$$\sin A = \frac{opposite}{hypotenuse} = \frac{a}{c} = \frac{perpendicular}{hypotenuse}$$

- **Cosine** function (cos), is defined as the ratio of the adjacent leg to the hypotenuse.

$$\cos A = \frac{adjascent}{hypotenuse} = \frac{b}{c} = \frac{base}{hypotenuse}$$

- **Tangent** function (tan), is defined as the ratio of the opposite leg to the adjacent leg.

$$\tan A = \frac{opposite}{adjascent} = \frac{a}{b} = \frac{a}{c} \times \frac{c}{b} = \frac{a}{c} \div \frac{b}{c} = \frac{\sin A}{\cos A} = \frac{perpendicular}{base}$$

The reciprocals of these functions are named the **Cosecant (cosec), Secant (sec),** and **Cotangent (cot)**, respectively:

$$\cos ecA = \frac{hypotenuse}{opposite} = \frac{c}{a}$$

$$\sec A = \frac{hypotenuse}{adjascent} = \frac{c}{b}$$

$$\cot A = \frac{adjascent}{opposite} = \frac{b}{a} = \frac{\cos A}{\sin A}$$

Let us look at figure 5.1 depicted again as below:

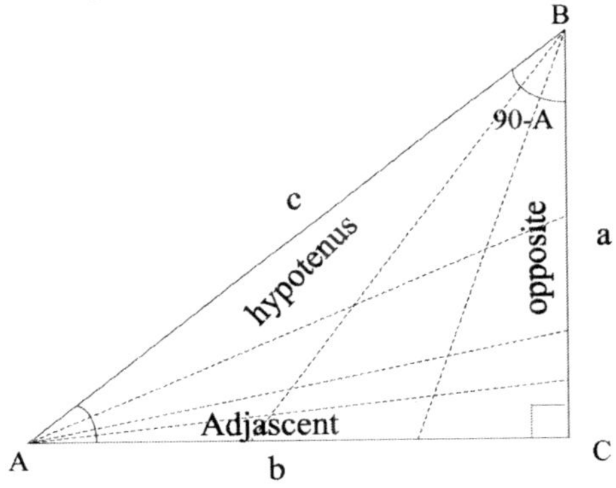

Note that as measure of angle A starts reducing, measure of opposite side BC keeps reducing till the hypotenuse coincides with adjacent side AC and goes equal to it at angle A = 0^0. Similarly as measure of complement of angle A i.e. angle B starts reducing, measure of side AC keeps reducing till the hypotenuse coincides with the altitudeBC and goes equal to it at angle A = 90^0 i.e. angle B (which is complement of A) = 0^0

Now refer to figure 5.1 and the following figures:

Angle A = 0^0

Figure 5.2.1

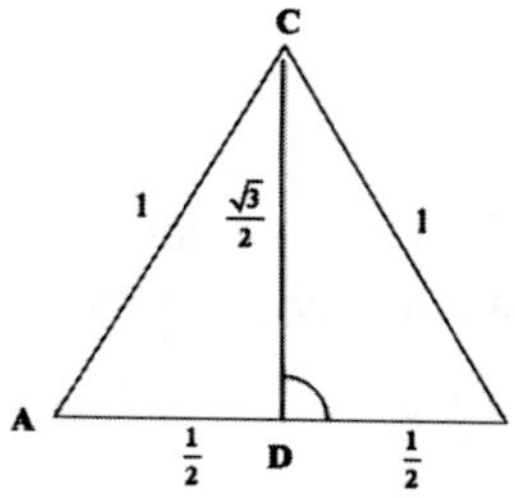

Equilateral triangle

Figure 5.2.2

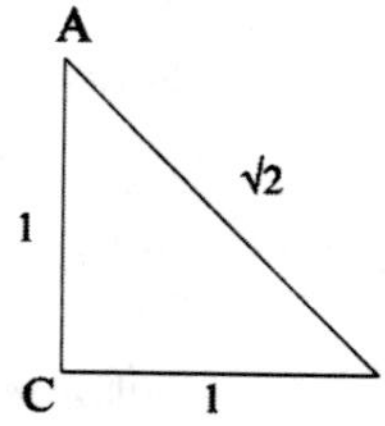

Right isosceles triangle

Figure 5.2.3

Angle A=90^0

Figure 5.2.4

Thus at angle A=0^0, point B coincides with point C (figure 5.2.1), thereby hypotenuse= adjacent side and the altitude goes 0. Thus

$\sin 0^0 = 0$ and $\cos 0^0 = \dfrac{AB}{AC} = 1$

Similarly at angle A=90^0, point A coincides with point C (figure 5.2.4), thereby hypotenuse = altitude and the adjacent side goes 0.

Thus $\cos 90^0 = 0$ and $\sin 90^0 = \dfrac{BA}{BC} = 1$

Consider the equilateral triangle of side length 1 as shown at figure 5.2.2 and recall that in an equilateral triangle, perpendicular drawn from the apex bifurcates the base as well as the apex angle, angle C in this case. Hence AD=DB= ½ and $\angle ACD = \angle BCD = 30^0$ and each angle of the triangle =60^0 Thus $\cos 60^0 = \sin 30^0 = \dfrac{AD}{AC} = \dfrac{1}{2}$ and $\cos 30^0 = \sin 60^0 = \dfrac{CD}{AC} = \dfrac{\sqrt{3}}{2}$

In the Right isosceles triangle figure 5.2.3, both acute angles= 45^0. Since the perpendicular sides are of length 1 each, the hypotenuse AB $= \sqrt{2}$.

$\therefore \sin 45^0 = \cos 45^0 = \dfrac{AC}{AB} = \dfrac{BC}{AB} = \dfrac{1}{\sqrt{2}}$

Tables are available for trigonometric function values for various angles. Values of trigonometric function for angles described above are summarized in the Table below

Angle value		Values					
in degrees	in radians	sine	co-sine	Tan-gent = sin/cos	Cose-cant =1/ sin	Secant = 1/cos	Cotangent = 1/tan
0	0	0	1	0	∞	1	∞
30	π/6	1/2	√3/2	1/Ö3	2	2/√3	√3
45	π/4	1/√2	1/√2	1	√2	√2	1
60	π/3	3/2	1/2	√3	2/√3	2	1/√3
90	π/2	1	0	∞	1	∞	0

These values should be memorized for ease and quicker problem solving.

If angle measure in degrees is ***x***, measure in radians **t** will be

$$\mathbf{t} = \frac{x \times \pi}{180}$$

Thus $30^0 = \frac{\pi}{6}$, $45^0 = \frac{\pi}{4}$, $60^0 = \frac{\pi}{3}$ and $90^0 = \frac{\pi}{2}$

Conversely $x = \frac{t \times 180}{\pi}$

The cosine, cotangent, and cosecant are so named because they are respectively the sine, tangent, and secant of the complementary angle abbreviated to "co-". We can therefore also note that:

$$\cos A = \sin\left(\frac{\pi}{2} - A\right)$$

$$\cot A = \tan\left(\frac{\pi}{2} - A\right)$$

$$\operatorname{cosec} A = \sec\left(\frac{\pi}{2} - A\right)$$

These relationships can easily be seen in the Table above. For example value of Cos 30^0 is same as value of Sin 60^0, value of cosec 30^0 is same as value of Sec 60^0 and so on.

The inverse functions of Sine, Cosine and Tangent are called the **arcsine**, **arccosine**, and **arctangent**, respectively.

Thus **arcsine**$\frac{a}{c}$ = **A, arccosine**$\frac{b}{c}$ = **A** and so on.

There are arithmetic relations between these functions, which are known as trigonometric identities.

5.2 Trigonometric identities

Let us consider Pythagorus theorem in relation to figure 5.1 above.

Here $a^2 + b^2 = c^2$,

Dividing both sides by c^2,

$\frac{a^2}{c^2} + \frac{b^2}{c^2} = 1$, which means

$\mathbf{\sin^2 A + \cos^2 A = 1}$ ..(1)

This is a fundamental trigonometric identity known as Pythagorean Identity. By manipulating this identity we arrive at the following additional identities:

$\sin^2 A = 1 - \cos^2 A$ ..(2)

$\cos^2 A = 1 - \sin^2 A$ ...(3)

$1 + \tan^2 A = \sec^2 A$ (By dividing both sides of equation 1 by $\cos^2 A$)

$1 + \cot^2 A = \text{cosec}^2 A$ (By dividing both sides of equation 1 by $\sin^2 A$)

You don't need to memorize these derived identities. These different forms of ***Pythagorean Identities*** and how these are arrived at should only be understood for instant recall.

5.3 Trignometric values for angles larger than 90°

So far, we have considered trigonometric values for only acute angles. Let us now see what happens when the angle exceeds 90^0. Let us assume angle A to be acute. Then angle A+90 or 180-A will lie in 2nd quadrant. Similarly angle 180+A or 270-A will lie in 3rd quadrant and 270+A or –A will lie in 4th quadrant

Value of trigonometric functions in different quadrants will depend upon sign of x and y coordinates. Since x-coordinates are negative in quadrant 2 and 3 and y-coordinates are negative in quadrant 3 and 4, sign of trigonometric functions in different quadrants will be as below:

Function	Quadrant 1	Quadrant 2	Quadrant 3	Quadrant 4
sine	+ve	+ve	-ve	-ve
cosine	+ve	-ve	-ve	+ve
tangent	+ve	-ve	+ve	-ve

Now we will explore values of trigonometric functions in different quadrants.

Refer to the figure 5.3 below:

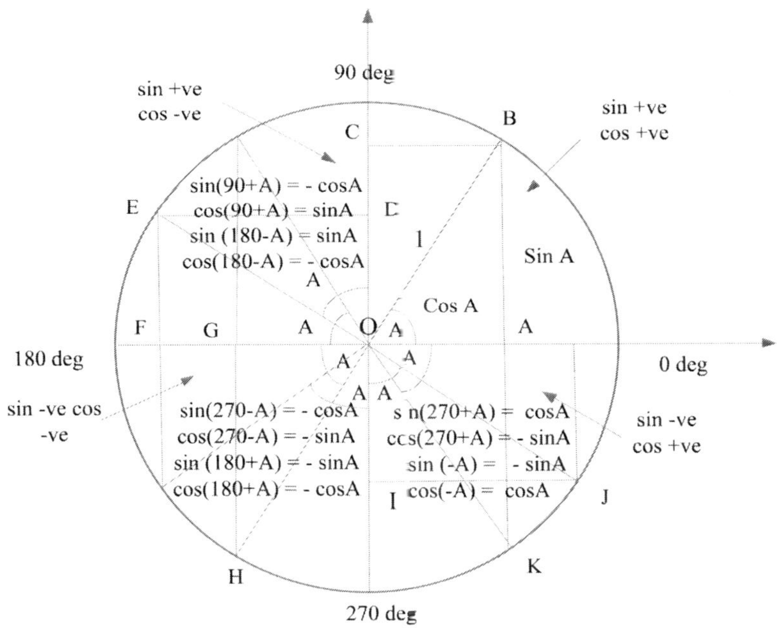

Figure 5.3: Sign and value of sin and cos in different quadrants

We always measure angles counter-clockwise starting at positive side of x-axis as 0^0 Line segment OE thus depicts an angle of 90 + A. If we measure with reference to negative x-axis, this angle (line segment OE) will measure 180° - (90° + A) = 90° - A. This is nothing but complement of A in absolute terms.

Sine of the angle 90 + A is (EF/OE) which is equivalent to Sin (90 - A) with reference to 180^0 line that also equals (BC/OB), which is same as Cos A (OA/OB). Similarly value of Cos (90° + A) (OF/OE) is equivalent to SinA but since value of Cos function in 2nd quadrant is negative; it is termed as – (Sin A). ***This concept is very important to understand.***

Now if we go to the third quadrant, for value 180° + A, the situation is mirror image of quadrant 1. Hence values of Sin and Cos will be same as in the first quadrant but with negative sign as both values are negative. Situation in quadrant 4 for angle of say 270° + A is mirror image of quadrant 2, values of Sin and Cos are same as in 2nd quadrant but with opposite sign i.e. Sin (270° + A) is value of positive CosA and corresponding Cos (270° + A) is negative SinA. Angles beyond 360^0 start repeating from quadrant 1.

As a rule of thumb, remember that angles measured anti clockwise from y-axis result in value of sine function to be equivalent to cosine function and vice versa. Sign of value of these functions in different quadrants is as given in the table above.

Recall from Chapter 3 that if we measure angles clockwise, measure of angles is negative. Angle A in 4th quadrant is measured clockwise in Figure 5.4 below:

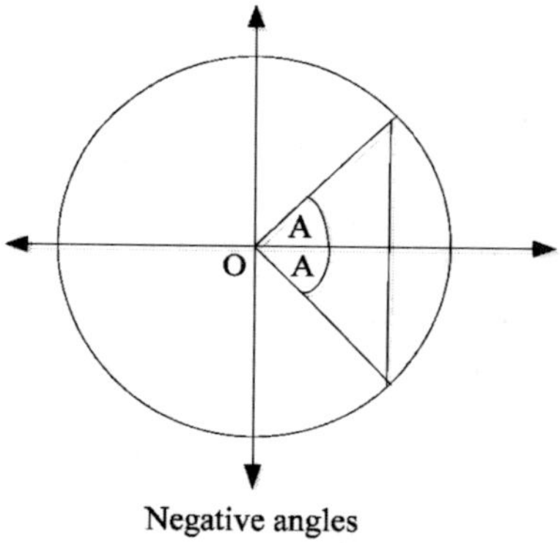

Figure 5.4

This angle is mirror image of angle A in the 1st quadrant. Hence numerical values of Sin and Cos are same as in first quadrant. However, Sin in 4th quadrant is negative and Cos is positive. ***Hence sin (-A) = - sin A and cos (-A) = cos A. Consequently tan (-A) = -tan A.*** cosec, sec and cot have same sign as sin, cos and tan respectively being reciprocal of these trigonometric functions.

A pictorial view of values of sin and cos functions for some of the angles in different quadrants is given in figure 5.5 below (coursey: Wikipedia.org). Complete understanding of the reason for +ve or –ve sign of a given trigonometric function in different quadrants is extremely important to avoid error during problem solving.

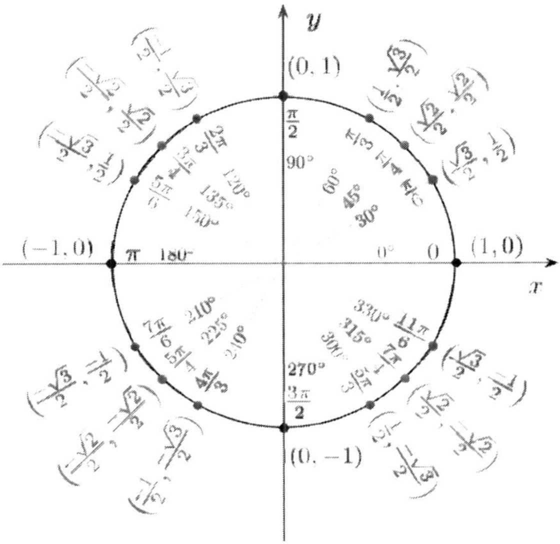

Figure 5.5

There is no need to memorize signs or values in different quadrants. Once you understand the concept, you should be able to visualize/imagine the sign and value of trigonometric functions of all these 5 angle values in different quadrants by knowing the quadrant in which the angle lies. Basic values of trigonometric functions of all these 5 angles are already given in the Table in section 5.1.

5.4 Important formulae relating two angles

5.4.1. *Sum and difference formulae:*

In case we know trigonometric value(s) of two angles say A and B, we can determine all trigonometric values of angles A ± B by following identities:

$$\sin(A \pm B) = \sin A \cos B \pm \cos A \sin B$$

$$\cos(A \pm B) = \cos A \cos B \mp \sin A \sin B$$

$$\tan(A \pm B) = \frac{\tan A \pm \tan B}{1 \mp \tan A \tan B}$$

$$\cot(A \pm B) = \frac{\cot A \cot B \mp 1}{\cot B \pm \cot A}$$

These identities are also known as *Angle transformation Formulae.*

Proof for Sin (A+B)

Consider figure 5.6 below.

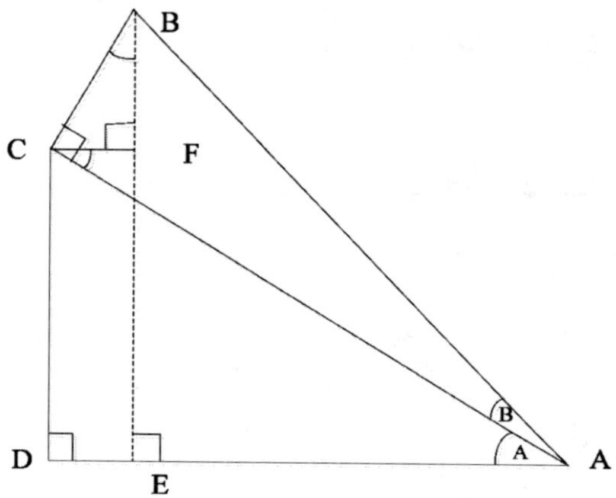

Figure 5.6

ΔADC is a Right angled triangle. BC is perpendicular to AC, BE is drawn perpendicular to AD and CF is drawn perpendicular to BE. Hence BE is parallel to CD as both these lines are perpendicular to the same line AD. Similarly CF and AD are parallel. CDEF is, therefore, a rectangle.

∴ CD=EF and CF=DE

Also ∠ACF=∠CAD=A, being alternate angles.

Thus, ∠FCB= 90-A and ∠FBC=A, being complementary angles.

ΔADC, ΔACB, ΔABE and ΔBFC are all Right angled triangles. These triangles yield following trigonometric relations:

$$\frac{CD}{AC}=\sin A\,,\ \frac{AD}{AC}=\cos A \qquad \ldots(1)$$

$$\frac{BC}{AB}=\sin B\,,\frac{AC}{AB}=\cos B \qquad \ldots(2)$$

$$and\frac{CF}{BC}=\sin A\ \frac{BF}{BC}=\cos A \qquad \ldots.(3)$$

$$\sin(A+B)=\frac{BE}{AB}=\frac{BF+EF}{AB}=\frac{BF+CD}{AB}=\frac{BF}{AB}+\frac{CD}{AB}..........(4)$$

$$and\cos(A+B)=\frac{AE}{AB}=\frac{AD-DE}{AB}=\frac{AD-CF}{AB}=\frac{AD}{AB}-\frac{CF}{AB}.......(5)$$

Multiplying sin A and cos B from equations 1 and 2 above

$$\frac{CD}{AB}=\sin A\cos B$$

Similarly multiplying cos A, sin B from equations 2 and 3 above

$$\frac{BF}{AB}=\cos A\sin B$$

Substituting in equation 4,

Sin (A+B) = sin A cos B + cos A sin B

Similarly, Multiplying cos A and cos B from equations 1 and 2 and Sin A, sin B from equation 2 and 3, we get

$$\frac{AD}{AB}=\cos A\cos B$$

$$\frac{CF}{AB}=\sin A\sin B$$

Substituting in equation 5,

cos(A+B) = cos A cos B -sin A sin B

An important hint: *To avoid any confusion of sign between the two terms of the formulae for sin (A+B) and cos (A+B), remember the following:*

Since value of sin function increases with increase in value of the angle in quadrant 1, formula for sin (A+B) has a +ve sign between the two components sinAcosB and cosAsinB. Similarly, since value of cos function decreases with increase in value of the angle in quadrant 1, formula for cos (A+B) has a -ve sign between the two components cosAcosB and sinAsinB.

We can easily find the formulae for sin (A - B) and cos (A - B) by substituting –B for B in the above formulae for sin (A + B) and cos (A +B) as below:

$$\sin(A-B)=\sin A\cos(-B)+\cos A\sin(-B)$$

Since cos(-B) = cos B and sin(-B) = - sin B

Sin (A-B) = sin A cos B – cos A sin B; and

cos (A-B) = cos A cos B + sin A sin B

We can algebraically calculate values of Tan (A $\pm$ B) by dividing formula for sin (A $\pm$ B) by cos (A $\pm$ B) as below

$$\tan(A \pm B) = \frac{\sin A \cos B \pm \cos A \sin B}{\cos A \cos B \mp \sin A \sin B}$$

Dividing numerator and denominator by cos A cos B

$$\tan(A \pm B) = \frac{\tan A \pm \tan B}{1 \mp \tan A \tan B}$$

$\cot(A \pm B)$ can also be found by taking reciprocal of $\tan(A \pm B)$ and expressing it in terms of cotA and cot B as given below:

$$\cot(A \pm B) = \frac{1 \mp \tan A \tan B}{\tan A \pm \tan B}$$

Dividing RHS by tan A tan B, we get

$$\cot(A \pm B) = \frac{\cot A \cot B \mp 1}{\cot B \pm \cot A}$$

5.4.2. *Double and triple angle formulae:*

In case angle A=B, the above sum angle formulae yield formulae for double angles in terms of given angle. From formula for sin (A+B) and cos (A+B) we get,

$$\text{Sin}\, 2A = 2 \text{Sin}\, A \text{Cos}\, A$$

$$\text{Cos}\, 2A = \text{Cos}^2 A - \text{Sin}^2 A = 2\text{Cos}^2 A - 1 = 1 - 2\text{Sin}^2 A \Rightarrow$$

$$\cos^2 A = \frac{1+\cos 2A}{2} \ \& \ \sin^2 A = \frac{1-\cos 2A}{2}$$

We can also say Cos A= $\pm\sqrt{\frac{1+\cos 2A}{2}}$, sin A= $\pm\sqrt{\frac{1+\cos 2A}{2}}$, and

Tan A= $\pm\sqrt{\frac{1-\cos 2A}{1+\cos 2A}}$

Similarly, $\tan 2A = \dfrac{2\tan A}{1-\tan^2 A}$

Using Sum Angle Formulae, we can also deduce formulae for triple angles, which are handy in solving some of the problems.

sin 3A= sin (2A+A)

= sin 2Acos A+ cos 2Asin A

$= 2\sin A\cos A \times \cos A + (\cos^2 A - \sin^2 A) \times \sin A$

$= 2\sin A\cos^2 A + \sin A\cos^2 A - \sin^3 A$

$= 3\sin A\cos^2 A - \sin^3 A$

$= 3\sin A(1-\sin^2 A) - \sin^3 A$

$\text{Sin}\,3A = 3\text{Sin}\,A - 4\text{Sin}^3 A$

In the same way, readers are encouraged to deduce formulae for cos 3A and tan 3A given below on their own; it should be fun:

Cos 3A= $4\cos^3 A$-3cos A

Tan 3A= $\dfrac{3\tan A - 4\tan^3 A}{1-3\tan^2 A}$

5.4.3. *Half angle formulae:*

The double angle formulae can also be used for deriving half angle formulae

Using double angle formulae, we can also say that

$$\sin A = 2\sin\frac{A}{2}\cos\frac{A}{2} \qquad ...(1)$$

and $\cos A = \cos^2\dfrac{A}{2} - \sin^2\dfrac{A}{2} = 2\cos^2\dfrac{A}{2} - 1 = 1 - 2\sin^2\dfrac{A}{2}$ ...(2)

From equation (2),

$\cos^2\dfrac{A}{2} = \dfrac{1+\cos A}{2}$ and

$\sin^2\dfrac{A}{2} = \dfrac{1-\cos A}{2}$. Or

$\cos\frac{A}{2} = \pm\sqrt{\frac{1+\cos A}{2}}$ and

$\sin\frac{A}{2} = \pm\sqrt{\frac{1-\cos A}{2}}$, yielding half angle formulae.

We can also see that $\tan\frac{A}{2} = \pm\sqrt{\frac{1-\cos A}{1+\cos A}}$

5.4.4 *Product to sum formulae*:

Using sum and difference formulae for Cos, we can see that

2 cos A cos B= cos (A+B) + cos (A-B) or

$cos\ A\ cos\ B = \frac{1}{2}[\cos(A+B)+\cos(A-B)]$.

Similarly $sin\ A\ sin\ B = \frac{1}{2}[\cos(A-B)-\cos(A+B)]$

And using sum and difference formulae for sin, we can find that

$sin\ A\ cos\ B = \frac{1}{2}[\sin(A+B)+\sin(A-B)]$

$cos\ A\ sin\ B = \frac{1}{2}[\sin(A+B)-\sin(A-B)]$

5.4.5. *Sum to Product formulae:*

Referring to the product to sum formulae above, let us assume that

$\angle A = \frac{\alpha+\beta}{2}$ and $\angle B = \frac{\alpha-\beta}{2}$

So A+B = α and A-B $=\beta$

Using these values in the formula for cos A cos B above

$\cos\frac{\alpha+\beta}{2}\cos\frac{\alpha-\beta}{2} = \frac{1}{2}[\cos\alpha+\cos\beta)$ or

$$\boldsymbol{cos\ \alpha + cos\ \beta = 2cos\ \frac{\alpha+\beta}{2}\ cos\ \frac{\alpha-\beta}{2}}$$

Similarly,

$$\boldsymbol{cos\ \alpha - cos\ \beta = -\ 2sin\frac{\alpha+\beta}{2}\ sin\frac{\alpha-\beta}{2}}$$

$$\boldsymbol{sin\ \alpha + sin\ \beta = 2\ sin\ \frac{\alpha+\beta}{2}\ cos\ \frac{\alpha-\beta}{2}}$$

$$\sin\alpha - \sin\beta = 2\cos\frac{\alpha+\beta}{2}\sin\frac{\alpha-\beta}{2}$$

Readers are encouraged to try and verify some of these formulae by substituting known values of trigonometric functions given above. Let us take a couple of examples:

Example 1:

$$\cos(A-B) = \cos A\cos B + \sin A\sin B$$

Let us assume $\angle A=60^0$ and $\angle B=30^0$

$$\cos(60^0 - 30^0) = \cos 30^0 = \cos 60^0 \cos 30^0 + \sin 60^0 \sin 30^0$$

$$= \frac{1}{2}\times\frac{\sqrt{3}}{2}+\frac{\sqrt{3}}{2}\times\frac{1}{2}=\frac{\sqrt{3}}{4}+\frac{\sqrt{3}}{4}=\frac{\sqrt{3}}{2}$$

$= \cos 30^0$----------------*Hence proved.*

Example 2:

$$\tan 2A = \frac{2\tan A}{1-\tan^2 A}$$

Let us assume $\angle A=45^0$

$$\tan 2A = \tan 90^0 = \frac{2\tan 45^0}{1-\tan^2 45^0}$$

$$= \frac{2\times 1}{1-(1)^2} = \frac{2}{0} = \infty$$

$$= \tan 90^0$$

This will help recall the identities in case of any confusion at the time of problem solving. As the above example clearly shows that the denominator is $1-\tan^2 A$ **not** $1+\tan^2 A$, a general confusion for many students.

5.4.6 *A special formula (Sum of cosines/sines of angles in arithmetic progression)*

This formula will give us sum of cosine of n angles in arithmetic progression say C.

Let **a** be the first term, **d** the common difference and **n** the number of terms of the AP. The required formula can be expressed as

$$C = \sum_{n=0}^{N-1} \cos(a+nd)$$

Proof:

Multiplying both sides by $2\sin\frac{d}{2}$ (assuming $\sin\frac{d}{2} \neq 0$), we get

$$2\sin\frac{d}{2} \times C = \sum_{n=0}^{N-1} 2\cos(a+nd)\sin\frac{d}{2}$$

Using formula $2\cos\alpha\sin\beta = \sin(\alpha+\beta) - \sin(\alpha-\beta)$, we get

$$2\sin\frac{d}{2} \times C = \sum_{n=0}^{N-1}\left[\sin\left(a+\left(n+\frac{1}{2}\right)d\right) - \sin\left(a+\left(n-\frac{1}{2}\right)d\right)\right]$$

Expanding the RHS, we get

$$2\sin\frac{d}{2} \times C =$$

$$\sin\left(a+\frac{d}{2}\right) - \sin\left(a-\frac{d}{2}\right) + \sin\left(a+\frac{3d}{2}\right) - \sin\left(a+\frac{d}{2}\right) + \ldots\ldots\ldots.$$

$$+\sin\left(a+\left(N-\frac{3}{2}d\right)\right) - \sin\left(a+\left(N-\frac{5}{2}d\right)\right) + \sin\left(a+\left(N-\frac{1}{2}d\right)\right) - \sin\left(a+\left(N-\frac{3}{2}d\right)\right)$$

We can notice that alternate terms cancel out leaving us with

$$2\sin\frac{d}{2} \times C = \sin\left(a+\left(N-\frac{1}{2}d\right)\right) - \sin\left(a-\frac{d}{2}\right)$$

Using the reverse formula, $\sin C - \sin D = 2\cos\frac{C+D}{2}\sin\frac{C-D}{2}$ we get

$$C = \frac{2\cos\left(a+\frac{(N-1)}{2}d\right)\sin\left(\frac{Nd}{2}\right)}{2\sin\frac{d}{2}} = \frac{\cos\left(a+\frac{(N-1)}{2}d\right)\sin\left(\frac{Nd}{2}\right)}{\sin\frac{d}{2}},$$

provided $\sin\frac{d}{2} \neq 0$

In case, $\sin\frac{d}{2} = 0$, C = N $\cos a$.

This is so because $\sin\frac{d}{2}=0$ implies that $d=2\pi$ and

Since $\cos(a+2n\pi)=\cos a$ for all integer values of n, $C=\sum_{n=0}^{N-1}\cos(a+nd)$ reduces to

$C=\cos a+\cos a+\cos a+.......$ N terms = $N\cos a$

This is an important formula. Students can try and prove similar formula for sum of sine of angles in AP with similar first term and common difference denoted as say S as

S= $N\sin a$ when $\sin\frac{d}{2}=0$

$$S=\frac{\sin\left(a+\frac{(N-1)}{2}d\right)\sin\left(\frac{Nd}{2}\right)}{\sin\frac{d}{2}}, \text{ otherwise.}$$

A solved example 10 later in the chapter will elucidate practical use of this formula.

5.5 Triangle identities:

Triangle identities that relate sides and angles of a given triangle are discussed below. In these identities, A, B and C are the angles of a triangle and a, b and c are the lengths of sides of the triangle opposite the respective angles (as shown in Figure 5.7 below).

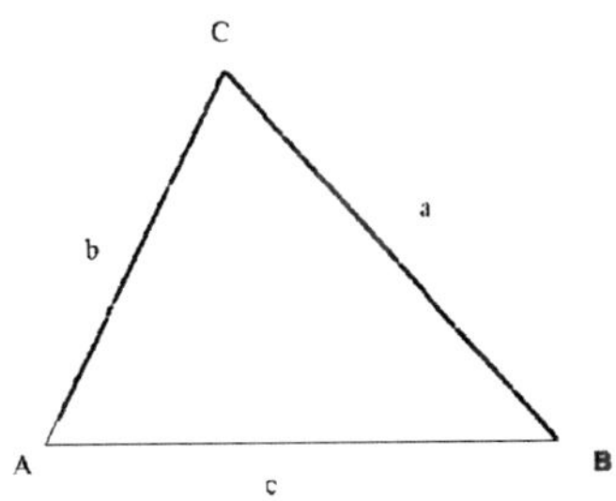

Figure 5.7

5.5.1 Law of sines:

Length of sides of a triangle is proportional to the angle opposite to it. Thus angle opposite to the largest side is the largest and vice versa. Law of sines depicts this relationship as below:

$$\frac{a}{\sin A}=\frac{b}{\sin B}=\frac{c}{\sin C}=2R=\frac{abc}{2\Delta},$$

where Δ is area of the triangle and R is radius of the circumscribed circle to the triangle. And values of R and Δ are:

$$R=\frac{abc}{\sqrt{(a+b+c)(a-b+c)(a+b-c)(b+c-a)}}$$

Area of the triangle $\Delta=\frac{1}{2}ab\sin C$

Proof for Law of Sines:

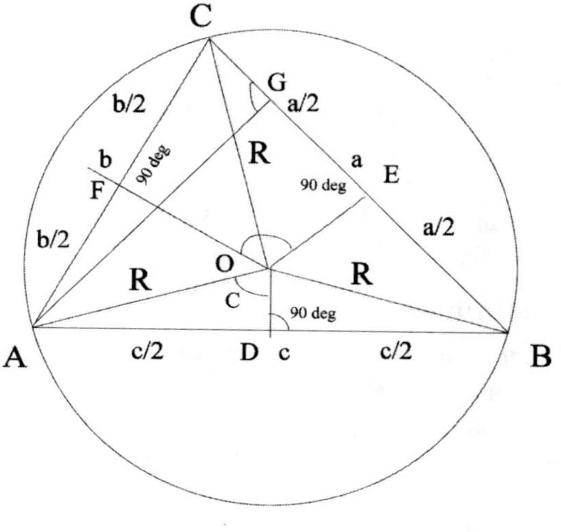

Figure 5.8

Refer to Figure 5.8. OD, OE and OF are respectively perpendicular bisectors of AB, BC and CA. O being point of intersection of perpendicular bisectors of the three sides is circumcentre of the triangle ABC.

$\therefore$ OA= OB = OC = R, circumradius of the triangle.

And $\angle$AOB = 2C, being the angle subtended at the centre by the same arc AB that subtends $\angle$C at the circle boundary.

Line AG is drawn perpendicular to side BC from A. From Δ AGC,

AG = bSin C

$\therefore$ Δ (Area of the triangle ABC) $=\frac{1}{2}ab\text{Sin}\,C$

It may further be noticed that

AD=DB, BE=CE and CF=AF.

It follows that set of triangles AOD/BOD, BOE/COE and COF/AOF are congruent.

$\therefore \angle$AOD $=\angle$C

In ΔAOD, Sin C$=\dfrac{c/2}{R}=\dfrac{c}{2R}$ or $\dfrac{c}{\operatorname{Sin} C}=2R$

Similarly in ΔBOE and ΔCOF,

$\dfrac{a}{\operatorname{Sin} A}=2R$ and $\dfrac{b}{\operatorname{Sin} B}=2R$

Hence $\dfrac{a}{\sin A}=\dfrac{b}{\sin B}=\dfrac{c}{\sin C}=2R$

So sin C$=\dfrac{c}{2R}$

Now area of the triangle $\Delta=\dfrac{1}{2}ab\operatorname{Sin} C=\dfrac{1}{2}ab\times\dfrac{c}{2R}=\dfrac{abc}{4R}$

Or $2R=\dfrac{abc}{2\times\Delta}$

$\therefore$ Law of sines is $\dfrac{a}{\sin A}=\dfrac{b}{\sin B}=\dfrac{c}{\sin C}=2R=\dfrac{abc}{2\times\Delta}$

Knowing side lengths of a triangle, we can find its area using Heron's theorem (to be proved later in this chapter) as below:

$\Delta=\sqrt{s(s-a)(s-b)(s-c)}$, where s is semi perimeter of the triangle $=\dfrac{a+b+c}{2}$

Substituting value of s, the above formula reduces to

$$\Delta=\frac{\sqrt{(a+b+c)(a+b-c)(a-b+c)(b+c-a)}}{4}$$

Since $\Delta=\dfrac{abc}{4R}$, $\dfrac{abc}{4R}=\dfrac{\sqrt{(a+b+c)(a+b-c)(a-b+c)(b+c-a)}}{4}$

Transposing, R$=\dfrac{abc}{\sqrt{(a+b+c)(a+b-c)(a-b+c)(b+c-a)}}$

5.5.2 Law of cosines:

Law of cosines is an extension of Pythagoras theorem to any arbitrary triangle shown at figure 5.7 in section 5.5 above.

As per this law,

$$c^2 = a^2 + b^2 - 2ab\cos C\,;$$

$$b^2 = a^2 + c^2 - 2ac\cos B$$

$$a^2 = b^2 + c^2 - 2bc\cos A$$

$$\text{or}\cos C = \frac{a^2 + b^2 - c^2}{2ab};\cos B = \frac{a^2 + c^2 - b^2}{2ac};\cos A = \frac{b^2 + c^2 - a^2}{2bc}$$

Readers are encouraged to prove law of Cosines on their own.

Hint: Draw a perpendicular from point C to the base AB meeting it at D as in the figure 5.9 below. This divides the triangle ABC into two right angled triangles. Now use Pythagoras theorem to prove the law of Cosines. Proof can be seen at end of this chapter in case needed.

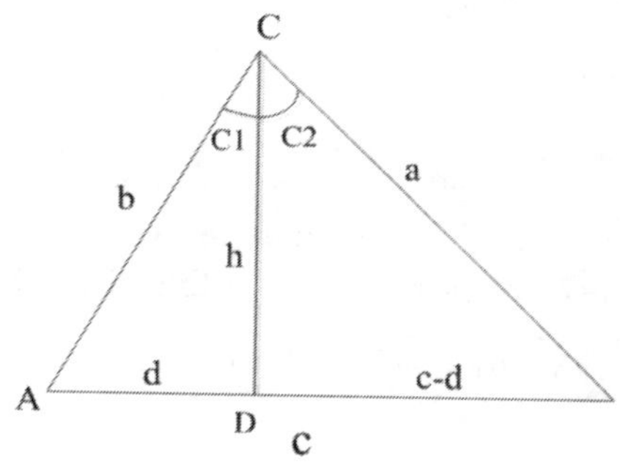

Figure 5.9

5.5.3 Heron's formula

The laws of cosine and sine may also be used to prove heron's formula for area of any arbitrary triangle given below:

Area of any triangle = $\sqrt{s(s-a)(s-b)(s-c)}$ where s = semi perimeter of the triangle = $\dfrac{(a+b+c)}{2}$

Proof:

From law of cosines, $\cos C = \dfrac{a^2 + b^2 - c^2}{2ab}$, and

From law of sines, area of triangle = $\frac{1}{2}ab\sin C$

We know that $\text{Sin}^2 C$ = 1- $\cos^2 C$ = (1+cos C)× (1-cos C)

Substituting value of cos C in the above equation,

$$\text{Sin}^2 C = \left(1+\frac{a^2+b^2-c^2}{2ab}\right)\times\left(1-\frac{a^2+b^2-c^2}{2ab}\right)$$

$$= \frac{(a+b)^2-c^2}{2ab}\times\frac{c^2-(a-b)^2}{2ab}$$

$$= \frac{(a+b+c)(a+b-c)(c+a-b)(c-a+b)}{4a^2b^2} \Rightarrow$$

$$\text{Sin } C = \frac{\sqrt{(a+b+c)(a+b-c)(a-b+c)(b+c-a)}}{2ab}$$

Or $ab\sin C = \frac{\sqrt{(a+b+c)(a+b-c)(a-b+c)(b+c-a)}}{2}$

Hence Area of the triangle

$$= \frac{1}{2}ab\sin C = \frac{\sqrt{(a+b+c)(a+b-c)(a-b+c)(b+c-a)}}{4}$$

$$= \sqrt{\frac{(a+b+c)(a+b-c)(a-b+c)(b+c-a)}{16}}$$

$$= \sqrt{\frac{(a+b+c)}{2}\times\frac{(b+c-a)}{2}\times\frac{(a-b+c)}{2}\times\frac{(a+b-c)}{2}} \quad \ldots\text{(A)}$$

Since s = $\frac{(a+b+c)}{2}$; Then

$$(s-a) = \frac{(b+c-a)}{2}, (s-b) = \frac{(a-b+c)}{2}, (s-c) = \frac{(a+b-c)}{2}$$

Substituting the values in equation (A) above, we get Heron's formula as below:

Area of the triangle = $\sqrt{s(s-a)(s-b)(s-c)}$

Heron's formula is very useful in calculating area of an arbitrary triangle, knowing length of its sides.

One can answer virtually all questions about arbitrary triangles by using the laws of sines and the ***laws of cosines. These laws can be used to compute the remaining angles and sides of any triangle as soon as two sides and their included angle or two angles and a side or three sides are known. These laws are useful in all branches of geometry, since every polygon may be described as a finite combination of triangles.***

5.6 Solved examples:

1. Prove that $\cos^2(A-B)+\cos^2 B-2\cos(A-B)\cos A\cos B=\sin^2 A$

Solution

Taking $\cos(A-B)$ common on the Left hand Side (LHS) we get

$$\text{LHS} =\cos(A-B)\left[\cos(A-B)-2\cos A\cos B\right]+\cos^2 B$$

$$=\cos(A-B)\left[\cos A\cos B+\sin A\sin B-2\cos A\cos B\right]+\cos^2 B$$

$$=\cos(A-B)\left[\sin A\sin B-\cos A\cos B\right]+\cos^2 B$$

$$=\left[\sin^2 A\sin^2 B-\cos^2 A\cos^2 B\right]+\cos^2 B=\sin^2 A\sin^2 B+\cos^2 B\left(1-\cos^2 A\right)$$

$$=\sin^2 A\sin^2 B+\sin^2 A\cos^2 B=\sin^2 A(\sin^2 B+\cos^2 B)=\sin^2 A$$

……Hence proved

2. If $\tan\frac{\alpha}{2}$ and $\tan\frac{\beta}{2}$ are roots of the equation $8x^2-26x+15=0$ find $\cos(\alpha+\beta)$

Solution

Using properties of roots of a quadratic equation, we get

$$\tan\frac{\alpha}{2}+\tan\frac{\beta}{2}=\frac{26}{8} \text{ and } \tan\frac{\alpha}{2}\tan\frac{\beta}{2}=\frac{15}{8}$$

Using angle sum property, we get

$$\tan\frac{\alpha+\beta}{2}=\frac{\tan\frac{\alpha}{2}+\tan\frac{\beta}{2}}{1-\tan\frac{\alpha}{2}\tan\frac{\beta}{2}}=\frac{\frac{26}{8}}{1-\frac{15}{8}}=-\frac{26}{7}$$

Using the property $1+ \tan^2\theta=\sec^2\theta$, we get

$\sec^2\frac{\alpha+\beta}{2}=1+\left(-\frac{26}{7}\right)^2=\frac{49+676}{49}=\frac{725}{49}$ leading to,

$\cos^2\frac{\alpha+\beta}{2}=\frac{49}{725}$;

using the identity $2\cos^2\theta-1=\cos 2\theta$ we now get

$$\cos(\alpha+\beta)=\frac{2\times 49}{725}-1=\frac{98-725}{625}=\frac{-627}{725}$$

3. If $\sin^{-1}x+\sin^{-1}y=\frac{2\pi}{3}$ and $\cos^{-1}x-\cos^{-1}y=\frac{\pi}{3}$; find x and y

Solution:

Let $\sin a=x$ and $\sin b=y$ which will also mean

$\cos\left(\frac{\pi}{2}-a\right)=x$ and $\cos\left(\frac{\pi}{2}-b\right)=y$

The above statements give us $\sin^{-1}x=a$, $\sin^{-1}y=b$, $\cos^{-1}x=\frac{\pi}{2}-a$ and $\cos^{-1}y=\frac{\pi}{2}-b$

So according to the given data,

$$a+b=\frac{2\pi}{3} \qquad \text{....(1)}$$

and $\frac{\pi}{2}-a-\left(\frac{\pi}{2}-b\right)=\frac{\pi}{3}$; leading to $b-a=\frac{\pi}{3}$ ......(2)

Adding (1) and (2), we get $2b=\left(\frac{2\pi}{3}+\frac{\pi}{3}\right)=\pi$ or $\boldsymbol{b}=\frac{\pi}{2}$

Substituting in equation (1),we get $a=\left(\frac{2\pi}{3}-\frac{\pi}{2}\right)=\frac{\pi}{6}$

Thus x = $\sin\frac{\pi}{6}=\frac{1}{2}$ and y $=\sin\frac{\pi}{2}=1$

4. If $\cos 25^0-\sin 25^0=k$, find $\cos 50^0$

Solution:

Squaring both sides,

$\cos^2 25^0+\sin^2 25^0-2\sin 25^0\cos 25^0=k^2$

Or $1-\sin 50^0=k^2$ leading to $\sin 50^0=1-k^2$ implying that k is positive and < 1

$$\cos 50^0=\sqrt{1-\sin^2 50^0}=\sqrt{1-(1-k^2)^2}=\sqrt{1-(1+k^4-2k^2)}=\pm\sqrt{2k^2-k^4}$$

Since value of $\cos 50^0$ is positive, we get value of

$\cos 50^0 = k\sqrt{2-k^2}$, which is a positive number as k is<1.

5. $12\cot^2\theta - 31\operatorname{cosec}\theta + 32 = 0$, find the value of $\sin\theta$

Solution:

We know that $1+\cot^2\theta = \operatorname{cosec}^2\theta$ or $\cot^2\theta = \operatorname{cosec}^2\theta - 1$

Substituting in the given equation, we get

$12\operatorname{cosec}^2\theta - 12 - 31\operatorname{cosec}\theta + 32 = 0$ or

$12\operatorname{cosec}^2\theta - 31\operatorname{cosec}\theta + 20 = 0$

Solving this quadratic equation, we get

$$\operatorname{cosec}\theta = \frac{31 \pm \sqrt{(31^2 - 4\times 12\times 20)}}{2\times 12}$$

$$= \frac{31 \pm \sqrt{(961-960)}}{24} = \frac{31\pm 1}{24} = \frac{4}{3}, \frac{5}{4}$$

So, $\sin\theta = \frac{3}{4}$ or $\frac{4}{5}$

6. Prove that $\tan 81^0 - \tan 63^0 - \tan 27^0 + \tan 9^0 = 4$

Solution:

Using identity $\tan\theta = \frac{\sin\theta}{\cos\theta}$, we get

$$\left(\frac{\sin 81^0}{\cos 81^0} + \frac{\sin 9^0}{\cos 9^0}\right) - \left(\frac{\sin 63^0}{\cos 63^0} + \frac{\sin 27^0}{\cos 27^0}\right) =$$

$$\left(\frac{\sin 81^0 \cos 9^0 + \cos 81^0 \sin 9^0}{\cos 81^0 \cos 9^0} - \frac{\sin 63^0 \cos 27^0 + \cos 63^0 \sin 27^0}{\cos 63^0 \cos 27^0}\right)$$

Using standard trigonometric formulae, the above expression reduces to,

$$\left(\frac{\sin 90^0}{\sin 9^0 \cos 9^0} - \frac{\sin 90^0}{\sin 27^0 \cos 27^0}\right) = \frac{2}{\sin 18^0} - \frac{2}{\sin 54^0} = 2\times\left(\frac{\sin 54^0 - \sin 18^0}{\sin 54^0 \sin 18^0}\right):$$

$$= \frac{2\times 2\cos 36^0 \sin 18^0}{\cos 36^0 \sin 18^0} = 4$$

Hence proved.

7. If $\sin\theta_1+\sin\theta_2+\sin\theta_3+\ldots\ldots\ldots+\sin\theta_n=n,$ find the value of $\sin\theta_1+\cos\theta_2+\sin\theta_3+\cos\theta_4\ldots\ldots\ldots\ldots+\sin\theta_{n-1}+\cos\theta_n$ given that $\theta_1,\theta_2,\theta_3\ldots\ldots\ldots\ldots\ldots\theta_n$ lie in first quadrant.

Solution:

Since $\sin\theta_1+\sin\theta_2+\sin\theta_3+.\ \ldots\ldots\ldots+\sin\theta_n=n,$

it implies that $\sin\theta_1=\sin\theta_2=\sin\theta_3=\ldots\ldots\ldots\ldots=\sin\theta_n=1$,

that is $\theta_1=\theta_2=\theta_3=\ldots\ldots\ldots\ldots\ldots\ \ldots\ldots\ldots\theta_n=\frac{\pi}{2}$

and $\cos\theta_2=\cos\theta_4=\ldots\ldots\ldots\ldots=\cos\theta_n=0$

Leading us to the solution that

$\sin\theta_1+\cos\theta_2+\sin\theta_3+\cos\theta_4\ldots\ldots+\sin\theta_{n-1}+\cos\theta_n$

$=\sin\theta_1+\sin\theta_3+\ldots\ldots\ldots+\sin\theta_{n-1}=\frac{n}{2}$, if n is even

and $\frac{n+1}{2}$, if n is odd.

8. Find how many solutions are possible for the following pair of equations:

$2\sin^2\theta-\cos 2\theta=0$ and $2\cos^2\theta-3\sin\theta=0$ in the interval $[0,2\pi]$

Solution:

From first equation we get $2\sin^2\theta-(1-2\sin^2\theta)=0$ leading to $4\sin^2\theta-1=0$ or $\sin^2\theta=\frac{1}{4}$ giving us two possible values of $\sin\theta$ as $\pm\frac{1}{2}$

For $\sin\theta=\frac{1}{2}$, as well as $-\frac{1}{2}$, two possible values of $\cos\theta$ are $\pm\frac{\sqrt{3}}{2}$,

If we substitute these values in equation (2), we will find that $\sin\theta=\frac{1}{2}$ will satisfy the equation but $\sin\theta=-\frac{1}{2}$ does not. Hence we have two possible solutions as $\sin\theta=\frac{1}{2}$ and $\cos\theta=\pm\frac{\sqrt{3}}{2}$, which will correspond to the angles $\frac{\pi}{6}$ and $\frac{5\pi}{6}$

9. Find the value of sin 15⁰

Solution:

Let $\angle a = 45^0$ and $\angle b = 30^0$, then we can find sin 15^0 as

$$\sin(45^0 - 30^0) = \sin 45^0 \cos 30^0 - \cos 45^0 \sin 30^0 =$$

$$\frac{1}{\sqrt{2}} \times \frac{\sqrt{3}}{2} - \frac{1}{\sqrt{2}} \times \frac{1}{2} = \frac{1}{2\sqrt{2}}\left(\sqrt{3}-1\right) \text{ or } \sin 15^0 = \sqrt{2} \times \left(\frac{\sqrt{3}-1}{4}\right)$$

10. Find value of $\sin\frac{\pi}{n} + \sin\frac{3\pi}{n} + \sin\frac{5\pi}{n} + \ldots\ldots\ldots\ldots$ upto n terms

Solution:

We know that sum of sine of angles in AP '**S**' with first term **a** and common difference **d** is given as

$$S = \frac{\sin\left(a + \frac{(N-1)}{2}d\right)\sin\left(\frac{Nd}{2}\right)}{\sin\frac{d}{2}} \text{ where there are total N terms}$$

Substituting a $= \frac{\pi}{n}$, d = $\frac{2\pi}{n}$, N = n, we get

$$S = \frac{\sin\left(\frac{\pi}{n} + \frac{(n-1)}{2}\frac{2\pi}{n}\right)\sin\left(\frac{n \times 2\pi}{2n}\right)}{\sin\frac{2\pi}{2n}} = \frac{\sin\left(\frac{\pi}{n}(1+n-1)\right)\sin\pi}{\sin\frac{\pi}{n}} = \frac{\sin^2\pi}{\sin\frac{\pi}{n}} = 0$$

11. $\sin x + \cos x = \frac{1}{5}$. Find tan 2x where $\frac{\pi}{2} < x < \frac{3\pi}{4}$

Solution:

Squaring both sides, $\sin^2 x + \cos^2 x + 2\sin x\cos x = \frac{1}{25}$

Or $1 + \sin 2x = \frac{1}{25}$ or $\sin 2x = \frac{1}{25} - 1 = \frac{-24}{25}$

$$\cos 2x = \pm\sqrt{1 - \sin^2 2x} = \pm\sqrt{1 - \frac{576}{625}} = \pm\sqrt{\frac{49}{625}} = \pm\frac{7}{25}$$

Since angle 2x is more than π but less than $3\pi/2$, $\cos 2x = -\frac{7}{25}$

Thus $\tan 2x = \frac{\sin 2x}{\cos 2x} = \frac{-24/25}{-7/25} = \frac{24}{7}$

12. Find maximum value of the function A $=3\cos x-4\sin x$ (alert: knowledge of derivation is essential for this question)

Solution:

The function will have maximum value when its differential = 0

So differentiating the function and equating it to zero, we get

$-3\sin x-4\cos x=0$ or $\tan x=\frac{-4}{3}$ leading to $\sec^2 x=1+\frac{16}{9}=\frac{25}{9}$

Or $\cos^2 x=\frac{9}{25}$ and $\sin^2 x=1-\frac{9}{25}=\frac{16}{25}$ leading to $\cos x=\pm\frac{3}{5}$ and $\sin x=\pm\frac{4}{5}$

Since $\tan x=\frac{-4}{3}$, x lies in 2nd or 4th quadrant.

The given function will have maximum value when cos x is positive and sin x is negative. Hence using $\cos x=\frac{3}{5}$ and $\sin x=-\frac{4}{5}$ we get

$A=3\cos x-4\sin x=\ 3\times\frac{3}{5}-4\times\frac{-4}{5}=\frac{9}{5}+\frac{16}{5}=5$

13. Find general solution of the equation $\sin x+\cos x=1$

Solution:

Multiplying both sides by $\frac{1}{\sqrt{2}}$ we get,

$\frac{1}{\sqrt{2}}\sin x+\frac{1}{\sqrt{2}}\cos x=\frac{1}{\sqrt{2}}$.

Since sine is positive in 1st and 2nd quadrant only, the above equation leads to

$\sin\left(x+\frac{\pi}{4}\right)=\sin\frac{\pi}{4}$ or $\sin\frac{3\pi}{4}$ or $\sin\frac{9\pi}{4}$ or $\sin\frac{11\pi}{4}$ and so on.

The above equation can be rewritten as

$\sin\left(x+\frac{\pi}{4}\right)=\sin\left(n\pi+(-1)^n\frac{\pi}{4}\right)$

$\therefore\ x=n\pi-\frac{\pi}{4}+(-1)^n\frac{\pi}{4}$ yielding

$x=n\pi-\frac{\pi}{2}$ ***for n being odd*** and $x=n\pi$ ***for n being even***

14. If $\tan A = 2\tan B + \cot B$, find the value of $2\tan(A-B)$

Solution:

From the given statement, we get $\tan A - \tan B = \tan B + \cot B$

Now $2\tan(A-B) = 2\times\dfrac{\tan A - \tan B}{1+\tan A\tan B} = 2\times\dfrac{\tan B+\cot B}{1+\tan A\tan B}$

$$= 2\times\frac{\tan B+\cot B}{1+(2\tan B+\cot B)\tan B}$$

simplifying further we get $2\tan(A-B) = 2\times\dfrac{\tan B+\cot B}{1+(2\tan^2 B+1)}$

$$= 2\times\frac{\tan B+\cot B}{2+2\tan^2 B} = \frac{\tan B + {}^{1}\!/_{\tan B}}{1+\tan^2 B} = \frac{1+\tan^2 B}{\tan B(1+\tan^2 B)} = \frac{1}{\tan B}$$

15. Find the value of sin 18^0, sin 36^0, sin 54^0 and sin 72^0

Solution:

Let θ = 18^0, then $5\theta = 90^0 = \dfrac{\pi}{2}$

So $2\theta + 3\theta = \dfrac{\pi}{2}$ or $2\theta = \dfrac{\pi}{2} - 3\theta$ leading to

$\sin 2\theta = \cos 3\theta \Rightarrow 2\sin\theta\cos\theta = 4\cos^3\theta - 3\cos\theta$ leading further to

$\Rightarrow 2\sin\theta = 4\cos^2\theta - 3 \Rightarrow 4 - 4\sin^2\theta - 3 - 2\sin\theta = 0$

Simplifying further we get

$4\sin^2\theta + 2\sin\theta - 1 = 0 \Rightarrow$

$$\sin\theta = \frac{-2\pm\sqrt{4+4\times4}}{8} = \frac{-2\pm\sqrt{20}}{8} = \frac{-1\pm\sqrt{5}}{4}$$

Since **sin 18⁰** is positive, we get $\sin 18^0 = \dfrac{-1+\sqrt{5}}{4}$

We know that $\cos 2\theta = 1 - 2\sin^2\theta \Rightarrow \cos 36^0 = 1 - 2\sin^2 18^0$

Thus $\cos 36^0 = 1 - 2\left(\frac{-1+\sqrt{5}}{4}\right)^2 = 1 - \frac{12-4\sqrt{5}}{16} = \frac{4+4\sqrt{5}}{16} = \frac{1+\sqrt{5}}{4}$

$$\sin 36^0 = \sqrt{1-\cos^2 36^0} = \sqrt{1-\left(\frac{1+\sqrt{5}}{4}\right)^2} = \sqrt{1-\left(\frac{6+2\sqrt{5}}{16}\right)} =$$

$$= \sqrt{\frac{10-2\sqrt{5}}{16}} = \sqrt{\frac{5-\sqrt{5}}{8}}$$

sin 54⁰= cos (90-54)⁰ = **cos 36⁰** $= \frac{1+\sqrt{5}}{4}$

Sin 72⁰ = cos 18⁰. Hence

$$\text{Sin } 72^0 = \sqrt{1-\sin^2 18^0} = \sqrt{1-\left(\frac{-1+\sqrt{5}}{4}\right)^2} = \sqrt{1-\left(\frac{6-2\sqrt{5}}{16}\right)}$$

$$= \sqrt{\frac{10+2\sqrt{5}}{16}} = \sqrt{\frac{5+\sqrt{5}}{8}}$$

5.7 Practice problems:

1. Find least positive non integer solution of

$$\sin \pi(x^2+x) - \sin \pi x^2 = 0$$

2. If tan A $= \frac{1-\cos B}{\sin B}$, find the value of tan 2A.
3. Find the value of $\cos\alpha \cos 2\alpha \cos 4\alpha \times .... \cos 2^{n-1}\alpha$ where n is a natural number.
4. Find the value of

$$3(\sin x - \cos x)^4 + 6(\sin x + \cos x)^2 + 4(\sin^6 x + \cos^6 x)$$

5. Sin x + cos x = $1/5$, find value of tan 2x ; $\pi/2 < x < 3\pi/4$

6. Find the value of $\dfrac{\cos^2 33^0 - \cos^2 57^0}{\sin 21^0 - \cos 21^0}$

7. If A lies in second quadrant and 3tan A + 4 = 0, find the value of $2\cot A - 5\cos A + \sin A$

8. Find the number of solutions of the following pair of equation in the interval $[0, 2\pi]$

9. General value of θ satisfying the equation $2\sin^2\theta - 3\sin\theta - 2 = 0$

10. If α lies in the third quadrant, find the value of $\sqrt{\dfrac{1-\cos\alpha}{1+\cos\alpha}} + \sqrt{\dfrac{1+\cos\alpha}{1-\cos\alpha}}$

11. Find value of $\sin 12^0 \sin 48^0 \sin 54^0$

12. If $\alpha + \beta = \dfrac{\pi}{2}$ and $\beta + \gamma = \alpha$, find $\tan\alpha$

13. Find $\cot\theta - 2\cot 2\theta$ for all real values of θ

14. If $\tan x + \cot x = 2$, find the value of $\sin^{2n} x + \cos^{2n} x$

15. Find the value of

$$\tan^{-1} \tfrac{1}{3} + \tan^{-1} \tfrac{2}{9} + \tan^{-1} \tfrac{4}{33} \ldots\ldots\ldots\ldots\ldots \tan^{-1} \frac{2^{n-1}}{1+2^{2n-1}}$$

Answers to Practice Problems:

1. $\dfrac{\sqrt{p}-1}{4}$, where p is an odd integer.
2. tan B
3. $\dfrac{\sin 2^n x}{2^n \sin x}$
4. 13
5. $\dfrac{24}{7}$
6. $-\dfrac{1}{\sqrt{2}}$
7. $\dfrac{23}{10}$
8. 2
9. $n\pi + (-1)^n \dfrac{7\pi}{6}$
10. $-\dfrac{2}{\sin\alpha}$
11. $\dfrac{1}{8}$
12. $\tan\beta + 2\tan\gamma$
13. $\tan\theta$
14. $\dfrac{1}{2^{n-1}}$
15. $\dfrac{\pi}{4}$

Appendix

Proof of law of cosines:

In the figure below, $\angle C = \angle C_1 + \angle C_2$,

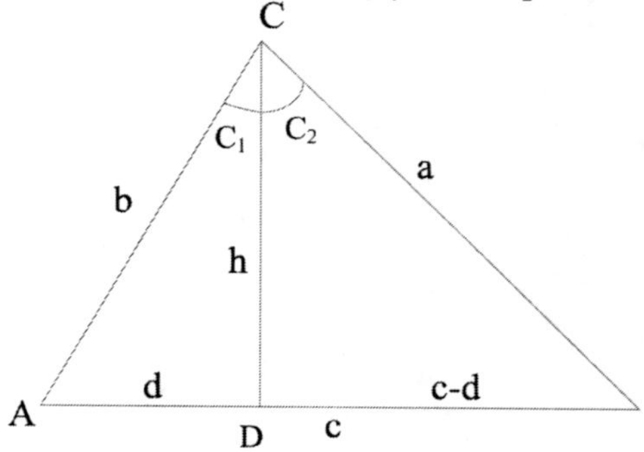

$$h^2 = b^2 - d^2 = a^2 - (c-d)^2$$

$\therefore \cos C = \cos C_1 \cos C_2 - \sin C_1 \sin C_2$

$$= \frac{h}{b} \times \frac{h}{a} - \frac{d}{b} \times \frac{c-d}{a}$$

$$= \frac{h^2 - d(c-d)}{ab} = \frac{h^2 - cd + d^2}{ab} = \frac{2h^2 - (2cd - 2d^2)}{2ab}$$

Substituting values of h^2 in equation for Cos C above,

$$\cos C = \frac{b^2 - d^2 + a^2 - (c-d)^2 - 2(cd - d^2)}{2ab}$$

$$= \frac{b^2 + a^2 - d^2 - (c-d)(c-d+2d)}{2ab}$$

$$= \frac{b^2 + a^2 - d^2 - (c-d)(c+d)}{2ab} = \frac{b^2 + a^2 - (d^2 + c^2 - d^2)}{2ab} = \frac{a^2 + b^2 - c^2}{2ab}$$

Same way, formula for cos A and cos B can be proved.

6

MENSURATION

Mensuration is a branch of mathematics that deals with measurement of geometric shapes. It involves derivation and use of algebraic formulae to measure lengths, areas and volumes and various other parameters of geometric figures.

Overview

Let us recall algebraic formulae for perimeter and area of different geometrical figures from Chapter 3.

6.1 Triangles

6.1.1 Perimeter

Perimeter equals sum of 3-sides of a triangle and it logically follows that:

Perimeter of Equilateral triangle = 3 x side length (since each side is equal)

Perimeter of Isosceles triangle = 2 x equal side + Base

6.1.2 Area

Area of a triangle is arrived at by multiplying base length and altitude divided by 2. Altitude is the height of a triangle and for most purposes it is a straight line drawn perpendicular to the base from the vertex of the triangle.

Thus area of Right triangle is simply $= \frac{1}{2} \times base \times height$

For other types of triangles, a formula can be derived where the variables can be replaced by values as applicable. Let's see how these formulae are arrived at and how these can be recalled and used to solve area for any type of triangle: Algebraic proof for Area of Equilateral and Isosceles triangles are as below. *Readers may refer to the Chapter 5-Trigonometry for proof of the Heron's formula for area of the Scalene triangle.*

6.1.2.1 Equilateral Triangle:

Refer to the equilateral triangle ABC in figure 6.1. Recall that a perpendicular drawn from the vertex of an equilateral triangle (the altitude) bisects the base.

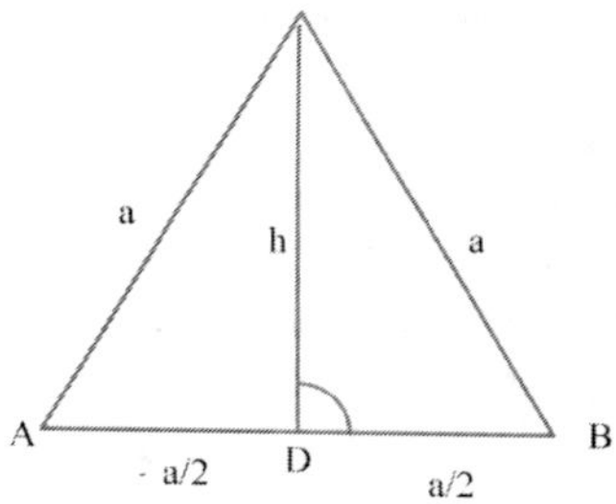

Figure 6.1

Now area of this triangle = $\frac{1}{2} \times base \times height = \frac{1}{2} ah$

By Pythagoras theorem, $h^2 = a^2 - (a/2)^2 = a^2 - \frac{a^2}{4} = \frac{3a^2}{4}$

Or $h = \frac{\sqrt{3}}{2} a$

Substituting the value of h in terms of a in the formula:

Area of an equilateral triangle $= \frac{1}{2} ah = \frac{1}{2} \times a \times \frac{\sqrt{3}}{2} a = \frac{\sqrt{3}}{4} a^2$

Thus Area of equilateral triangle = $\frac{\sqrt{3}}{4} a^2$

6.1.2.2 Isosceles Triangle:

Refer to the Isosceles triangle PQR in figure 6.2 below where equal side length is **a,** base length **b** and height **h**.

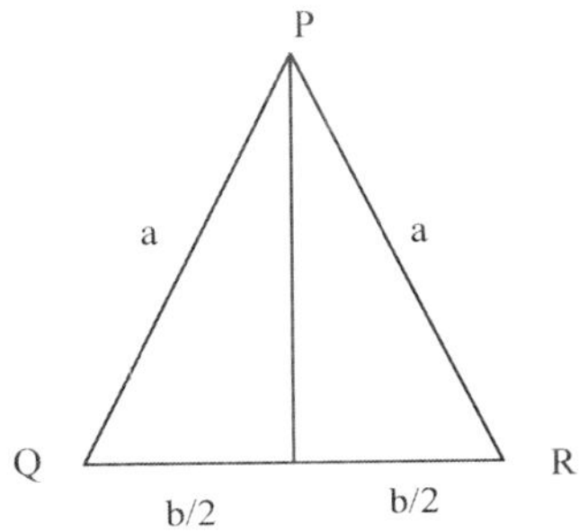

Figure 6.2

A perpendicular drawn from the vertex of an isosceles triangle (the altitude) also bisects the base. By the same logic as in case of equilateral triangle,

$$h^2 = a^2 - (b/2)^2 = a^2 - \frac{b^2}{4} = \frac{4a^2 - b^2}{4}$$

Or $h = \dfrac{\sqrt{4a^2 - b^2}}{2}$

Substituting the value of h in terms of a (equal side) and b (base):

Area of Isoceles Triangle = $\dfrac{1}{2}$ x b x h = $\dfrac{1}{2}$ x b x $\dfrac{\sqrt{4a^2 - b^2}}{2}$

Hence area of an isosceles triangle = $b \times \dfrac{\sqrt{4a^2 - b^2}}{4}$

6.1.2.3 Scalene triangle:

In a scalene triangle, if we don't know base length and height of the triangle, we need to know at least 3 parameters – length of two sides and the measure of the included angle or length of all three sides.

Case 1 : Length of two sides say a and b and the measure of the included angle C are known

Refer to the figure 6.3 below:

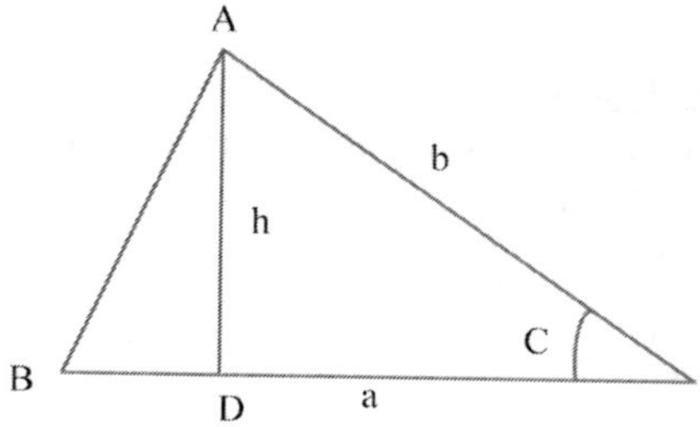

Figure 6.3

Height of the triangle ABC = h = b sin C

$\therefore$ **Area of the triangle =** $\frac{1}{2}a \times h = \frac{1}{2}ab\sin C$

Case 2 : Length of all three sides say a, b and c are known

Refer to figure 6.4 below:

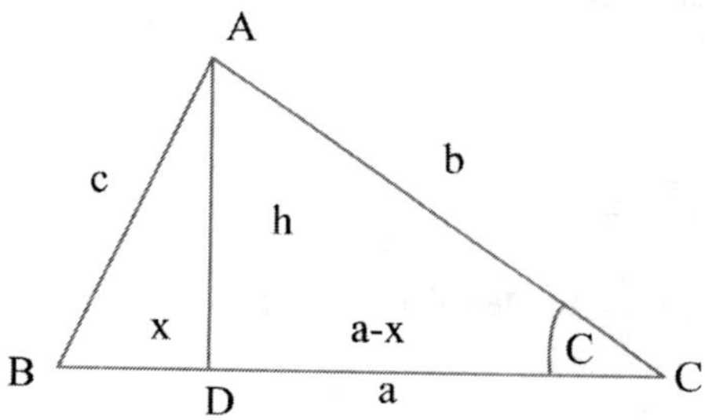

Figure 6.4

Here dimensions of the three sides a, b, c are known. Let us denote length of line BD = x , yielding CD = a – x.

Considering, ΔABD, $x^2 = c^2 - h^2$ or $x = \sqrt{c^2 - h^2}$ and

Considering, ΔADC, $h^2 = b^2 - (a-x)^2 = b^2 - a^2 - x^2 + 2ax$

Substituting values of x^2 and x, we get

$$h^2 = b^2 - a^2 - (c^2 - h^2) + 2a\sqrt{c^2 - h^2}$$

$$\Rightarrow 2a\sqrt{c^2 - h^2} = h^2 - b^2 + a^2 + c^2 - h^2 = a^2 + c^2 - b^2$$

Squaring both sides

$$c^2 - h^2 = \frac{(a^2 + c^2 - b^2)^2}{4a^2} \Rightarrow h^2 = c^2 - \left\{\frac{(a^2 + c^2 - b^2)}{2a}\right\}^2$$

$$\text{Or } h^2 = \left\{c - \frac{(a^2 + c^2 - b^2)}{2a}\right\}\left\{c + \frac{(a^2 + c^2 - b^2)}{2a}\right\}$$

$$= \left\{\frac{2ac - (a^2 + c^2 - b^2)}{2a}\right\}\left\{\frac{2ac + (a^2 + c^2 - b^2)}{2a}\right\}$$

$$= \left\{\frac{b^2 - (a-c)^2}{2a}\right\}\left\{\frac{(a-c)^2 - b^2)}{2a}\right\}$$

$$= \left\{\frac{(b - a + c)(b + a - c)}{2a}\right\}\left\{\frac{(a + c + b)(a + c - b)}{2a}\right\}$$

$$= \left\{\frac{(a+b+c)(a+b+c-2a)(a+b+c-2b)(a+b+c-2c)}{(2a)^2}\right\}$$

Perimeter P = $(a+b+c)$ leading to

$$h^2 = \left\{\frac{P(P-2a)(P-2b)(P-2c)}{(2a)^2}\right\}$$

$$\text{Or } h = \left\{\frac{\sqrt{P(P-2a)(P-2b)(P-2c)}}{2a}\right\}$$

Area of the triangle =

$$\frac{1}{2}ah = \frac{1}{2}a \times \frac{\sqrt{P(P-2a)(P-2b)(P-2c)}}{2a} = \sqrt{\frac{P(P-2a)(P-2b)(P-2c)}{16}}$$

If s is semi-perimeter, P = 2s, substituting in the above equation, we get

$$\text{Area of the triangle} = \sqrt{\frac{2s(2s-2a)(2s-2b)(2s-2c)}{16}}$$

$$= \sqrt{\frac{2s \times 2(s-a) \times 2(s-b) \times 2(s-c)}{16}} = \sqrt{s(s-a)(s-b)(s-c)}$$

This is another Prooffor Heron's formula.

6.1.3 Properties of Means of a triangle

6.1.3.1 The three means divide the given triangle into 6 triangles of equal area.

Reference figure 6-5 below where AD, BE and CF are the three medians.

Since AD is mean of the triangle, BD = CD. Similarly, AF = BF and AE = CE

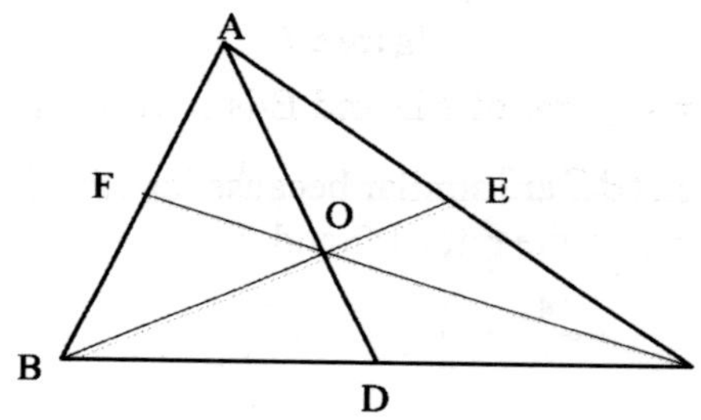

Figure 6.5

We know that triangles with equal base and common vertex have equal area.

∴ Area of Δ AOF = Area Δ BOF, Area of Δ AOE = Area Δ COE and Area of Δ COD = Area Δ BOD.

Let us denote these areas as x, y and z respectively.

We can now see that Area of Δ ABE = Area Δ CBE leading to 2x + y = 2z +y, implying that x=z

If we consider triangles ABD and ACD, we find 2x + z = 2y + z

Since x = z, 2y + z = 3z leading to y = z.

Thus x = y = z, implying that the medians divide the given triangle into 6 equal triangles.

6.1.3.2 Line joining mid-points of two sides is parallel to the third and equal to half of it.

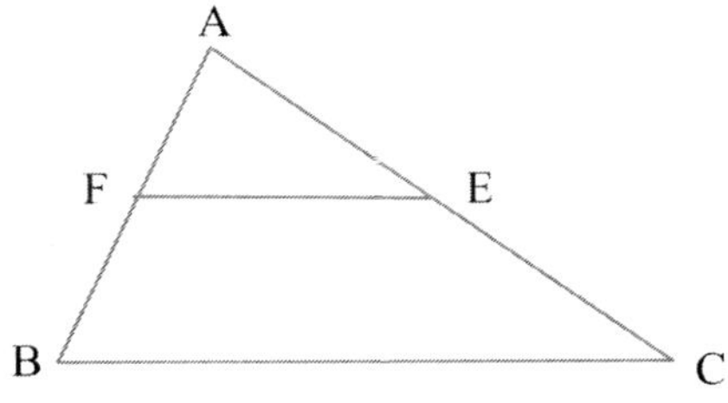

Figure 6.6

Given: F is mid-point of AB and E is mid-point of AC.

ΔAFE and ΔABC are similar because length of AF/AB and AE/AC are proportional in the ratio 1:2 and included angle A is common.

$\therefore$ FE/BC also have the proportion 1:2. Thus FE = $\frac{1}{2}BC$ and all three angles of ΔAFE and ΔABC are correspondingly equal.

Since $\angle AFE = \angle ABC$ and $\angle AEF = \angle ACB$, it implies that these two angles being corresponding angles of line FE and BC, it implies that the two lines FE and BC are parallel.

6.1.4 Radius of Circumcircle

Formula for circumradius of a triangle is derived as below:

In figure 6.7 below, BD and AF are perpendicular bisector of AC and BC respectively meeting at O. O is the circumcentre of the triangle. ABC there be BO = R (radius of the circumcircle. BD = height of the triangle from B denoted as h_b. leading to

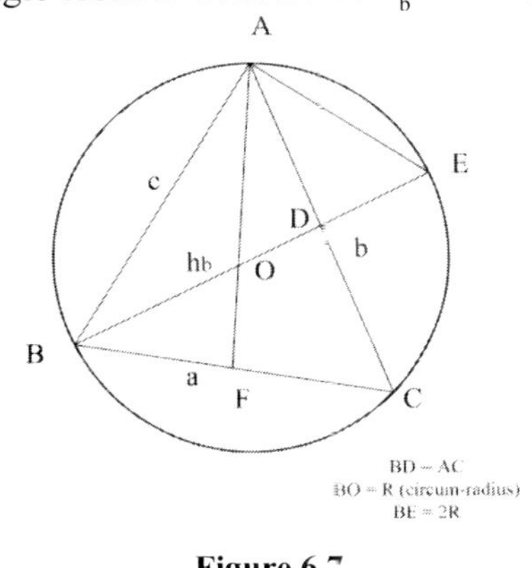

Figure 6.7

Area of the $\Delta ABC = \frac{1}{2} h_b \times b$

BD is extended to meet the circle at E. $\therefore$ BE = 2R

Now ΔABE and ΔDBC are similar because $\angle A$ being angle in the semi-circle & BD being perpendicular to AC, leading to $\angle A = \angle D = 90^0$ and $\angle BCA = \angle BEA$ (being angles in the same quadrant),

$$\frac{BE}{BC} = \frac{AB}{DB} \Rightarrow \frac{2R}{a} = \frac{c}{h_b}$$

Or $R = \frac{ac}{2h_b} = \frac{abc}{2 \times bh_b} = \frac{abc}{4 \times \frac{1}{2} bh_b} = \frac{abc}{4 \times \text{Area of the } \Delta ABC}$

Another way is to use trigonometric parameters.

From ΔABE, we find AB =BE Sin E

Since $\angle BCA = \angle BEA$, AB = BE Sin C or c = 2R Sin C

$$\Rightarrow \frac{c}{\text{Sin}\, C} = 2R$$

Similarly it can be proved that $\frac{a}{\text{Sin}\, A} = \frac{b}{\text{Sin}\, B} = 2R \Rightarrow$ $\frac{a}{\text{Sin}\, A} = \frac{b}{\text{Sin}\, B} = \frac{c}{\text{Sin}\, C} = 2R$, which is Law of Sines for triangles as proved in the preious chapters.

6.1.5 Radius of Incircle:

Formula for radius of the incircle of a triangle is derived as below:

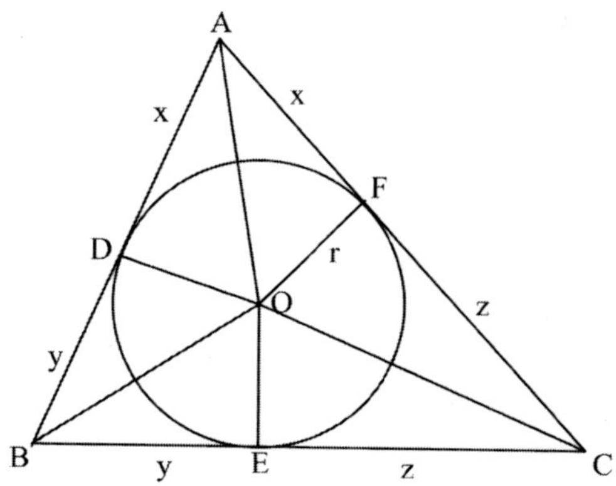

Figure 6.8

Recall that if a point lies on the angle bisector, it is equidistant from the arms of the angle. In the diagram above, since arms of the angle are sides of the triangle, the point of intersection of angle bisectors O is equidistant from the sides of the triangle. Hence ***all three perpendicular lengths OD/OE/OF are equal.***

When we draw circle with the above perpendicular length as radius and centre O, the circle touches all three sides of the triangle. O is thus the ***Incentre*** *which is nothing* but *centre of the circle that can be inscribed in the triangle touching all three sides of the triangle.*

It can be seen that height of the three Δs AOB, AOC and BOC equals Radius of Incircle r

Area of these 3 triangles each $= \frac{1}{2} \times r \times$ side length.

$\Rightarrow$ Area $\Delta ABC =$ Sum of the areas of the three triangles $= \frac{1}{2} \times r \times$ perimeter of the triangle.

Thus Radius of Incircle r is given by the formula

$$r = \frac{2 \times \Delta}{Perimeter}$$

Looking back at the formula for radius of the Circumcircle $R = \frac{abc}{4 \times \Delta}$, we can see that

$$r \times R = \frac{2 \times \Delta}{Perimeter} \times \frac{abc}{4 \times \Delta} = \frac{abc}{2 \times perimeter}$$

It may also be noticed that ΔAOD and ΔAOF are congruent (as OD = OF = r, AO is common and $\angle ADO = \angle AFO = 90^0$), leading to the conclusion that $AD = AF$. Similarly $CE = CF$ and $BD = BE$. If these dimensions are termed as $x, z \,\&\, y$ respectively, we have perimeter of the $\Delta = 2(x + y + z)$ or semi-perimeter $s = x + y + z$

Using Heron's formula, Area of the ΔABC works out to be = $\sqrt{(x + y + z)(xyz)}$ as three side lengths are $(x + y)$, $(y + z)$, $(z + x)$.

Also Area of the ΔABC = Area of $(\Delta AOB + \Delta AOC + \Delta BOC)$

$= \frac{r}{2} \times 2(x + y + z)$

$\therefore r \times (x + y + z) = \sqrt{(x + y + z)(xyz)}$

$r = \sqrt{\dfrac{xyz}{x+y+z}}$; this is another formula for r

6.1.6 Orthocentre

As mentioned in Chapter 3, section 3.2.6.3, method for locating parameters for orthocenter and its relationship with Circumcenter and centroid are dealt with in section 7.2.9 Chapter 7.

6.2 Quadrilaterals:

6.2.1 Perimeter:

Perimeter of a Quadrilateral equals sum of its four sides. Following logically from this definition we can derive the formula for a few types of quadrilaterals as follows:

Square = 4a (where a is any of the 4 equal sides)

Rectangle/Parallelogram/kite = 2 (a + b) (where a and b are the 2 different lengths of the given figure)

Irregular quadrilateral = a + b + c + d, where a, b, c, d are lengths of the 4 sides.

6.2.2 Area:

6.2.2.1 Area of a square = a^2 (where a is any side of the square)

6.2.2.2 Area of a rectangle is very evident as length multiplied by width.

Thus Area of a rectangle = a $\times$ b (where a and b are the 2 different sides of the rectangle). Diagonals of a rectangle are equal with length= $\sqrt{a^2 + b^2}$.

Another interesting formula for area of rectangle in terms of semi-perimeter and length of diagonal is

Area = $\frac{1}{2}(s^2 - d^2)$ where s is semi-perimeter and d is length of the diagonal.

6.2.2.3 Area of parallelogram: Let us now consider the case of a parallelogram shown in figure 6.9 below.

Line segments AE and BF are perpendicular to line CD. Since side AB is parallel to side CD, height between these two lines will

be equal. Hence AE=BF=h

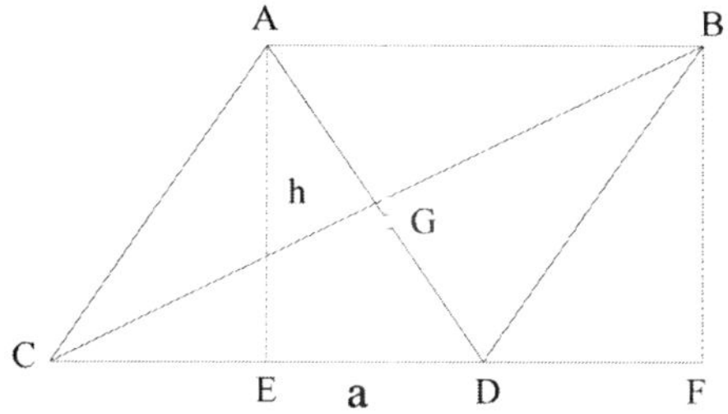

Figure 6.9

Since AC=BD, the two Right angled triangles ACE and BDF are congruent. Hence ΔBDF can be replaced by ΔACE, yielding area of the parallelogram as area of the rectangle AEBF= AB$\times$AE

Thus Area of the parallelogram = $a \times h$

It may be noticed that

Area of ΔABC= $\frac{1}{2} BC \times AG$ and

Area of ΔBCD= $\frac{1}{2} BC \times GD$

Further recall that diagonals of a parallelogram bisect each other at right angles. Thus

AG=GD

$\therefore$ Area of the parallelogram= Area of ΔABC+ Area of ΔBCD

$$= \frac{1}{2} BC \times (AG + GD) = \frac{1}{2} BC \times AD$$

Thus Area of the parallelogram is also = $\frac{1}{2} diagonal1 \times diagonal2$

This is a very important formula. Since a Rhombus is a special case of a parallelogram, it applies to Rhombus as well.

6.2.2.4 Area of Kite

Diagonals of a Kite and a square also intersect at right angles.

The logic given above applies as much to the kite as a square. Hence

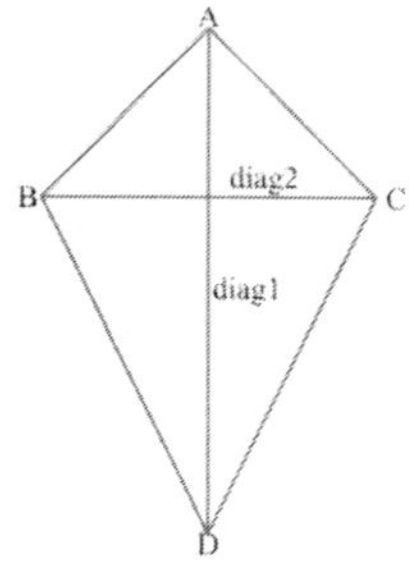

Figure 6.10: Kite

> **Area of the kite and a square also =** $\frac{1}{2} diagonal1 \times diagonal2$
> **Both diagonals of a square being equal, it also implies that Area of square =** $\frac{1}{2} \times diag^2$

6.2.2.5 Area of part of a figure

Most problems on mensuration will seek area of part of the given figure. For example, there is a footpath along boundary of a rectangular field (figures 6.11 and 6.12) and you need to find area of the footpath.

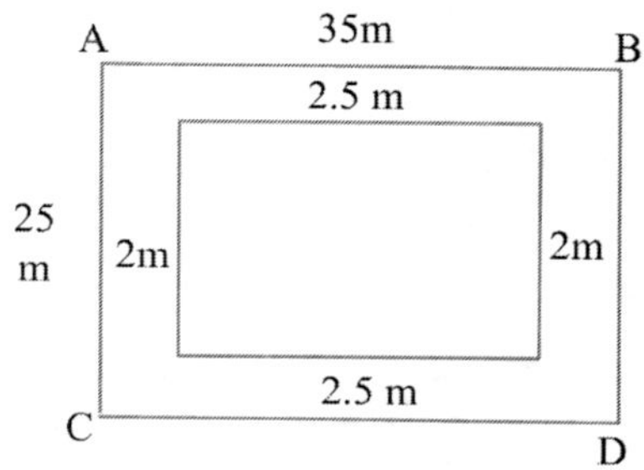

Figures 6.11

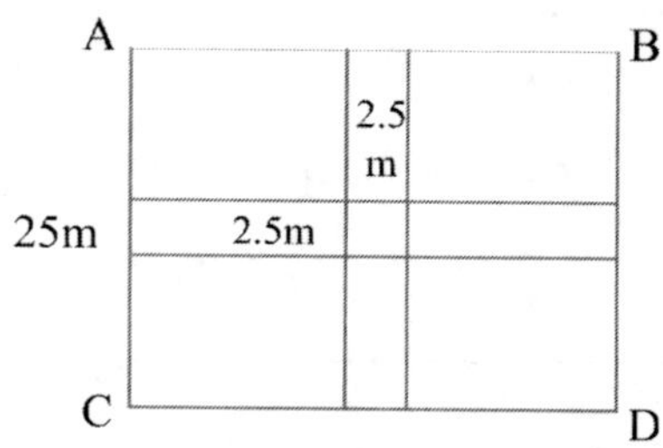

Figure 6.12

There can be two approaches to such a question:

1. Path along AB and CD can be considered to extend right from A to B and C to D. Hence length of these footpaths will be 35 m each. The other footpaths along sides AC and BD will then be (25-2× 2.5) = 20m each

Thus area of footpath = $35 \times 2 \times 2.5 + 20 \times 2 \times 2 = 255 sqm$

2. The other approach is to calculate area of two rectangular fields, one outer and the other inner (excluding the footpath). By this approach

   Area of footpath $= 35 \times 25 - 31 \times 20 = 875 - 620 = 255 sqm$

It may be noticed that approach 2 is easier and less prone to error.

There could be similar problem where foot paths are in the centre of the field crossing vertically, as in figure 6.12. Here also you can calculate by two methods- one by treating footpath parallel to side AB/CD as full and the other footpath as shorter than side length by width of the footpath and the second by considering footpath area fully along length and width of the field and subtracting area of square where footpaths cross as that area gets counted twice. The second approach is easy.

Thus area of footpath in this case

$$= (35 + 25) \times 2.5 - (2.5 \times 2.5) = 150 - 6.25 = 143.75 sqm$$

Students are encouraged to try different problems by evaluating the easier approach.

6.2.2.6 Area of an irregular Quadrilateral

Finding area of an irregular quadrilateral is not very straight forward. We can find area if we know all 4 sides and one of the angles (SASSS) or 2 angles, the common side and adjoining 2 sides (SASAS). Let us consider the two cases: *(courtesy mathguide.com)*

Case-1 : (SASSS)

Refer figure 6.13 below. We know side lengths a,b,c,d and angle C.

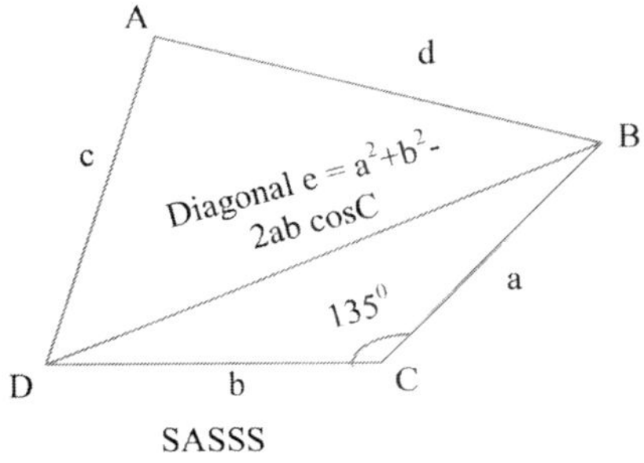

Figure 6.13

Recall Laws of cosine and Heron's formula to solve this case. By law of Cosine, we can find length e of the diagonal BD as

Diagonal e = $\sqrt{a^2 + b^2 - 2ab\ Cos\ C}$

Knowing length e and all 4 sides, area of both ΔABD and Δ BCD can be found using Heron's Formula.

Let us consider a=30, b=33, c= 35, d=42 and angle C= 135^0

Diagonal $e^2 = 30^2+33^2 - 2\times30\times33\cos135^0$

$= 900+1089+1980\times\frac{1}{\sqrt{2}} = 3388.86$

e = 58.21

Area of Δ ADB

$$s = \frac{35+42+58.21}{2} = 67.6$$

$$\text{Area} = \sqrt{67.6\times32.6\times25.6\times9.39}$$

$$= 727.83$$

Area of Δ BCD

$$s = \frac{30+33+58.21}{2} = 60.6$$

$$\text{Area} = \sqrt{60.6\times30.6\times27.6\times2.39}$$

$$= 349.74$$

Total area = 727.83 + 349.74 = 1077.57

Case-2: (SASAS)

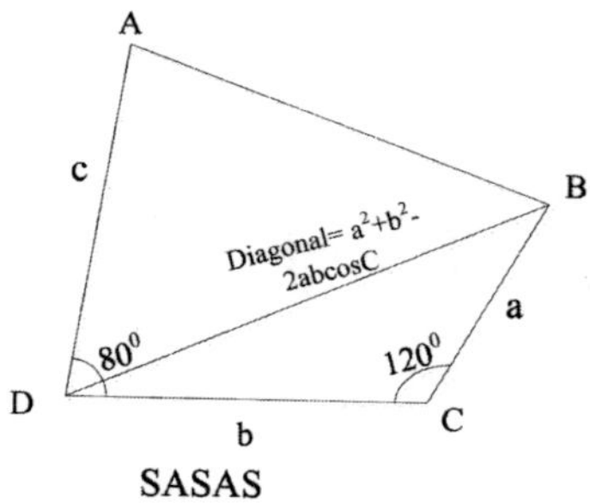

Figure 6.14

We need to use Heron's formula, Laws of Sine as well as Law of Cosine in this case.

As in the previous case we will fine length of diagonal e by law of cosine. This will give us area of Δ *BCD*.

However, to find area of ΔABD, we need to know $\angle ADB$.

Knowing $\angle ADC$, we need to find $\angle BDC$ to achieve this, $\angle BDC$ is found by using Law of Sines for Δ *BCD* as below:

$$\frac{a}{\sin\angle BDC} = \frac{e}{\sin 120^0}$$

$$\sin\angle BDC = \frac{a\sin 120^0}{e}$$

Thereafter we can find $\angle ADB$.

Area of $\Delta ADB = \frac{1}{2}c \times e\sin\angle ADB$

Now let us assume values of a, b and c as 20, 30 and 32 to find actual area.

$$\text{Diagonal } e^2 = 20^2+30^2- 2\times 20\times 30\cos 120^0$$

$$= 400+900+1200\times\frac{1}{2} = 1900$$

$$e = 43.59$$

$$\sin\angle BDC = \frac{20\sin 120^0}{43.59} = \frac{20\times 0.866}{43.59} = 0.3973$$

$$\angle BDC = 23.41^0$$

Hence $\angle ABD = 80 - 23.41 = 56.59^0$

Area of Δ ABD using Heron's formula

$$s = \frac{20+30+43.59}{2} = 46.79$$

$$\text{Area} = \sqrt{46.79\times 26.79\times 16.79\times 3.2}$$

$$= 259.51$$

Area of Δ ABD

$$= \frac{1}{2}c\times e\sin\angle ADB$$

$$= \frac{1}{2} \times 32 \times 43.59 \sin 56.59^0$$

$$= 582.59$$

Total area = 259.51+ 582.59=842.10

6.3 Area of a regular Polygon:

6.3.1 Area of pentagon:

Formula for internal angle of an n sided polygon, as given in Chapter 3 = $\frac{(n-2)\times\pi}{n}$

And if we divide the polygon into n equal triangles by connecting each vertex of the polygon to centre of the polygon, vertex angle of each of these triangles at centre of the polygon will = $\frac{2\pi}{n}$

Thus each internal angle of a regular pentagon

$=\frac{(5\text{-}2)\times 180}{5} = \frac{540}{5} = 108°$ and angle of vertex of each of the triangle $= \frac{360^5}{5} = 72°$

Refer to figure 6.,15 of a regular pentagon below. OF is a perpendicular drawn from center O to the base AB. Length of this vertical is h. Now Area of the

$$\Delta AOB = \frac{1}{2} s \times h$$

$$\& \frac{\frac{s}{2}}{h} = \tan 36° \text{ or } h = \frac{s}{2\tan 36^0}$$

Substituting value of h, Area of the triangle AOB= $\frac{s^2}{4\tan 36°}$

And Area of the pentagon as $\frac{5s^2}{4\tan 36°}$

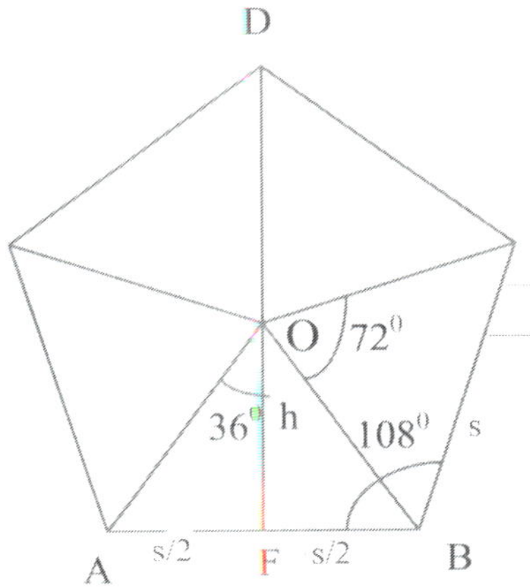

Figure 6.15

We can also generalize that Area of a regular polygon of n sides with side length s is given by $\frac{ns^2}{4\tan\frac{\pi}{n}}$

It is not essential to know value of tan36⁰ as the same can be easily found from trigonometric tables. However it is interesting and easy to derive this value by using known trigonometric formulae as below: (courtesy math-only-math.com).

Let us assume angle A= 36^0

$\therefore$ 5A= 180^0

Or 2A=180^0-3A

Sin 2A= Sin (180^0-3A) = Sin 3A

Using formulae for Sin 2A and Sin 3A,

2 Sin A Cos A = 3 Sin A-4 Sin^3A=Sin A (3-4 Sin^2A)

Since Sin $36^0 \neq$ 0, Sin A can be removed, leading to

2 Cos A = 3-4 Sin^2A= 4 Cos^2A-1

or 4 Cos^2A- 2 CosA-1=0

Solving it as a quadratic equation,

Cos A= $\frac{2 \pm \sqrt{4+16}}{8}$

Since A= 36^0 and Cos 36^0 has a positive value

$$\text{Cos } 36^0 = \frac{2+\sqrt{20}}{8} = \frac{1+\sqrt{5}}{4}$$

$$\text{Cos}^2\ 36^0 = \left(\frac{1+\sqrt{5}}{4}\right)^2 = \frac{6+2\sqrt{5}}{16} = \frac{3+\sqrt{5}}{8}$$

$$\text{Sin}^2\ 36^0 = 1 - \text{Cos}^2\ 36^0 = 1 - \frac{6+2\sqrt{5}}{16} = \frac{5-\sqrt{5}}{8}$$

$$\therefore\ \text{Tan}^2\ 36^0 = \frac{5-\sqrt{5}}{3+\sqrt{5}} = \frac{(5-\sqrt{5})(3-\sqrt{5})}{(3+\sqrt{5})(3-\sqrt{5})} = \frac{20-8\sqrt{5}}{4} = 5-2\sqrt{5}$$

$$\text{Tan } 36^0 = \sqrt{5-2\sqrt{5}}$$

$$\therefore \textbf{Area of Pentagon} = \frac{5s^2}{4\tan 36^0} = \frac{5s^2}{4\times\sqrt{5-2\sqrt{5}}}$$

Using general formula for area of polygon,

$$\textbf{Area of Hexagon} = \frac{6s^2}{4\tan\frac{180^0}{6}} = \frac{6s^2}{4\tan 30^0} = \frac{6s^2\times\sqrt{3}}{4}$$

$$\textbf{Area of Octagon} = \frac{6s^2}{4\tan\frac{180^0}{8}} = \frac{6s^2}{4\tan 22.5^0}$$

Value of tan 22.5^0 can be easily calculated as below:

Let A= 22.5^0

Tan 2A= Tan 45^0=1

$$\text{Now Tan 2A} = \frac{2\,\text{Tan}\,A}{1-\text{Tan}^2 A}$$

Substituting value of Tan 2A,

1-Tan^2A= 2 TanA

Tan^2A+2 TanA-1=0

Solving it as a quadratic equation,

$$\text{Tan A} = \frac{-2\pm\sqrt{4+4}}{2} = \frac{-2\pm 2\sqrt{2}}{2} = -1\pm\sqrt{2}$$

Since Tan 22.5^0 has a positive value,

Tan $22.5^0 = \sqrt{2} - 1$

Hence **Area of Octagon=** $\dfrac{8s^2}{4(\sqrt{2}-1)}$

Same way, we can calculate area of any regular polygon of n sides of length s each by using generic formula for area of a regular polygon as $\dfrac{ns^2}{4\tan\dfrac{\pi}{n}}$

6.4 Circles:

6.4.1 Circumference and area

Recall the discussion about circles in Chapter 3. Let us assume radius of the circle AC as **r** and diameter AB as **d** as shown below.

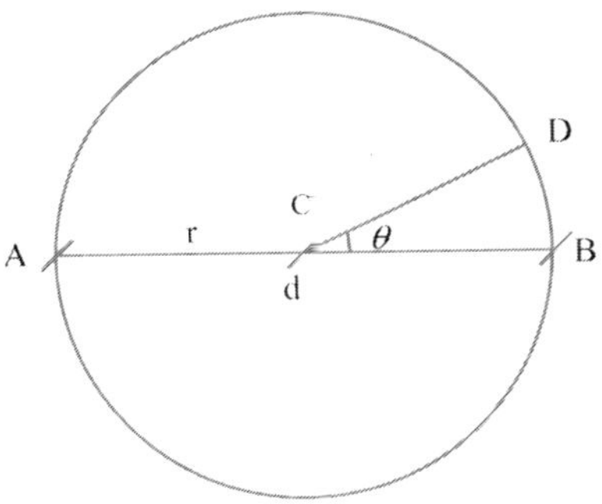

Figure 6.16

Entire length of the circle or Circumference of the circle is related to its radius. Larger the radius, larger is the circumference. Also recall that angle subtended by the complete circle at the centre is 360^0 or 2π in radians.

Length of arc BD of the circle is directly related to the angle θ subtended at the centre in radians and is given as $r\theta$. Hence

Circumference of the circle = 2π **r** or π **d** and

Circumference of the semi-circle= π **r**

Similarly **area** of an arc is also directly proportional to the angle θ subtended at the centre in radians and is given as $\frac{1}{2}\theta r^2$. Therefore

Area of the circle = πr^2 or $\frac{\pi d^2}{4}$ and

Area of the semi-circle = $\frac{1}{2}\pi r^2$

6.4.2 Length of a chord and area of triangle formed by it at the centre

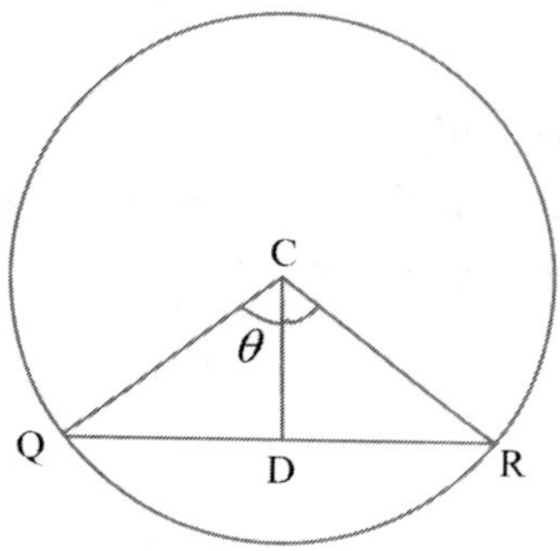

Figure 6.17

We can see that end points of the chord QR are equidistant from the centre C. Hence ΔCQR is an isosceles triangle and perpendicular CD from centre C bisects the chord QR as well as $\angle QCR\,(\theta)$.

Length of the chord QR= $2r\sin\frac{\theta}{2}$ **and**
Area of the Δ *CQR* =
$\frac{1}{2}QR\times CD=\frac{1}{2}\times 2r\sin\frac{\theta}{2}\times r\cos\frac{\theta}{2}=\frac{1}{2}r^2\sin\theta$

We can find any related area or arc length using the above basic understanding of the circles.

6.5 Cuboids,Cylinders, Cones and Sphere:

6.5.1 Cuboids

A cuboid is a solid shape like a dice with rectangular base, vertical edges on all sides of the base rectangle and the top face.. Thus it has six faces (top/ bottom and 4 vertical faces) like a cube but with all 6 rectangular faces. Special case of a Cuboid is a Cube where all dimensions are equal.

Total surface area of a cube with side length 'l' = $\mathbf{6l^2}$

Volume of this cube = $\mathbf{a^3}$

Similarly **Total surface area** of a cuboid with length, width and height as l, b and h respectively = **2lb + 2h(l + b) = 2(lb + lh + bh)**

And **Volume** of this cuboid = **lbh**

6.5.2 Cylinders:

The cylinder has parallel congruent circular bases.Volume of a cylinder is product of the base area times the height. Since the base of a cylinder is always a circle, Formulae for surface area and volume of cylinderare as below:

The total surface area of the cylinder is given by **A**= $2\pi r(r+h)$ where **r** is radius of the base and **h** is height and **Curved surface area** of the cylinder is given by **A**= $2\pi rh$

Volume of the Cylinder $V = \pi r^2 h$ or $\frac{\pi d^2 h}{4}$

You can solve problems relating to hollow cylinders using the above formulae appropriately.

A metallic wire is also technically a cylinder and knowing its length and diameter, the above formulae can be used to find its volume.

6.5.3 Cone:

A cone is a geometrical figure with one curved surface and one circular surface at the bottom. The top of the curved surface is called the vertex of the cone. An edge that joins the curved surface with the circular surface is called the curved edge of the cone.

If the vertex of the cone lies above the center of the circular base, it is called a Right circular cone. Oblique cones do not have the apex directly above the center of the base. The curved height or slant height of the cone is called the lateral height of the cone.

The cone may be hollow or solid. The curved surface area of the cone is calculated by the formula $A = \pi rs$ where r is the radius of the base, and s is the slant height of the cone.

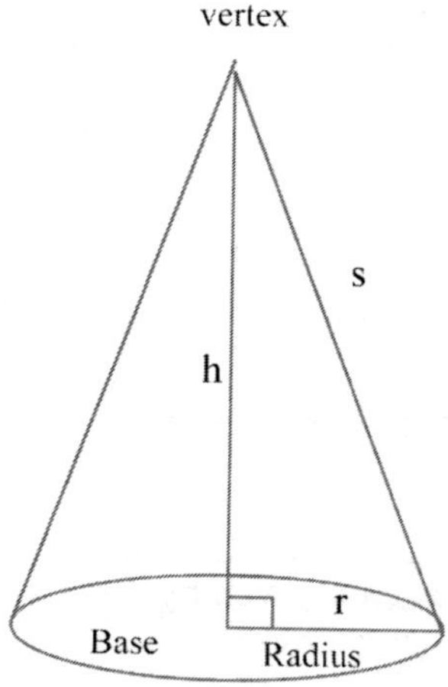

Figure 6.18

Slant height s is related to radius and vertical height as $s^2 = r^2 + h^2$. Formula for Curved surface area can be derived as below:

If we were to cut the cone up one side along a vertical slant line and roll it out flat, it would look something like sector of a circle with radius s (the slant height of the cone) and arc length is circumference of the base of the cone i.e. $2\pi r$ as shown below.

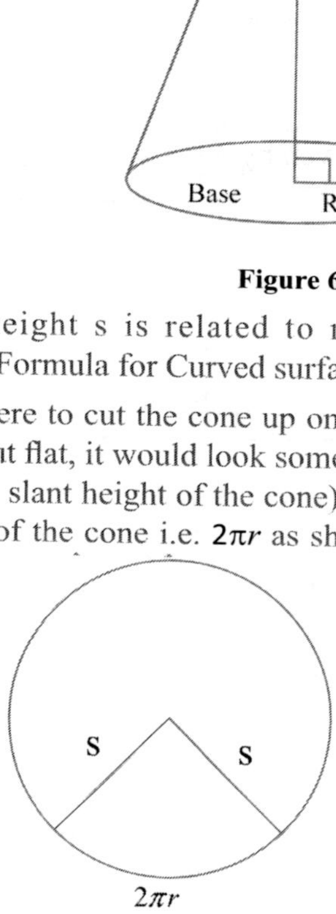

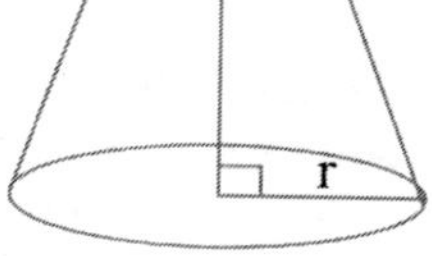

Figure 6.19 **Figure 6.20**

This section is actually part of a larger circle that has a radius of s, (To flatten it, the cone was cut along the red lines, the length of this cut is the slant height of the cone.).

We know that area of sector of a circle = $\frac{\theta s^2}{2}$ where θ is the angle in radians subtended by the shaded arc at the center (as we know that area of whole circle = πs^2 with an angle of 2π at the centre)

Further, length of arc $2\pi r = s\theta$

Substituting value of $s\theta$ in the above expression for area of the sector, we get

Curved surface area of the cylinder = area of sector of a circle $= \frac{2\pi rs}{2} = \pi rs$

Thus Curved surface area is given by **A= π rs** and **The total surface area** of the cone is given by A= $\pi r^2 + \pi rs = \pi r(r + s)$

Derivation for Volume of a cone requires knowledge of integration. You can see this derivation in Chapter 9, section 9.9.2. For the present please note that

Volume of the Cone is $\frac{1}{3}$ **π r² h** where h is vertical height of the Cone.

Important: **For ease of recall, please note that volume of a Cone is 1/3 of the volume of a cylinder with same radius and height.**

There could also be problems relating to area/volume/surface area of a partial cone, say cone truncated at the top. Let us consider the following figure 6.21

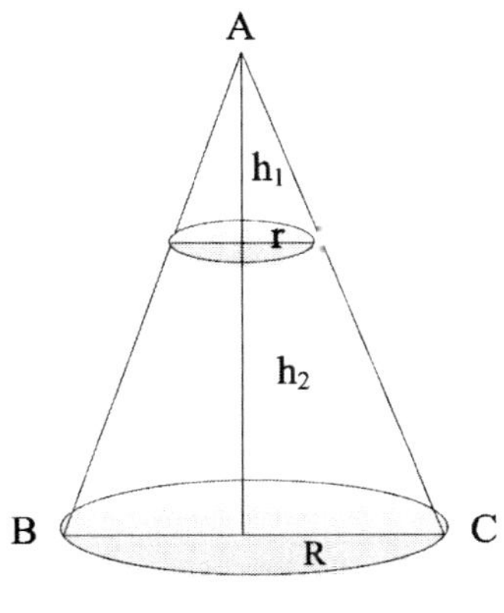

Figure 6.21

We may need to find height of liquid if the cone is partially filled or we may need to find surface area of a cone that has been truncated at the top say h_1 height from the top.

We notice that radius of the cone keeps reducing as the height increases. Thus volume of a partially filled cone = full volume of the cone – volume of the cone not filled. We can find these two volumes if we know height of each of these sections. Total height of the cone is h_1+h_2. Reduced radius r and large radius R is in proportion of height h_1 and full height (h_1+h_2). Thus

$$\frac{r}{R}=\frac{h_1}{h_1+h_2}$$

Using the values of r, R, h_1, h_2, any mensuration problem related to cones can be solved.

6.5.4 Sphere:

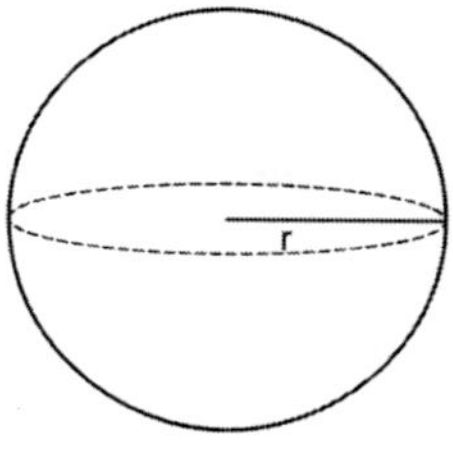

Figure 6.22.1: Sphere

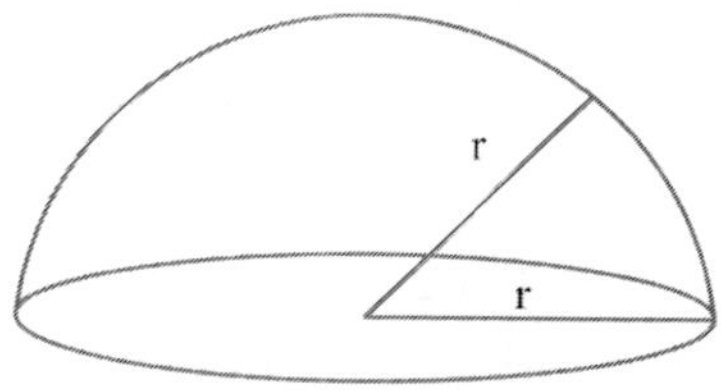

Figure 6.22.2: Hemisphere

A sphere is a space/shape defined by all the points that are at the same distance from the center point. Like a circle, a sphere has a radius and a diameter. The shape of the earth is like a large sphere.

Volume of a Sphere is given by **V** = $\frac{4}{3}\pi r^3$ where **r** is radius of the sphere.

Volume of a hemi-sphere withradius **r** is given by **V** = $\frac{2}{3}\pi r^3$

Surface area of the Sphere withradius **r**is given by **A**= $4\pi r^2$ and **Surface area** of a hemi-sphere with radius **r** including the flat surface = $3\pi r^2$

An easy way to remember volume and surface area of a sphere is that surface area is same and volume is 4/3 times that of a cylinder with equal base radius and height r.

Important Note: While calculating area and volume in any question, make sure that units of different dimensions are same. Don't mix millimeters with centimeter, decimeters or meters and inches with feet etc.

There would also be problems of solid objects with mixed features like partly cylindrical and partly conical (at the ends) or partly cylindrical and partly hemispherical etc. Such problems will require complete understanding of the structure and its dimensions and application of one or more of the above formulae. ***Readers are expected to look for unknown problems of different kind on mensuration on the internet and solve them to gain confidence.*** A few examples are given below for practice

6.6 Solved examples:

1. The length, breadth and height of a cuboid are 20 cm, 15 cm, 10 cm respectively. Find its total surface area.

Solution:

Total surface area of a cuboid = 2× Area of base + perimeter of the base × height.

Hence total surface area

$$= 2\times 20\times 15 + 2\times(20+15)\times 10 = 600 + 700 = 1300\text{ cm}^2$$

2. In a building there are 24 cylindrical pillars with each having a radius 28 cm and height 4 m. Find the cost of painting the curved surface area of all pillars at the rate of Rs. 8 per sq.m.

Solution:

Curved surface area of a cylindrical object = $2\pi rh$ where ***r*** is radius and ***h*** is height.

Hence with given data, surface area of one pillar =

$$2\times\frac{22}{7}\times\frac{28}{100}\times 4 = \frac{176}{25}\text{ sq. m}$$

Total surface area of all pillars = $\frac{176}{25}\times 24 = 168.96$ sq. m

Cost of painting in Rs. = 168.96 × 8 = 1351.68

3 The lateral surface area of a hollow cylinder is 4224 cm^2. It is cut along its height and formed a rectangular sheet of width 33 cm. find the perimeter of rectangular sheet.

Solution:

When the hollow cylinder is cut along its height, length of the sheet will equal perimeter of the base of the cylinder and width will equal height of the cylinder.

As per given data, height of the cylinder works out to be = 33 cm

Perimeter of the cylinder = lateral surface area/ height.

Hence Perimeter of the cylinder = length of the sheet = $\frac{4224}{33} = 128$ cm

$\therefore$ Perimeter of rectangular sheet = $2 \times (33 + 128) = 322$ cm

4. A roller takes 750 complete revolutions to move once over section of a road. Find the area of road if the diameter of the roller is 84 cm and width is 1 m.

Solution:

Area covered by 1 complete revolution = Lateral surface area of the roller

So lateral surface area of the roller = $\pi \times diameter \times length$

$$= \frac{22}{7} \times \frac{84}{100} \times 1 = \frac{66}{25} \text{ sq. m}$$

$\therefore$ Area of road = $\frac{66}{25} \times 750 = 1980$ sq. m

5. If the area of a rhombus be 24 cm^2 and one of its diagonals be 4 cm, find the perimeter of the rhombus.

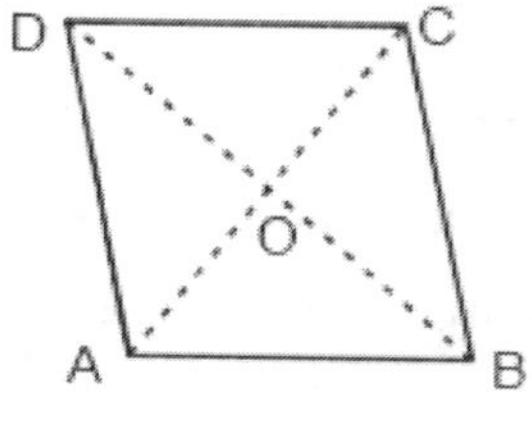

Solution:

We know that diagonals of a rhombus intersect at right angles and area of a rhombus equals $\frac{1}{2}\times$ product of the two diagonals.

Hence the other diagonal = $\frac{24\times 2}{4}=12$

Side of the rhombus = length of the diagonal of right triangle comprising side lengths equal to half the diagonal lengths.

Hence Side of the rhombus = $\sqrt{2^2+6^2}=\sqrt{40}=2\sqrt{10}$

And perimeter of the rhombus = $4\times 2\sqrt{10}=8\sqrt{10}$

6. Find the area of a regular octagon each of whose sides measures 4 cm.

Solution:

We have learnt that Area of a regular polygon = $\frac{ns^2}{4\tan\frac{\pi}{n}}$, where n = no. of sides, s = side length

$\therefore$ Area of the given regular octagon = $\frac{8\times 4^2}{4\tan\frac{\pi}{8}}=\frac{8\times 4^2}{4\tan\frac{\pi}{8}}=\frac{32}{\tan 22.5^0}$

$\tan 45^0=\frac{2\tan 22.5^0}{1-\tan^2 22.5^0}$ leading to $1-\tan^2 22.5^0=2\tan 22.5^0$

If we substitute $\tan 22.5^0=$ x, we get a quadratic equation $x^2+2x-1=0$

Solving this quadratic we get $x=\frac{-2\pm\sqrt{4+4}}{2}=-1\pm\sqrt{2}$

Since $\tan 22.5^0$ is positive, we get $\tan 22.5^0=\sqrt{2}-1$

$\therefore$ Area of the given regular octagon =

$\frac{32}{\sqrt{2}-1}=32\times(\sqrt{2}+1)=32\times 2.414=77.248\,\text{cm}^2$

7. Find the total surface of a hollow cylinder open at ends, if the length is 12 cm, the external diameter 10 cm and thickness 2 cm.

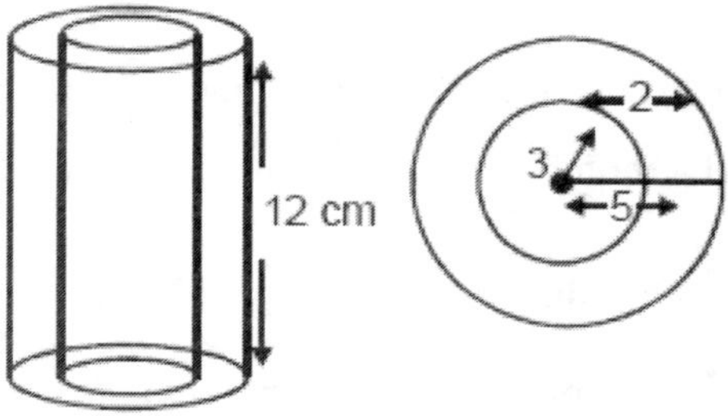

Clearly, the external radius r_1 =5 cm and internal radius r_2 =3 cm.

Total surface area of a hollow cylinder open at ends will be $2\times$ area of the edge + internal and external surface areas.

$\therefore$ Total surface area

$$= 2\pi(r_1^2 - r_2^2) + 12\times(2\pi r_1 + 2\pi r_2) = 2\pi\times(25-9) + 24\pi(5+3)$$

$$= 32\pi + 192\pi = 224\times\frac{22}{7} = 704 \text{ cm}^2$$

8. If radius of the base of a cone is 120 deci-metre and its vertical height is 8 m. Find the
   (i) curved surface area
   (ii) total surface area

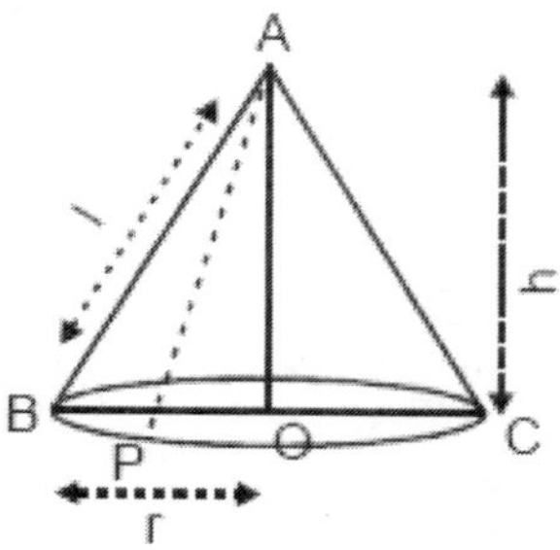

Solution:

Let vertical height of the cone be h = 8 m.

Radius r of base of the cone = $\frac{120}{2} = 60$ dm = 6 m

We know that slant height l = $\sqrt{h^2 + r^2} \Rightarrow$ leading to $l = \sqrt{6^2 + 8^2} = 10$ m

(i) Curved surface area

$$= \pi rl = \pi \times 10 \times 6 = 60\pi = 60 \times 3.14 = 188.4m^2$$

(ii) Total surface area

$$= \pi r(r + l) = 6\pi(6 + 10) = 96\pi = 96 \times 3.14 = 301.44m^2$$

9. The surface area of a sphere is 2464 dm^2. Find its diameter.

Solution:

Surface area of a sphere with radius r = $4\pi r^2$

Thus radius r works out to be

$$r = \sqrt{\frac{2464}{\pi}}\ dm = \sqrt{\frac{2464 \times 7}{22}} = \sqrt{112 \times 7} = \sqrt{784} = 28\ dm = 2.8\ m$$

10. A cuboid is of dimensions (60*50*30)cm. How many small cubes with side 6 cm can be placed in the given cuboid?

Solution:

Here since the small cube is a solid object, maximum number of cubes can be accommodated by organizing small cubes in rows and columns. For this we need to know how many small cubes can be accommodated in each dimension of the cuboid box.

In the given data, we can see that two dimensions can accommodate exactly 10 and 5 cubes respectively but the third dimension can accommodate 8 cubes leaving out vacant space of 2 cms on the width side.

Hence the cuboid can accommodate $10 \times 5 \times 8 = 400$ cubes

11. A 5 cubic centimeter cube is painted on all its side. If it is sliced into 1 cubic centimeter cubes, how many 1 cubic centimeter cubes will have exactly one of their sides painted?

Solution:

The question implies that the larger cube is sliced into $5 \times 5 \times 5 = 125$ smaller cubes. Each face of the larger cube will yield 25 smaller cubes. We can now visualize that there will be 4 corner pieces (with 3 sides painted), 12 edge pieces (with 2 sides painted), and remaining 9 pieces with one face painted.

Since there are 6 faces of a cube, ***we will have*** $9 \times 6 = 54$ ***pieces with one face painted.***

12. The area of a square field is 24200 sq m. How long will a lady take to cross the field diagonally at the rate of 6.6 km/hr?

Solution:

Area of a square with side a = a^2, and its diagonal^2 = $2a^2$

$\therefore$ diagonal^2 = $2 \times 24200 = 48400 \Rightarrow$ diagonal = $\sqrt{48400}$ = 220m

Time taken to cross the field = $\frac{220}{6600} \times 60 = 2$ minutes.

13. The circumference of the front wheel of a cart is 30 ft long and that of the back wheel is 36 ft long. What is the distance travelled by the cart, when the front wheel has done five more revolutions than the rear wheel?

Solution:

Let number of revolutions done by rear wheel = a

Number of revolutions done by front wheel = a+5

Since the distance covered by both wheels is same, we get

$36a = 30(a+5) = 30a + 150 \Rightarrow 6a = 150 \Rightarrow a = 25$

The distance covered by the cart = $36 \times 25 = 900$ feet

14. If the sides of a triangle measure 72, 75 and 21, what is the measure of its in-radius?

Solution:

We know that 24, 7, 25 is a Pythagorean triplet. Thus 72, 21and 75 (each being multiple of 3) is also a Pythagorean triplet. Therefore, the given triangle is a right angled triangle with 75 as the diagonal.

Area of the triangle = $\frac{1}{2} \times 72 \times 21 = 36 \times 21$

And its perimeter = $75 + 21 + 72 = 168$

Diameter of incircle $= \frac{2 \times Area}{Perimeter} = \frac{2 \times 36 \times 21}{84} = 18$

$\therefore$ Radius of the in-circle $= \dfrac{18}{2} = 9$

15. If volume of the following solid objects is same, which one will have least surface area?
    (a) Cube
    (b) Hemi-sphere
    (c) Cylinder with height same as diameter
    (d) Sphere

Solution:

Let volume = V

(a) For Cube, side length $= \sqrt[3]{V}$ and Total surface area $= 6a^2 = 6V^{2/3}$

(b) For hemisphere, radius $= \left(\dfrac{3}{2\pi} \times V\right)^{1/3}$
And Total surface area =

$$3\pi r^2 = 3\pi \times \left(\frac{3}{2\pi}V\right)^{2/3} = \left((3\pi)^{3/2} \times \frac{3}{2\pi}\right)^{2/3} \times V^{2/3} =$$

$$\left(\sqrt{3\pi} \times \frac{9}{2}\right)^{2/3} \times V^{2/3} = \left(\frac{243\pi}{4}\right)^{1/3} \times V^{2/3}$$

(c) For cylinder with given volume V, for least surface area, the height $= 2\times$ radius r.

$$\therefore \text{V} = \pi r^2 \times 2r = 2\pi r^3 \Rightarrow r = \left(\frac{V}{2\pi}\right)^{1/3}$$

And Total surface area =

$$2\pi r(r + 2r) = 6\pi r^2 = 6\pi \times \left(\frac{V}{2\pi}\right)^{2/3} = \left(\frac{(6\pi)^{3/2}}{(2\pi)}\right)^{2/3} \times V^{2/3}$$

$$= \left(\frac{\sqrt{216\pi}}{2}\right)^{2/3} \times V^{2/3} = (54\pi)^{1/3} \times V^{2/3}$$

(d) Volume of sphere $V = \frac{4}{3}\pi r^3 \Rightarrow r = \left(\frac{3V}{4\pi}\right)^{1/3}$

And Total surface area $= 4\pi r^2 = 4\pi \times \left(\frac{3}{4\pi}V\right)^{2/3}$

$$= 4\pi \times \left(\frac{3}{4\pi}\right)^{2/3} V^{2/3}$$

$$= \left((4\pi)^{3/2} \times \frac{3}{4\pi}\right)^{2/3} \times V^{2/3} = \left(8 \times \frac{3}{4}\sqrt{\pi}\right)^{2/3} \times V^{2/3} = (36\pi)^{1/3} \times V^{2/3}$$

By comparison we find that $(36\pi)^{1/3} < (54\pi)^{1/3} < \left(\frac{243\pi}{4}\right)^{1/3} < 6$ (as $6 = 216^{1/3}$) leading to the conclusion that for the given objects with same volume, sphere has the minimum surface area.

6.7 Practice problems:

1. Four horses are tethered at 4 corners of a square field of side 70 metres so that they can just about reach one another. Find the area left ungrazed by the horses.
2. A square sheet of paper is converted into a cylinder by rolling it along its length. What is the ratio of the base radius to the side of the square?
3. The surface area of the three adjacent faces of a cuboid are 6, 15, 10 sq.cm respectively. Find the volume of the cuboid.
4. If the diagonal and the area of a rectangle are 25 m and 168 m^2, what is the length of the rectangle?
5. A regular hexagon is inscribed in a circle of radius r cm. What is the perimeter of the regular hexagon?
6. Reference figure below. ABCD has area equal to 28. BC is parallel to AD. BA is perpendicular to AD. If BC is 6 and AD is 8, then what is CD?

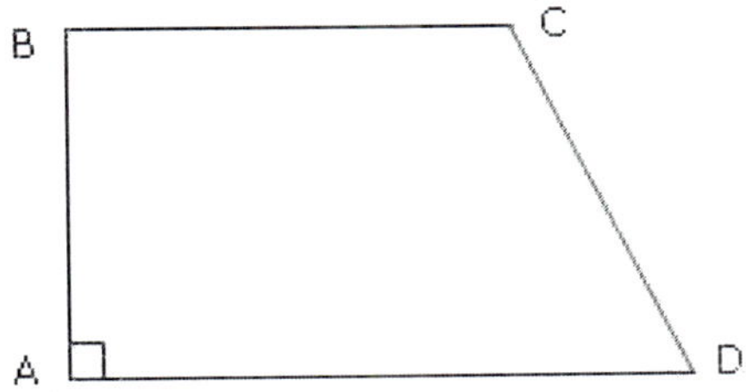

7. A solid metal cylinder of 10 cm height and 14 cm diameter is melted and re-cast into two cones in the proportion of 3 : 4 (volume), keeping the height 10 cm. What would be the percentage change in the flat surface area before and after?

8. A circular road is constructed outside a square field. The perimeter of the square field is 200 ft. If the width of the road is $7\sqrt{2}$ ft. and cost of construction is Rs. 100 per sq. ft. Find the lowest possible cost to construct 50% of the total road.

9. In the figure below, CD = BF = 10 units and ∠CED = ∠BAF = 30°. What would be the area of triangle AED?

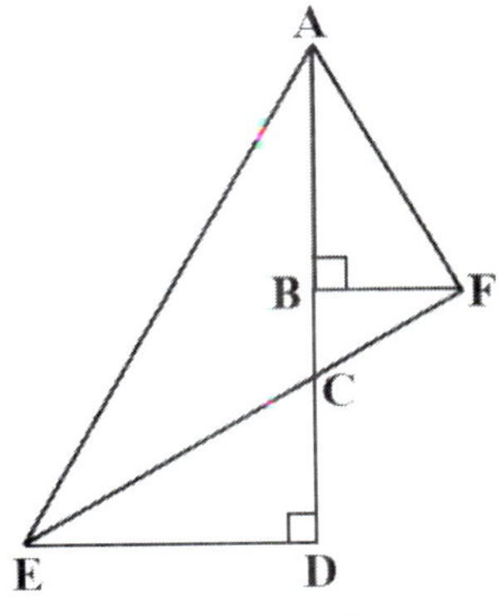

10. The length, breadth and height of a room are in the ratio 3:2:1. If the breadth and height are halved while the length is doubled, then what will be the change in the total area of the four walls of the room.

11. In the figure given below AB and BC are adjacent sides of a square and the arc ADC is drawn with B as centre and BC as radius. D is a point on the arc and DEFG is a square such that, FE, if extended, passes through C while FG, if extended, passes through A. What is the ratio of the area of a square of

side AB and the square DEFG?

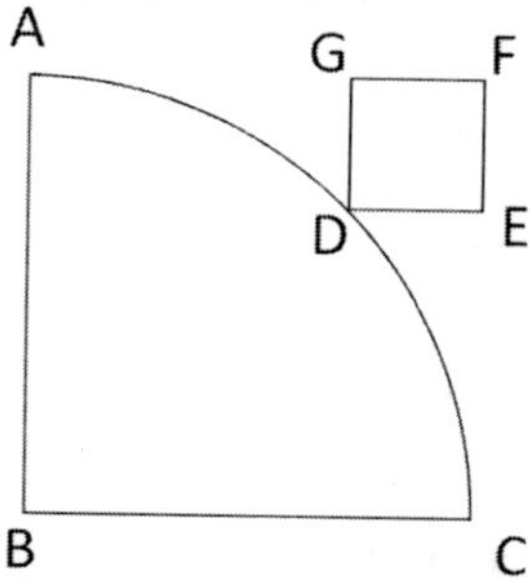

12. One of the angles of a triangle is 60^0 and length of its two sides is 6 cm and 7 cm. Find length of the third side of the triangle.
13. The external length, breadth and height of a closed box are 10 cm, 9 cm and 7 cm respectively. The total inner surface area of the box is 262 sq. cm. If the walls of the box are of uniform thickness, find the thickness.
14. $\triangle ABC$ is an equilateral triangle of side 14 cm as shown in the figure below. A semi-circle on BC as diameter is drawn to meet AB at D and AC at E. Find the area of the region between sector BD, Sector CE and the adjoining sides of the triangle.

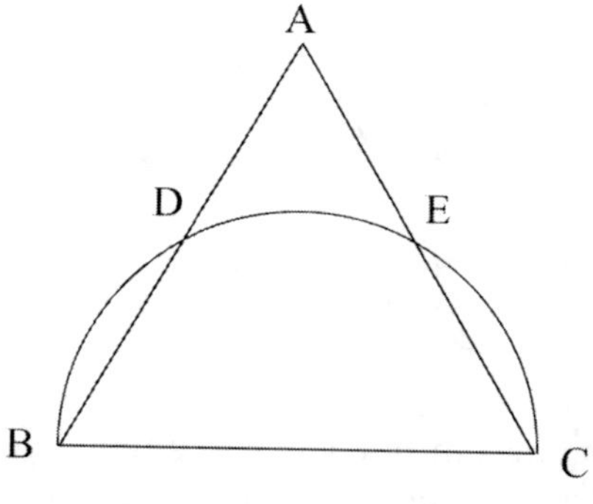

15. A sphere of radius r is cut by a plane at a distance of h from its centre, thereby breaking this sphere into two different pieces. The cumulative surface area of these two pieces is 25% more than that of the sphere. What is the value of h?

Answers to Practice Problems:

1. 1050 sq. m
2. $\frac{1}{2\pi}$
3. 30 cc
4. 24 m
5. 6r
6. $2\sqrt{5}$
7. 50%
8. Rs.1,25,400
9. 50 ($\sqrt{3}$ + 4)
10. Decrease by 30%
11. $2:(3-2\sqrt{2})$
12. $\sqrt{43}$ or $3+\sqrt{22}$
13. 1
14. $49\left(\frac{\pi}{3}-\frac{\sqrt{3}}{2}\right)$
15. $\frac{r}{\sqrt{2}}$

7

COORDINATE GEOMETRY

7.1 Overview

Coordinate geometry is one of the most important and exciting ideas of mathematics. The number plane (Cartesian plane) is divided into four quadrants by two perpendicular axes called the x–axis (horizontal line) and the y–axis (vertical line). These axes intersect at a point called the origin denoted by value zero for both x and y coordinates. There are different names for the co–ordinate system like

- rectangular coordinate system
- xy–coordinate system or
- xy–coordinate plane

Refer to figure 7.1. Basic properties of a coordinate system are as below:

- The Horizontal or x–axis extends infinitely in both directions – left and right. Values to left of the origin are negative and to right of the origin are positive.
- The Vertical or y–axis extends infinitely in both directions – up and down. Values above the origin are positive and below the origin are negative.

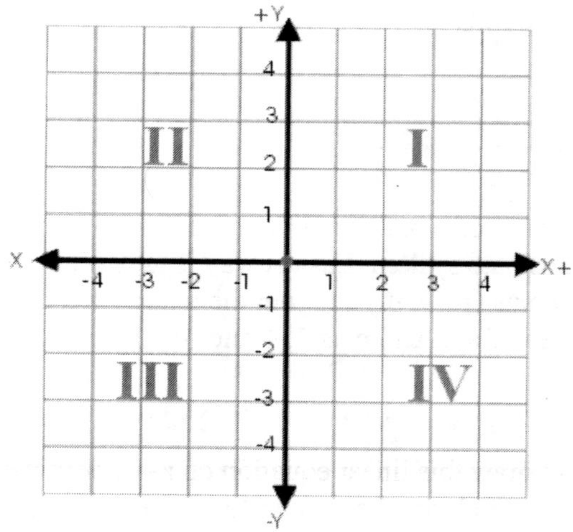

Figure 7.1

- The two axes divide the coordinate area into 4 quadrants – I where both X and Y are positive, II where X is negative and Y is positive, III where X and Y both are negative and IV where X is positive and Y is negative
- Every point in the plane has a unique pair of coordinates that defines its location by two numbers. The first number is the x–coordinate and the second number is the y–coordinate. The coordinates are taken to be real numbers in elementary mathematics. You can easily remember it by recalling that x comes before y in the English alphabet. Any point on the coordinate plane can then be depicted as (x,y)
- If you trace the point to the two axes, it will form a rectangle/ square

The use of a coordinate system allows problems in geometry to be translated into problems about numbers and vice versa; this is the basis of analytical geometry.

7.2 Different forms of linear eqations

7.2.1 General form:

General form of a linear equation in variables x and y is **ax+by+c=0**

By rearranging we get,

$$by = -c - ax$$

$$y = \frac{-(ax+c)}{b} = \frac{-a}{b}x - \frac{c}{b}$$

Knowing value of constants a, b and c, we can draw such a line on the x–y coordinate system. Easiest way is to mark points on x–axis (for y=0) and y axis (for x=0) and join them and extend on both sides. Assuming values of a, b and c as 3, 4 and –8, the equation reduces to

$$y = \frac{-3}{4}x + 2 \qquad ...(1)$$

We can also draw this linear equation on x–y coordinate system by assuming different values of x and finding corresponding values of y.

> **Example 1:** Let values of x be 4, –8 and 12. Corresponding values of y will be –1, 8, –7. Now let us plot these three points (4, –1), (–8, 8) and (12, –7) on a graph paper and join as in figure 7.2 below:

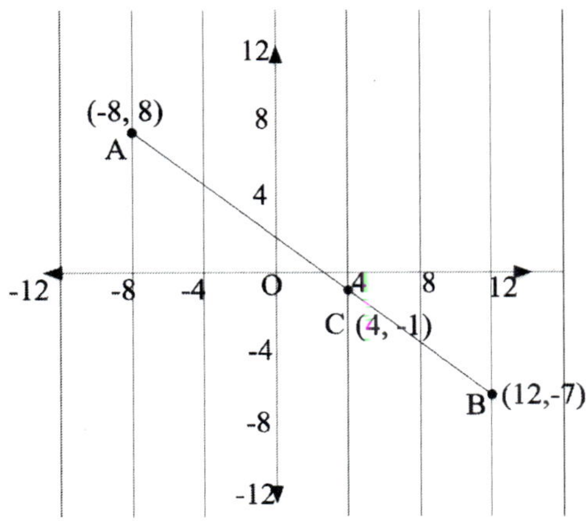

Figure 7.2

We can now find coordinates of any point lying on this line. For example when x=0, y=2 and when y=0, x= – 8/3 and when x=8, y=–4

7.2.2 Slope intercept form

Slope of a straight line refers to its steepness. It is denoted by ratio of vertical distance (change in y coordinate) to horizontal distance (change in x–coordinate) between any two points on the line and is a constant. Vertical distance between any two points with coordinates (x_1, y_1) and (x_2, y_2) is $y_2 - y_1$ and horizontal distance is $x_2 - x_1$.

Thus slope of the line $= \frac{y_2 - y_1}{x_2 - x_1}$. Be careful to be consistent with which co–ordinates are set 1 and which co–ordinates are set 2

Considering points A and B from above where A is (– 8,8) and B is (12,–7), slope of the line in the above graph= $\frac{-7-8}{12-(-8)} = \frac{-15}{20} = \frac{-3}{4}$

It will be noticed that the slope is nothing but the quotient of x in the equation of the line $y = \frac{-3}{4}x + 2$ given at (1) above.

Thus generic equation of a straight line can also be written as y= mx+c, where m is the slope of the line and c is y intercept.
If instead we know x–intercept say d, the equation is**y= m(x–d).**
These equations are known as Slope intercept form

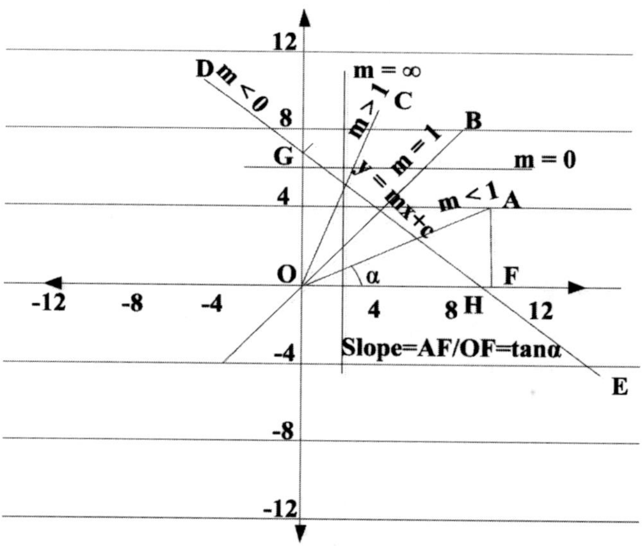

Different slope types

Figure 7.3

Slope is a very important concept and is used extensively in solving practical problems. Figure 7.3 depicts lines with different slopes. As the magnitude of the slope increases, the line gets steeper. In line OA change in value of y is less than the corresponding change in value of x, hence slope m is < 1. Similarly lines OB and OC has slope =1 and >1. Whenvalue of y decreases with increase in value of x and vice versa as in line DE; the value of slope is negative. Line parallel to x–axis has slope = 0, whereas line parallel to y–axis has slope = ∞. It will also be seen in triangle OAF that OA makes an angle of α to the x–axis and as x–coordinate increases from 0 at O to value x_1 at F, y–coordinate increases from 0 to y_1 of point A. Thus,

$$slope = \frac{AF}{OF} = \frac{y_1}{x_1} = \tan\alpha$$

Thus slope is also defined as tangent value of angle the line makes to the x–axis.

Since for angles <90⁰, value of tangent is positive and for angles >90⁰, it is negative, value of slope tells us whether the line makes an acute angle or obtuse angle to the x–axis. We can also find angle between two intersecting lines by finding slope and consequently angles subtended by the two lines (say α_1 and α_2) to the x–axis. Angle between the lines is then $\alpha_1-\alpha_2$. Since slope of two parallel lines is same, angle between two parallel lines is 0^0. Similarly if two lines intersect at right angle, $\alpha_1-\alpha_2= 90^0$.

Hence $\alpha_1 = 90^0+\alpha_2$

$\tan\alpha_1 = \tan(90^0+\alpha_2) = -\cot\alpha_2 = \dfrac{-1}{\tan\alpha_2}$, or

$\tan\alpha_1 \tan\alpha_2 = -1$

This implies that product of slope of two perpendicular lines = –1

This also implies that if generic equation of a line is ax+by+c=0, equation of the line perpendicular to it will be bx-ay+k=0 where k is any parameter. And if equations of two perpendicular lines are $ax+by+c=0$ ***and*** $a'x+b'y+c'=0$ ***then*** $aa'+bb'=0$

Line OB passes through the centre, implying thereby that when x=0, y is also 0. Equation of such a line is y=mx. Equation x = k simply implies that for any value of y, x has to be equal to k. Thus, it will result in a vertical line at value k on the x–axis. Similarly, equation y = k simply implies that for any value of x, y has to equal k. such a line is a horizontal line at value k on the y–axis.

7.2.3 Slope and one point form

If instead of x or y intercept, we know slope m and one point say (x_1, y_1) on the line. Equation of the line is

$$y - y_1 = m(x - x_1)$$

Here slope m = tan θ where θ is the angle made by the line with x–axis.

Now if another line passing through point (x_1, y_1) makes an angle α with the given line $y - y_1 = m(x - x_1)$ angle made by that line with x–axis will be θ–α or θ+α. Therefore equation of intersecting line will be $y - y_1 = \tan(\theta - \alpha)(x - x_1)$ or $y - y_1 = \tan(\theta + \alpha)(x - x_1)$

7.2.4 Two point form

We can also easily write equation for a straight line if we know coordinates of two points on the line say (x_1, y_1) and (x_2, y_2). Simply calculate slope of the equation by using formula given above and write equation of the line by substituting value of slope in the formula as

$$y - y_1 = \frac{y_2 - y_1}{x_2 - x_1}(x - x_1)$$ **. This is known as Two Point form**

7.2.5 Intercept form

If we know both intercepts (say **a** and **b**) on x–axis and y–axis respectively, we can write equation of the line as below:

Suppose there is a line 2x + 5y = 10. We can calculate value of **a** and **b** by substituting y=0 and x=0 respectively. In this case, y intercept will be 2 and x–intercept will be 5.

We can rewrite the equation as $\frac{x}{5} + \frac{y}{2} = 1$. Denominator of x and y represent the x and y intercepts respectively.

Intercept form of equation of straight line is given as $\frac{x}{a} + \frac{y}{b} = 1$

7.2.6 Normal or Perpendicular form:

Refer to Figure 7.4 below. Here length of the perpendicular from origin to the line is **p** and angle formed by the perpendicular to the y–axis is α as shown. m and n are the y and x intercept respectively. By intercept form, equation of such a line is

$$\frac{x}{n}+\frac{y}{m}=1$$

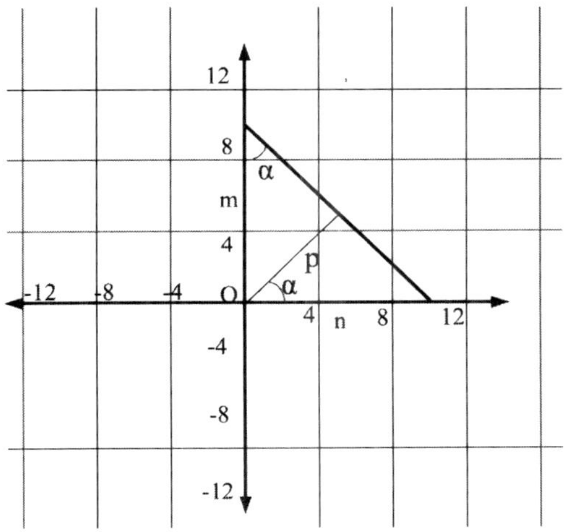

Figure 7.4

We can see that angle formed by the line to the y–axis is also α. Accodingly,

$$\frac{p}{m}=\sin\alpha \text{ and } \frac{p}{n}=\cos\alpha$$

$$\therefore \text{ n} = \frac{p}{\sin\alpha} \text{ and m} = \frac{p}{\cos\alpha}$$

Substituting in intercept form, equation of straight line in perpendicular form is,

$$\frac{x\cos\alpha}{p}+\frac{y\sin\alpha}{p}=1 \; \textit{or} \; x\cos\alpha + y\sin\alpha = p.$$

This is also known as Normal form of the equation.

Drawing lines on a Graph is a useful technique to solve problems related to linear equations in two variables. For example if we draw two straight lines. Their point of intersection (with coordinates x_1 and y_1) lies on both lines and is hence a solution to the two simultaneous linear equations depicting the two lines.

7.2.7 Other important formulae related to straight lines

7.2.7.1 Section Formula Internal/ External:

Knowing coordinates of end points of a line, we can find co–ordinates of any other point (internal or external) on the line. Refer to figure 7.5.1 below for internal division. Point P (x,y) divides the line AB in the ratio m:n.

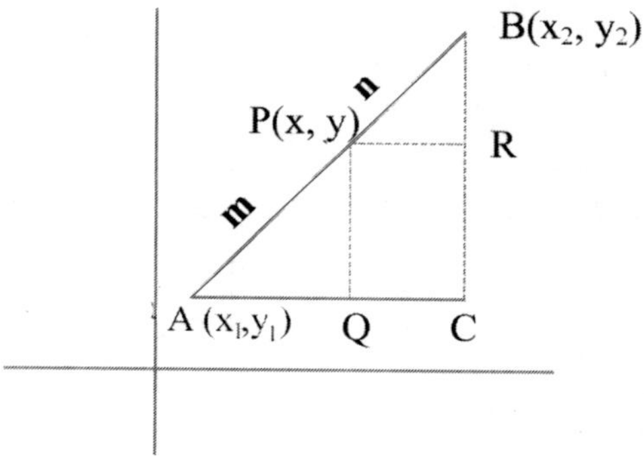

Figure 7.5.1

Internal division:

Consider ΔAPQ and ΔBPR. These are similar triangles. Hence

$\frac{AP}{BP} = \frac{AQ}{PR} = \frac{BR}{PQ}$; by substituting values

$$\frac{m}{n} = \frac{x - x_1}{x_2 - x} = \frac{y - y_1}{y_2 - y}$$

Cross multiplying and rearranging we get,

$$x = \frac{mx_2 + nx_1}{m+n} \ \& \ y = \frac{my_2 + ny_1}{m+n}$$

External division:

Refer to figure 7.5.2 below for external division.

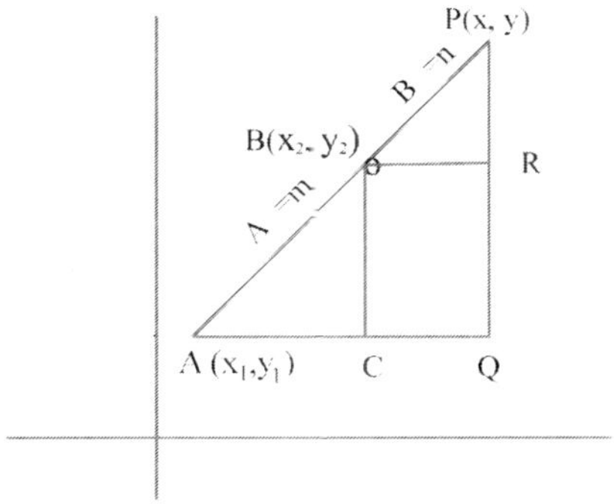

Figure 7.5.2

Consider ΔAPQ and ΔBPR. These are similar triangles. Hence

$$\frac{AP}{BP}=\frac{AQ}{BR}=\frac{PQ}{PR}$$; by substituting values

$$\frac{m}{n}=\frac{x-x_1}{x-x_2}=\frac{y-y_1}{y-y_2}$$

Cross multiplying and rearranging we get,

$$x=\frac{mx_2-nx_1}{m-n}$$

$$y=\frac{my_2-ny_1}{m-n}$$

For easy recall, if the point of division P is internal i.e. point P is towards A from B, n is positive and if the point of division P is external i.e. point P is away from A and beyond B, n is negative and section formula remains same.

We can see that x,y coordinates of midpoint of a straight line from point (x_1, y_1) and (x_2, y_2) are $\left(\frac{x_1+x_2}{2},\frac{y_1+y_2}{2}\right)$

7.2.7.2 Distance between 2 points

Consider the figure 7.6 below. We need to find distance between two arbitrary points A,B with coordinates (x_1,y_1) and (–x_2,y_2) respectively. It will be noticed that the lines indicating x and y coordinates of the two points form a right angled triangle ABC. Length of the perpendicular length AC of this triangle is $(y_1 - y_2)$ andthe horizontal length BC of this triangle is absolute value of x_1+ absolute value of x_2. Thus length BC will be treated as $(x_1 + x_2)$. Distance AB between the two given points is the length of the hypotenuse of this right angled triangle. Using Pythagoras theorem,

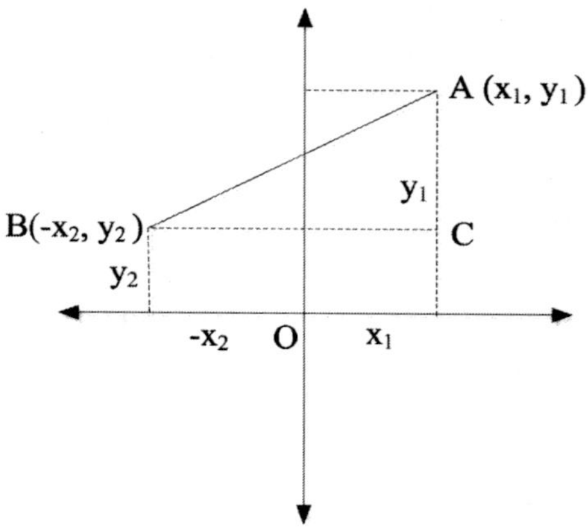

Figure 7.6

Distance between the two points AB $=\sqrt{(x_1+x_2)^2+(y_1-y_2)^2}$

Since we are squaring both entities, x_1 and x_2 are interchangeable and so are y_1 and y_2. Value of distance is always positive.

7.2.7.3 Area of a triangle

We can find area of any arbitrary triangle by coordinate geometry knowing coordinates of the three vertices. One way could be by finding length of the three sides and using Heron's formula. But that will be somewhat tedious.

Let us refer to the figure 7–7 below: Area of ΔABC can be seen as sum of area of two trapeziums ABEF and ACGF less area of trapezium BEGC. Knowing coordinates of points A, B and C, area of the three Trapezium are as below:

Area of Trapezium ABEF $=\frac{1}{2}\times(y_1+y_2)\times(x_1-x_2)$. Similarly

Area of Trapezium ACGF $=\frac{1}{2}\times(y_1+y_3)\times(x_3-x_1)$

Area of Trapezium EFGC = $\frac{1}{2} \times (y_3 + y_2) \times (x_3 - x_2)$

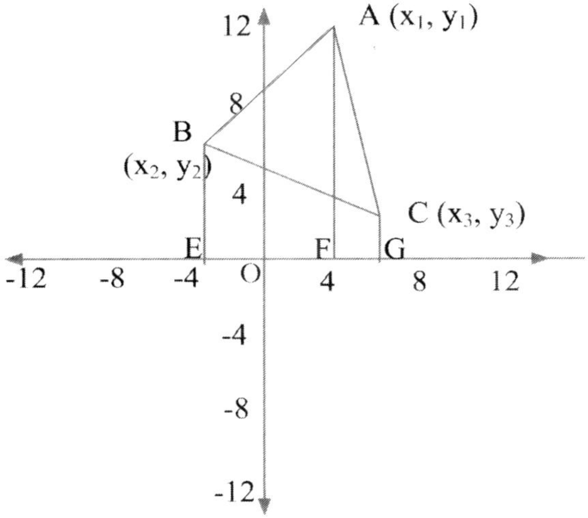

Figure 7.7

Area of ΔABC = $\frac{1}{2} \times (y_1 + y_2) \times (x_1 - x_2) + \frac{1}{2} \times (y_1 + y_3) \times (x_3 - x_1) - \frac{1}{2} \times (y_3 + y_2) \times (x_3 - x_2)$

By rearrangement, Area of ΔABC =

$$\frac{1}{2}\left[x_1(y_2 - y_3) + x_2(y_3 - y_1) + x_3(y_1 - y_2)\right]$$

Value of the expression $\left[x_1(y_2 - y_3) + x_2(y_3 - y_1) + x_3(y_1 - y_2)\right]$ can yield positive or negative value depending upon location of the three points on the coordinate axis. Hence area of the triangle should be taken as modulus $\frac{1}{2}\left|x_1(y_2 - y_3) + x_2(y_3 - y_1) + x_3(y_1 - y_2)\right|$

Let us verify with a simple example of a Right angled triangle with base 10 and height 8. Area of such a triangle is 40. Let us assume one end of the base is at a point having coordinates (1, 1). Thus coordinates of the other two points will be (11,1) and (11,9)

Area of such a triangle as per the above formula = $\frac{1}{2}[1\times(1-9)+11\times(9-1)+11\times(1-1)]=40$. Hence the formula is verified.

7.2.74 Distance of a line from a point (x_1, y_1)

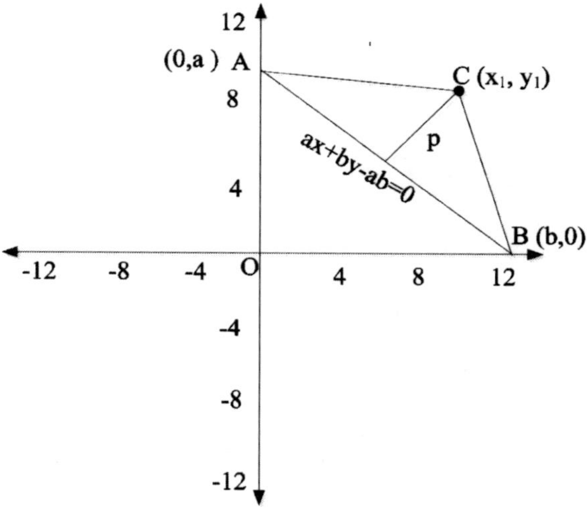

Figure 7.8

For simplicity, let equation of the line be $ax + by - ab = 0$. The intercepts of this line on x–axis and y–axis will respectively be b & a respectively leading to coordinates of points A and B as (0, a) and (b, 0) respectively.

Let us calculate area of triangle ABC by two methods– one using coordinates of the three points and the other, by length of line AB and perpendicular from C to AB.

Case 1:

Using formula for area of the triangle derived in 7.2.7.3

$$\text{Area} = \frac{1}{2}\left|0\times(0-y_1)+b\times(y_1-a)+x_1\times(a-0)\right|$$

$$= \frac{1}{2}\left|ax_1+by_1-ab\right|$$

Case 2:

Area= $\frac{1}{2} \times p \times \sqrt{a^2 + b^2}$

Equating the two

$\frac{1}{2} \times p \times \sqrt{a^2 + b^2} = \frac{1}{2}(ax_1 + by - ab)$; or

$$p = \left| \frac{(ax_1 + by_1 - ab)}{\sqrt{a^2 + b^2}} \right|.$$

In case equation of the line is ax+by+c = 0, the distance p will be

$$\left| \frac{(ax_1 + by_1 + c)}{\sqrt{a^2 + b^2}} \right|$$

7.2.7.5 Distance between two parallel lines:

Case 1: Equation of lines in generic form:

When two straight lines are parallel, their slopes are same. Hence any line parallel to ax+by+c=0 is of the type ax+by+d=0, where d is also known as parameter of the line.

Reference figure 7.9 below. Lines AB and CD are parallel. X–intercept of line AB on x–axis $= -\frac{c}{a}$

Using the formula derived in 7.2.7.4 for perpendicular distance **p** from point B (–c/a,0) to line ax+by+d=0

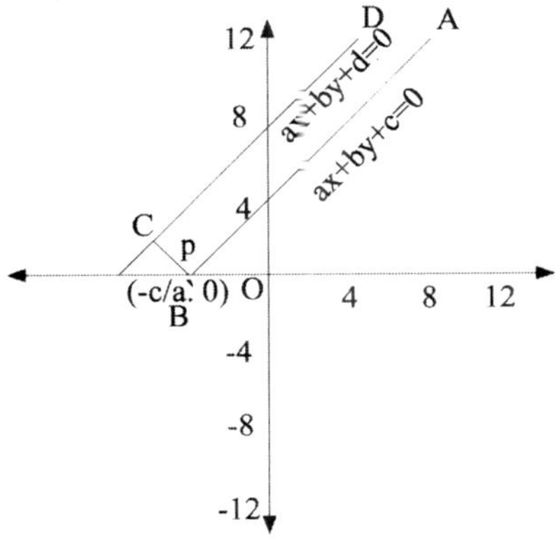

Figure 7.9

we can see that distance between the parallel lines

$$p=\frac{|a(-c/a)+d|}{\sqrt{a^2+b^2}}=\frac{|d-c|}{\sqrt{a^2+b^2}}$$

Case 2: Equation of lines in slope intercept form:

Let equations of the parallel lines be $y = m_1x + c_1$, and $y = m_1x + c_2$

Comparing these equations with the generic equations in Case 1,

$$m_1=\frac{-a}{b}, c_1=\frac{-c}{b} \;\&\; c_2=\frac{-d}{b}$$

Dividing numerator and denominator of the expression for p derived above by –b, we get,

$$p=\left|\frac{\frac{-d}{b}-\frac{-c}{b}}{\sqrt{\frac{a^2}{b^2}+1}}\right|=\left|\frac{c_2-c_1}{\sqrt{m_1^{\,2}+1}}\right|$$

7.2.8 A few interesting observation regarding straight lines:

7.2.8.1 Position of a point (x_1, y_1) relative to the line ax + by + c = 0

If $ax_1 + by_1 + c = 0$ is of the same sign as c, the point (x_1, y_1) lie on the origin side of ax + by + c = 0. But if $ax_1 + by_1 + c = 0$ is of the opposite sign as that of c, the point (x_1, y_1) lie on the non–origin side of ax + by + c = 0

7.2.8.2 Point of intersection of 2 lines $a_1x + b_1 y + c_1 = 0$ and $a_2x + b_2 y + c_2 = 0$

We find the point of intersection by equating the two equations as below:

Multiplying equation of line $\mathbf{a_1x + b_1 y + c_1 = 0}$ by b_2 and $\mathbf{a_2x + b_2 y + c_2 = 0}$ by b_1 and subtracting we get $x(a_1b_2-a_2b_1)+(c_1b_2-c_2b_1)=0$ leading to $x=\frac{b_1c_2-b_2c_1}{(a_1b_2-a_2b_1)}$

Multiplying equation of line $\mathbf{a_1x + b_1 y + c_1 = 0}$ by a_2 and $\mathbf{a_2x + b_2 y + c_2 = 0}$ by a_1 and subtracting we get $y(b_1a_2-b_2a_1)+(c_1a_2-c_2a_1)=0$ leading to $y=\frac{c_1a_2-c_2a_1}{(a_1b_2-a_2b_1)}$

Thus point of intersection of lines $a_1x + b_1y + c_1 = 0$ and $a_2x + b_2y + c_2 = 0$ is $\left(\frac{b_1c_2 - b_2c_1}{(a_1b_2 - a_2b_1)}, \frac{c_1a_2 - c_2a_1}{(a_1b_2 - a_2b_1)}\right)$

T 7.2.8.3 The ratio in which a given line divides the line segment joining two points

Let the given line $ax + by + c = 0$ divide the line joining point A (x_1, y_1) & B (x_2, y_2) in the ration m:n, then $\frac{m}{n} = \frac{ax_1 + by_1 + c}{ax_2 + by_2 + c}$.

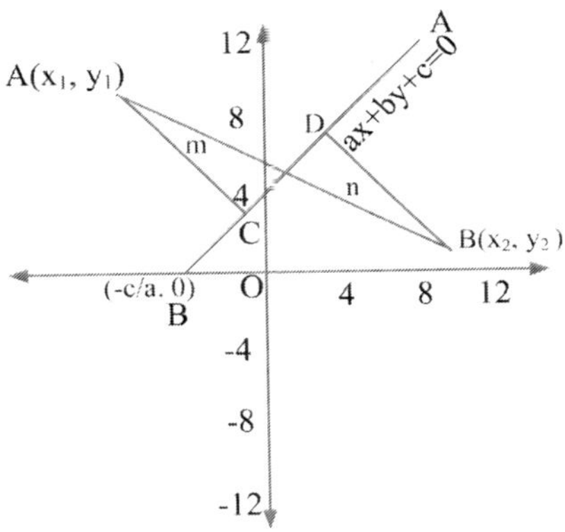

Figure 7.10

The above relationship can be easily visualized by considering two similar triangles AEC and BED where AC and BD are perpendiculars to the given line implying that ratio of the perpendicular distances is also m:n.

Now using the equation for distance of a known point from a line, derived in section 7.2.7.4, ratio of the perpendicular distance of the two given points (x_1,y_1), (x_2, y_2) from the given line are

$$AC = \left|\frac{(ax_1 + by_1 + c)}{\sqrt{a^2 + b^2}}\right| \text{ and } BD = \left|\frac{(ax_2 + by_2 + c)}{\sqrt{a^2 + b^2}}\right|$$

yielding $\frac{m}{n} = \frac{ax_1 + by_1 + c}{ax_2 + by_2 + c}$

If A & B are on the same side of the given line, then $\frac{m}{n}$ is negative but if A & B are on the opposite side of the given line, then

$\frac{m}{n}$ is positive. This has already been discussed in the section formula at 7.2.7.1 above.

7.2.8.4 Angle between two straight lines in terms of their slope:

If m_1 & m_2 are slopes of two intersecting straight lines ($(m_1 \times m_2 \neq -1)$ & θ is the acute angle between them, then $\tan\theta = \left|\frac{m_1 - m_2}{1 + m_1 m_2}\right|$

Note: If m_1, m_2 & m_3 are slopes of three lines L_1=0, L_2=0 & L_3=0, where $m_1 > m_2 > m_3$, then the interior angles of the ΔABC formed by these lines are given by

$$\tan A = \left|\frac{m_1 - m_2}{1 + m_1 m_2}\right|, \quad \tan B = \left|\frac{m_2 - m_3}{1 + m_2 m_3}\right| \ \& \ \tan C = \left|\frac{m_3 - m_1}{1 + m_3 m_1}\right|$$

7.2.8.5 Area of a parallelogram knowing perpendicular distance between the parallel lines and angle between the adjacent sides or equations of the sides

As shown in figure 7.11 below, let pairs of parallel lines be $y = m_1x + c_1$, $y = m_1x + c_2$, $y = m_2x + d_1$ & $y = m_2x + d_2$ and let angle between the adjacent sides be θ. p_1 & p_2 are perpendiculars on the two non–parallel sides.

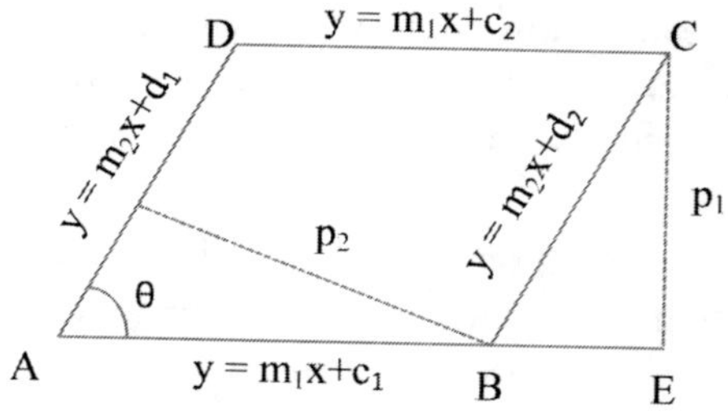

Figure 7.11

Area of the parallelogram = AB × p_1.

We can see that $\sin\theta = \frac{p_2}{AB}$ leading to $AB = \frac{p_2}{\sin\theta}$

Thus Area of the parallelogram $= p_1 \times AB = \frac{p_2 \times p_1}{\sin\theta}$

Reference section 7.2.7.5, distance $\mathbf{p}_1$ between two parallel lines $y = m_1x + c_1$ and $y = m_1x + c_2$ is given as $\left|\frac{c_2 - c_1}{\sqrt{m_1^2 + 1}}\right|$. Similarly $\mathbf{p}_2 = \left|\frac{d_2 - d_1}{\sqrt{m_2^2 + 1}}\right|$

We have seen in section 7.8. that angle between two lines knowing their slopes is given by the formula $\tan\theta = \left|\frac{m_1 - m_2}{1 + m_1m_2}\right|$

We need value of sin θ to find area of the parallelogram. So let us find value of sin θ as below:

$$1 + \tan^2\theta = \sec^2\theta \text{ leading to } \cos^2\theta = \frac{1}{1 + \tan^2\theta}$$

$$\Rightarrow \sin^2\theta = 1 - \frac{1}{1 + \tan^2\theta} = \frac{\tan^2\theta}{1 + \tan^2\theta} = \frac{(m_1 - m_2)^2}{(1 + m_1m_2)^2 + (m_1 - m_2)^2}$$

$$= \frac{(m_1 - m_2)^2}{\left(1 + m_1^2m_2^2 + 2m_1m_2\right) + \left(m_1^2 + m_2^2 - 2m_1m_2\right)}$$

$$= \frac{(m_1 - m_2)^2}{\left(1 + m_1^2m_2^2 + m_1^2 + m_2^2\right)} = \frac{(m_1 - m_2)^2}{\left(1 + m_1^2\right)\left(1 + m_2^2\right)} \Rightarrow$$

$$\sin\theta = \frac{(m_1 - m_2)}{\sqrt{1 + m_1^2}\sqrt{1 + m_2^2}}$$

Substituting values of p_1, p_2 and $\sin\theta$ in the expression for Area of the parallelogram derived above, we find

Area of the parallelogram

$$= \frac{p_1 \times p_2}{\sin\theta} = \frac{\left|\frac{c_2 - c_1}{\sqrt{m_1^2 + 1}}\right| \times \left|\frac{d_2 - d_1}{\sqrt{m_2^2 + 1}}\right|}{\frac{(m_1 - m_2)}{\sqrt{1 + m_1^2}\sqrt{1 + m_2^2}}} = \left|\frac{(c_2 - c_1)(d_2 - d_1)}{(m_1 - m_2)}\right|$$

7.2.9 Method to find location of Orthocenter, Centroid, Circumcenter and parameters of Euler line

As mentioned in section 3.2.8.3, there is no direct formula for location of Orthocenter of a triangle. Let us now see how to calculate location of Orthocenter, Centroid and Circumcenter and verify that these points lie on a straight line known as Euler line. We will further go on to verify distance between these points and their relationship.

Figure 7.12 below shows an arbitrary triangle with known coordinates of its vertices:

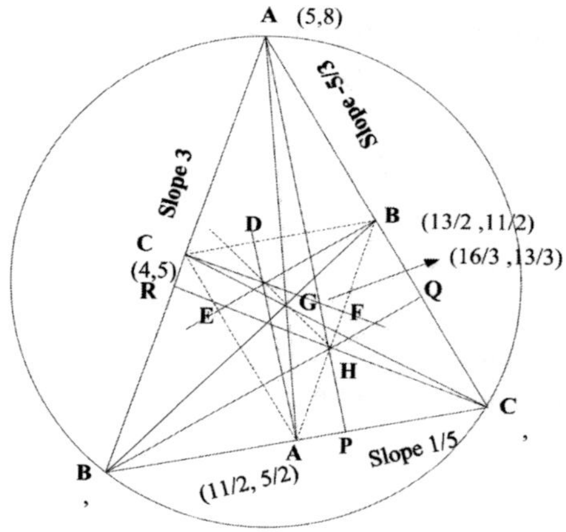

Figure 7.12

Let us find coordinates of the Centroid (denoted as G in the above diagram)

Centre points of the 3 sides BC (A), CA (B) and AB (C) are

$$\frac{5+6}{2},\frac{2+3}{2}\left(\frac{11}{2},\frac{5}{2}\right),\ \frac{8+5}{2},\frac{3+8}{2}\left(\frac{13}{2},\frac{11}{2}\right) \text{ and } \frac{5+3}{2},\frac{8+2}{2}(4,5)$$

respectively.

Equations of the medians using two point form of equation are as below:

$$\mathrm{AA}' \Rightarrow y-8=\frac{\frac{5}{2}-8}{\frac{11}{2}-5}\times(x-5)=-11(x-5), \text{ leading to}$$

$$y+11x-63 = 0 \quad \text{... (1)}$$

$$\text{BB}' \Rightarrow y-2=\frac{2-\frac{11}{2}}{3-\frac{13}{2}}\times(x-3)=1\times(x-3)\text{, leading to}$$

$$y-x+1 = 0 \quad \text{... (2)}$$

$$\text{And CC}' \Rightarrow y-3=\frac{3-5}{8-4}\times(x-3)=-\frac{1}{2}\times(x-8)\text{, leading to}$$

$$2y+x-14 = 0 \quad \text{...(3)}$$

Now, point of intersection of AA′ and BB′ can be found by equating equations (1) and (2).

Subtracting equation (2) from equation (1), we get, $12x-64=0 \Rightarrow x=\frac{16}{3}$

Substituting value of x in equation (2), $y=x-1 \Rightarrow y=\frac{16}{3}-1=\frac{13}{3}$, implying that coordinates of the point of intersection of AA′ and BB′ are $\left(\frac{16}{3},\frac{13}{3}\right)$

Substituting these coordinate values in equation 3, we find

$2\times\frac{13}{3}+\frac{16}{3}-14 = 0$ implying thereby that ***the three medians meet at the point G whose coordinates are*** $\left(\frac{16}{3},\frac{13}{3}\right)$

Coordinates of point A and A′ are (5, 8) and $\left(\frac{11}{2},\frac{5}{2}\right)$ respectively. Assuming that point G divides line AA′ in ratio $m_1:m_2$, we can find the ratio using section formula as below:

$$\frac{m_2\times 5+m_1\times\frac{11}{2}}{m_1+m_2}=\frac{16}{3}\Rightarrow 30m_2+33m_1=32m_1+32m_2\Rightarrow m_1=2m_2$$

or $\frac{m_1}{m_2}=\frac{2}{1}$. This ratio can be verified for the other two medians as well.

In generic terms, if we consider coordinates of the three vertices as $A(x_1,y_1)$, $B(x_2,y_2)$ and $C(x_3,y_3)$. Coordinates of mid point of BC is $\left(\frac{x_2 + x_3}{2}\right), \left(\frac{y_2 + y_3}{2}\right)$. Since Centroid divides the line joining A to mid point of BC in the ratio 2:1, it implies that coordinates of centroid will be

$$\left(\frac{\frac{x_2 + x_3}{2} \times 2 + x_1}{2+1}, \frac{\frac{y_2 + y_3}{2} \times 2 + y_1}{2+1}\right) \Rightarrow \left(\frac{x_1 + x_2 + x_3}{3}, \frac{y_1 + y_2 + y_3}{3}\right).$$

Hence we can see that coordinates of centroid are average of the coordinates of the three vertices.

Triangle $\Delta A'B'C'$ formed by mid points of the three sides of ΔABC. is similar to the given triangle as line joining mid points of the two sides is parallel to the third side and half of it in length. This further implies that medians of the $\Delta ABC'$ are also medians of $\Delta A'B'C'$. Thus G is also Centroid of $\Delta A'B'C'$. $\Delta A'B'C'$ is also known as ***Medial triangle.***

Similarly let us find out coordinates of the Circumcenter, which is the point of intersection of perpendicular bisectors of the triangle.

Perpendicular bisectors of $\Delta ABC'$ are A′D, B′E and C′F. To find equation of these bisectors, we need to know the mid points of the three sides and slope of their perpendiculars. We know that slope of a perpendicular line $= -\frac{1}{m}$ where m is slope of the side of the triangle being bisected.

Coordinates of the mid points and Slope of different sides of the triangle are given in the figure 7.12 above.

Hence equation of the perpendicular bisectors are as below:

A′D$\Rightarrow y - \frac{5}{2} = -5 \times \left(x - \frac{11}{2}\right)$, leading to

$y + 5x - 30 = 0$ ... (4)

B′E$\Rightarrow y - \frac{11}{2} = \frac{3}{5} \times \left(x - \frac{13}{2}\right)$, leading to

$5y - 3x - 8 = 0$ ...(5)

And C′F$\Rightarrow y - 5 = \frac{-1}{3} \times (x - 4)$, leading to

$$3y + x - 19 = 0 \quad ...(6)$$

Equating equations (4) and (5) we get $28x - 142 = 0 \Rightarrow x = \frac{71}{14}$ and

$$y = 30 - 5x \Rightarrow y = 30 - \frac{5 \times 71}{14} = \frac{420 - 355}{14} = \frac{65}{14}$$

Substituting these coordinate values in equation 6, we find

$3 \times \frac{65}{14} + \frac{71}{14} - 19 = 0$, confirming thereby that ***coordinates of the point of intersection of the three perpendicular bisectors i.e the Circumcenter denoted as O are*** $\left(\frac{71}{14}, \frac{65}{14}\right)$

Another interesting fact can be noticed that the perpendicular bisectors of ΔABC ***are altitudes of medial triangle*** $\Delta A'B'C'$ ***(all three sides of*** $\Delta A'B'C'$ ***being parallel to the sides of*** ΔABC***. Thus Circumcenter O of*** ΔABC ***is Orthocenter of*** $\Delta A'B'C'$.

Now let us find coordinates of the Orthocenter of ΔABC

Slope of altitude of a side of the triangle is same as slope of its perpendicular bisector. Hence we can easily find equations of the three altitudes AP. BQ and CR as below:

AP $\Rightarrow y - 8 = -5 \times (x - 5)$, leading to

$$y + 5x - 33 = 0 \quad ... (7)$$

BQ $\Rightarrow y - 2 = \frac{3}{5} \times (x - 3)$, leading to

$$5y - 3x - 1 = 0 \quad ...(8)$$

And CR $\Rightarrow y - 3 = \frac{-1}{3} \times (x - 8)$, leading to

$$3y + x - 17 = 0 \quad ...(9)$$

Equating equations (7) and (8) we get $28x - 164 = 0 \Rightarrow x = \frac{41}{7}$ and Substituting value of x in equation (8),

$5y = 3x + 1 \Rightarrow 5y = 3 \times \frac{41}{7} + 1 = \frac{130}{7} \Rightarrow y = \frac{26}{7}$, implying that coordinates of the point of intersection of AP and BQ are $\left(\frac{41}{7}, \frac{26}{7}\right)$

Substituting these coordinate values in equation 9, we find

$3\times\frac{26}{7}+\frac{41}{7}-17=0$, confirming thereby that ***coordinates of the point of intersection of the three altitudes i.e the Orthocenter denoted as H are*** $\left(\frac{41}{7},\frac{26}{7}\right)$

To sum up coordinates of Circumcenter (G), Centroid (O) and Orthocenter (H) are $\left(\frac{71}{14},\frac{65}{14}\right)$, $\left(\frac{16}{3},\frac{13}{3}\right)$ ***and*** $\left(\frac{41}{7},\frac{26}{7}\right)$ ***respectively***

Now let us check whether these three points lie on a straight line.

One method to check is to see whether slope of line joining Circumcenter to Centroid is same as the slope of line joining Centroid to the Orthocenter.

Second method is to check area of triangle formed by these three points. If area = 0, the points lie on a straight line.

Method 1:

Slope of line joining Circumcenter to Centroid

$$=\frac{\frac{65}{14}-\frac{13}{3}}{\frac{71}{14}-\frac{16}{3}}=\frac{\frac{195-182}{42}}{\frac{213-224}{42}}=\frac{13}{-11}=-\frac{13}{11}$$

Slope of line joining Centroid to the Orthocenter

$$=\frac{\frac{13}{3}-\frac{26}{7}}{\frac{16}{3}-\frac{41}{7}}=\frac{\frac{91-78}{21}}{\frac{112-123}{21}}=\frac{13}{-11}=-\frac{13}{11}$$

As the slope is same, the points lie on the straight line.

Method 2:

As shown in section 2.3, area of a triangle with vertices (x_1, y_1), (x_2, y_2), (x_3, y_3) equals

$$\frac{1}{2}\left[x_1(y_2-y_3)+x_2(y_3-y_1)+x_3(y_1-y_2)\right]$$

So substituting coordinate values for ***Circumcenter, Centroid and Orthocenter,*** we get Area

$$= \frac{1}{2}\times\left\{\frac{71}{14}\left(\frac{13}{3}-\frac{26}{7}\right)+\frac{16}{3}\left(\frac{26}{7}-\frac{65}{14}\right)+\frac{41}{7}\left(\frac{65}{14}-\frac{13}{3}\right)\right\}$$

$$=\frac{1}{2}\times\left\{\frac{71}{14}\times\frac{13}{21}+\frac{16}{3}\times-\left(\frac{13}{14}\right)+\frac{41}{7}\times\frac{13}{42}\right\}=\frac{1}{2}\times\left\{\frac{71\times13}{42\times7}-\frac{16\times13}{42}+\frac{41\times13}{7\times42}\right\}$$

$$=\frac{1}{2}\times\left\{\frac{(71+41)\times13}{42\times7}-\frac{16\times13}{42}\right\}=\frac{1}{2}\times\left\{\frac{112\times13}{42\times7}-\frac{16\times13}{42}\right\}=0$$

This again proves that the three points ***Circumcenter, Centroid and Orthocenter of any triangle lie on a straight line known as Euler Line after Swiss mathematician Leonhard Euler.***

It is surprising that these three points of any triangle lie on a straight line. But let us now see conceptually why it is so. Refer to figure 7.12 again.

Focus your attention on the centroid *G*. For each point, like *A* on one side of it, there is another, like *A′* on the other side of it but half as far away (as seen while discussing properties of medians). On one side is *B*, the other *B′*; on one side *C*, the other *C′*. In fact, this correspondence sends the whole triangle *ABC* to the smaller, but similar, triangle *A′B′C′*, called the *medial triangle*. The sides of the medial triangle *ABC* are parallel and half the length of the sides of the original triangle *ABC*. Thus G is the Centroid of the medial triangle as well.

You can see from the figure that this correspondence sends the altitudes of the original triangle, which are *AP*, *BQ*, and CR, to the altitudes of the medial triangle, which are *A′D*, *B′E*, and *C′F*. Since the altitudes of the original triangle meet at the orthocenter *H′* of the original triangle, the altitudes of the medial triangle will meet at its orthocenter *H* which you can see in the figure is labelled *O* because this orthocenter *O* of the medial triangle is the circumcenter of the original triangle! Thus, this correspondence sends *H* to *O*, that is, *H* and *O* are on the opposite sides of the centroid ***G****, and* ***O* is half as far away from *G as H*** is.

We can verify this fact as below:

Distance between Centroid and Circumcenter (Orthocenter of the Medial triangle)

$$= \sqrt{\left(\frac{65}{14}-\frac{13}{3}\right)^2+\left(\frac{71}{14}-\frac{16}{3}\right)} = \sqrt{\left(\frac{13}{42}\right)^2+\left(\frac{-11}{42}\right)^2} = \frac{1}{42}\times\sqrt{290}$$

Distance between Centroid and Orthocenter

$$= \sqrt{\left(\frac{13}{3}-\frac{26}{7}\right)^2+\left(\frac{16}{3}-\frac{41}{7}\right)} = \sqrt{\left(\frac{13}{21}\right)^2+\left(\frac{-11}{21}\right)^2} = \frac{1}{21}\times\sqrt{290}$$

This proves another interesting fact that **Orthocenter** is twice as far from the **Centroid** as the **Circumcenter**.

In an equilateral triangle all 3 sides and angles being equal and because of symmetry all four points i.e circumcentre, incentre, orthocentre and centroid coincide at the same point.

7.3 Conic Sections:

We discussed straight lines in detail so as to have a very clear understanding of basic concepts of coordinate geometry. Any line segment that does not have constant slope is called non linear. Slope in case of non linear curves keeps changing all the time.

Let us consider two straight lines– one vertical (m) in figure 7.13 below and another (n) at an angle α to the vertical. If we rotate line n around the vertical line, we end up getting two cones one above the other with their vertex V facing each other.

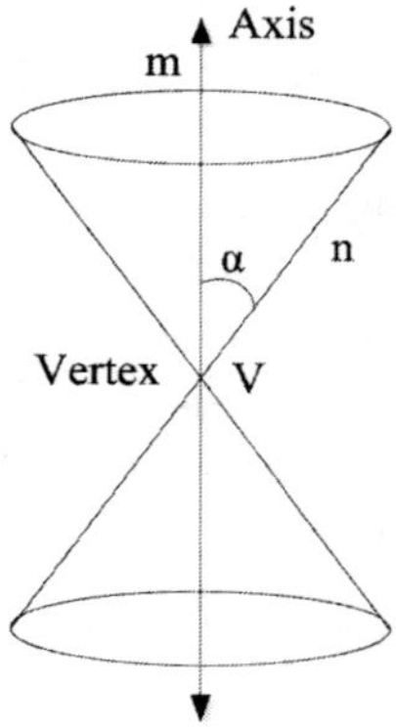

Figure 7.13

Now if we slice either of these cones with a plain surface at an angle of 90^0 to the vertical axis, we will get a circular section as shown in figures 7.14. And if we slice either of these cones with a plain surface at an angle β to the vertical axis, we will get an elliptical section as shown in figures 7.15

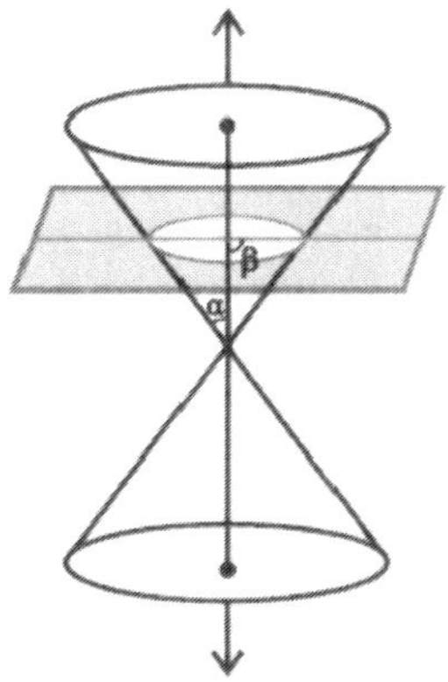

Figure 7.14

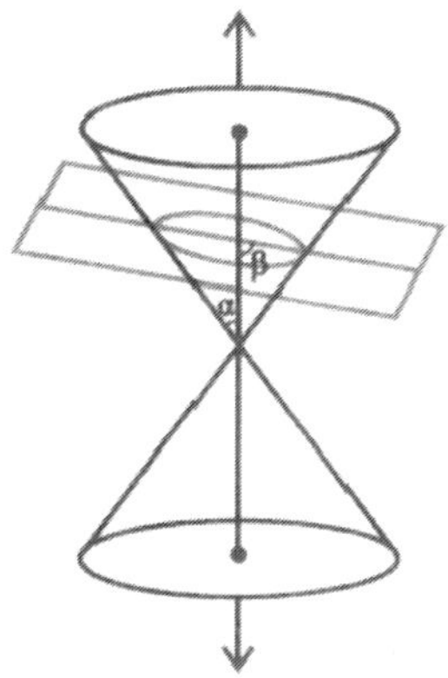

Figure 7.15

Now if we change angle β of the plain surface such that it slices only part of one of the cones as shown in figure 7.16 below, we get a conical section called **Parabola**. And if we reduce angle β further to a value less than angle α as shown in figure 7.17, we will get a conical section known as **Hyperbola**. We will discuss properties of these conical shapes and derive coordinate equations for depicting the same in the following sections.

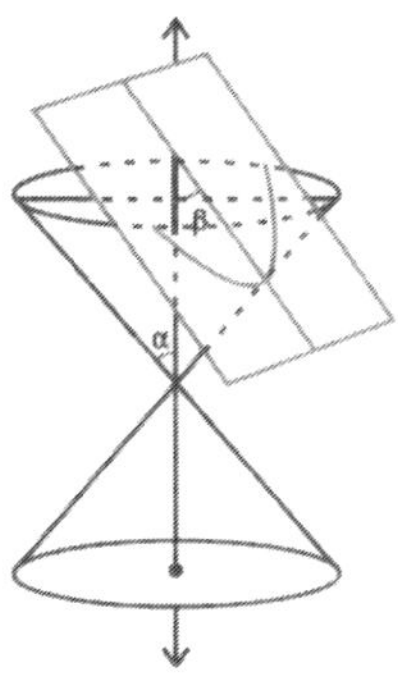

Figure 7.16

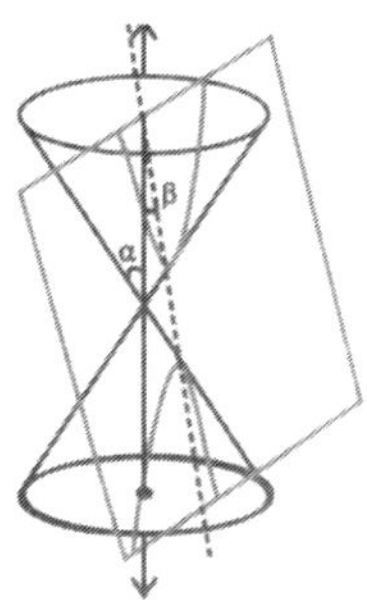

Figure 7.17

7.3.1 Circle

Let us consider circle in some detail. Refer to figure 7.18. As we know all points on the circle are equidistant from centre of the circle. This distance is denoted by radius, marked **r** here.

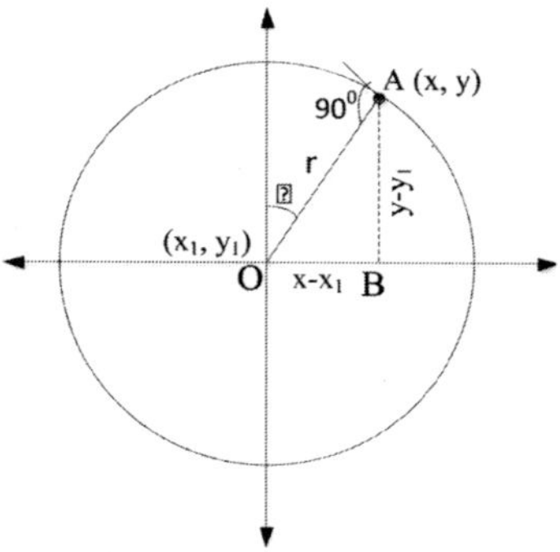

Figure 7.18

An arbitrary point A is marked on the circle boundary. Let its coordinates be (x, y). Let coordinates of centre of the circle be (x_1, y_1). Now consider a horizontal line passing through centre O of the circle and a vertical line from point A to the horizontal line meeting it at point B. OAB forms a right angled triangle. Radius of the circle is its hypotenuse. By Pythagoras theorem,

$$r^2 = (x-x_1)^2 + (y-y_1)^2$$

This is the generic equation of a circle in coordinate geometry. The same formula can also be arrived at by using distance formula. In case centre of the circle is located at the centre of the co–ordinate system i.e. coordinates of the centre are (0,0), equation of the circle reduces to

$$r^2 = x^2 + y^2$$

These are quadratic equations in x and y variables. It would be noticed that slope of the curvature of the circle keeps varying all the time. Slope at a point on circle boundary is proportionate to the angle θ that radius at that point makes with the vertical axis and is equal to tan θ. Thus slope is maximum (tan 90^0) at the point where circle boundary meets x–axis and is minimum (tan 0^0) at the point where circle boundary meets y–axis.

Understanding of these properties of the circle is useful to solve/comprehend problems related to circular motion and its other connected parameters. Recall that we have already covered area and parameter of the circle in chapter on mensuration.

7.3.2 Ellipse:

Ellipse is a sort of stretched circle. Refer to figure 7.19. As per one definition, an ellipse is the locus of a point with a fixed distance ratio of less than one between the focus and a fixed line called the directrix. For simplicity, the ellipse here is centered on the centre of the coordinate axis. We will deal with other variants later in this chapter. It has the following properties:

1. Ellipse is a set of all points in a plane, sum of whose distances from two fixed points (known as focal points or foci) is a given positive constant that is greater than the distance between the fixed points.
2. It has a center like that in a circle
3. There are two axes– a Major axis along the longer side and a Minor axis along the shorter side. Length of both these axes from the center to its edge is equal and is denoted as **a** & **b** respectively.Major and Minor axis together are called Principal Axes of the ellipse. Points where major axis meets the ellipse are called vertices– A & B. Coordinates of the vertices for major axis along x–axis are (a, 0) and (–a, 0) and for major axis along y–axis are (0, a) and (0, –a).
4. There are two focal points, shown as F_1 and F_2. ***Sum of the distance of any point on the ellipse to the two focal points equals length of the major axis (2a).***

   Thus $F_1P + F_2P = F_1A + F_2A = 2a$
5. Distance from center of the ellipse to the two focal points (written as foci) OF_1 and OF_2 is equal and denoted by **c**.

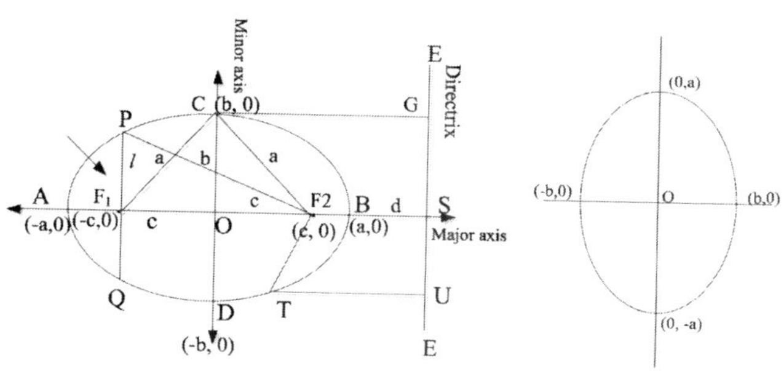

Figure 7.19

6. Ratio of the distance from a point on the ellipse to one of the foci and to the perpendicular distance from that point to a line EE′, outside the ellipse and perpendicular to the major axis called the directrix, is constant. In figure above, $\frac{TF_2}{TU}$ = constant. This ratio is known as eccentricity of the ellipse. Ellipse being a stretched circle, its eccentricity is also defined as its deviation from being a circle and is always less than 1. Circle has zero eccentricity. Similarly all conic sections have eccentricity.

The equation of an ellipse is simplest if thecenter of the ellipse is at the origin and the foci areon the x–axis or y–axis. Two such possible orientations are shown in Figure 7.20 and 7.21 below.

Now let us try and understand ellipse at a visual level. Let us consider a circle with radius 1. Equation of such a circle centered at origin is $x^2 + y^2 = 1$. In order to convert such a circle into an ellipse we need to elongate it along x–axis or y–axis. In case we elongate it along x–axis by say a factor **a** > **1**, keeping distance of boundary along y–axis same,x coordinates of the ellipse will be larger than those of the circle by a factor of **a**. The equation of ellipse will change to $\left[\frac{x}{a}\right]^2 + \left[\frac{y}{1}\right]^2 = 1$.

An example with factor a=3 is shown below (figure 7–20).

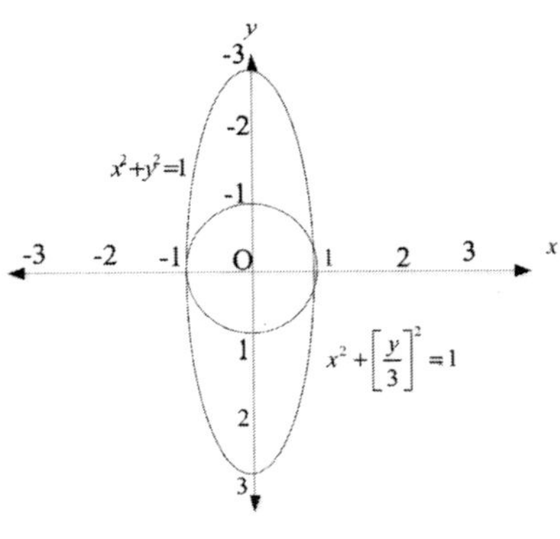

Figure 7.20

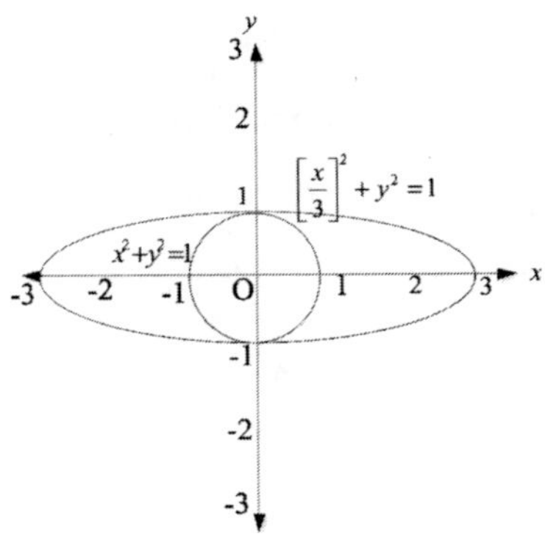

Figure 7.21

Similarly if we elongate the ellipse along y–axis by a factor **b** > **1**, y coordinates of the ellipse will be larger than those of the circle by a factor of **b**. The equation of ellipse will change to $\left[\frac{x}{1}\right]^2 + \left[\frac{y}{b}\right]^2 = 1$.

An example with factor b=3 is shown in figure 7.21 above. This implies that equation of an ellipse is $\left[\frac{x}{a}\right]^2 + \left[\frac{y}{b}\right]^2 = 1$, where a is the length of semi–major axis and b is the length of semi–minor axis, ie. a and b are the distances of points where the ellipse meets x=axis and y–axis respectively from center of the ellipse.

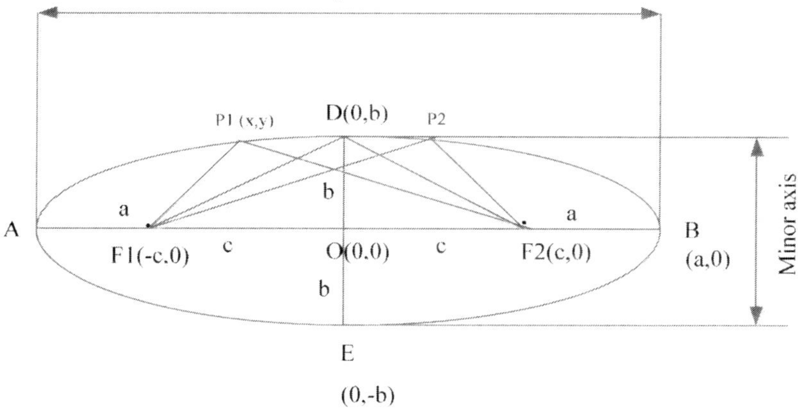

Figure 7.22

Alternatively, let us try and derive generic equation of an ellipse by considering basic property of the ellipse. Different parameters of the ellipse are as depicted in Figure 7.22 above. Coordinates of the center of the ellipse are (0,0), as considered above.

Sum of the distance of point B from the two foci is

$F_1B+F_2B = F_1O+OB+F_2B = c + a + (a–c) = 2a$, which is length of the major axis, a constant.

It also follows that distance of point D from the two foci $DF_1=DF_2=OA=OB=a$

$\therefore\ DF_1+DF_2 = 2\times\sqrt{b^2+c^2} = 2a \Rightarrow a = \sqrt{b^2+c^2}$

Squaring both sides, $a^2 = b^2 + c^2$ or $a^2 - c^2 = b^2$; this is a very **important identity.**

Let us now consider sum of the distances of an arbitrary point P_1 with coordinates (x,y) to the two foci. It is equal to:

$F_1P_1+F_2P_1 = \sqrt{(x+c)^2+y^2}+\sqrt{(x-c)^2+y^2}$, which in turn equals 2a, as seen above. Thus $\sqrt{(x+c)^2+y^2} = 2a-\sqrt{(x-c)^2+y^2}$

squaring both sides

$$(x+c)^2+y^2 = 4a^2+(x-c)^2+y^2-4a\times\sqrt{(x-c)^2+y^2}$$

Or $4a\times\sqrt{(x-c)^2+y^2} = 4a^2-4xc$ or $\sqrt{(x-c)^2+y^2} = a-\frac{xc}{a}$; squaring again

$x^2+c^2-2xc+y^2 = a^2+\frac{x^2c^2}{a^2}-2xc$; cancelling common terms and transposing

$y^2 = (a^2-c^2)+x^2\left(\frac{c^2}{a^2}-1\right)$. Simplifying further,

$y^2 = (a^2-c^2)-\frac{x^2}{a^2}(a^2-c^2)$ leading to

$y^2 = (a^2-c^2)\times\left(1-\frac{x^2}{a^2}\right)$ Or $\frac{y^2}{a^2-c^2}+\frac{x^2}{a^2} = 1$,

transposing and simplifying further

Since $a^2 - c^2 = b^2$, ***the generic equation of ellipse is***

$$\frac{x^2}{a^2}+\frac{y^2}{b^2} = 1$$

This is the equation of the ellipse, as deduced earlier. **Note *again that if a=b=r, the equation reduces to the equation of a circle. Circle is also often called a special case of ellipse with same distancefrom center to all points. It also follows that apoint (x_1, y_1) lies outside/ inside/ or on the ellipse if*** $\frac{x_1^2}{a^2}+\frac{y_1^2}{b^2} > / < or = 1$

And in case center of the ellipse lies at an arbitrary point (x_1, y_1), formula for the ellipse changes to $\frac{(x-x_1)^2}{a^2}+\frac{(y-y_1)^2}{b^2} = 1$, as in case of a circle. This is the equation of an ellipse oriented horizontally. In case the ellipse is oriented vertically, association of major and minor axis reverses, yielding equation of the ellipse as $\frac{(y-y_1)^2}{a^2}+\frac{(x-x_1)^2}{b^2}=1$

An important point to note is that axes of the ellipse are parallel to x–axis and y–axis respectively. We shall consider the case of conics where the axis of the conic are oriented ina direction different than the coordinate axis later in section 7.4 .

7.3.2.1 Basic Properties of ellipse

7.3.2.1.1 Eccentricity

Eccentricity of an ellipse is already defined above. Refer figure 7.23 below.

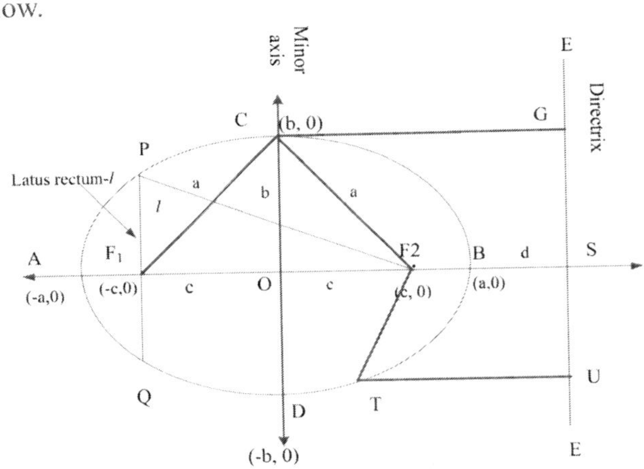

Figure 7.23

Considering point T, eccentricity $e=\frac{F_2T}{TU}$

Similarly for points B and C, $e=\frac{F_2B}{BS}=\frac{a-c}{d}$ ....(1)

and also $\frac{F_2C}{CG}=\frac{a}{a+d}$ ......(2) respectively.

Equating the two, $\frac{a-c}{d}=\frac{a}{a+d}$; cross multiplying

$$(a-c)\times(a+d)=ad$$

$$a(a-c)+ad-cd=ad$$

$$a(a-c)=cd \Rightarrow d=\frac{a(a-c)}{c}$$

Substituting value of d in equation (1) above

$$e=\frac{a-c}{d}=(a-c)\div\frac{a(a-c)}{c}\Rightarrow$$

$$\boldsymbol{e=\frac{(a-c)\times c}{a(a-c)}=\frac{c}{a}}$$ ***or c = ae ⇒ coordinates of the foci are*** $(ae,0)\;\&\;(-ae,0)$

Thus eccentricity of an ellipse is also the ratio of the distances from the centre of the ellipse to one of the foci and to one of the vertices of the ellipse along its major axis. Since the focus is at a distance of c from the center, in terms of the eccentricity, the focus is at a distance of ***ae*** from the center.

In simple terms, eccentricity denotes flatness of the ellipse. All ellipses have eccentricity values greater than or equal to zero, and less than one (0 < e <1)

Equivalently, we can also say that $e=\frac{\sqrt{a^2-b^2}}{a}\Rightarrow a^2-b^2=a^2e^2$ or $1-\frac{b^2}{a^2}=e^2$

Value of e = 0 denotes a circle as it implies that $1-\frac{b^2}{a^2}=0$ ***or b = a, reducing the ellipse to a circle. It also implies that c = 0 meaning thereby that the two foci collapse to the center and distance from center to the boundaries equals a=b, as radius.***

We can also calculate distance d of the directrix from nearest vertex, in terms of eccentricity. As already proved,

$$d = \frac{a(a-c)}{c} = a\left[\frac{a}{c} - 1\right] = a\left[\frac{1}{e} - 1\right] = a\left[\frac{1-e}{e}\right] \text{ or}$$

Distance of directrix from centre = a + d = $\frac{a}{e} = \frac{a^2}{c}$

Equation of directrix will be x= $\pm\frac{a}{e}$ *or* $\pm\frac{a^2}{c}$

7.3.2.1.2 Focal chord and Double ordinate

Any chord of the ellipse passing through the focus is called **Focal Chord**.And any chord perpendicular to the axisand whose end points lie on the ellipse is called **Double Ordinate**.

7.3.2.1.3 Latus rectum

Latus rectum of an ellipse is a Double Ordinate passing through any of the foci. Refer to figure 7.23.

Here in ΔF_1PF_2 , $F_1P^2+F_1F_2^2= PF_2^2$

Let Latus rectum PQ be 2**l**.

We know that F_1F_2=2c and F_1P+PF_2= 2a (proved earlier). Hence

$$PF_2 = \sqrt{l^2 + 4c^2}$$

$F_1P+PF_2 = l + \sqrt{l^2 + 4c^2}$ = 2a. Squaring

$$l^2+l^2+4c^2+2l\times \sqrt{l^2 + 4c^2} = 4a^2$$

$$2l^2+2l\times \sqrt{l^2 + 4c^2} = 4(a^2-c^2);\ \text{since } a^2-c^2=b^2$$

$$2l^2+2l\times \sqrt{l^2 + 4c^2} = 4b^2$$

$$2l(l + \sqrt{l^2 + 4c^2}) = 4b^2$$

$$2l \times 2a = 4b^2 \text{ or } 2l = \frac{2b^2}{a} = \frac{(2b)^2}{2a}$$

Moreover, ratio of PF_2 to distance of P to the directrix = eccentricity e

Hence latus rectum ***2l =2e*** $\times$ ***distance to the nearest.***

$\therefore$ ***Latus rectum*** $2l = \frac{2b^2}{a} = \frac{(2b)^2}{2a} = \frac{(minoraxis)^2}{majoraxis}$ $=2e\times$ ***distance to the nearest directrix***

7.3.2.2 Advanced properties of ellipse

7.3.2.2.1 Auxiliary circle of the ellipse

A circle drawn on major axis as diameter as shown in Figure 7–24 below is called Auxiliary circle. Equation of the circle is $x^2 + y^2 = a^2 \Rightarrow \frac{x^2}{a^2} + \frac{y^2}{a^2} = 1$

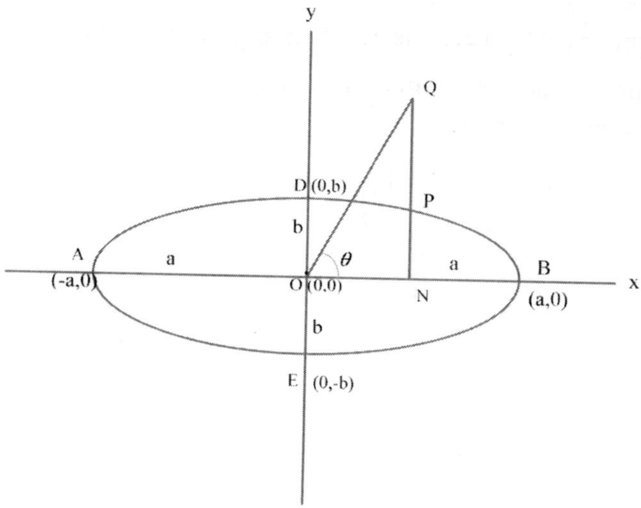

Figure 7.24

Let a perpendicular drawn on a point N on the major axis meet the ellipse and the circle at points P and Q respectively. These are called Corresponding points on the ellipse and the Auxiliary circle. θ is called Eccentric angle of the point P on ellipse $(0 \le \theta \le \pi)$

Let $ON = x,\ PN = y_1\ \&\ QN = y$.

Since P lies on the ellipse and Q on the circle, we will get

$\frac{x^2}{a^2} + \frac{y^2}{a^2} = 1$ and $\frac{x^2}{a^2} + \frac{{y_1}^2}{b^2} = 1$

From the above two equations, we get

$$\frac{y_1^2}{b^2}=\frac{y^2}{a^2}\Rightarrow\frac{y_1^2}{y^2}=\frac{b^2}{a^2}\Rightarrow\frac{PN}{QN}=\frac{b}{a}$$

Thus if from each point on a circle a perpendicular is drawn on a fixed diameter, then the locus of points dividing the perpendiculars in a given ratio is an ellipse. This is another definition of the ellipse. The circle is the Auxiliary Circle.

7.3.2.2.2 Parametric representation of the ellipse

The equations $x=a\cos\theta$ & $y=b\sin\theta$ together represent the ellipse $\frac{x^2}{a^2}+\frac{y^2}{b^2}=1$ where θ is eccentric angle and is called parameter of the ellipse. Note that if $P(\theta)\equiv(a\cos\theta, b\sin\theta)$ lies on the ellipse, $Q(\theta)\equiv(a\cos\theta, a\sin\theta)$ lies on the auxiliary circle.

7.3.2.2.3 Line intersection and tangents of an ellipse

Line $y=mx+d$ ...(1) can meet the ellipse $\frac{x^2}{a^2}+\frac{y^2}{b^2}=1$ ..... (2) at one, two or no points. We can find coordinates of the points of intersection by substituting value of y from eq (1) into eq (2). So

$$\frac{x^2}{a^2}+\frac{(mx+d)^2}{b^2}=1\Rightarrow\frac{x^2}{a^2}+\frac{m^2x^2+d^2+2mxd}{b^2}=1 \text{ or}$$

$$\left(\frac{b^2+a^2m^2}{a^2b^2}\right)x^2+\frac{2md}{b^2}x+\left(\frac{d^2-b^2}{b^2}\right)=0 \quad ......(3)$$

There will be two points of intersection if discriminant of the above quadratic equation is positive, one point of intersection implying thereby the point where the line will touch the ellipse when discriminant of the equation is 0 and no point of intersection if discriminant is negative.

$$\text{Discriminant}=\left(\frac{2md}{b^2}\right)^2-4\left(\frac{b^2+a^2m^2}{a^2b^2}\right)\left(\frac{d^2-b^2}{b^2}\right)$$

$$=4\left(\frac{a^2m^2d^2-b^2d^2-a^2d^2m^2+b^4+a^2b^2m^2}{a^2b^4}\right)$$

$$= 4\left(\frac{b^2\left(-d^2+b^2+a^2m^2\right)}{a^2b^4}\right) = 4\left(\frac{\left(-d^2+b^2+a^2m^2\right)}{a^2b^2}\right)$$

Hence sign of $-d^2+b^2+a^2m^2$ determines whether points of intersection are real (two)/ coincident (one) or imaginary (none). This leads to the conclusion that points of intersection are real (two)/ coincident (one) or imaginary (none) depending upon $b^2+a^2m^2 >= or < d^2$.

$\therefore$ ***The line is tangent to the ellipse if***

$$a^2m^2+b^2=d^2 \Rightarrow d=\pm\sqrt{a^2m^2+b^2}$$

Given the above condition, equation of the tangent to the ellipse is $y=mx\pm\sqrt{a^2m^2+b^2}$ ....(4) for all values of m.

Now, we can find point of contact of the tangent at the ellipse using

$$\text{equation(3)as } x = -\frac{\frac{2md}{b^2}}{2\times\left(\frac{b^2+a^2m^2}{a^2b^2}\right)} = -\frac{a^2md}{b^2+a^2m^2} = -\frac{a^2md}{d^2} = -\frac{a^2m}{d}$$

And substituting in the equation of the ellipse, we find

$$a^2y^2 = a^2b^2-x^2b^2 = b^2\left(a^2-x^2\right) \Rightarrow a^2y^2 = b^2\left(a^2-\left(-\frac{a^2m}{d}\right)^2\right)$$

$$= a^2b^2\left(\frac{d^2-a^2m^2}{d^2}\right)$$

$$\text{Or } y^2 = b^2\left(\frac{d^2-a^2m^2}{d^2}\right) = \frac{b^2\times b^2}{d^2} \Rightarrow y = \frac{b^2}{d}$$

Thus coordinates of the point of contact of the tangent are $\left(\frac{-a^2m}{d}, \frac{b^2}{d}\right)$.

Here d refers to the intercept of the tangent at the y–axis.

It may be noted that there are two tangents to the ellipse for each value of m implying thereby that there is a pair of parallel tangents for each value of m.

The examples of such pairs of tangents are PM/NQ and QM/NP. Eccentric angles for the points of contact of the parallel tangents QM/NP as shown in figure 7.25 are θ and π + 0, thus difference of eccentric angles for points of contact of parallel tangents is π. Conversely if the difference of eccentric angles for two points on an ellipse is π, tangents drawn at these points are parallel.

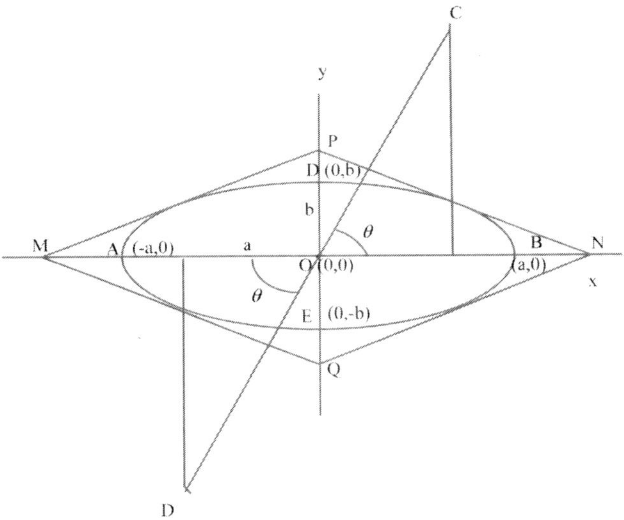

Figure 7.25

Let us find points of intersection of the tangent with slope m at x–axis and y–axis. Thus by substituting x = 0 in equation (4), we get points of intersection on y–axis as $y = \pm\sqrt{a^2m^2 + b^2}$ and by substituting y = 0 in equation (4), we get points of intersection on x–axis as $x = \mp\frac{\sqrt{a^2m^2 + b^2}}{m}$. If m is positive, slope of PM and NQ is ***m*** and slope of PN and MQ is ***–m.*** Coordinates of points P, M, N, Q are $\left(0, \sqrt{a^2m^2 + b^2}\right)$, $\left(-\frac{\sqrt{a^2m^2 + b^2}}{m}, 0\right)$, $\left(\frac{\sqrt{a^2m^2 + b^2}}{m}, 0\right)$

$\left(0,-\sqrt{a^2m^2+b^2}\right)$ respectively.

Length of the tangent PM $=\left|\sqrt{\left(\sqrt{a^2m^2+b^2}\right)^2+\left(-\frac{\sqrt{a^2m^2+b^2}}{m}\right)^2}\right|$

$$=\left(a^2m^2+b^2\right)\left|\sqrt{1+\frac{1}{m^2}}\right|=\left|\frac{a^2m^2+b^2}{m}\right|\left(\sqrt{1+m^2}\right)$$

For a given value of slope, all 4 tangents will have the same length. Hence PMNQ is a Rhombus.

7.3.2.2.4 Generic equation of tangent at point (x_1, y_1)

Equation of a line at point (x_1, y_1) is $y-y_1=m\left(x-x_1\right)$

Substituting values of coordinates of the point of contact of the tangent derived in section 7.3.2.2.3 above,

$$\frac{y_1}{x_1}=\frac{\frac{b^2}{d}}{\frac{-a^2m}{d}}=-\frac{b^2}{a^2m}\Rightarrow m=-\frac{b^2x_1}{a^2y_1} \qquad \text{.... (5)}$$

$\therefore$ Equation of the tangent is

$$y-y_1=-\frac{b^2x_1}{a^2y_1}\left(x-x_1\right)\Rightarrow a^2yy_1+b^2xx_1=a^2{y_1}^2+b^2{x_1}^2$$

Since point (x_1, y_1)lies on the ellipse, $a^2{y_1}^2+b^2{x_1}^2=a^2b^2$ reducing to the

Equation of tangent at (x_1, y_1) as

$$a^2yy_1+b^2xx_1=a^2b^2\Rightarrow\frac{xx_1}{a^2}+\frac{yy_1}{b^2}=1$$

7.3.2.2.5 Equation of tangent in parametric form

As seen in section 7.3.2.2.2 above, a generic point P on the ellipse in parametric form is $P(\theta)\equiv\left(a\cos\theta,b\sin\theta\right)$. Substituting parametric coordinates of the point of contact of tangent in equation of tangent at 7.3.2.2.4 above, we get

Equation of tangent at point $P(\theta)$ as

$$\frac{xa\cos\theta}{a^2}+\frac{yb\sin\theta}{b^2}=1 \Rightarrow \frac{x\cos\theta}{a}+\frac{y\sin\theta}{b}=1$$

7.3.2.2.6 Tangent touching latus rectum

As a special case, if the tangent touches the ellipse at the latus rectum, value of slope of that tangent can be found as below.

We know that Coordinates of end points of latus rectum are $\left(\pm c, \pm\frac{b^2}{a}\right)$ where c is distance of foci from the center. Considering one of the end points as $\left(-c, \frac{b^2}{a}\right)$

Slope ***m*** of the tangent as seen in equation (5) above $=-\frac{b^2 x_1}{a^2 y_1}=-\frac{b^2(-c)}{a^2\times\frac{b^2}{a}}=\frac{c}{a}$

$\therefore$ Points of intersection of tangents to the two axis of the ellipse and consequently vertices of the rhombus will be $\left(0, \pm\sqrt{a^2\times\frac{c^2}{a^2}+b^2}\right) \Rightarrow \left(0, \pm\sqrt{c^2+b^2}\right) \Rightarrow (0, a)$, $(0, -a)$ and

$$\left(\pm\frac{\sqrt{a^2\times\frac{c^2}{a^2}+b^2}}{\frac{c}{a}}, 0\right), \Rightarrow \left(\pm\frac{\sqrt{c^2+b^2}}{\frac{c}{a}}, 0\right) \Rightarrow \left(-\frac{a^2}{c}, 0\right), \left(\frac{a^2}{c}, 0\right).$$

This leads to length of the two diagonals of the Rhombus as $\frac{2a^2}{c}$ & $2a$

$\therefore$ ***Area of the Rhombus*** $=\frac{1}{2}\times diag1\times diag2$

$$=\frac{1}{2}\times\frac{2a^2}{c}\times 2a=\frac{2a^3}{c}=\frac{2a^2}{c/a}=\frac{2a^2}{e}$$

7.3.2.2.7 Locus of point of intersection of tangents which meet at right angles is a circle called Director Circle.

We have seen at 7.3.2.2.3 above that equation of tangent to an ellipse with slope m is

$$y = mx \pm \sqrt{a^2m^2 + b^2} \Rightarrow y - mx = \pm\sqrt{a^2m^2 + b^2} \qquad .....(1)$$

For a tangent to be perpendicular to it, its slope will be $-\frac{1}{m}$, leading to equation of that tangent to be

$$y = -\frac{1}{m}x \pm \sqrt{\frac{a^2}{m^2} + b^2} \Rightarrow my + x = \pm\sqrt{a^2 + m^2b^2} \qquad ...(2)$$

If coordinates of the point of intersection are (r, s), the above two equations of tangents will yield

$s - mr = \pm\sqrt{a^2m^2 + b^2}$ and $ms + r = \pm\sqrt{a^2 + m^2b^2}$

Squaring and adding the two equations, we get

$$\left(s^2 + m^2r^2 - 2smr\right) + \left(m^2s^2 + r^2 + 2smr\right) = \left(a^2m^2 + b^2\right) + \left(a^2 + m^2b^2\right) \Rightarrow$$

$$s^2\left(1 + m^2\right) + r^2\left(1 + m^2\right) = \left(1 + m^2\right)\left(a^2 + b^2\right) \Rightarrow s^2 + r^2 = a^2 + b^2$$

In general form, we can see that locus of point of intersection of the tangents is a circle with the equation $x^2 + y^2 = a^2 + b^2$

Thus point of intersection of tangents to an ellipse with equation $\frac{x^2}{a^2} + \frac{y^2}{b^2} = 1$ ***which meet at right angles lies on a circle with center at the origin and radius*** $\sqrt{a^2 + b^2}$

7.3.2.2.8 Chord of contact

If two tangents are drawn to an ellipse from an external point say $T(x_0, y_0)$, and these meet the ellipse at two points say P_1 and P_2, as shown in figure 7.26 below. The line joining points of contact of the two tangents P_1P_2 is called ***Chord of contact of point T.***

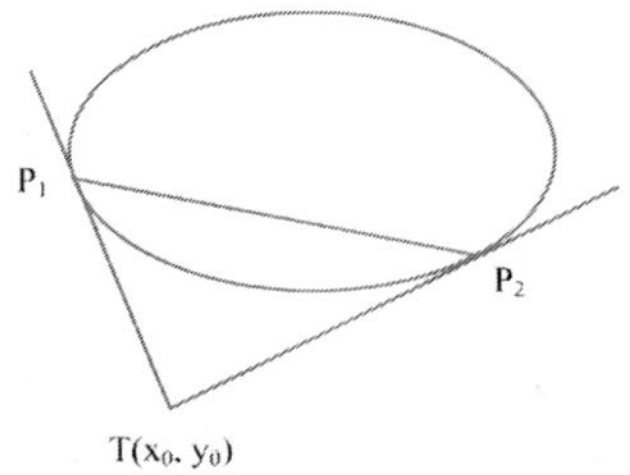

Figure 7.26

Let us now derive equation of the chord of contact. Let coordinates of points P_1 and P_2 be (x_1, y_1) and (x_2, y_2). As we know equation of tangent at point(x_1, y_1) is $\frac{xx_1}{a^2}+\frac{yy_1}{b^2}=1$

Similarly equation of tangent at point (x_2, y_2) will be $\frac{xx_2}{a^2}+\frac{yy_2}{b^2}=1$

Since both the tangents pass through point T, equation of the tangents can be rewritten as $\frac{x_0x_1}{a^2}+\frac{y_0y_1}{b^2}=1$ and $\frac{x_0x_2}{a^2}+\frac{y_0y_2}{b^2}=1$

Since Chord of contact touches both these tangents at points with coordinates (x_1, y_1) and (x_2, y_2), we can write equation of the chord of contact as $\frac{x_0x}{a^2}+\frac{y_0y}{b^2}=1$

7.3.2.2.9 Equation of the chord with a given mid-point

Refer figure 7.27. P_1P_2 is the chord with mid-point M. Let coordinates of mid–point be (x_0,y_0). Let distance to the ellipse from the mid–point be r and angle made by the chord with x–axis be θ. With reference to the x–axis, slope of line P_1MP_2 is tan θ.

Coordinates of point P_2 work out to be $\left(x_0+r\cos\theta, y_0+r\sin\theta\right)$.

Similarly, coordinates of point P_1 work out to be $\left(x_0-r\cos\theta, y_0-r\sin\theta\right)$

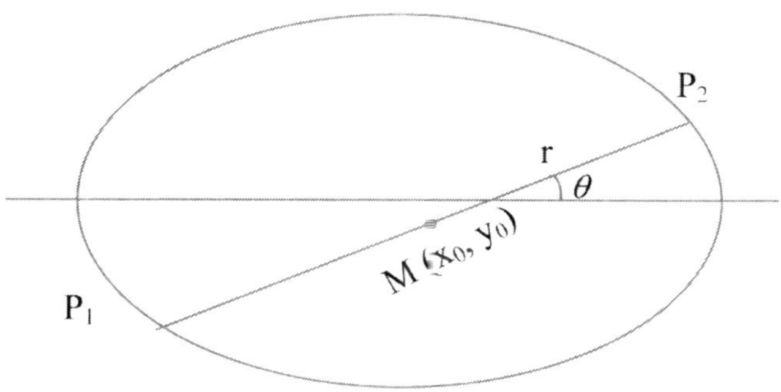

Figure 7.27

Substituting coordinates of point P_2 in the equation of the ellipse, we get

$$\frac{\left(x_0+r\cos\theta\right)^2}{a^2}+\frac{\left(y_0+r\sin\theta\right)^2}{b^2}= \Rightarrow$$

$$r^2\left(b^2\cos^2\theta + a^2\sin^2\theta\right) + r\left(2b^2x\cos\theta + 2a^2y\sin\theta\right) + \left(b^2x_0^{\,2} + a^2y_0^{\,2} - a^2b^2\right) = 0$$

This is a quadratic equation in variable r which will yield two values of r say $r_1$and r_2. Since distance of the two end points of the chord is equal and in coordinate form distance to the lower point P_1 is negative vis a vis the upper point $P_{2.}$ Hence sum of $r_1$and r_2 is 0.

From the above quadratic equation, it implies that $-\dfrac{2b^2x_0\cos\theta + 2a^2y_0\sin\theta}{\left(b^2\cos^2\theta + a^2\sin^2\theta\right)} = 0 \Rightarrow b^2x_0\cos\theta + a^2y_0\sin\theta = 0$ ***or***

$$b^2x_0\cos\theta = -a^2y_0\sin\theta \Rightarrow \tan\theta = -\frac{b^2x_0}{a^2y_0}$$

Equation of the chord will be

$$y - y_0 = -\frac{b^2x_0}{a^2y_0}\left(x - x_0\right) \Rightarrow a^2yy_0 - a^2y_0^{\,2} = -b^2xx_0 + b^2x_0^{\,2}$$

Dividing both sides by a^2b^2 and rearranging, we get

$$\frac{xx_0}{a^2} + \frac{yy_0}{b^2} = \frac{x_0^{\,2}}{a^2} + \frac{y_0^{\,2}}{b^2}$$

This is equation of the Chord with mid–point (x_0,y_0). It may also be noticed that LHS is exactly same as Chord of contact of two tangents from an external point with coordinates (x_0,y_0) and RHS is equation of the ellipse if point (x_0,y_0) lies on it. This equality is also written as $T = S_0$.

7.3.2.2.10 Equation of the Normal to an ellipse

(a) at a point (x_1, y_1)

If slope of tangent is m, slope of a normal $= \dfrac{-1}{m}$

$\therefore$ Using slope of tangent derived in eq (5) at 7.3.2.2.4 above,

slope of a normal $= \dfrac{a^2y_1}{b^2x_1}$

Hence equation of the normal at point (x_1, y_1) is

$$y - y_1 = \frac{a^2 y_1}{b^2 x_1}(x - x_1) \Rightarrow b^2 x_1 y - b^2 x_1 y_1 = a^2 x y_1 - a^2 x_1 y_1$$

$$\Rightarrow a^2 x y_1 - b^2 x_1 y = x_1 y_1 \left(a^2 - b^2\right) \Rightarrow \frac{a^2 x}{x_1} - \frac{b^2 y}{y_1} = a^2 - b^2$$

$$= c^2 = a^2 \times \frac{c^2}{a^2} = a^2 e^2$$

(b) in parametric form

If point on the ellipse is $(a\cos\alpha, b\sin\alpha)$, equation of the normal at that point will be $\frac{a^2 x}{a\cos\alpha} - \frac{b^2 y}{b\sin\alpha} = c^2 \Rightarrow ax\sec\alpha - by\,\mathrm{cosec}\,\alpha = a^2 - b^2$

(c) in slope form with slope m

We know that equation of tangent with slope t is $y = tx \pm \sqrt{a^2 t^2 + b^2}$ with point of contact at the ellipse as $\left(\frac{-a^2 t}{d}, \frac{b^2}{d}\right)$ where $d = \pm\sqrt{a^2 t^2 + b^2}$

Slope of the normal $m = -\frac{1}{t}$, therefore equation of the normal at point $\left(\frac{-a^2 t}{d}, \frac{b^2}{d}\right)$ will be $y - \frac{b^2}{d} = m\left(x + \frac{a^2 t}{d}\right)$

Since $m = -\frac{1}{t}$ and $d = \sqrt{a^2 t^2 + b^2}$, substituting these values, we get equation of the normal as

$$y - \frac{b^2}{\sqrt{a^2 t^2 + b^2}} = m\left(x + \frac{a^2 t}{\sqrt{a^2 t^2 + b^2}}\right) \Rightarrow y - \frac{b^2}{\sqrt{\frac{a^2}{m^2} + b^2}} =$$

$$m\left(x - \frac{\frac{a^2}{m}}{\sqrt{\frac{a^2}{m^2} + b^2}}\right) \Rightarrow$$

$$y = mx - \left(\frac{a^2 \alpha_1}{\sqrt{\frac{a^2}{m^2} + b^2}} - \frac{b^2}{\sqrt{\frac{a^2}{m^2} + b^2}} \right) \Rightarrow y = mx - \frac{\left(a^2 - b^2\right)m}{\sqrt{a^2 + m^2 b^2}}$$

7.3.2.2.11 A normal at a point bisects angle made by the two focal radii at that point

Reference figure 7.28 below. Let coordinates of point P be (x_1, y_1). PQ is normal at point P. The two focal radii r_1 and r_2 subtend angles θ_1 & θ_2 respectively and the Normal PQ subtend angle θ_3 with the x–axis. Let angle between r_1/r_2 and the normal be α_1/α_2 respectively.

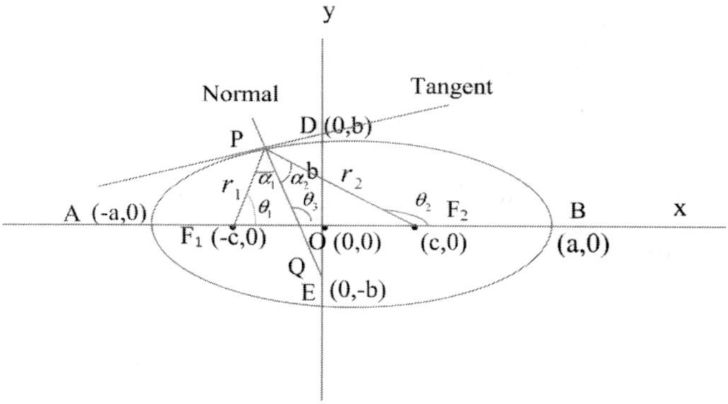

Figure 7.28

We can find angles α_1 & α_2 by using slopes of r_1, r_2 and the normal.

Slope of $r_1 = \frac{y_1}{x_1 + c}$, Slope of $r_2 = \frac{y_1}{x_1 - c}$

We have seen at 7.3.2.2.4 above that slope of the tangent at point $(x_1, y_1) = -\frac{b^2 x_1}{a^2 y_1}$

Hence slope of the normal $= \frac{a^2 y_1}{b^2 x_1}$

We have earlier seen that angle say θ between the two lines with slopes m_1, m_2 is $\tan\theta = \frac{m_1 - m_2}{1 + m_1 m_2}$

Using this formula and slopes of different lines given above, we can find that

$$\tan\alpha_1 = \frac{\frac{a^2y_1}{b^2x_1} - \frac{y_1}{x_1+c}}{1+\frac{a^2y_1}{b^2x_1}\times\frac{y_1}{x_1+c}} = \frac{a^2x_1y_1 + a^2cy_1 - b^2x_1y_1}{b^2x_1^2 + b^2cx_1 + a^2y_1^2}$$

Since point (x_1, y_1) lies on the ellipse, $b^2x_1^2 + a^2y_1^2 = a^2b^2$
Substituting in the above we get

$$\tan\alpha_1 == \frac{x_1y_1\left(a^2-b^2\right)+a^2cy_1}{a^2b^2+b^2cx_1} = \frac{x_1y_1c^2+a^2cy_1}{b^2\left(a^2+cx_1\right)} = \frac{cy_1\left(a^2+cx_1\right)}{b^2\left(a^2+cx_1\right)} = \frac{cy_1}{b^2}$$

Similarly angle between r_2 and the normal α_2 is

$$\tan\alpha_2 = \frac{\frac{y_1}{x_1-c} - \frac{a^2y_1}{b^2x_1}}{1+\frac{a^2y_1}{b^2x_1}\times\frac{y_1}{x_1-c}} = \frac{b^2x_1y_1 - a^2x_1y_1 + a^2cy_1}{b^2x_1^2 - b^2cx_1 + a^2y_1^2} = \frac{x_1y_1\left(b^2-a^2\right)+a^2cy_1}{a^2b^2-b^2cx_1}$$

$$= \frac{-x_1y_1c^2+a^2cy_1}{b^2\left(a^2-cx_1\right)} = \frac{cy_1\left(a^2-cx_1\right)}{b^2\left(a^2-cx_1\right)} = \frac{cy_1}{b^2}$$

Thus angles α_1 & α_2 are equal. ***Hence it is proved that a normal at a point bisects angle made by the two focal radii at that point***

7.3.2.2.12 Equation of the chord of an ellipse with eccentric angles **α & β *is given as*** $\frac{x}{a}\cos\frac{\alpha+\beta}{2}+\frac{y}{b}\sin\frac{\alpha+\beta}{2}=\cos\frac{\alpha-\beta}{2}$. Reference figure 7.29 below:

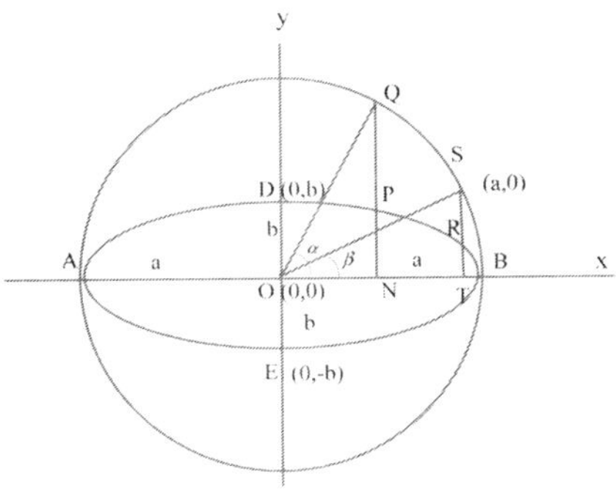

Figure 7.29

Perpendiculars QPN and SRT form eccentric angles of α & β respectively. We need to find equation of chord PR. OQ and OS each = **a,** being radii of the auxiliary circle. Thus coordinates of points Q and S are (a cos α, a sin α) and (a cos β, a sin β).

We have earlier seen in property (i) above that $\frac{PN}{QN}=\frac{b}{a}$

Since QN = $a\sin\alpha$, $PN=QN\times\frac{b}{a}=a\sin\alpha\times\frac{b}{a}=b\sin\alpha$

$\therefore$ Coordinates of point P are $(a\cos\alpha, b\sin\alpha)$. Similarly, coordinates of point R are $(a\cos\beta, b\sin\beta)$. So, equation of chord PR is

$y-b\sin\alpha=\frac{b\sin\alpha-b\sin\beta}{a\cos\alpha-a\cos\beta}(x-a\cos\alpha)$, leading to

$$(y-b\sin\alpha)(a\cos\alpha-a\cos\beta)=(b\sin\alpha-b\sin\beta)(x-a\cos\alpha)\Rightarrow$$

$$ay(\cos\alpha-\cos\beta)-ab\sin\alpha(\cos\alpha-\cos\beta)=xb(\sin\alpha-\sin\beta)-$$

$$ab\cos\alpha(\sin\alpha-\sin\beta)\Rightarrow$$

$$-2ay\left(\sin\frac{\alpha+\beta}{2}\sin\frac{\alpha-\beta}{2}\right)-2bx\left(\cos\frac{\alpha+\beta}{2}\sin\frac{\alpha-\beta}{2}\right)=$$

$$ab\{\sin\alpha(\cos\alpha-\cos\beta)-\cos\alpha(\sin\alpha-\sin\beta)\}$$

$$\Rightarrow -2\sin\frac{\alpha-\beta}{2}\left(ay\sin\frac{\alpha+\beta}{2}+bx\cos\frac{\alpha+\beta}{2}\right)=$$

$$-ab\{\sin\alpha\cos\beta-\cos\alpha\sin\beta\}$$

$$\Rightarrow -2\sin\frac{\alpha-\beta}{2}\left(ay\sin\frac{\alpha+\beta}{2}+bx\cos\frac{\alpha+\beta}{2}\right)=$$

$$-ab\{\sin(\alpha-\beta)\}=-2ab\left(\sin\frac{\alpha-\beta}{2}\cos\frac{\alpha-\beta}{2}\right)$$

Cancelling out, $\left(-2\sin\frac{\alpha-\beta}{2}\right)$, we get

$$\left(ay\sin\frac{\alpha+\beta}{2}+bx\cos\frac{\alpha+\beta}{2}\right)=ab\left(\cos\frac{\alpha-\beta}{2}\right)$$

Dividing both sides by ab, we get

$$\frac{x}{a}\cos\frac{\alpha+\beta}{2}+\frac{y}{b}\sin\frac{\alpha+\beta}{2}=\cos\frac{\alpha-\beta}{2}$$ ....Hence proved

7.3.2.2.13 Point of intersection of tangents drawn at points with eccentric angles α *&* β *is*

$$\left(a\frac{\cos\frac{\alpha+\beta}{2}}{\cos\frac{\alpha-\beta}{2}},b\frac{\sin\frac{\alpha+\beta}{2}}{\cos\frac{\alpha-\beta}{2}}\right)$$

As seen above in figure 7.29, points P and R have eccentric angles α & β respectively. We have further seen that coordinates of point P with eccentric angle α are $(a\cos\alpha,b\sin\alpha)$. Similarly, coordinates of point R with eccentric angle β are $(a\cos\beta,b\sin\beta)$

As shown in property 7.2.3.2.4 above, generic equation of tangent at point (x_1,y_1) is $\frac{xx_1}{a^2}+\frac{yy_1}{b^2}=1$. Thus equation of tangent at point P is

$$\frac{x\times a\cos\alpha}{a^2}+\frac{y\times b\sin\alpha}{b^2}=1\Rightarrow\frac{x\cos\alpha}{a}+\frac{y\sin\alpha}{b}=1 \quad ....(1)$$

Similarly equation of tangent at point R is

$$\frac{x\cos\beta}{a}+\frac{y\sin\beta}{b}=1 \quad ......(2)$$

We can find point of intersection of the above tangents by equating their equations. So, Multiplying eq (1) by $\sin\beta$ and eq (2) by $\sin\alpha$ and subtracting, we get

$$\frac{x\cos\alpha\sin\beta}{a}-\frac{x\cos\beta\sin\alpha}{a}=\sin\beta-\sin\alpha\Rightarrow$$

$\frac{x}{a}\sin(\beta-\alpha)=\sin\beta-\sin\alpha$ leading to

$$\frac{x}{a}=\frac{\sin\beta-\sin\alpha}{\sin(\beta-\alpha)}=\frac{2\cos\frac{\beta+\alpha}{2}\sin\frac{\beta-\alpha}{2}}{2\cos\frac{\beta-\alpha}{2}\sin\frac{\beta-\alpha}{2}}=\frac{\cos\frac{\beta+\alpha}{2}}{\cos\frac{\beta-\alpha}{2}}\Rightarrow$$

$$x = \frac{a\cos\frac{\alpha+\beta}{2}}{\cos\frac{\alpha-\beta}{2}}$$

To find y coordinate, multiplying eq (1) by cos β and eq (2) by cos α and subtracting, we get

$$\frac{y\sin\alpha\cos\beta}{b} - \frac{y\sin\beta\cos\alpha}{b} = \cos\beta - \cos\alpha \Rightarrow$$

$$\frac{y}{b}\sin(\alpha-\beta) = \cos\beta - \cos\alpha \text{ leading to}$$

$$\frac{y}{b} = \frac{\cos\beta - \cos\alpha}{\sin(\alpha-\beta)} = \frac{-2\sin\frac{\beta+\alpha}{2}\sin\frac{\beta-\alpha}{2}}{2\sin\frac{\alpha-\beta}{2}\cos\frac{\alpha-\beta}{2}} = \frac{\sin\frac{\beta+\alpha}{2}}{\cos\frac{\alpha-\beta}{2}} \Rightarrow$$

$$y = \frac{b\sin\frac{\alpha+\beta}{2}}{\cos\frac{\alpha-\beta}{2}}$$

From the above, coordinates of point of intersection are $\left(\frac{a\cos\frac{\alpha+\beta}{2}}{\cos\frac{\alpha-\beta}{2}}, \frac{b\sin\frac{\alpha+\beta}{2}}{\cos\frac{\alpha-\beta}{2}}\right)$

7.3.3 Parabola

A parabola is the set of all points in a plane that are equidistant from a fixed line, called Directrix and a fixed point (not on the line) called Focus in the plane. Refer to the figure 7.30. Here by definition $P_1B_1 = P_1F$ and $P_2B_2 = P_2F$. This is similar to the definition of eccentricity in an ellipse. Since the two distances are equal, eccentriciy in case of parabola is 1.

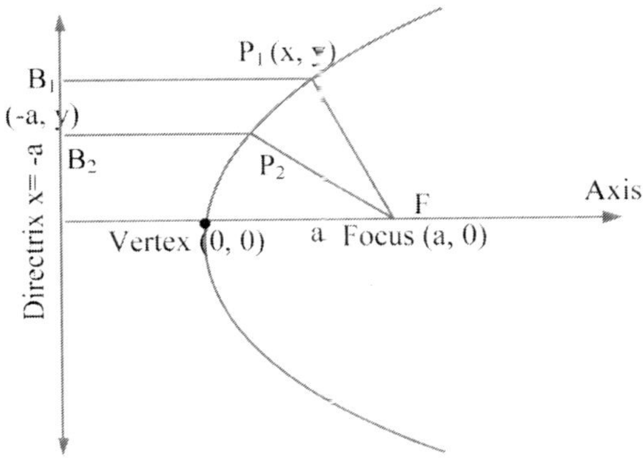

Figure 7.30

As shown, the parabola is assumed to be symmetrical about x–axis, vertex lies on center of the co–ordinate axis and distance of the Focus from vertex is **a**. Coordinates of the Focus will then be (a, 0) and equation of the Directrix will be x= –a.Vertex is thus mid– point of the distance from Focus to the Directrix.

Focus has another property that all lines joining the parabola parallel to the axis, converge on to the focus. This is known as Reflective property. We will discuss this property in section 7.4.3.1.2

Using formula for distance between two points,

$P_1B_1 = (x-(-a))^2 + (y-y)^2 = (x+a)^2$ and

$P_1F = (x-a)^2 + (y-0)^2 = (x-a)^2 + y^2$

$\therefore (x-a)^2 + y^2 = (x+a)^2$ or $y^2 = (x+a)^2 - (x-a)^2 = 4ax$

Thus generic equation of a parabola symmetrical about x–axis and vertex at origin is $y^2 = 4ax$

Here we notice that the x coordinate is always positive. Even though y coordinate is negative in quadrant 4, it does not matter in the equation as y coordinate is squared.

However, when parabola lies in quadrant 2 and 3 with Focus (–a, 0) and Directrix x=a, x coordinate being negative and y^2 being positive, the equation of parabola is $y^2 = -4ax$

Similarly, when parabola is symmetrical about y–axis and is located in quadrant 1/2 or quadrant 3/4, coordinates of focus will be (0,–a) or (0, a) and equation for Directrix y=–a and y=a respectively. Hence equation of parabola is $x^2 = 4ay$ or $x^2 = -4ay$ respectively

If the vertex is located at an arbitrary point say (g, h) and a is distance of the Focus from vertex, we can similarly find equation of the parabola symmetrical to x–axis as $(y-h)^2 = 4a(x-g)$ and symmetrical to y–axis as $(x-g)^2 = 4a(y-h)$

7.3.3.1 Properties of Parabola

Here we will assume parabola symmetrical to the x–axis having vertex at (0,0) with equation $y^2 = 4\,ax$ in quadrant 1 and 4. Properties will be similar for parabola symmetrical to x–axis in quadrant 2 and 3 and those symmetrical to y–axis. Properties will just be listed for these cases.

7.3.3.1.1 Tangents to a parabola

(a) at a given point

Let us try and find equation of tangent to parabola $y^2 = 4ax$ at a point (x_1, y_1)

Let there be two points $P(x_1, y_1)$ and $Q(x_2, y_2)$ on the parabola as shown below:

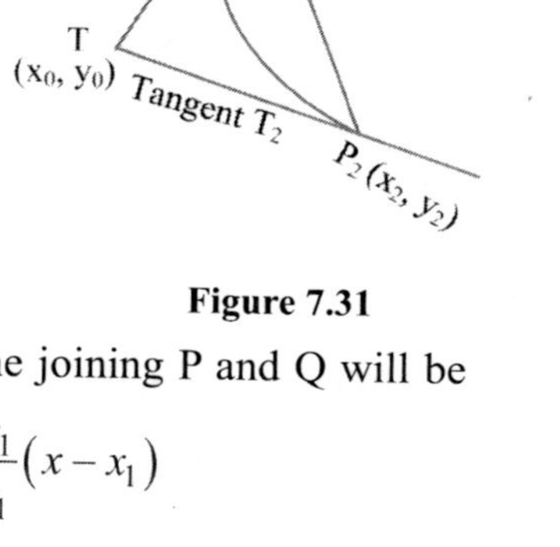

Figure 7.31

Equation of line joining P and Q will be

$$y - y_1 = \frac{y_2 - y_1}{x_2 - x_1}(x - x_1) \qquad \text{.....(1)}$$

where $\frac{y_2 - y_1}{x_2 - x_1}$ is slope of this line.

Since points P and Q lie on the parabola, $y_1^2 = 4ax_1$ ....(2) and $y_2^2 = 4ax_2$ ....(3)

Subtracting (2) from (3),

$$y_2^2 - y_1^2 = 4a(x_2 - x_1) \Rightarrow \frac{y_2 - y_1}{x_2 - x_1} = \frac{4a}{y_2 + y_1}$$

Substituting in equation (1), we get equation for line PQ as

$$y - y_1 = \frac{4a}{y_2 + y_1}(x - x_1) \qquad \ldots(4)$$

For line PQ to be tangent to the parabola, point Q must approach point P leading to $y_2 \to y_1$ & $x_2 \to x_1$

Substituting $y_2 = y_1$ & $x_2 = x_1$ in equation (4), we get slope of the line as

$$\frac{4a}{y_2 + y_1} = \frac{4a}{2y_1} = \frac{2a}{y_1}$$

$$\therefore y - y_1 = \frac{2a}{y_1}(x - x_1) \Rightarrow yy_1 = y_1^2 + 2ax - 2ax_1 \Rightarrow yy_1 - 2ax + 2ax_1 = y_1^2$$

Subtracting $4ax_1$ on both sides, we get

$$yy_1 - 2ax + 2ax_1 - 4ax_1 = y_1^2 - 4ax_1 \Rightarrow yy_1 - 2a(x + x_1) = y_1^2 - 4ax_1 \qquad ..(1)$$

This equation is also written as T = S where T= $yy_1 - 2a(x + x_1)$ and S = $y_1^2 - 4ax_1$

Since $y_1^2 - 4ax_1 = 0$ for point (x_1, y_1) lying on the parabola,

> ***Equation of the tangent can be written as*** $yy_1 - 2a(x + x_1) = 0$ ***(T=0) at the point*** (x_1, y_1)

(b) in slope form

General equation of a straight line with slope m is

$$y = mx + c \qquad \ldots..(1)$$

Substituting in the equation for the parabola, we get

$$(mx + c)^2 = 4ax \Rightarrow m^2x^2 + 2x(mc - 2a) + c^2 = 0 \qquad \ldots(2)$$

We can find x–coordinates (abscissa) of the points of intersection of this line on the parabola from the above quadratic equation. However since the above line needs to only touch the parabola, value of x–coordinate should be only a single value leading to the conclusion that the discriminant of the quadratic equation (2) should be zero.

Hence

$$2^2(mc-2a)^2-4m^2c^2=0 \Rightarrow m^2c^2+4a^2-4amc-m^2c^2=0 \Rightarrow c=\frac{a}{m}$$

Thus equation of the tangent in slope form is $y=mx+\frac{a}{m}$ ......(3)

We also know from section 7.3.3.1.1 that slope of the tangent at point (x, y) is $m=\frac{2a}{y} \Rightarrow y=\frac{2a}{m}$.

Substituting in equation (3), we get

$$\frac{2a}{m}=mx+\frac{a}{m} \Rightarrow m^2x=a \Rightarrow x=\frac{a}{m^2}$$

> Thus ***Equation of the tangent in slope form is*** $y=mx+\frac{a}{m}$ ***that meets the parabola at the point*** $\left(\frac{a}{m^2},\frac{2a}{m}\right)$. ***We can also assume,*** $\frac{1}{m}=t$ ***leading to the point of contact as*** $\left(at^2,2at\right)$ ***and Equation of the tangent as*** $yt=x+at^2$ ***'t' is known as parameter of the parabola.***

Equation of the chord joining two points with parameters t_1 and t_2 is

$$y-2at_1=\frac{2a(t_1-t_2)}{a\left(t_1^2-t_2^2\right)}\left(x-at_1^2\right) \Rightarrow y\left(t_1+t_2\right)-2at_1\left(t_1+t_2\right)=2x-2at_1^2$$

> ***Or Equation of the chord joining two points with parameters t_1 and t_2 is*** $2x-y\left(t_1+t_2\right)+2at_1t_2=0$

If two chords joining points t_1t_2 and t_3t_4 intersect on the x–axis at point (c, 0), we can see that $2c+2at_1t_2=0 \Rightarrow t_1t_2=\frac{-c}{a}$ ***and*** $2c+2at_3t_4=0 \Rightarrow t_3t_4=\frac{-c}{a}$

Thus $t_1t_2=t_3t_4=\frac{-c}{a}$

(c) from an external point and its Chord of contact

If two tangents are drawn to a parabola from an external point say T, these will meet the parabola at two points say P_1 and P_2, as shown below. The line joining the points of contact of the two tangents P_1P_2 is called ***Chord of contact of point T.***

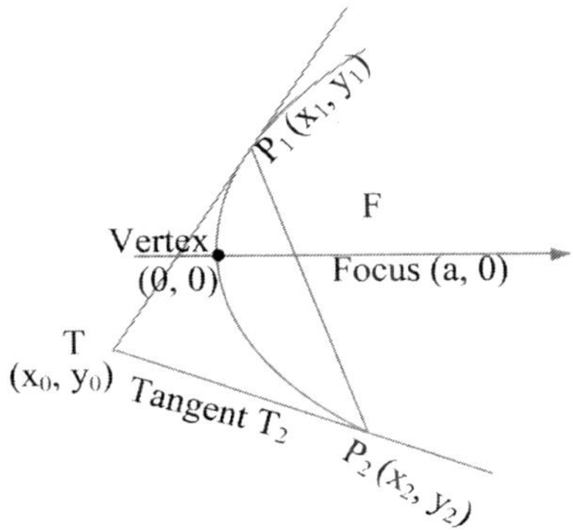

Figure 7.32

Let coordinates of points P_1 and P_2 be (x_1, y_1) and (x_2, y_2) and that of the point of intersection Tbe (x_0, y_0) and let $y_0 = mx_0 + \frac{a}{m}$ be any tangent to the parabola $y^2 = 4ax$ passing through the point of intersection (x_0, y_0) with slope m.

We get $x_0m^2 - y_0m + a = 0$ ...(1)

If m_1 and m_2 are slopes of the two tangents intersecting at point T, the two roots of equation (1) will give the two slopesm_1 and m_2 leading to

$$m_1 + m_2 = \frac{y_0}{x_0} \text{ and } m_1m_2 = \frac{a}{x_0} \qquad ...(2)$$

A few important observations are:

Slope of the tangent from point P_2 (m_2) is negative and that from point P_1 (m_1) is positive. Hence product $m_1m_2$is always negative.

The point of intersection will lie outside the parabola and hence will have x–coordinate negative and y–coordinate may be positive or negative.

Given slope of the tangents being m_1 & m_2, equations of the two tangents will be

$y_1 = m_1x_1 + \frac{a}{m_1}$ and $y_2 = m_2x_2 + \frac{a}{m_2}$ with their points of contact at the parabola as derived in section 7.3.3.1.2 $\left(at_1^2, 2at_1\right) \& \left(at_2^2, 2at_2\right)$ where $t_1 = \frac{1}{m_1}$ and $t_2 = \frac{1}{m_2}$

Substituting values of m_1 & m_2 in equation (2) above,

$$\frac{1}{t_1}+\frac{1}{t_2}=\frac{y_0}{x_0}\Rightarrow\frac{t_1+t_2}{t_1t_2}=\frac{y_0}{x_0} \text{ and } \frac{1}{t_1t_2}=\frac{a}{x_0} \text{ leading to}$$

$$\frac{a(t_1+t_2)}{x_0}=\frac{y_0}{x_0}\Rightarrow y_0=a(t_1+t_2) \text{ and } x_0=at_1t_2 \qquad ....(3)$$

Substituting back coordinate values in terms of (x_1, y_1) and (x_2, y_2), we get $y_0=at_1+at_2=\dfrac{y_1+y_2}{2}$ and $x_0=\sqrt{at_1^{\,2}\times at_2^{\,2}}=\sqrt{x_1x_2}$

Arithmetically x_0 could be positive or negative but since point of intersection lies outside the parabola, x_0 will always be negative and x_1 as well as x_2 are always positive. Therefore,the coordinates of point of intersection T are $\left(-\sqrt{x_1x_2}, \dfrac{(y_1+y_2)}{2}\right)$.

We can note that x–coordinate and y–coordinate of point of intersectionof the two tangents are negative geometric mean of the x–coordinates and arithmetic mean of y–coordinates of the two points of contact respectively.

Given the points of contact of the two tangents on the parabola being $(at_1^{\,2}, 2at_1)\,\&\,(at_2^{\,2}, 2at_2)$,

Equation of Chord of contact will be

$$y-2at_1=\frac{2at_2-2at_1}{at_2^{\,2}-at_1^{\,2}}\left(x-at_1^{\,2}\right)\Rightarrow y-2at_1=\frac{2}{t_1+t_2}\left(x-at_1^{\,2}\right)$$

$$\text{Or } y=2at_1+\frac{2x}{t_1+t_2}-\frac{2at_1^{\,2}}{t_1+t_2}\Rightarrow y=\frac{2x}{t_1+t_2}+2at_1\left(1-\frac{t_1}{t_1+t_2}\right)\Rightarrow$$

$$y=\frac{2ax}{a(t_1+t_2)}+\frac{2a(at_1t_2)}{a(t_1+t_2)}$$

If we Substitute coordinates of the point of intersection of the tangents $(at_1t_2, a(t_1+t_2))$ derived in equation (3) above, we get

$$y=\frac{2ax}{y_0}+\frac{2ax_0}{y_0}$$

Leading to the equation of chord of contact of point (x_0, y_0) as $yy_0 = 2a(x + x_0) \Rightarrow yy_0 - 2a(x + x_0) = 0$

Interestingly, this equation is exactly same as the equation of tangent to the parabola from an external point (x_0, y_0). Essential difference is that point (x, y) in case of the tangent lies anywhere on the parabola whereas in case of chord of contact it lies anywhere on the Chord of contact.

(c) Chord of a parabola with given midpoint coordinates as (x_0, y_0)

Coordinates of the midpoint of the chord with end point coordinates as $\left(at_1^2, 2at_1\right) \& \left(at_2^2, 2at_2\right)$ are $x_0 = \dfrac{a\left(t_1^2 + t_2^2\right)}{2}$ and $y_0 = a(t_1 + t_2)$.

Slope of the chord is $\dfrac{2}{(t_1 + t_2)}$, which equals $\dfrac{2a}{y_0}$.

Therefore, equation of the required chord with mid–point (x_0, y_0) is

$$y - y_0 = \frac{2a}{y_0}(x - x_0) \Rightarrow yy_0 - y_0^2 = 2a(x - x_0) \Rightarrow yy_0 - y_0^2 + 4ax_0$$

$$= 2a(x - x_0) + 4ax_0$$

$$\Rightarrow yy_0 - 2a(x + x_0) = y_0^2 - 4ax_0 \qquad \ldots(1)$$

LHS of this equation is again similar to the equation of the Chord of contact of an external point (x_0, y_0) with the exception that coordinates used are of the mid–point in place of point of intersection.

We have already seen above that equation of a tangent at point (x_0, y_0) on the parabola is $yy_0 - 2a(x + x_0) = 0$ …… (2) and for any generic point (x, y), $y^2 - 4ax = 0$. Substituting in equation (2) we get $yy_0 - 2a(x + x_0) = y^2 - 4ax$ ……….(3)

Multiplying equations (1) and (3), we get

$$\left(yy_0 - 2a(x + x_0)\right)^2 = \left(y_0^2 - 4ax_0\right)\left(y^2 - 4ax\right)$$

LHS represents product of equation of tangent T to the parabola at point (x_0, y_0) and RHS represents two parabolic curves – one with generic point (x,y) on the parabola and the other with point of intersection of tangents drawn from chord of contact/ mid–point of chord of the parabola (x_0, y_0).

$\Rightarrow T^2 = SS_0$. where T is equation of the tangent/ chord of contact, $S = y^2 - 4ax$ and $S_0 = y_0^{\ 2} - 4ax_0$

(d) Another interesting property is that locus of mid-point of any system of parallel chords of a parabola is called its DIAMETER and is parallel to the axis of the parabola and meets it at a point the tangent at which is parallel to the given system of parallel chords.

Refer to figure 7.33 below showing a parabola symmetrical about x-axis with equation $y^2 = 4ax$

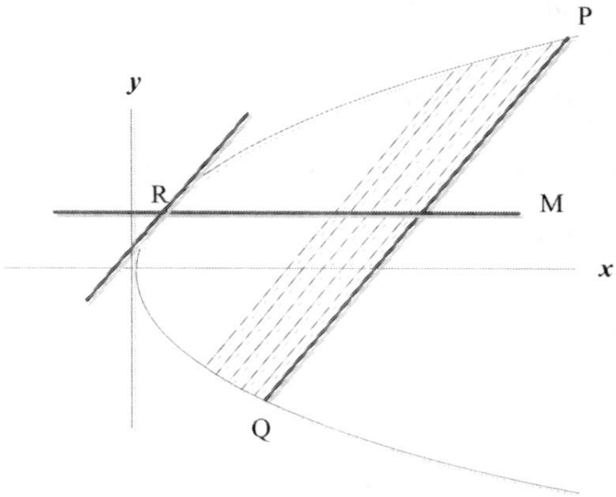

Figure 7.33

Since chord PQ and all other chords are parallel, they make the same angle with x-axis. Hence let their slope be m leading to equation of chord PQ and any other parallel chord as $y = mx + c$ where c is different and m is same for all the chords. Let RM be the line joining mid-points of the chord joining points P $(at_1^{\ 2}, 2at_1)$ and Q $(at_2^{\ 2}, 2at_2)$ that meets the parabola at point R (h, k).

Hence $k = \dfrac{2a(t_1 + t_2)}{2} = a(t_1 + t_2) \Rightarrow (t_1 + t_2) = \dfrac{k}{a}$ ......(1)

And slope of this chord as derived above is

$$m = \frac{2}{(t_1 + t_2)} \Rightarrow (t_1 + t_2) = \frac{2}{m}$$

Substituting value of $(t_1 + t_2)$ in equation (1), we get

$$\frac{2}{m} = \frac{k}{a} \Rightarrow k = \frac{2a}{m}$$

Since slope m of all parallel lines is equal, ordinate of mid-point of all lines will be equal.

This leads us to the equation of the DIAMETER (locus of mid-point of system of parallel chords with slope m) as $y = \frac{2a}{m}$, a line parallel to the x-axis.

Since this straight line meets the parabola at P, it yields $\left(\frac{2a}{m}\right)^2 = 4ah \Rightarrow h = \frac{a}{m^2}$

Tangent at this point $\left(\frac{a}{m^2}, \frac{2a}{m}\right)$ is

$$y\frac{2a}{m} - 2a\left(x + \frac{a}{m^2}\right) = 0 \Rightarrow 2amy - 2am^2x - 2a^2 = 0$$

$$y = mx + \frac{a}{m}$$

Thus equation of the tangent at point P has the same slope m as the system of parallel chords and is therefore parallel to the given set of chords...........***Hence proved.***.

7.3.3.1.2 *Normals to the parabola*

Like tangents, we can find equations of the normal (line segment perpendicular to the tangent) to a parabola from a given point in different forms.

Slope of the normal at point P (x, y) will be

$$= -\frac{1}{slope\ of\ tangent} = \frac{-1}{2a/y} = \frac{-y}{2a},$$

(a) at a given point (x_1, y_1)

Equation of the normal at point (x_1, y_1) will be
$y - y_1 = -\frac{y_1}{2a}(x - x_1) \Rightarrow 2ay + xy_1 - y_1(2a - x_1) = 0$

(b) in slope form

If slope of the normal is considered m, slope of the tangent will be $\frac{-1}{m}$ which would imply that point of contact of the normal will be

$\dfrac{a}{\left(-\frac{1}{m}\right)^2} = am^2$ and $\dfrac{2a}{\left(-\frac{1}{m}\right)} = -2am$ leading us to

Equation of the Normal with slope m as
$y + 2am = m(x - am^2) \Rightarrow y = mx - 2am - am^3$ ***at*** $(am^2, -2am)$

(c) Parametric equation of the normal

Slope m of the normal with reference to the parameter t is, m = –t

Hence point of contact of the normal with slope m will be $\left(at^2, 2at\right)$ ***leading to***

Equation of the Normal with parameter t as
$y - 2at = -t(x - at^2) \Rightarrow y + tx = 2at + at^3$ ***at*** $\left(at^2, 2at\right)$

(d) Point of intersection of normal drawn at two points of the parabola

If normal is drawn to a parabola at each of the points with parameter t_1 and t_2, their point of intersection can be found as below:

$y + t_1x - 2at_1 - at_1^{\,3} = y + t_2x - 2at_2 - at_2^{\,3}$. leading to

$$x\left(t_1 - t_2\right) = a\left(2\left(t_1 - t_2\right) + \left(t_1^{\,3} - t_2^{\,3}\right)\right) = a\left(t_1 - t_2\right)\left(2 + t_1^{\,2} + t_2^{\,2} + t_1t_2\right)$$

Or $x = a\left(2 + t_1^{\,2} + t_2^{\,2} + t_1t_2\right)$

Substituting in the equation for the normal say at t_1,

$y + at_1\left(2 + t_1^{\,2} + t_2^{\,2} + t_1t_2\right) = 2at_1 + at_1^{\,3}$ giving value of

$$y = -at_1\left(t_2^{\,2} + t_1t_2\right) = -at_1t_2\left(t_1 + t_2\right)$$

So the point of intersection is $\left(a\left(2 + t_1^{\,2} + t_2^{\,2} + t_1t_2\right), -at_1t_2\left(t_1 + t_2\right)\right)$

(e) If Normal to a parabola at a point with parameter t_1 meets the parabola again at point with parameter t_2, then t_2 in terms of t_1 is $t_2 = -\left(t_1 + \dfrac{2}{t_1}\right)$.

Equation of normal at point t_1 is

$$y + t_1x = 2at_1 + at_1^{\,3} \Rightarrow y - 2at_1 = t_1\left(at_1^{\,2} - x\right)$$

For point t_2 to lie on the above normal, gives us

$$2at_2 - 2at_1 = t_1\left(at_1^2 - at_2^2\right) \Rightarrow 2a\left(t_2 - t_1\right) = at_1\left(t_1^2 - t_2^2\right)$$

$$\Rightarrow 2 = -t_1\left(t_1 + t_2\right)$$

Rearranging we get value of t_2 in terms of t_1 as

$$t_2 = \frac{-2}{t_1} - t_1 \Rightarrow t_2 = -\left(t_1 + \frac{2}{t_1}\right)$$

(f) If Normal to a parabola at points with parameter t_1 & t_2 meets the parabola again at point with parameter t_3, then $t_1t_2=2$, $t_3 = -(t_1+t_2)$ and line joining t_1 & t_2 passes through a fixed point (–2a, 0)

For point with parameter t_3 to lie on normal at points with parameter t_1 & t_2, we can use the result arrived above to get

$$t_3 = -\left(t_1 + \frac{2}{t_1}\right) \ldots\ldots (1) \quad \& \quad t_3 = -\left(t_2 + \frac{2}{t_2}\right) \ldots\ldots\ldots (2)$$

Subtracting (1) and (2),

$$-\left(t_1 + \frac{2}{t_1}\right) = -\left(t_2 + \frac{2}{t_2}\right) \Rightarrow t_1 - t_2 = \frac{2}{t_2} - \frac{2}{t_1} = \frac{2\left(t_1 - t_2\right)}{t_1t_2}$$

Cancelling out $t_1 - t_2$ (since $t_1 \neq t_2$) gives $1 = \frac{2}{t_1t_2} \Rightarrow t_1t_2 = 2$

Adding (1) and (2),

$$2t_3 = -\left(t_1 + \frac{2}{t_1}\right) - \left(t_2 + \frac{2}{t_2}\right) = -\left(\left(t_1 + t_2\right) + \frac{2\left(t_1 + t_2\right)}{t_1t_2}\right)$$

Since $t_1t_2 = 2$, we get

$$2t_3 = -\left(\left(t_1 + t_2\right) + \frac{2\left(t_1 + t_2\right)}{2}\right) = -2\left(t_1 + t_2\right) \Rightarrow t_3 = -\left(t_1 + t_2\right)$$

Equation of line joining normal at t_1 and t_2

$$y - 2at_1 = \frac{2at_1 - 2at_2}{at_1^2 - at_2^2}\left(x - at_1^2\right) \Rightarrow y - 2at_1 = \frac{2}{t_1 + t_2}\left(x - at_1^2\right)$$

Or $y\left(t_1 + t_2\right) - 2at_1\left(t_1 + t_2\right) = 2x - 2at_1^2 \Rightarrow y\left(t_1 + t_2\right) = 2x + 2at_1t_2 \ldots(3)$

We have seen above that $t_1t_2 = 2$ which implies that t_1 & t_2 are of the same sign i.e. they lie on the same side of x–axis. The above line once extended will cross the x–axis at some point. At that point y = 0.

Substituting this value of y and of t_1t_2 above in the equation (3) above, we get

$$2x + 2a \times 2 = 0 \Rightarrow x = -2a$$

Thus the line joining normal at t_1 and t_2 passes through point (–2a, 0)

(g) If a normal to the parabola $y^2 = 4ax$ passes through point (h,k), find equation of the normal and prove that

(i) Sum of the slopes of three concurrent normals is 0

(ii) Sum of the ordinates of three co–normal points is zero

(iii) Centroid of the Δ formed by three co–normal points lies on the x–axis

Equation of the normal at a given point in slope form as derived in property 2 above is $y = mx - 2am - am^3 \Rightarrow am^3 + m(2a - x) + y = 0$

And slope is related to the parameter of the point of normal as t = –m, giving point coordinates as $(am^2, -2am)$

Since the normal passes through point (h,k), the equation of the normal reduces to $am^3 + m(2a - h) + k = 0$,a cubic equation in m.

Any cubic equation has three roots. Hence three values of slopes say m_1, m_2, m_3 will satisfy this equation implying thereby that there will be three concurrent normals and coordinates of the three co–normal points will be

$$(am_1^2, -2am_1), (am_2^2, -2am_2), (am_3^2, -2am_3)$$

The three roots of the cubic equation will be related as below:

$m_1 + m_2 + m_3 = 0$ (since the m^2 term is 0)……(1)

$m_1m_2 + m_2m_3 + m_3m_1 = \dfrac{2a - h}{a}$ …..(2) and $m_1m_2m_3 = \dfrac{-k}{a}$ …(3)

(i) We can see from the equation (1) above that Sum of the slopes of three concurrent normals is 0

(ii) Since ordinate is $-2a \times slope$, we also observe that sum of the ordinates of three co–normal points is zero

(iii) Coordinates of the Centroid of a Δ formed by 3 points with coordinates (x_1, y_1), (x_2, y_2) and (x_3, y_3) is $\left(\frac{x_1+x_2+x_3}{3}, \frac{y_1+y_2+y_3}{3}\right)$

Substituting values of the coordinates of the three points, coordinates of the Centroid are

$$\left(\frac{a\left(m_1^2+m_2^2+m_3^2\right)}{3}, \frac{-2a\left(m_1+m_2+m_3\right)}{3}\right) \Rightarrow \left(\frac{a\left(m_1^2+m_2^2+m_3^2\right)}{3}, 0\right),$$

implying thereby that the centroid lies on the x–axis.

We know that

$$\left(m_1+m_2+m_3\right)^2 = m_1^2+m_2^2+m_3^2+2\left(m_1m_2+m_2m_3+m_3m_1\right)$$

Since $m_1+m_2+m_3=0$, it leads to

$$m_1^2+m_2^2+m_3^2+2\left(m_1m_2+m_2m_3+m_3m_1\right)=0 \text{ or}$$

$$m_1^2+m_2^2+m_3^2=-2\left(m_1m_2+m_2m_3+m_3m_1\right)=-2\left(\frac{2a-h}{a}\right)$$

Hence abscissa of the centroid is $\frac{a}{3}\times-2\left(\frac{2a-h}{a}\right)=\frac{2(h-2a)}{3}$

The centroid is thus $\left(\frac{2(h-2a)}{3}, 0\right)$

7.3.3.1.3 *Reflective property of parabola*

Parabola has the property that a line (ray) meeting the parabola at any point reflects to it focus. This property is used extensively in satellite communication by use of parabolic dish antennae.

Proof:

Refer to the diagram 7.34 below:

We can see line P_1P_2, parallel to the axis of the parabola meeting directrix at point P_2 (perpendicular to it) and the parabola at point P having generic coordinates $\left(at^2, 2at\right)$. F is focus of the parabola with PF as focal radius.

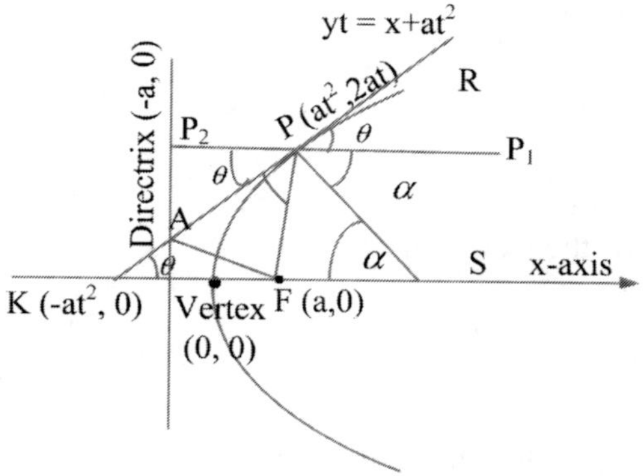

Figure 7.34

A tangent and a normal are drawn at P meeting x–axis at K and S respectively.

Equation of the tangent being $yt = x + at^2$, coordinates of point K are $\left(-at^2, 0\right)$ with slope $\frac{1}{t}$

Coordinates of point S can be found by equating slope of the normal to –t.

Thus $\frac{2at - 0}{at^2 - s} = -t \Rightarrow st = 2at + at^3 \Rightarrow s = a\left(2 + t^2\right)$

$\therefore$ Coordinates of point S (y = 0), are $\left(a\left(2+t^2\right), 0\right)$

We can further see that $\angle RPP_1$ is assumed as θ and $\angle SPP_1$ is assumed as α

$\angle P_2PK = \angle RPP_1 = \theta$ (vertically opposite angle) and $\angle P_2PK = \angle PKF = \theta$ (alternate angle)

And $\angle SPP_1 = \angle PKF = \alpha$ (alternate angle)

Let us now find lengths of the line segments PF, FK and FS

PF = $\sqrt{\left(at^2 - a\right)^2 + \left(2at\right)^2} = \sqrt{a^2t^4 + a^2 - 2a^2t^2 + 4a^2t^2} = \sqrt{\left(at^2 + a\right)^2} = a\left(t^2 + 1\right)$

FK = $a - (-at^2) = a + at^2 = a\left(t^2 + 1\right)$

FS = $a\left(2 + t^2\right) - a = a + at^2 = a\left(t^2 + 1\right)$

Thus PF = FK = FS implying thereby that ΔPFK and ΔPFS are isosceles Δs

This leads us to $\angle SPF = \alpha$ and $\angle KPF = \theta$

Hence $\angle P_2PK = \angle KPF = \theta$ and $\angle SPP_1 = \angle SPF = \alpha$

We can now conclude that:

1. ***If tangent and normal at any point P of a parabola intersect the axis at K and S, then FK = FS = PF where F is focus of the parabola.***
2. ***Tangent and normal at any point P of a parabola are the bisectors of the angle between Focal radius PF and perpendicular from P to the directrix.***
3. ***A circle drawn with focal radius as diameter will touch the tangent drawn at the vertex.***
4. ***A ray striking the parabola at any point (P) parallel to the axis of the parabola making an angle say α to the normal, will get reflected making the angle of reflection α to meet the focus of the parabola. Thus all rays striking the parabola at any point parallel to the axis of the parabola will converge at the focus.***

7.3.3.1.4 *A few other interesting features are that:*

(a) A circle on any focal radii of a point P (at^2, 2at) as diameter touches the tangent drawn at the vertex and intercepts a chord of length $a\sqrt{t^2+1}$ on normal at point P

(b) Any tangent to a parabola and perpendicular to it from Focus meet on the tangent drawn at the vertex and its distance to mid–point of focal radius = half of Focal radius

(c) Point P, Focus (F), Point of intersection of the tangent at P and tangent at the vertex, R and intercepted chord on the normal at P, PM form a rectangle of side lengths $a\sqrt{t^2+1}$ and $at\sqrt{t^2+1}$

(d) Portion of the tangent between the curve and the directrix subtends a right angle at the focus

Refer to figure 7.35 below for all the above properties

Proof for (a):

PF is Focal radius and the diameter of the circle with center C. PR is tangent at P meeting y–axis (tangent at the vertex) at R. PS is normal at point P meeting the circle at M.

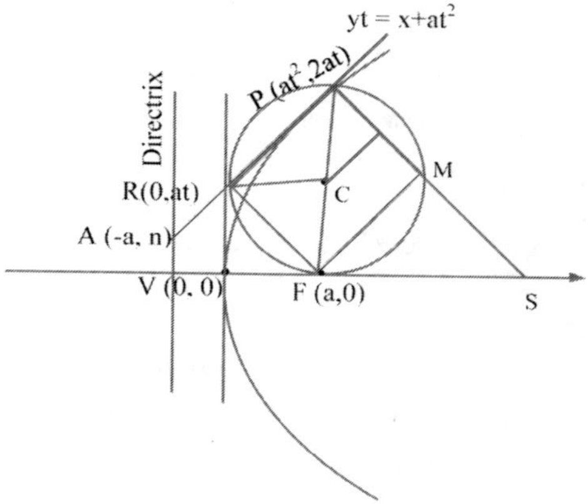

Figure 7.35

Length of focal radius

$$= \sqrt{\left(at^2 - a\right)^2 + \left(2at\right)^2} = a\sqrt{t^4 + 1 - 2t^2 + 4t^2} = a\left(t^2 + 1\right)$$

Coordinates of the center of the circle C are

$$\left(\frac{at^2 + a}{2}, \frac{2at}{2}\right) \Rightarrow \left(\frac{a\left(t^2 + 1\right)}{2}, at\right)$$

Slope of the tangent and the normal at P are $\frac{1}{t}$ and –t respectively.

Hence equation of the tangent and `normal at P are respectively

$$y - 2at = \frac{1}{t}\left(x - at^2\right) \Rightarrow yt = x + at^2$$

$$y - 2at = -t\left(x - at^2\right) \Rightarrow y + xt = 2at + at^3$$

Equation of the circle with given end points will be $(x-a)\left(x-at^2\right) + y(y-2at) = 0$.

We can find points of intersection of this circle with tangent at the vertex by substituting x = 0 (equation of tangent line at the vertex)

This leads us to $(-a)\left(-at^2\right) + y(y-2at) = 0 \Rightarrow y^2 - 2ayt + a^2t^2 = 0$ or $(y-at)^2 = 0$ implying thereby that ***the circle meets the tangent at***

vertex, only at one point R (0, at), implying thereby that the circle just touches the tangent at the vertex.

We can now find length of the intercept on the normal PM. Since M lies in the semicircle, $\angle PMF$ is right angle. Hence FM will be parallel to the tangent with slope $\frac{1}{t}$ and its equation will be $y = \frac{1}{t}(x-a) \Rightarrow yt - x + a = 0$

We can find length of the chord PM as distance from P to the line FM as

$$\frac{\left|2at \times t - at^2 + a = 0\right|}{\sqrt{1+t^2}} = \frac{a\left(1+t^2\right)}{\sqrt{1+t^2}} = a\sqrt{1+t^2}$$ ...... ***Hence proved***

Proof for (b):

We have seen in (a) above that FR is perpendicular to the tangent at point P and R lies on the tangent drawn at the vertex. Since point P is any arbitrary point with parameter t, we conclude that ***any tangent to a parabola and perpendicular to it from Focus meet on the tangent drawn at the vertex.***

Further we have seen in (a) above that coordinates of C (mid–point of the focal radius) and point R are $\left(\frac{a\left(t^2+1\right)}{2}, at\right)$ ***and (0, at) respectively.*** Since ordinate of the center of the Focal radius C is same as that of the point R, distance CR = $\frac{a\left(t^2+1\right)}{2}$

We have seen above that length of Focal radius $= a\left(t^2+1\right)$

Hence any tangent to a parabola and perpendicular to it from Focus meet on the tangent drawn at the vertex and its distance to mid–point of focal radius = half of Focal radius

Proof for (c):

We have seen that $\angle PRF = \angle PMF = 90^0$ (being angles formed in the semi–circle) and $\angle RPM = 90^0$, normal being perpendicular to the tangent. Hence RPMF is a rectangle leading to PM = RF and PR = MF.

Length PM has already been derived in (a) above as $a\sqrt{1+t^2}$

Length MF can be found by the distance from focus F to the normal at point P, which has equation $y+xt-\left(2at+at^3\right)=0$

Using distance formula from F (a, 0) to the normal PM ($y+xt-\left(2at+at^3\right)=0$), we find

$$\text{MF} = \frac{\left|at-\left(2at+at^3\right)\right|}{\sqrt{1+t^2}} = \frac{\left|-at-at^3\right|}{\sqrt{1+t^2}} = \frac{at\left(1+t^2\right)}{\sqrt{1+t^2}} = at\sqrt{1+t^2}$$

Hence point P, Focus (F), Point of intersection of the tangent at P and tangent at the vertex, R and intercepted chord on the normal at P, PM form a rectangle of side lengths $a\sqrt{t^2+1}$ ***and*** $at\sqrt{t^2+1}$

Proof for (d):

The tangent at P meets the directrix at A (Figure 7.34 above).

Let coordinates of A be (–a, n). Substituting in the equation of the tangent, we get

$$nt = -a+at^2 \Rightarrow n = a\left(\frac{t^2-1}{t}\right)$$

Thus coordinates of point A are $\left(-a, a\left(\frac{t^2-1}{t}\right)\right)$

Let slope of the line segment PF be m_1 and slope of the line segment AF be m_2

$$\therefore m_1 = \frac{2at}{at^2-a} = \frac{2t}{t^2-1} \text{ and}$$

$$\therefore m_2 = \frac{a\left(\frac{t^2-1}{t}\right)}{-a-a} = \frac{t^2-1}{t}\times\frac{1}{-2} = -\frac{t^2-1}{2t}$$

$$m_1 \times m_2 = \frac{2t}{t^2-1}\times\frac{-\left(t^2-1\right)}{2t} = -1$$

Thus line segments PF and AF are perpendicular. ***In other words, portion of the tangent between the curve and the directrix (PA) subtends a right angle at the focus F.***

7.3.3.1.5 *Focal Chord of a parabola*

A chord passing through the focal point is called Focal chord of the parabola. Refer figure 7.36 below.

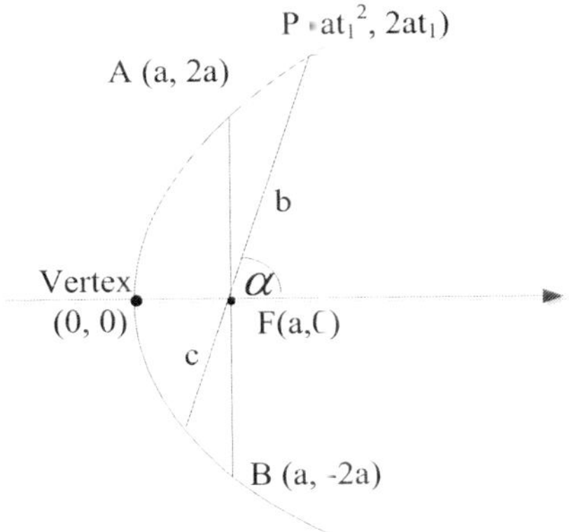

Figure 7.36

Since the chord passes through the focus (a, 0) of the parabola $y^2 = 4ax$, equation of the chord is $y - 0 = m(x-a) \Rightarrow y = m(x-a)$

Given points of intersection of the chord with the parabola as $\left(at_1^2, 2at_1\right) \& \left(at_2^2, 2at_2\right)$, equation of the focal chord is

$$y = \frac{2a(t_1 - t_2)}{a(t_1^2 - t_2^2)}(x-a) \Rightarrow y = \frac{2}{(t_1 + t_2)}(x-a) \Rightarrow$$

$$y = \frac{2a}{y_0}(x-a) \Rightarrow yy_0 = 2a(x-a),$$

where y_0 will be ordinate of the mid-point of the focal chord or point of intersection of tangents drawn from the points of contact of the focal chord with the parabola.

If we compare above equation of the Focal Chord with equation of the chord in the form of Chord of contact ($yy_0 - 2a(x + x_0) = 0$), we find that $x - a = x + x_0 \Rightarrow x_0 = -a$

This implies that tangents drawn from end point of any Focal Chord will meet on the line x = –a, which is nothing but the Directrix.

We also know that slope of the tangent at point with parameter t is $\frac{1}{t}$ and (as seen in section 7.4.3.1.3) that abscissa of the point of intersection of the tangents from points with parameters $t_1 \& t_2$ is at_1t_2. Thus in case of Focal Chord, $at_1t_2 = -a \Rightarrow t_1t_2 = -1 \Rightarrow \frac{1}{t_1} \times \frac{1}{t_2} = -1$

This shows that two tangents from ends of any focal chord meet at right angles at the Directrix. Thus locus of the point of intersection of perpendicular tangents is the Directrix.

It can also be concluded that a circle drawn with any focal chord as diameter will touch the directrix and is also called Director Circle of the parabola.

7.3.3.1.6 *Length of focal chord given angle made with x–axis*

Let angle formed with x–axis be α as shown in figure 7.36 above.

Equating slopes of line segment PF and FQ, we get

$$\frac{2at_1}{at_1^2 - a} = \frac{2at_2}{at_2^2 - a} \Rightarrow \frac{t_1}{t_1^2 - 1} = \frac{t_2}{t_2^2 - 1}$$

Cross multiplying and rearranging we get

$$t_1t_2^2 - t_2t_1^2 = t_1 - t_2 \Rightarrow t_1t_2\left(t_2 - t_1\right) \Rightarrow t_1t_2 = -1$$

$t_1t_2 = -1$ is an important relation. We can now treat parameters of the points of contact of a focal chord as $t \& \frac{-1}{t}$ leading to coordinates of the points of contact as

$$\left(at^2,\ 2at\right) \text{ and } \left(\frac{a}{t^2}, \frac{-2a}{t}\right)$$

Length of the Focal chord

$$PQ = PF + FQ = \sqrt{(2at)^2 + \left(at^2 - a\right)^2} + \sqrt{\left(\frac{-2a}{t}\right)^2 + \left(a - \frac{a}{t^2}\right)^2}$$

$$= a\left(\sqrt{4t^2 + t^4 + 1 - 2t^2} + \sqrt{\frac{4}{t^2} + 1 + \frac{1}{t^4} - \frac{2}{t^2}}\right)$$

$$= a\left(\sqrt{\left(t^2 + 1\right)^2} + \sqrt{\left(\frac{1}{t^2} + 1\right)^2}\right) = a\left(t^2 + 2 + \frac{1}{t^2}\right)$$

$$= a\left(t + \frac{1}{t}\right)^2 \quad ...(1)$$

Since angle formed to the x–axis is α, slopes of line segment PF = $\tan\alpha$

$$\therefore \tan\alpha = \frac{2t}{t^2 - 1} \Rightarrow t^2 \tan\alpha - 2t - \tan\alpha = 0$$

Solving this quadratic equation, we get

$$t = \frac{2 \pm \sqrt{4 + 4\tan^2\alpha}}{2\tan\alpha} = \frac{1 \pm \sec\alpha}{\tan\alpha}$$

Substituting one of the values of t in equation (2), we get

Length of the Focal chord

$$PQ = a\left(\frac{1+\sec\alpha}{\tan\alpha} + \frac{\tan\alpha}{1+\sec\alpha}\right)^2$$

$$= a\left(\frac{1+\sec^2\alpha + 2\sec\alpha + \tan^2\alpha}{\tan\alpha(1+\sec\alpha)}\right)^2 = a\left(\frac{2\sec\alpha(1+\sec\alpha)}{\tan\alpha(1+\sec\alpha)}\right)^2$$

∴ ***Length of the Focal chord with angle α to x–axis***

$$= a\left(\frac{2\sec\alpha}{\tan\alpha}\right)^2 = 4a\,\mathrm{cosec}^2\alpha$$

The same result will be achieved by using the other value of t.

7.3.3.1.7 *Angle between the two tangents*

As noted above slope m_1 is positive and slope m_2 is negative. Slope corresponds to the angle made by the tangent to the horizontal. Hence angle made by tangent with slope m_1 (say α_1) is positive and that with slope m_2 (say α_2) is negative. Since angle between the two tangents (say θ) is numerical sum of the two angles and since angle α_2 has negative value, which leads to $\theta = \alpha_1 - \alpha_2$.

Using angle sum property,

$$\tan\theta = \tan(\alpha_1 - \alpha_2) = \frac{\tan\alpha_1 - \tan\alpha_2}{1 + \tan\alpha_1\tan\alpha_2} = \frac{m_1 - m_2}{1 + m_1 m_2}$$

Depending upon values of m_1 and m_2, angle θ may be acute, obtuse or right angle.

7.3.3.1.8 *Latus Rectum*

In case the Chord of contact is perpendicular to the x–axis at the focal point, Coordinates of the end points of the chord will be (a, y) and (a,–y).This chord is known as ***Latus Rectum of the parabola*** as shown in figure 7.37 below.

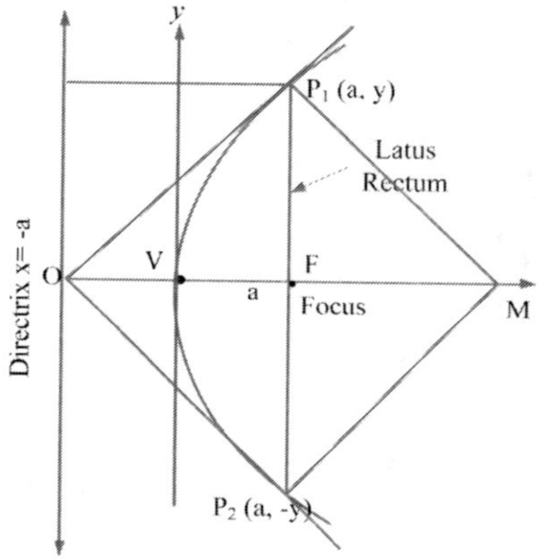

Figure 7.37

Length of Latus rectum will be 2y. Now substituting coordinates of point P_1 in the equation of parabola, we find $y^2 = 4a \times a \Rightarrow y = \pm 2a$ Thus y = P_1F = P_1B_1 = 2a. Points of contact of the Latus Rectum with parabola will thus be (a, 2a) and (a, –2a) and ***Length of Latus rectum=P_1P_2= 4a. It also equals twice the distance from focus to the Directrix.***

7.3.3.1.8.1 *Properties of Latus rectum:*

(a) ***Points of intersection of tangents and normal at end points of the Latus rectum form a square with Latus rectum as one of its diagonals.***

Slope of the tangents drawn from end points of the Latus Rectum will equal $\pm\frac{2a}{y} = \pm\frac{2a}{2a} = \pm 1$

This shows that each tangent drawn at end points of the Latus Rectum will subtend an angle of 45⁰ at the point of intersection as well as with the latus rectum . Coordinates of the point of intersection will be $\left(-\sqrt{a\times a},\dfrac{2a-2a}{2}\right)=(-a,0)$ implying that the point of intersection lies on the junction of the x–axis and the Directrix and the Latus rectum subtends an angle of 90^0 at the junction.

Now normal at points P_1 and P_2 being perpendicular to the tangents, will also subtend an angle of 45^0 with the latus rectum and will also meet on the x–axis at point M at a distance of 2a from the focusfor similar reasons. Thus all 4 angles at points P_1, P_2, O and M are of 90^0 each, thereby P_1P_2OM is a square. Point of intersection of the tangents and normals are respectively (–a, 0) and (3a, 0). Side length of the square is $\sqrt{(2a)^2+(2a)^2}=2\sqrt{2}a$ and diagonal length P_1P_2 (latus rectum) = OM = 4a

(b) ***Semi Latus Rectum of the parabola is harmonic mean between the two segments of any focal chord.***

Proof:

Let the two segments of a focal chord be b and c respectively as shown in Figure 7.36 above. And let h be the harmonic mean of b and c.

$$\therefore \frac{1}{h}=\frac{1}{2}\left(\frac{1}{b}+\frac{1}{c}\right)=\frac{c+b}{2bc}\Rightarrow h=\frac{2bc}{b+c}$$

As seen in section 7.4.3.1.6, relationship between parameters of the points of contact P and Q of a focal chord t_1, t_2 are related as $\boldsymbol{t_1 t_2 = -1}$.

Length of segment PF (b) =

$$\sqrt{\left(at_1^{\,2}-a\right)^2+(2at_1)^2}=a\sqrt{t_1^{\,4}+1-2at_1^{\,2}+4at_1^{\,2}}=a\left(t_1^{\,2}+1\right)$$

Length of segment QF (c) =

$$\sqrt{\left(a-at_2^{\,2}\right)^2+(-2at_2)^2}=a\sqrt{t_2^{\,4}+1-2at_2^{\,2}+4at_2^{\,2}}=a\left(t_2^{\,2}+1\right)$$

$$\therefore h = \frac{2bc}{b+c} = \frac{2\left[a\left(t_1^2+1\right)\times a\left(t_2^2+1\right)\right]}{a\left(t_1^2+1\right)+a\left(t_2^2+1\right)} = \frac{2a^2\left[\left(t_1^2t_2^2+t_1^2+t_2^2+1\right)\right]}{a\left(t_1^2+t_2^2+2\right)}$$

Since $t_1t_2 = -1$, $h = \frac{2bc}{b+c} = \frac{2a^2\left[\left(1+t_1^2+t_2^2+1\right)\right]}{a\left(t_1^2+t_2^2+2\right)} = 2a =$ Semi Latus rectum.

Hence proved.

(c) Two parabolas are said to be equal if they have equal Latus Rectum.

Another common term is ***Double ordinate*** which is line segment perpendicular to axis of parabola. In case the double ordinate is at the focus of parabola, it is same as the ***Latus rectum. Tangents at end points of any Double Ordinate will also make equal angles and the point of intersection will lie on mid–point of the Directrix. Angle will be obtuse if the chord is positioned between the Focus and vertex and acute if the chord is positioned beyond the Focus.***

7.3.3.2 Other interesting properties of the Parabola

Length of the Chord intercepted by the parabola $y^2 = 4ax$ on line $y = mx + c$ is $\left(\frac{4}{m^2}\right)\sqrt{a\left(1+m^2\right)\left(a-mc\right)}$

Proof:

To find points of intersection of the line with the given parabola, let us substitute y from the line to the parabola leading to

$$\left(mx+c\right)^2 = 4ax \Rightarrow m^2x^2 + (2mc-4a)x + c^2 = 0$$

Roots of this equation will give x–coordinates of the two points of intersection say (x_1,y_1) and (x_2,y_2)

$$\therefore x_1 + x_2 = \frac{2(2a-mc)}{m^2} \text{ and } x_1x_2 = \frac{c^2}{m^2} \quad ...(1)$$

$$\text{Length of the chord} = \sqrt{\left(y_2-y_1\right)^2+\left(x_2-x_1\right)^2} \quad ...(2)$$

Since line y = mx + c touches the parabola at points (x_1,y_1) and (x_2,y_2)

$y_1 = mx_1 + c$ and $y_2 = mx_2 + c$, giving us $y_2 - y_1 = m(x_2 - x_1)$

Substituting in equation (2),

Length of the chord =

$$\sqrt{(m(x_2 - x_1))^2 + (x_2 - x_1)^2} = (x_2 - x_1)\sqrt{1+m^2} \quad \text{...(3)}$$

$$(x_2 - x_1) = \sqrt{(x_2 + x_1)^2 - 4x_2x_1} = \sqrt{\left(\frac{2(2a - mc)}{m^2}\right)^2 - \frac{4c^2}{m^2}} =$$

$$\frac{2}{m^2}\sqrt{(2a - mc)^2 - c^2m^2}$$

$$= \frac{2}{m^2}\sqrt{\left(4a^2 + m^2c^2 - 4amc\right) - c^2m^2} = \frac{2}{m^2}\sqrt{4a^2 - 4amc} = \frac{4}{m^2}\sqrt{a(a - mc)}$$

Going back to equation (3), Length of the chord

$$= \frac{4}{m^2}\sqrt{a(a - mc)} \times \sqrt{1+m^2}$$

$$= \frac{4}{m^2}\sqrt{a(a - mc)\left(1 + m^2\right)} \,..$$

Hence proved

(b) Area of a triangle formed by any three points on a parabola is twice the area of the triangle formed by tangents at these points

Proof:

Let parameters of the three points on the parabola be t_1, t_2 & t_3.

$\therefore$ Coordinates of the three points will be

$$\left(at_1^2, 2at_1\right), \left(at_2^2, 2at_2\right), \left(at_3^2, 2at_3\right)$$

And coordinates of the points of intersection of the three tangents will be

$$\left(-at_1t_2, a(t_1 + t_2)\right),\ \left(-at_2t_3, a(t_2 + t_3)\right) \text{ and } \left(-at_3t_1, a(t_3 + t_1)\right)$$

Area of a triangle formed by points $(x_1, y_1), (x_2, y_2), (x_3, y_3) =$

$$\frac{1}{2}\left|x_1(y_2 - y_3) + x_2(y_3 - y_1) + x_3(y_1 - y_2)\right|$$

Substituting coordinate values of the 3 points on the parabola, we get Area of triangle formed by these 3 points =

$$\frac{1}{2}\left|at_1^2(2at_2 - 2at_3) + at_2^2(2at_3 - 2at_1) + at_3^2(2at_1 - 2at_2)\right|$$

$$=\frac{2a^2}{2}\left|t_1^{\,2}(t_2-t_3)+t_2^{\,2}(t_3-t_1)+t_3^{\,2}(t_1-t_2)\right|=$$

$$a^2\left(t_1t_2(t_1-t_2)+t_2t_3(t_2-t_3)+t_3t_1(t_3-t_1)\right)\text{....(1)}$$

Substituting coordinate values of the 3 points of intersection of the tangents, we get area of triangle formed by the 3 tangents equals

$$\frac{1}{2}\left|-at_1t_2\left(a(t_2+t_3)-a(t_3+t_1)\right)-at_2t_3\left(a(t_3+t_1)-a(t_1+t_2)\right)-\right.$$

$$\left.at_3t_1\left(a(t_1+t_2)-a(t_2+t_3)\right)\right|$$

$$=\frac{1}{2}\left|-a^2t_1t_2(t_2-t_1)-a^2t_2t_3(t_3-t_2)-a^2t_3t_1(t_1-t_3)\right| \qquad \text{...(2)}$$

Comparing equations (1) and (2), clearly area in eq (1) = 2 times area in Eq (2)

Hence Area of triangle formed by any three points on the parabola = 2 times Area of triangle formed by tangents at these three points

7.3.4 Hyperbola

Among conic sections, Hyperbola is somewhat tricky. Its components and coordinate formula resemble those of an ellipse but with fine differences. In case of an ellipse, it is easy to visualize parameters **a, b** and **c**, but in hyperbola, while **a** (distance from center to the vertex) and **c** (distance from center to the focal point) can be understood, parameter **b** (half length of minor axis in ellipse) is more abstract here. It is also a little complex to arrive at formulae for different parameters like directrix, eccentricity, latus rectum etc. Hyperbola is therefore treated in detail for ease in understanding.

7.3.4.1 Definition

(a) Refer to figure 7.38. Hyperbola is a set of two conical curves that are mirror image of each other with reference toan axis called Conjugate axis. Each curve has a Focal point. Thus there are two foci marked F’ and F. The mid–point of the line segment joining the foci is called Center of the Hyperbola, marked O. The line through the foci is called the transverse axis and the line through the center and perpendicular to the transverse axis is called the conjugate axis. Distance of the focal point from center is assumed as **c** as in the ellipse.

(b) For simplicity, let us assume a hyperbola centered at the center of coordinate axis. Thus coordinates of the two foci are (–c, 0) and (c, 0). Let coordinates of P be (x, y).

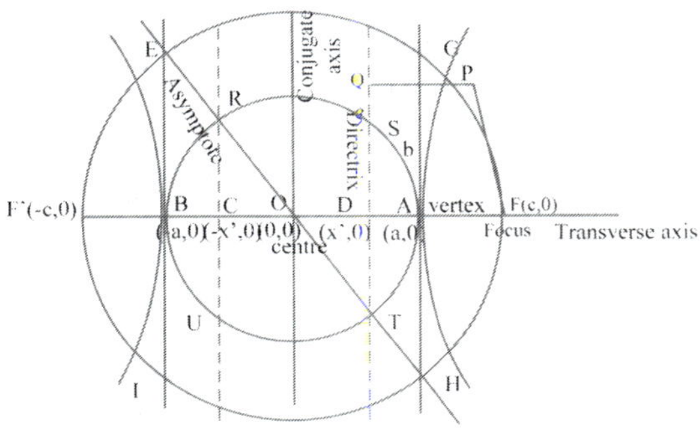

Figure 7.38

(c) The points where the Transverse axis meets the two curves are known as Vertex, marked A & B. Distance from the centre to each vertex is assumed as **a,** similar to the case of ellipse**.** Coordinates of the vertices will thus be (–a,0) and (a,0)

(d) If a circle is drawn with O as center & **c** as radius and two tangents are drawn, one on each vertex, the circle meets these tangents at 4 points, marked E, G, H, I. Vertical distance from the transverse axis to the points (like segment BE) where the circle meets the tangent is assumed as **b** (this is at variance to the case of ellipse). Parameter **b** in hyperbola is known as half–length of conjugate axis Note that length OE=**c**, being radius of the circle. In Δ OBE where BE=**b** and OB=**a**, $a^2+b^2=c^2$, the identity same as in an ellipse. Thus coordinates of point E,G,H,I are ***(–a,b), (a,b), (a,–b) and (–a,–b)*** respectively.

(e) Straight lines joining opposite points are called Asymptotes. Asymptotes as they extend to infinity will approach nearer and nearer to the curve but never touch it. Equation for the tangents are $x = -a$ and $x = a$ respectively. Slope of the asymptotes is $\frac{-b}{a}$ and $\frac{b}{a}$ respectively. Since the asymptotes are passing through the centre, equations for the asymptotes EH and GI are, therefore $y=\frac{-b}{a}x$ and $y=\frac{b}{a}x$ respectively.

f. Now if we draw a circle (called Auxiliary circle) with center of hyperbola as center and radius **a**, the asymptotes meet the auxiliary circle at R, S, T and U. Distance of all these 4 points is **a** from centre **O,** being radius of the auxiliary circle. Lines drawn from these points parallel to the conjugate axis are known as Directrices. Distance from centre to the point where a Directrix meets the transverse axis is assumed as **x′**. Thus equation of the directrices are x = –**x′** and x = **x′**.

Δs OCR and OBE are similar. Hence

$$\frac{OC}{OB} = \frac{OR}{OE}$$

or $\frac{x'}{a} = \frac{a}{c}$. Thus $x' = \frac{a^2}{c}$

$\therefore$ Equation of directrices are $x = -\frac{a^2}{c}$ and $x = \frac{a^2}{c}$ respectively

7.3.4.2 Properties

(a) ***Difference of the distance of any point on the hyperbola from the two foci is always constant.*** The term “difference” means the distance to the farther point (F’) minus the distance to the closer point (F).

As seen in figure 7.38, AF′= c + a and AF= c – a.

Thus, AF′–AF= c + a – (c –a) = c + a –c +a = 2a

(b) ***As in case of an ellipse, distance from any point on the hyperbola to the focus and perpendicular distance of that point to the respective directrix is in a fixed ratio. This ratio is eccentricity of the hyperbola and is denoted by e.***

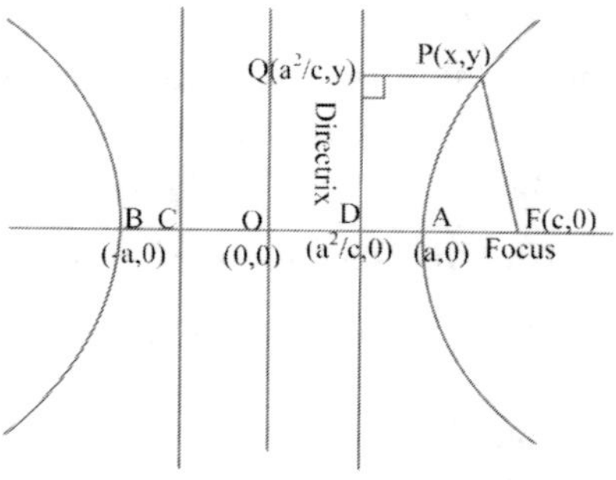

Figure 7.39.

Refer to figure 7.39. P is any point on the hyperbola and PF is drawn to the Focus and PQ is drawn perpendicular to the Directrix, $\frac{PF}{PQ}=e.$

By definition, $\frac{FA}{AD}$ also equals e.

FA= $c-a$, and AD = $a-\frac{a^2}{c}$

$\therefore$ e= $\frac{c-a}{a-\frac{a^2}{c}}=\frac{c-a}{\frac{a}{c}(c-a)}=\frac{1}{\frac{a}{c}}=\frac{c}{a}$; or **$c=ae$.**

This also leads to $e^2=\frac{c^2}{a^2}=\frac{a^2+b^2}{a^2}=1+\frac{b^2}{a^2}$

> ***Thus eccentricity of a hyperbola is also the ratio of the distances from the centre of the hyperbola to one of the foci and to the corresponding vertex of the hyperbola, same as in an ellipse.***
>
> ***Note that value of e for hyperbola is always >1***

7.3.4.3 Equation for hyperbola

Any of the above two properties can be used to derive formula for the hyperbola.

Consider property (a) above and any arbitrary point P (x,y) on the hyperbola. As per the property,

PF' – PF = 2a

PF'= $\sqrt{(c+x)^2+y^2}$ and PF = $\sqrt{(c-x)^2+y^2}$

PF' – PF $=\sqrt{(c+x)^2+y^2}-\sqrt{(c-x)^2+y^2}=2a$ ... (1)

Multiplying by $\sqrt{(c+x)^2+y^2}+\sqrt{(c-x)^2+y^2}$ on both sides and transposing,

$$2a\times\left(\sqrt{(c+x)^2+y^2}+\sqrt{(c-x)^2+y^2}\right)=(c+x)^2+y^2-(c-x)^2-y^2=4cx$$

$$\sqrt{(c+x)^2+y^2}+\sqrt{(c-x)^2+y^2}=\frac{2cx}{a} \quad ...(2)$$

Adding equations (1) and (2),

$\sqrt{(c+x)^2+y^2}=\frac{1}{2}\times\left(2a+\frac{2cx}{a}\right)=\left(a+\frac{cx}{a}\right)$; squaring both sides

$$c^2 + x^2 + 2cx + y^2 = a^2 + \frac{c^2x^2}{a^2} + 2cx \text{ or}$$

$$x^2\left(1 - \frac{c^2}{a^2}\right) + y^2 = a^2 - c^2; \text{ simplifying further}$$

$$\frac{x^2}{a^2}(a^2 - c^2) + y^2 = a^2 - c^2 \text{ or}$$

$\dfrac{x^2}{a^2} + \dfrac{y^2}{a^2 - c^2} = 1$; substituting value of $a^2 – c^2 = – b^2$ from the common identity mentioned above,

> Equation of the hyperbola reduces to
> $$\frac{x^2}{a^2} - \frac{y^2}{b^2} = 1$$

Considering property (b) above,

$$\frac{PF}{PQ} = e$$

PF = $\sqrt{(c - x)^2 + y^2}$ and PQ = $x - \dfrac{a^2}{c}$

$$e = \frac{\sqrt{(c - x)^2 + y^2}}{x - \dfrac{a^2}{c}} \text{ or}$$

$\sqrt{(c - x)^2 + y^2} = ex - \dfrac{a^2e}{c} = ex - a$; squaring both sides

$$(c - x)^2 + y^2 = e^2x^2 + a^2 - 2aex$$

$c^2 + x^2 - 2cx + y^2 = \dfrac{c^2x^2}{a^2} + a^2 - 2cx$; transposing and cancelling common term,

$$x^2\left(1 - \frac{c^2}{a^2}\right) + y^2 = a^2 - c^2; \text{ simplifying further}$$

$$\frac{x^2}{a^2}(a^2 - c^2) + y^2 = a^2 - c^2 \text{ or}$$

$\dfrac{x^2}{a^2} + \dfrac{y^2}{a^2 - c^2} = 1$; substituting value of $a^2 – c^2 = – b^2$

Equation of the hyperbola again reduces to

$$\frac{x^2}{a^2}-\frac{y^2}{b^2}=1$$

7.3.4.4 Tangent and Normal at a point

Let us find equation of tangent and normal at a point (x_1, y_1).

Concept of differentiation is needed for finding slope of a curve. Refer Chapter 8 for differentiation concepts before going through the following derivation.

Differentiating the equation of the hyperbola $\frac{x^2}{a^2}-\frac{y^2}{b^2}=1$, we get

$$\frac{2x}{a^2}-\frac{2y\frac{dy}{dx}}{b^2}=0 \Rightarrow \frac{dy}{dx}=\frac{b^2x}{a^2y}$$

So slope of the tangent,m,at point$(x_1, y_1)=\frac{b^2x_1}{a^2y_1}$ leading to equation of the tangent as

$$y-y_1=\frac{b^2x_1}{a^2y_1}(x-x_1) \Rightarrow a^2yy_1-b^2xx_1=a^2y_1^{\ 2}-b^2x_1^{\ 2} \text{ or}$$

Dividing by $-a^2b^2$, we get equation of the tangent as

$$\frac{x_1x}{a^2}-\frac{y_1y}{b^2}=\frac{x_1^{\ 2}}{a^2}-\frac{y_1^{\ 2}}{b^2}=1$$

Slope of normal at (x_1, y_1) = $-\frac{1}{m}=-\frac{a^2y_1}{b^2x_1}$ leading to equation of the tangent as

$$y-y_1=-\frac{a^2y_1}{b^2x_1}(x-x_1) \Rightarrow b^2x_1y-b^2x_1y_1=-a^2xy_1+a^2x_1y_1 \text{ or}$$

$$a^2y_1x+b^2x_1y-x_1y_1\left(a^2+b^2\right)=0$$

7.3.4.5 Latus rectum

The chord of the hyperbola through its focus and perpendicular to the transverse axis (or parallel to the directrix) is called the latus rectum of thehyperbola, shown as FE in figure 7.40. Let it be denoted as ***2l***.

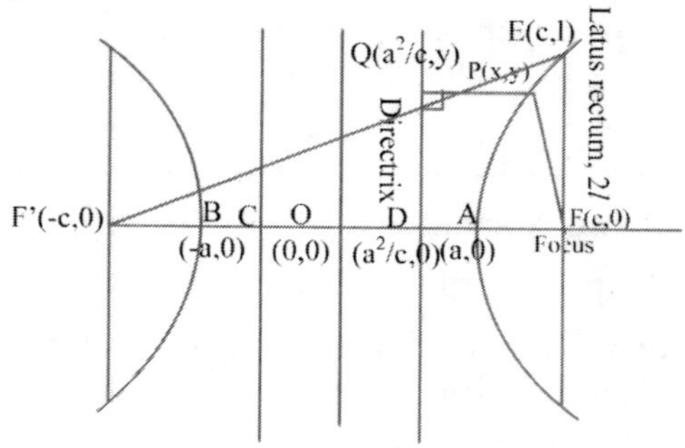

Figure 7.40

As per property of hyperbola,

F′E–FE= 2a

$$F'E = \sqrt{(2c)^2 + l^2} \text{ and } FE = l,$$

Hence $\sqrt{(2c)^2 + l^2} - l = 2a$ or

$\sqrt{(2c)^2 + l^2} = 2a + l$; squaring both sides

$$4c^2 + l^2 = 4a^2 + l^2 + 4al$$

$4(c^2 - a^2) = 4al$; ***since*** $c^2 - a^2 = b^2$

$4b^2 = 4al$; ***or***

Latus rectum $2l = \dfrac{2b^2}{a}$

7.4 Generic equation of Conics:

We have studied equations for different type of conics. A generic expression of a conic section in x–y coordinate system is defined as $Ax^2 + Bxy + Cy^2 + Dx + Ey + F = 0$ .... (1) where $A \neq 0$, $B \neq 0$ & $C \neq 0$.

Result of $B^2 - 4AC$ determines the type of the conic section obtained. It is important to know the differences in the equations to help quickly identify the type of conic that is represented by a given equation.

- If B^2-4AC is < 0 and if a conic exists, it will be an ellipse. Further if B^2-4AC is less tha 0 and if B = 0 and A = C, the equation will be a circle.
- If $B^2-4AC = 0$ and if a conic exists, it will be a parabola.
- If B^2-4AC is > 0 and if a conic exists, it will be bea hyperbola. As a special case
- if A + C =0, then we have a rectangular Hyperbola.

Through rotation of axes causing a change of coordinates, the above generic equation changes to

$$A'x'^2 + B'x'y' + C'y'^2 + D'x' + E'y' + F' = 0 \qquad ...(2)$$

This equation can be put into a standard form, where there is no x′y′ term, which is easier to work with. Angle at which the coordinate axes need to be rotated can be termed as say θ .

We will work on this concept, which is quite important for tackling problems offered in competitive examinations.

Refe to figure 7.41 below. The x and y axes are simultaneously rotated through an angle θ to their new position as x′/y′ axes.

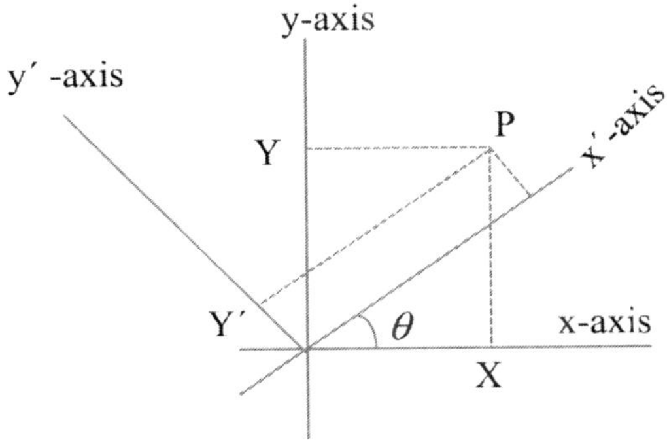

Figure 7.41

Considering polar coordinates of point P in x/y axes as (r, α) we get $x = r\cos\alpha$ & $y = r\sin\alpha$

Due to rotation of the axes by an angle θ, the polar coordinates with reference to x'/y' axes will be (r, α– θ)leading to

$x' = r\cos(\alpha - \theta) = r\cos\alpha\cos\theta + r\sin\alpha\sin\theta$.....(3) &

$y' = r\sin(\alpha - \theta) = r\sin\alpha\cos\theta - r\cos\alpha\sin\theta$......(4)

Substituting values of x and y above we get transformed coordinates x′/y′ as

$$x' = x\cos\theta + y\sin\theta \quad ......(5)\ \&$$
$$y' = -x\sin\theta + y\cos\theta ......(6)$$

Multiplying eq (5) by cos θ and eq (6)by sin θ, we get

$$x'\cos\theta = x\cos^2\theta + y\cos\theta\sin\theta \quad ......(7)\ \&$$
$$y'\sin\theta = -x\sin^2\theta + y\cos\theta\sin\theta ......(8)$$

Subtracting eq (8) from eq (7), we get

$$x'\cos\theta - y'\sin\theta = x\left(\cos^2\theta + \sin^2\theta\right) = x$$

Similarly multiplying eq (5) by sin θ and eq (6)by cos θ, we get

$$x'\sin\theta = x\sin\theta\cos\theta + y\sin^2\theta \quad ......(9)\ \&$$
$$y'\cos\theta = -x\cos\theta\sin\theta + y\cos^2\theta ......(10)$$

Adding eq (9) to eq (10), we get

$$x'\sin\theta + y'\cos\theta = y\left(\cos^2\theta + \sin^2\theta\right) = y$$

So we get x/y coordinates in terms of x′/y′ coordinates as

$$x = x'\cos\theta - y'\sin\theta \quad \&$$
$$y = x'\sin\theta + y'\cos\theta$$

We can substitute values of x and y in the generic equation (1) to get

$$A(x'\cos\theta - y'\sin\theta)^2 + B(x'\cos\theta - y'\sin\theta)(x'\sin\theta + y'\cos\theta) + C(x'\sin\theta + y'\cos\theta)^2$$
$$+D(x'\cos\theta - y'\sin\theta) + E(x'\sin\theta + y'\cos\theta) + F = 0 \Rightarrow$$

$$(x')^2\left(A\cos^2\theta + B\sin\theta\cos\theta + C\sin^2\theta\right) + x'y'\left[2(C-A)\sin\theta\cos\theta + B\left(\cos^2\theta - \sin^2\theta\right)\right]$$

$$+(y')^2\left(A\sin^2\theta - B\sin\theta\cos\theta + C\cos^2\theta\right) + x'(D\cos\theta + E\sin\theta) + y'(-D\sin\theta + E\cos\theta) + F = 0..(11)$$

Comparing equation (11) to equation (2) we get

$$A' = A\cos^2\theta + B\sin\theta\cos\theta + C\sin^2\theta$$

$$B' = 2(C-A)\sin\theta\cos\theta + B\left(\cos^2\theta - \sin^2\theta\right) = (C-A)\sin 2\theta + B\cos 2\theta$$

$$C' = A\sin^2\theta - B\sin\theta\cos\theta + C\cos^2\theta$$

$$D' = D\cos\theta + E\sin\theta$$

$$E' = -D\sin\theta + E\cos\theta$$

And $F' = F$

For eliminating $x'y'$ term, value of B' must equal zero.

Hence $(C-A)\sin 2\theta + B\cos 2\theta = 0 \Rightarrow B\cos 2\theta = (A-C)\sin 2\theta$

$$\Rightarrow \cot 2\theta = \frac{A-C}{B}$$

We thus can find value of angle θ using the above equation for conversion of given generic equation to a more manageable equation. This is a very important result.

7.5 Practical applications of conic sections

We see conic sections everyday but often just don't notice them. They can be man–made or natural. Conics are found in architecture, physics, astronomy and navigation. Bridges, buildings and statues use conics as support systems. Hence mastering this topic is very important in later part of scientific education. Here are some real life applications and occurrences of conic sections:

- There are many examples of circular objects in day to day life like pizza, iris of the eye, CD/DVDs, soccer ball, clock or wheels etc.
- The paths of the planets around the sun are ellipses with the sun at one focus.
- Tilt a glass of water, surface of the liquid acquires elliptical outline. Rugby ball is elliptical in shape. Even our eye is an ellipse.
- Conics can be seen in wide variety in many buildings, churches and arches. The bottom part of the Eiffel Tower seems as if it is showing a "negative parabola". The Eiffel Tower was built and designed this way so that it could support wind pressure and so be more stable. Parabolic curve is used by engineers in designing some suspension bridges.

- Parabolas are formed when a football is kicked, a baseball is hit, a basketball hoop is made, dolphins jump and much more. A parabola is also formed when a comet shoots across the sky. Trajectory of objects thrown or shot near the earth's surface follow a parabolic path.
- Parabolic mirrors are used to converge light beams at the focus of the parabola. This principle is used in design of car headlights and in spotlights because it aids in concentrating the light beam. Parabolic microphones perform a similar function with sound waves
- Solar ovens use parabolic mirrors to converge light beams to use for heating
- The best roller coasters are parabolic. When on a roller coaster, it feels like you're defeating or going against the force of gravity. When a coaster falls from the peak of a parabola, it is rejecting air resistance.
- Paths some of the astronomical objects take around the sun are hyperbolic (they do not revolve around the sun over and over, they approach, get close then leave in a hyperbolic path)
- Light from a Table lamp open on top and bottom, forms a hyperbola.
- Nuclear cooling towers have hyperbolic shape to make them most efficient and to enable it to withstand high wind pressure.
- Radio waves intersect each other in hyperbola. Hyperbolas are used in a navigation system known as LORAN (long range navigation)
- Hyperbolic as well as parabolic mirrors and lenses are used in systems of telescopes

7.6 Summary of different formulae for ready recall

A number of formulae have been derived for straight lines and different conic sections in this chapter. Some of these are quite similar. A good understanding of the formulae helps in avoiding mistakes in problem solving. Different formulae are summarized in the Table below for ease of recall.

7.6.1 Straight lines – Different forms of equations

General form	***ax + by + c = 0***	*a, b, c are arbitrary constant, +ve or –ve* *x–intercept= –c/a* *y–intercept= –c/b and* *slope= –a/b*
Slope– intercept form	***If** y–intercept (c) is known,* ***y = mx + c** or if* *y–intercept(d)is known* ***y = m(x–d)***	*m is slope*
Two point form	$y=\frac{y_2-y_1}{x_2-x_1}x+c$	*Points with coordinates*(x_1,y_1) *and* (x_2,y_2) *lie on the line and c is y–intercept*
Intercept form	$\frac{x}{a}+\frac{y}{b}=1$	*a and b are x–intercept and y–intercept respectively*
Normal form	$x \sin\alpha + x \cos\alpha = p$	*p is normal from origin to the line and α is angle the normal makes to the x–axis*
Section formula	$x=\frac{m_1x_2+m_2x_1}{m_1+m_2}$ $y=\frac{m_1y_2+m_2y_1}{m_1+m_2}$	*Points with coordinates*(x_1,y_1) *and* (x_2,y_2) *lie on the line and point with coordinates (x, y) divides the line in ratio* $m_1{:}m_2$

7.6.2 Straight lines – other formulae

Distance between two points	$\sqrt{(x_1-x_2)^2+(y_1-y_2)^2}$	*End points have coordinates*(x_1,y_1) *and* (x_2,y_2)
Distance **p** of a line **ax+by+c=0** from a point (x_1, y_1)	$p=\frac{(ax_1+by_1-ab)}{\sqrt{a^2+b^2}}$	
Distance **p** between two parallel lines **ax+by+c=0** and **ax+by+d=0**	$p=\frac{[d-c]}{\sqrt{a^2+b^2}}$	

7.6.3 Area of a triangle with (x_1, y_1), (x_2, y_2) and (x_3, y_3) as the vertices

$$\text{Area of triangle} = \frac{1}{2}[x_1(y_2-y_3)+x_2(y_3-y_1)+x_3(y_1-y_2)]$$

$$\text{Coordinates of Centroid are} \left(\frac{x_1+x_2+x_3}{3},\frac{y_1+y_2+y_3}{3}\right)$$

7.6.4 Conic sections

Conic section	Generic equation	Formula for eccentricity (e)	Latus rectum *(2l)*	Equation of directrix
Circle	$x^2 + y^2 = r^2$ *with center at origin or* $r^2=(x-x_1)^2+(y-y_1)^2$ *with origin at* (x_1, y_1)	$e = 0$	Not applicable	Not applicable
Ellipse	$\frac{x^2}{a^2}+\frac{y^2}{b^2}=1$, where center is at origin and ***a*** *is semi–major axis along x–axis &* ***b*** *is semi minor axis along y–axis and* $\frac{(x-x_1)^2}{a^2}+\frac{(y-y_1)^2}{b^2}=1$ *where center is at*(x_1, y_1) *In case major axis is along y–axis, the formula changes to* $\frac{x^2}{b^2}+\frac{y^2}{a^2}=1$	$e= c/a$ *and is always* <1	$2l=2b^2/a$	$x=\pm \frac{a^2}{c}$
Parabola	*Equations of the parabola with vertex at (0,0) are* $y^2= 4ax$, *if parabola lies in Quadrant 1&4 and symmetrical about x–axis* $y^2= -4ax$ *if parabola lies in Quadrant 2&3 and symmetrical about x–axis* $x^2= 4ay$, *if parabola lies in Quadrant 1&2 and symmetrical about y–axis and* $x^2= -4ay$, *if parabola lies in Quadrant 3&4 and symmetrical about y–axis* *Equations of the parabola with vertex at (g,h) and symmetrical about x–axis is* $(y-h)^2 = 4a(x-g)$ *and for symmetrical about x–axis is* $(x-g)^2 = 4a(y-h)$	$e = 1$	$l = 4a$	$x = -a$ $x = a$ $y = -a$ $y = a$ $x = g \pm a$ $y = h \pm a$

Hyperbola	$\frac{x^2}{a^2}-\frac{y^2}{b^2}=1$, *where* *center is at origin and* ***a*** *is distance of vertex from the origin , c is distance of focal point from the origin and* ***b*** $=\sqrt{c^2-a^2}$. *Hyperbola is symmetrical about y–axis* $\frac{(x-x_1)^2}{a^2}+\frac{(y-y_1)^2}{b^2}=1$ *where center is at* (x_1, y_1) *If Hyperbola is symmetrical about x–axis, equation is* $\frac{y^2}{a^2}-\frac{x^2}{b^2}=1$	*e= c/a and is always >1 (c=* $\sqrt{a^2+b^2}$ *)*	$2l=2b^2/a$	$x=\pm\frac{a^2}{c}$ & *equation of asymptote is* $y=\pm\frac{b}{a}x$ $y=\pm\frac{a^2}{c}$ & *equation of asymptote is* $y=\pm\frac{a}{b}x$

7.6.5 Solved examples:

1. A Graph G has a line of symmetry of x $=-\frac{5}{2}$. Graph G passes through the point (3, 3). What is the x–coordinate of another point that must have a y–coordinate of 3?

   Line x = $-\frac{5}{2}$ is a vertical line with variable values of y coordinate.

   Being the line of symmetry, the other point will have to be at the same distance from this line as the given point (3, 3), but on the other side i.e. on negative x–coordinate side of this line. Distance of the point (3, 3). from line x = $-\frac{5}{2}$ equals length of a perpendicular drawn from point (3,3) to the vertical line x = $-\frac{5}{2}$. Such a perpendicular will be parallel to x–axis at a distance of 3 from the x–axis and its length $=3-\left(-\frac{5}{2}\right)=\frac{11}{2}$. Thus coordinates of the base of this perpendicular will be $\left(-\frac{5}{2},3\right)$

   Now for the other point to have same distance from the line of symmetry, it must lie on the same vertical line extended beyond

point $\left(-\frac{5}{2},3\right)$ at a distance of $\frac{11}{2}$.
Thus X–coordinate of that point (with same y–coordiane 3) will equal $\left(-\frac{5}{2}-\frac{11}{2}\right)=-8$

2. Reference figure below. Circle J has a center at point W and radius of r = 5. This circle intersects the y–axis at one intercept and the x–axis at two intercepts. What is the area of the triangle formed by these three points of intercept?

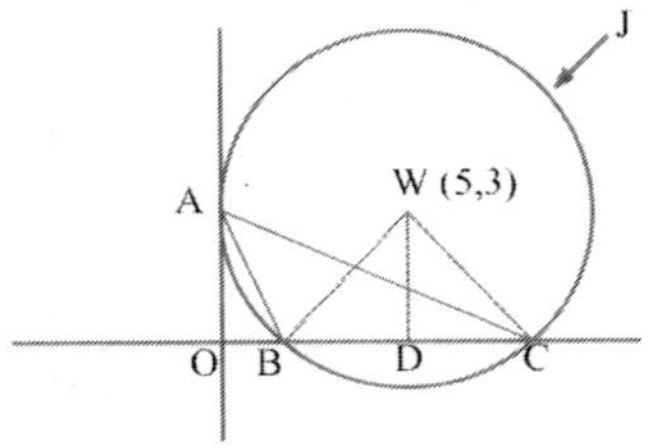

Since the co–ordinate of the center W of the circle is (5,3) and radius is 5, the circle will just touch the y–axis at point A and will intercept the x–axis at two points B & C. Now if we join W to B and C, we find WB = WC = 5 (radius of the circle)

Thus ΔWBC is an isosceles triangle. If we join midpoint D of BC to W, WD will be the perpendicular bisector of BC and equal 3 (y–coordinate of W).

$\therefore$ BD = $\sqrt{5^2-3^2}=4 \Rightarrow BC=8$

We can see that the y–axis is tangent to the circle. If we join, WA, it will be perpendicular to the y–axis, resulting in WAOD to be a rectangle yielding height of the ΔABC AO = WD = 3

So area of $\Delta ABC=\frac{1}{2}\times 8\times 3=12$

3. Given the vertices of parallelogram ABCD in the standard (x, y) coordinate plane below, what is the area of triangle ABC, in square units?

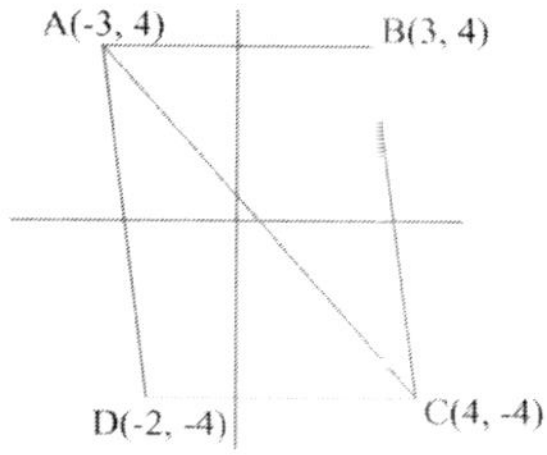

Line AB is parallel to x–axis. Hence perpendicular distance of line AB from point C = 4 – (–4) = 8 units and length of line AB = 3 – (–3) = 6 units.

$\therefore$ Area of $\Delta ABC = \frac{1}{2} \times 6 \times 8 = 24$ square units

4. In the standard (x, y) coordinate plane, what is the equation of the line that passes through the origin and the point (3, 4)?

   Generic equation of a line passing through two points with coordinates (x_1, y_1) and (x_2, y_2) is $y - y_1 = \frac{y_2 - y_1}{x_2 - x_1}(x - x_1)$.

5. What is the slope of the line that is perpendicular to the line determined by the equation 7x + 4y = 11?

   We know that Slope intercept form of generic equation of a straight line is y= mx+c, where m is the slope of the line and c is y intercept.

   The given equation can be reduced to $4y = -7x + 11 \Rightarrow$

   $y = \frac{-7}{4}x + \frac{11}{4}$ giving us slope of the line as $\frac{-7}{4}$

   Now slope of the line perpendicular to a given line = $\frac{-1}{m}$ where m is slope of the given line. We thus get slope of the perpendicular line = $\frac{-1}{-7/4} = \frac{4}{7}$

6. A system of linear equations is shown below.

   4y = 3x + 12 and

– 4y = –3x – 8

   Describe the relationship of this system of linear equations.

   Transforming the two equations in slope intercept form we get

   $y = \frac{3}{4}x + 3$ and $y = \frac{-3}{-4}x - \frac{8}{-4} \Rightarrow y = \frac{3}{4}x + 2$

Since slope of the two lines is same but intercept is different, it implies that these two lines are parallel to each other with positive slope.

7. Let A be (3,5) and B be (5,10). Find
   (i) the distance AB
   (ii) the midpoint P of AB
   (iii) the point Q which divides AB in the ratio 2 : 5.

(i) Distance between two points with coordinates $(x_{1,} y_1)$ and $(x_{2,} y_2) = \sqrt{(y_2 - y_1)^2 + (x_2 - x_1)^2}$

Therefore distance AB

$$= \sqrt{(10-5)^2 + (5-3)^2} = \sqrt{25+4} = \sqrt{29}$$

(ii) Midpoint of AB =

$$\left(\frac{x_1 + x_2}{2}, \frac{y_1 + y_2}{2}\right) \Rightarrow \left(\frac{3+5}{2}, \frac{5+10}{2}\right) \Rightarrow \left(4, \frac{15}{2}\right)$$

(iii) Coordinates of point which divides AB in the ratio $m_1 : m_2$ are

$$\left(\frac{m_2 x_1 + m_1 x_2}{m_1 + m_2}, \frac{m_2 y_1 + m_1 y_2}{m_1 + m_2}\right) \Rightarrow \text{coordinates of point Q as}$$

$$\left(\frac{3\times5+5\times2}{2+5}, \frac{5\times5+10\times2}{2+5}\right) \Rightarrow \left(\frac{25}{7}, \frac{45}{7}\right)$$

8. Find the coordinates of the point P which divides the interval A(3,7), B(1,4) externally in the ratio 4 : 3.

We know that coordinates of the point that divides a given line with end points $(x_{1,} y_1)$ and $(x_{2,} y_2)$ externally in the ratio m_1:m_2=

$$\left(\frac{m_1 x_2 - m_2 x_1}{m_1 - m_2}, \frac{m_1 y_2 - m_2 y_1}{m_1 - m_2}\right)$$

Hence coordinates of the point P are

$$\left(\frac{4\times1-3\times3}{4-3}, \frac{4\times4-3\times7}{4-3}\right) = (-5, -5)$$

9. Find the slope of the lines joining the following pairs of points:

(a) $\left(cp, \frac{c}{p}\right), \left(cq, \frac{c}{q}\right)$

(b) $(ap^2, 2ap), (aq^2, 2aq)$

(c) $(a\cos\theta, b\sin\theta), (a\cos\varphi, b\sin\varphi)$.

(a) $$Slope = \frac{c/q - c/p}{cq - cp} = \frac{(cp - cq)/pq}{cq - cp} = -\frac{1}{pq}$$

(b) $$Slope = \frac{aq^2 - ap^2}{2aq - 2ap} = \frac{q^2 - p^2}{2(q - p)} = \frac{q + p}{2}$$

(c) $$Slope = \frac{b\sin\varphi - b\sin\theta}{a\cos\varphi - a\cos\theta} = \frac{b}{a}\frac{\sin\varphi - \sin\theta}{\cos\varphi - \cos\theta}$$

$$= \frac{b}{a} \times \frac{2\cos\frac{\varphi+\theta}{2}\sin\frac{\varphi-\theta}{2}}{-2\sin\frac{\varphi+\theta}{2}\sin\frac{\varphi-\theta}{2}} = -\frac{b}{a} \times \cot\frac{\varphi+\theta}{2}$$

10. Find the equation of the line parallel to the x–axis which passes through the point where the lines $4x + 3y - 6 = 0$ and $x - 2y - 7 = 0$ meet.

Equation of a line parallel to x axis is $y = c$ where c is the y intercept.

Point of intersection of the lines $4x + 3y - 6 = 0$ and $x - 2y - 7 = 0$ can be found by solving the equations of the two lines. Thus by multiplying second equation by 4 and subtracting from the first equation, we get $3y + 8y = 6 - 28$ leading to $11y = -22$ or $y = -2$ and $x = 2y + 7 = -4 + 7 = 3$. Thus the point of intersection is $(3, -2)$.

-2 being the distance of this point from x–axis, equation of the line parallel to the x–axis and passing through this point is $y = -2$ or $y + 2 = 0$

11. In a triangle ABC, coordinates of B are (0,0). AB =2, angle ABC = $\pi/3$ and the mid point of BC has coordinates (2,0). Find centroid of the triangle.

Coordinates of Bare given as (0,0). Since coordinates of the mid point of BC (say B′) are (2,0). So line BC is along x–axis and hence coordinates of point C are (4,0).

Since AB = 2 and angle ABC = π/3, length of the perpendicular from A to the base BC meeting BC at say B′ =

$$\frac{AB'}{AB}=\sin\frac{\pi}{3}=\frac{\sqrt{3}}{2}\Rightarrow AB'=AB\times\frac{\sqrt{3}}{2}=2\times\frac{\sqrt{3}}{2}=\sqrt{3}\text{ and}$$

$$\frac{BB'}{AB}=\cos\frac{\pi}{3}=\frac{1}{2}\Rightarrow BB'=AB\times\frac{1}{2}=2\times\frac{1}{2}=1$$

Hence coordinatins of point A are $\left(1,\sqrt{3}\right)$.

We know that coordinates of centroid are average of coordinates of the three vertices. Hence coordinates of the centroid are $\left(\frac{1+0+4}{3},\frac{\sqrt{3}+0+0}{3}\right)\Rightarrow\left(\frac{5}{3},\frac{1}{\sqrt{3}}\right)$

(x_i,y_i) are vertices of an equilateral triangle such that

$$(x_1-2)^2+(y_1-3)^2=(x_2-2)^2+(y_2-3)^2=(x_3-2)^2+(y_3-3)^2$$

Find value of $2(x_1+x_2+x_3)+3(y_1+y_2+y_3)$

12. (x_i,y_i) are vertices of an equilateral triangle such that

$$(x_1-2)^2+(y_1-3)^2=(x_2-2)^2+(y_2-3)^2=(x_3-2)^2+(y_3-3)^2$$

Find value of $2(x_1+x_2+x_3)+3(y_1+y_2+y_3)$

$(x_1-2)^2+(y_1-3)^2$ implies square of the distance of point (x_1, y_1) from point (2,3).

From the given equality, it means that all three vertices have equal distance from point (2,3) Hence point (2,3) is circumcenter of the triangle.

Since circumcenter, incenter, orthocenter and centroid are the same point in an equilateral triangle, coordinates of the centroid are also (2,3). We know that coordinates of the centroid are arithmetic mean of coordinates of the three vertices, hence

$$\left(\frac{x_1+x_2+x_3}{3}\right)=2\ \ \&\ \ \left(\frac{y_1+y_2+y_3}{3}\right)=3\Rightarrow$$

$$x_1+x_2+x_3=6\ \ \&\ \ y_1+y_2+y_3=9$$

Thus $2(x_1 + x_2 + x_3) + 3(y_1 + y_2 + y_3) = 2 \times 6 + 3 \times 9 = 39$

13. A circle has center (1, 2) and radius $\sqrt{5}$

(a) Find the perpendicular distance from the center of the circle to the line with equation x +2y −10 = 0 and hence show this line is a tangent to the circle.

(b) Find the perpendicular distance from the center of the circle to the line with equation x +2y −12 = 0 and hence show the line does not meet the circle.

(c) Equation of the circle with given parameters is $(y-2)^2 + (x-1)^2 = \sqrt{5}^2 \Rightarrow y^2 - 4y + x^2 - 2x = 0$

Perpendicular distance p of the line x +2y −10 = 0 from center of circle (1, 2) is

$$p = \frac{|1 + 2 \times 2 - 10|}{\sqrt{1^2 + 2^2}} = \frac{5}{\sqrt{5}} = \sqrt{5}$$

Since the perpendicular distance of the line x +2y −10 = 0 equals radius of the circle, the line x +2y −10 = 0 touches the circle and is hence tangent to the circle.

Perpendicular distance p of the line x +2y −12 = 0 from center of circle (1, 2) is

$$p = \frac{|1 + 2 \times 2 - 12|}{\sqrt{1^2 + 2^2}} = \frac{7}{\sqrt{5}} = 1.4 \times \sqrt{5}$$

Since the perpendicular distance of the line x +2y −12 = 0 equals radius of the circle, the line x +2y −12 = 0 does not touch the circle.

14. Three points A(3, 6), B(7, 8) and C(5, 2) are the vertices of ΔABC. Find the area of this triangle.

Area of a triangle with vertices $(x_1, y_1), (x_2, y_2), (x_3, y_3)$ equals

$$\frac{1}{2}\left[x_1(y_2 - y_3) + x_2(y_3 - y_1) + x_3(y_1 - y_2)\right]$$

Hence area of ΔABC =

$$= \frac{1}{2} \times |3(8-2) + 7(2-6) + 5(6-8)| = \frac{1}{2} \times |18 - 28 - 10| = 10$$

15. An ellipse is given by the equation $8x^2 + 2y^2 = 32$. Find
    (a) the major axis and the minor axis of the ellipse and their lengths,
    (b) the vertices of the ellipse,
    (c) and the foci of this ellipse.

The given equation can be re–written as $4x^2 + y^2 = 16$

$$\Rightarrow \frac{x^2}{4} + \frac{y^2}{16} = 1$$

Comparing it with the standard form $\frac{(x-g)^2}{a^2} + \frac{(y-h)^2}{b^2} = 1$, we get g = 0, h = 0, a = 2 and b = 4. Therefore

(a) The centre of the ellipse is clearly at centre of the axis (0,0). Major axis is along y–axis with its length = 2 × 4 = 8 and Minor axis is along x–axis with its length = 2× 2 = 4

(b) The vertices are (0, 4), (2, 0), (0, –4) and (–2, 0) as shown in the figure below:

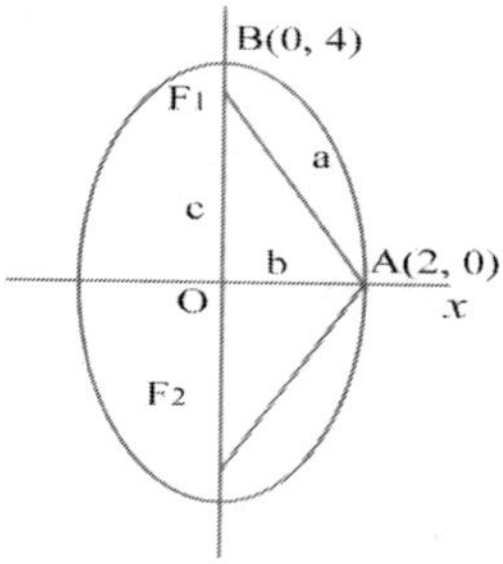

(c) Since sum of the distance for any point on the ellipse to the two focal points = length of the major axis, let us consider point A on the ellipse. Let the distance of the two foci from centre of the ellipse be denoted as c each, therefore distance of the point A from the two foci will equal length of semi–major axis each. Hence, $2^2 + c^2 = 4^2 \Rightarrow c = \pm\sqrt{12}$ giving us foci as ($\sqrt{12}$,0) and ($-\sqrt{12}$,0)

16. An ellipse is given by the equation $\frac{(x-1)^2}{9}+\frac{(y+4)^2}{16}=1$, find

(a) Its centre

(b) the major axis and the minor axis of the ellipse and their lengths,

(c) its vertices,

(d) and the foci.

Comparing the given equation with the standard form $\frac{(x-g)^2}{a^2}+\frac{(y-h)^2}{b^2}=1$, we get g = 1, h = –4, a = 3 and b = 4. Therefore

(a) The centre of the ellipse is at (1, – 4).

(b) Major axis is parallel to y–axis with its length = $2\times 4=8$ and Minor axis is parallel to x–axis with its length = $2\times 3=6$

(c) The vertices are (1, 0), (–2, –4), (1, –8) and (–5, –4)

(d) Considering distance of focal point from centre of the ellipse as c, and knowing semi–major axis as b and semi–minor axis as a, we get $a^2+c^2=b^2 \Rightarrow c^2=b^2-a^2=4^2-3^2 \Rightarrow c=\pm\sqrt{7}$, giving us foci as (1, –4+$\sqrt{7}$) and (1, –4–$\sqrt{7}$)

17. If a double ordinate of the parabola $y^2=4ax$ be of length 8a, find the angle between the lines joining the vertex of the parabola to the ends of this double ordinate.

Since double ordinate length is 8a, y–coordinate of the points of intersection of the double ordinate with the parabola (P_1, P_2) in the figure below will be $\pm 4a$. Let x–coordinate of the point of intersection of the double ordinate with the x–axis be x′.

$\therefore (4a)^2=4ax' \Rightarrow x'=4a$

This means coordinates of P_1 and P_2 are (4a, 4a) and (4a, –4a) yielding an angle of 45^0 from vertex to P_1 and P_2.

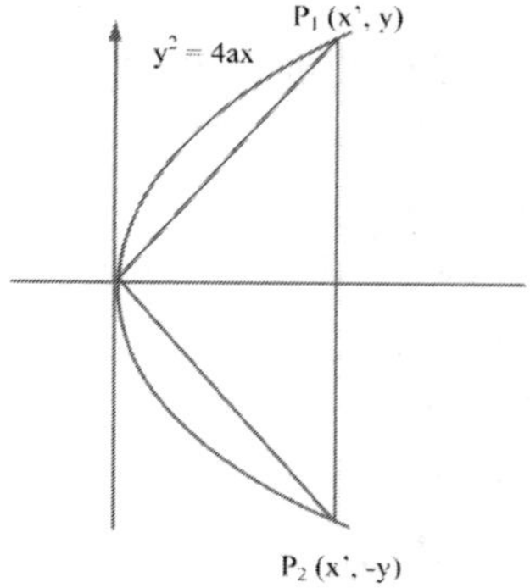

Thus angle between the lines joining the vertex of the parabola to the ends of this double ordinate is 90^0

18. Consider a branch of hyperbola $x^2-2y^2-2\sqrt{2}x-4\sqrt{2}y-6=0$ with vertex at point A. Let B be one of the end points of its Latus rectum. If C is the focus of the hyperbola nearest to the point A, find area of ΔABC.

The given equation $x^2-2y^2-2\sqrt{2}x-4\sqrt{2}y-6=0$ can be rewritten as

$$\left(x-\sqrt{2}\right)^2-2\left(y+\sqrt{2}\right)^2-4=0 \Rightarrow \frac{\left(x-\sqrt{2}\right)^2}{4}-\frac{\left(y+\sqrt{2}\right)^2}{2}=1,$$

yielding coordinates of the center of hyperbola as $\left(\sqrt{2},-\sqrt{2}\right)$

Parameters are a = 2, b = $\sqrt{2}$ yielding $c=\sqrt{a^2+b^2}=\sqrt{4+2}=\sqrt{6}$

Eccentricity $e=\frac{c}{a}=\frac{\sqrt{6}}{2}=\sqrt{\frac{3}{2}}$

In ΔABC, end point of latus rectum C is perpendicular to the focus and equals l. And distance of focus from vertex of the hyperbola = $\boldsymbol{c-a}$

$$\therefore \text{Area of } \Delta ABC = \frac{1}{2}\times(c-a)\times l=\frac{1}{2}\times(c-a)\times\frac{b^2}{a}$$

$$=\frac{1}{2}\times\left(\sqrt{6}-2\right)\times\frac{2}{2}=\sqrt{\frac{3}{2}}-1$$

19. Given a circle $2x^2 + 2y^2 = 5$ and a parabola $y^2 = 4\sqrt{5}x$ have a common tangent, find the value of slope m and equation of the tangent.

Tangent of a parabola with equation $y^2 = 4ax$ in slope form is $y = mx + \frac{a}{m}(m \neq 0)$. Since in the give equation of the parabola $a = \sqrt{5}$, the equation of the tangent is $y = mx + \frac{\sqrt{5}}{m}$ .... (1)

Equation of the circle can be re–written as $x^2 + y^2 = \frac{5}{2}$ implying that center of the circle is at (0, 0) and its radius is $\sqrt{\frac{5}{2}}$

Distance of the tangent from center of a circle = radius of the circle.

Since the tangent is common to the circle, the above equation (1) can be rewritten as $m^2x - my + \sqrt{5} = 0$

Using formula for distance to a line from an external point (centre of the circle in this case), we get

$$\frac{\sqrt{5}}{\sqrt{m^4 + m^2}} = \sqrt{\frac{5}{2}} \Rightarrow \frac{5}{\left(m^4 + m^2\right)} = \frac{5}{2} \Rightarrow m^2\left(m^2 + 1\right)$$

$$= 2 \Rightarrow m^4 + m^2 - 2 = 0$$

Solving this quadratic equation,

$$m^2 = \frac{-1 \pm \sqrt{1 - 4 \times 1 \times -2}}{2} = \frac{-1 \pm \sqrt{9}}{2} = \frac{-1 \pm 3}{2} = 1$$

because negative value of m^2 is not considered. Or $m = \pm 1$

Hence equation of the tangent is $y = \pm\left(x + \sqrt{5}\right)$

20. As shown in the diagram below, the base of a triangle passes through a fixed point P(a,b) and its other two sides are respectively bisected at right angles by the lines x + y = 0 and x = 9y. If the locus of the third vertex is a circle, then find equation of the circle.

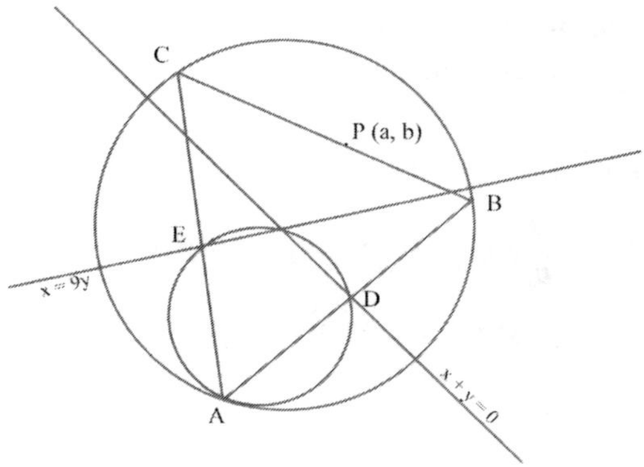

Since the two perpendicular bisectors are known, point of intersection of these perpendicular bisectors is circumcenter of the triangle.

So by solving x + y =0 and x = 9y, we get $x+\frac{x}{9}=0 \Rightarrow x=0$ and $y=0$

Thus the circumcenter is the origin (0 , 0)

Let coordinates of the vertices be A (x , y), B (x′, y′) and C (a′, b′).

Side AB is bisected by line x + y = 0

Coordinates of the mid–point of AB are $\left(\frac{x+x'}{2}, \frac{y+y'}{2}\right)$.

Since the mid–point lies on the line x + y = 0,

$\therefore\ x+x'+y+y'=0 \Rightarrow x'+y'=-x-y$ ...(1)

Slope of line x + y = 0 say m = –1

Therefore slope of line AB being perpendicular to the above line = $\frac{-1}{m}=\frac{-1}{-1}=1$

Using coordinates of points A and B, Slope of line AB = $\frac{y'-y}{x'-x}=1$, leading to $y'-x'=y-x$ ...(2)

Combining equations (1) and (2), we get

2y’ = –2x $\Rightarrow y'=-x$ and $x'=-y$

Hence coordinates of point B are (–y, –x)

Similarly let coordinates of point C be (a',b'), we find mid–point coordinates of AC as $\left(\frac{x+a'}{2},\frac{y+b'}{2}\right)$ leading to

$$x+a'=9(y+b') \text{ or } a'=9y+9b'-x \quad ...(3)$$

$$\text{Slope of line AC} = -\frac{1}{1/9}=-9$$

$$\therefore \frac{b'-y}{a'-x}=-9 \text{ leading to } a'=\frac{9x+y-b'}{9} \quad ...(4)$$

From equations (3) and (4),

$$9y+9b'-x=\frac{9x+y-b'}{9}\Rightarrow 82b'=18x-80y \text{ or}$$

$$b'=\frac{9x-40y}{41} \text{ and}$$

$$a'=\frac{9x+y-\dfrac{9x-40y}{41}}{9}=\frac{360x+81y}{9\times 41}=\frac{40x+9y}{41}$$

To sum up, coordinates of vertices B and C are B (–y, –x) and C $\left(\frac{40x+9y}{41},\frac{9x-40y}{41}\right)$

Since point P (a, b) lies on side BC, we can equate slopes of line BC and section BP Slope of BC =

$$\left(\frac{\dfrac{9x-40y}{41}+x}{\dfrac{40x+9y}{41}+y}\right)=\left(\frac{50x-40y}{40x+50y}\right)=\frac{5x-4y}{4x+5y}$$

and slope of BP $=\frac{b+x}{a+y}$ leading to $\frac{5x-4y}{4x+5y}=\frac{b+x}{a+y}$ or

$$5ax-4ay+5xy-4y^2=4bx+4x^2+5by+5xy$$

Simplifying, the above equation, we get

$$4x^2 + 4y^2 - x(5a - 4b) - y(-4a - 5b) = 0$$

Or $x^2 + y^2 - x\left(\frac{5a - 4b}{4}\right) - y\left(\frac{-4a - 5b}{4}\right) = 0$ ...(5)

Looking back on section 7.4.1, generic equation of a circle with center at (x_1, y_1) and radius r is $\mathbf{r^2=(x-x_1)^2+(y-y_1)^2}$. If the circle passes through the origin, substituting x=0, y =0, we get $r^2=x_1^2+y_1^2$.

For such a circle, the equation of the circle reduces to $(x-x_1)^2 + (y-y_1)^2 = x_1^2+y_1^2$ or

$x^2 + x_1^2 - 2xx_1 + y^2 + y_1^2 - 2yy_1 = x_1^2 + y_1^2$ leading to $x^2 + y^2 - x(2x_1) - y(2y_1) = 0$

Comparing equation 5 with the above equation, we find that equation 5 is of a circle with center at $\left(-\frac{4b-5a}{8}, -\frac{4a+5b}{8}\right)$ and passing through the vertex A (x, y) as well as the circumcenter (0, 0). The vertex A will follow the circle with above equation for all trianglessatisfying the given conditions where values of a, b and consequently center of the circle will change accordingly.

7.6.6 Practice problems

1. Find equation of the line passing through point (5,2) and perpendicular to the line joining points (2, 3) and (3, –1)

2. Show that tangent of an angle between the lines $\frac{x}{a} + \frac{y}{b} = 1$ and $\frac{x}{a} - \frac{y}{b} = 1$ equals $\frac{2ab}{a^2 - b^2}$

3. In the figure given below, find the point on segment JK that is four times as far from K as it is from J.

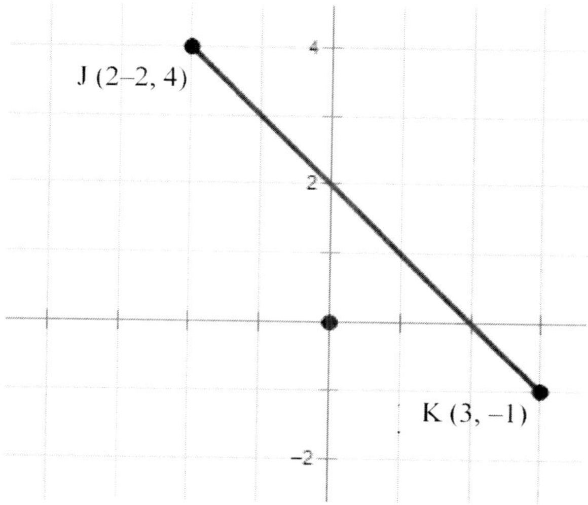

4. Line M has a y–intercept of –4, and its slope must be an integer–multiple of 1/7. Given that Line M passes below (4, –1) and above (5, –6), how many possible slopes could Line M have?

5. For what value of n would the following system of equations have an infinite number of solutions.

   $3a + b = 12$ and $12a + 4b = 3n$

6. In the standard (x, y) coordinate plane, if a square has the vertices (–2, –3), (2, –3) and (2, 1), what is the set of coordinates for the final vertex?

7. Suppose M is the midpoint of AB, where A is (8, 10) and B is (18, 20). Further suppose that P divides AB internally and Q divides AB externally in the ratio 2:3. Show that $MP \times MQ = MB^2$.

8. Find the equation of the line *l* through (1, 3) perpendicular to the line $2x + 3y = 12$. Find the equation of the line through (4, 5) parallel to *l*.

9. Use coordinate geometry to show that the altitudes of a triangle ABC are concurrent (at the orthocentre of the triangle).

10. Put $2x^2+2y^2+8x+24y+48=0$ into the standard circle form $(x-h)^2+(y-k)^2=r^2$, draw the circle on coordinate plane. Take a point on the circle and verify the equation.

11. To remove xy term from the second degree equation $5x^2+8xy+5y^2+3x+2y+5=0$ the coordinate axis are rotated through an angle θ, find value of θ

12. Let A(2,3) and B(–4,5) be two fixed points. Third point P moves such that area of the triangle PAB = 12 sq. units. Find locus of point P

13. The mid points of sides of a triangle are (5,0), (5,12), (0,12). Find coordinates of orthocentre of this triangle

14. Put the equation $\frac{4x^2}{9}+25y^2=1$ of an ellipse into standard form and find

    (a) Its center?

    (b) Length of the major axis?

    (c) Length of the minor axis?

15. An ellipse is defined by its parametric equations as follows

    x = 6 sin(t) and y = 4 cos(t). Find the centre, the major and minor axes and their lengths of this ellipse.

16. Find latus rectum of a parabola whose directrix is x + y –2 = 0 and focus is (3, –4).

17. A hyperbola having a transverse axis of length 2sinθ has same foci as an ellipse with the equation $3x^2+4y^2=12$. Find equation of the hyperbola.

18. Tangents are drawn to the hyperbola $4x^2 - y^2 = 36$ at the points P and Q. If these tangents meet at point T (0, 3), find area of $\Delta\,PQT$ in square units.

19. Let A (6, 3) be a point on the hyperbola $\frac{x^2}{a_1^2}-\frac{y^2}{b_1^2}=1$. If normal at point A meets the x–axis at (9, 0), then find eccentricity of the hyperbola.

20. Reflecting the point (2, –1) about y–axis, coordinate axes are rotated at 45° angle in negative direction without shifting the origin, find the new coordinates of the point

Answers to Practice Problems:

1. $4y - x - 3 = 0$
3. $(-1, 3)$
4. 8
5. 16
6. $(-2, 1)$.
8. $2y - 3x - 3 = 0$ and $2y - 3x + 2 = 0$
11. $\dfrac{\pi}{4}$
12. $x^2 + 9y^2 + 6xy - 66y - 22x - 23 = 0$
13. $(0,0)$
14. $(0,0)$. 3 and $\dfrac{2}{5}$
15. $(0, 0)$., 12, 8
16. $3\sqrt{2}$
17. $x^2 \cos ec^2\theta - y^2 \sec^2\theta = 1$
18. $45\sqrt{5}$
19. $\sqrt{\dfrac{3}{2}}$
20. $\left(\dfrac{-1}{\sqrt{2}}, \dfrac{-3}{\sqrt{2}}\right)$

8

LIMITS, CONTINUITY AND DERIVATIVES

Overview

In this chapter we will learn briefly the ideas behind the concepts of limits, continuity and derivatives. We will adopt an intuitive approach to understand these concepts.

8.1 The Idea of Limits of Functions

When we talked about algebra in Chapter 2, we saw algebraic expressions like x+y=5. The algebraic symbols x and y can take on any different values satisfying the expression and are thus called **variables**. In Chapter 7 on Co-ordinate Geometry, we learnt how to represent algebraic expressions on x-y coordinates. The graph on x-y coordinates of an algebraic expression indicates how value of one variable, say y, varies with values of the other variable, say x. Even though we draw a graph using values of the two variables at a few specific points, the graph can be used to understand how the two variables relate at any point.

The above expression x+y=5 can be expressed as y=5-x. Since value of y depends on value of x, we can also say that y is a function of x and express it as y= f(x)

The idea behind limits is to analyze what the function is "approaching" to when x "approaches" a specific value. The term "approaches" means it is very close to the specific value but not equal to that value. To start getting used to this idea, let's turn to this graph:

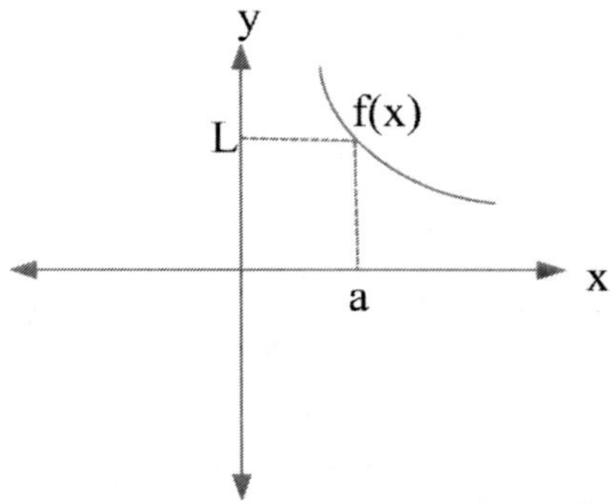

Figure 8.1

When x approaches the value "a" on the x axis, the function f(x) approaches "L" on the y axis. Notice a big hole in this graph at the point (a,L). This has been done deliberately to say that we don't necessarily know the value of function f at x=a.

Let's turn to the graph of a function whose expression we know:

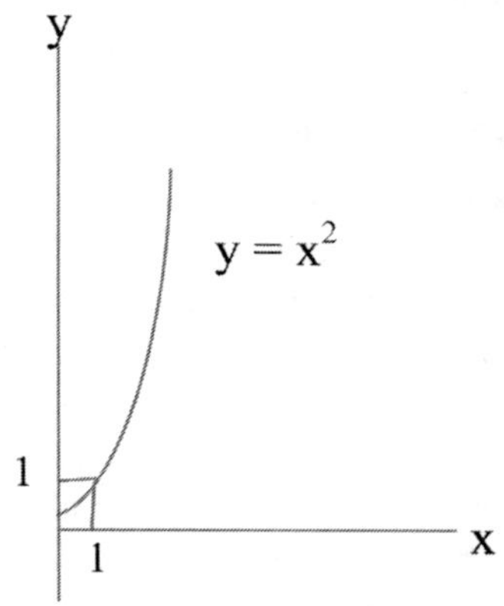

Figure 8.2

This is the function $f(x)=x^2$. Let's focus on the point (1, 1). We can see from the graph that when x approaches 1, the function f(x) approaches 1. When this happens, we say that

$$\lim_{x \to 1} x^2 = 1$$

This is read as "the limit, as x approaches 1, of x^2 equals 1".

Now let us consider another example. We want to find limit of function $f(x) = \frac{x^2-4}{x-2}$ as x approaches 2. In case we try to substitute value of x=2, both numerator and denominator equal to 0, thereby value of f(x) cannot be determined at x=2. However since we want to find limit of f(x) as x ***approaches*** 2, we can do the following:

$$f(x)= \frac{x^2-4}{x-2} = \frac{(x+2)(x-2)}{x-2} = x+2$$

We are able to cancel out factor (x-2) in numerator and denominator as we know that value of x-2 ***approaches close to 0 but not equal to zero***.

Thus, as x approaches 2, value of f(x) approaches 4 (i.e. very close to 4).

8.2 Existence of Limit

If we want to find limit of function f(x) as x approaches 2, we can determine the value of function as x approaches 2 from less than 2 (denoted as x approaches 2^-) or from more than 2 (denoted as x approaches 2^+). Limit is said to exist if values of function as x approaches 2^- and 2^+ are equal.

8.3 Why Limits are useful

One might ask what limit is useful for. Why would you need to know what the function is approaching to? We already know the function equals 1 when x equals 1, right?

Well, the point is that sometimes we don't care what the function is at x=1.

As an example, let's consider the following function:

$$f(x) = \begin{Bmatrix} x^2 if\ x \neq 1 \\ 0\ if\ x = 1 \end{Bmatrix}$$

The expression seems to be complex but this only means that this function equals x^2 when x is anything other than 1, and has a value 0 when x equals 1. This function is the same as the one we saw before, but in this case it has a «hole» at x=1. Let's see what the graph looks like:

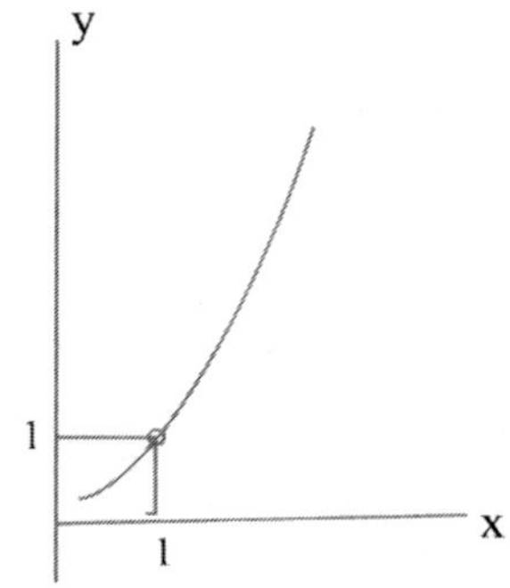

Figure 8.3

What does the function approach when x approaches 1? It also approaches 1, right? It doesn't matter that the function is other than 1 at that point! So, we state that

$$\lim_{x\to 1} f(x) = 1$$

In many real life situations, it is sufficient to know the value a function approaches to as the variable approaches to a specific value even if the value of the function at the specific point we care about is not defined, like 0/0 (which is complete junk), or useless, like zero or infinite.

8.4 Solving Limits

Here we will discuss everything we need to know about solving problems involving limits. All possible cases of problems are dealt. If the reader masters these techniques, he/she will be able to solve any type of problem involving limits.

8.4.1 Type 1: Limits by Direct Substitution

These are easiest problems. In these problems you only need to substitute the value to which the independent variable is approaching.

For example: $\lim_{x\to a} x^2$

Here we simply replace x by a to get

$$\lim_{x\to a} x^2 = a^2$$

This type is easy to solve.

8.4.2 Type 2: Limits by Factoring

In these limits, if you try to substitute as in the case discussed previously, you get an indetermination. For example:

$$\lim_{x \to 1} \frac{x^2 - 1}{x - 1}$$

If we simply substitute x by 1 in the expression we will get 0/0. So, we use our algebraic skills to simplify the expression as below:

$$\lim_{x \to 1} \frac{x^2 - 1}{x - 1} = \lim_{x \to 1} \frac{(x-1)(x+1)}{x-1} = \lim_{x \to 1}(x+1) = 2$$

Factor x – 1 is cancelled out in numerator and denominator as its value is very close to 0 but not equal to 0. It is easy to spot this type of problems: whenever you see a quotient of two polynomials, you may try this technique if there is an indetermination.

8.4.3 Type 3: Limits by Rationalization

These involve limits with square roots. For example:

$$\lim_{h \to 0} \frac{\sqrt{a+h} - \sqrt{a}}{h}$$

If we substitute value of h=0, we get 0/0 and we cannot factor this. In these limits, we apply an algebraic technique called rationalization. The trick is to multiply and divide the fraction by a convenient expression. (Remember that if you multiply and divide numerator and denominatorof a number by the same factor you get the same number). In this case we use the identity $\sqrt{a+h} + \sqrt{a}$ to rationalize as below:

$$\lim_{h \to 0} \frac{\sqrt{a+h} - \sqrt{a}}{h} = \lim_{h \to 0} \frac{\left(\sqrt{a+h} - \sqrt{a}\right)\left(\sqrt{a+h} + \sqrt{a}\right)}{h \times \left(\sqrt{a+h} + \sqrt{a}\right)} =$$

$$\lim_{h \to 0} \frac{a+h-a}{h \times \left(\sqrt{a+h} + \sqrt{a}\right)} = \frac{1}{2\sqrt{a}}$$

In the above example, h gets cancelled in numerator and denominator to eliminate reason for 0/0 result. So, whenever we see the difference or the sum of two square roots, we can apply the rationalization process. ***The two factors*** $\sqrt{a+h} - \sqrt{a}$ ***and*** $\sqrt{a+h} + \sqrt{a}$ ***are called conjugate expressions.***

Let us look at another example:

$$\lim_{x\to 1}\frac{1-\sqrt{x}}{1-x}$$

Here we use 1 + $\sqrt{x}$ as the rationalization factor leading to

$$\lim_{x\to 1}\frac{1-\sqrt{x}}{1-x}=\lim_{x\to 1}\frac{\left(1-\sqrt{x}\right)\left(1+\sqrt{x}\right)}{(1-x)\left(1+\sqrt{x}\right)}$$

On simplification,

$$\lim_{x\to 1}\frac{\left(1-\sqrt{x}\right)\left(1+\sqrt{x}\right)}{(1-x)\left(1+\sqrt{x}\right)}=\lim_{x\to 1}\frac{(1-x)}{(1-x)\left(1+\sqrt{x}\right)}$$

Here again factor (1-x) goes away and we get the desired result:

$$\lim_{x\to 1}\frac{(1-x)}{(1-x)\left(1+\sqrt{x}\right)}=\lim_{x\to 1}\frac{1}{\left(1+\sqrt{x}\right)}=\frac{1}{\left(1+\sqrt{1}\right)}=\frac{1}{2}$$

Another typical example is:

$$\lim_{x\to 0}\frac{\sqrt{x^2+p^2}-\sqrt{p}}{\sqrt{x^2+q^2}-\sqrt{q}}$$

where p and q are constants. In this case we have square roots both on the numerator and the denominator.

We get an indetermination if we substitute x by zero. This example is trickier, in the sense that we need to multiply by two expressions. We need to multiply and divide by the conjugate of the numerator and denominator and work algebraically.

This problem is good practice and it is recommended that readers try to solve on their own. In case you try and still can't solve it, solution is available at end of the chapter.

8.4.4 Type 4: Limits at Infinity

What are limits at infinity? Symbol for infinity is ∞. Infinity is not a number; it is more like a concept. When we say that x approaches infinity we mean that the variable x grows without bounds or where x takes values greater than any number you can come up with.

As an example, let us consider the example below:

$$\lim_{x\to\infty} \frac{1}{x}$$

Here you can't simply "plug" value of x as infinity because ∞ is not a number. However, we can guess what this limit will be using our intuitive understanding. As you divide 1 by very large numbers, you get very small fraction. This implies that 1 divided by x approaches 0 when x approaches infinity. Graph below depicts function $\frac{1}{x}$.

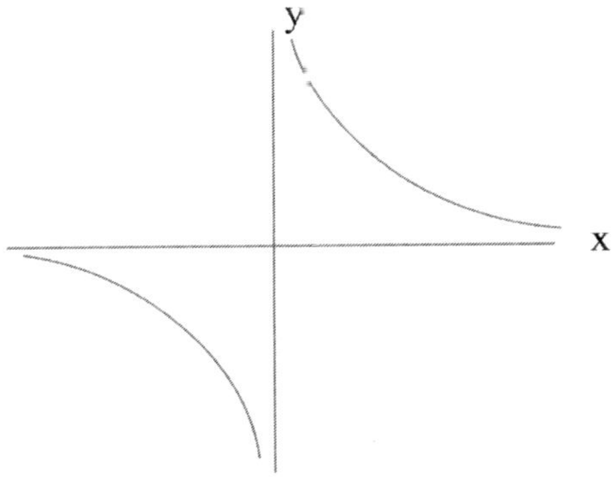

Figure 8.4

We can see that x could be approaching positive or negative infinity.

- x→ +∞ means that x is approaching a large positive number say in millions etc.
- x→ -∞ means that x is approaching a large negative number say in millions etc.
- x→ ∞ (without sign) means that x is taking big numbers, either positive or negative

In the graph above we can see that when x approaches very big numbers, either positive or negative, 1 divided by x approaches zero.

So: $\lim_{x\to\infty} \frac{1}{x} = 0$

This means that the two limits, when x→ +∞ and when x→ -∞, are equal to zero.

8.4.5 Limits theorems:

8.4.5.1 Squeeze theorem:

Let us consider three functions $u(x) \le z(x) \le v(x)$. That $u(x) \le z(x) \le v(x)$. That is, z(x) is always greater than or equal to u(x) and less than or equal to v(x). And also that limit of the two functions u(x) and v(x) as x→ a is same (L) i.e.

$$\lim_{x\to a} u(x) = L \text{ and } \lim_{x\to a} v(x) = L$$

The squeeze theorem states that given these conditions, we can say that:

$$\lim_{x\to a} z(x) = L$$

i.e. z(x) has been squeezed between u(x) and v(x)! Here's a graph that shows this:

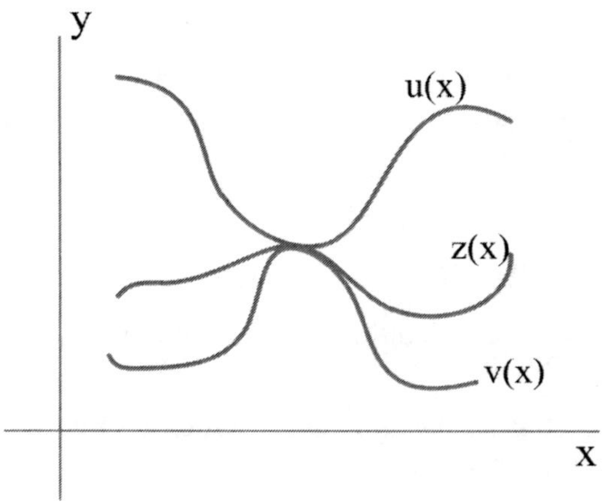

Figure 8.5

The point where u(x), z(x) and v(x) touch (or almost touch) is the point (a, L). Because z(x) is squeezed between the other two functions at that point, it also has to approach the same value at that point.

We will see that this theorem is actually useful to solve some limits.

8.4.5.2 Other Limits theorems:

$$\lim_{x\to c} \{f(x)+g(x)\} = \lim_{x\to c} f(x) + \lim_{x\to c} g(x)$$

$$\lim_{x\to c} \{f(x)g(x)\} = \lim_{x\to c} f(x) \times \lim_{x\to c} g(x)$$

8.4.5.3 Solving Limits at Infinity

Good thing about limits at infinity is that using a single technique we are able to solve almost any limit of this type. Let's look at this example:

$$\lim_{x\to\infty} \frac{4x^3 - 2x^2 + 1}{3x^3 - 5}$$

We cannot plug infinity and we cannot factor. So, now we'll use **the basic technique used to solve almost any limit at infinity**. It is a little algebraic trick.

Remember the property of fractions that you can divide both the numerator and denominator by the same number and the fraction remains the same, provided the number is not 0.

So, if we divide both the numerator and denominator of this limit function by x to the greatest exponent found in the function, (In this case it is x^3) we get:

$$\lim_{x\to\infty} \frac{\frac{4x^3}{x^3} - \frac{2x^2}{x^3} + \frac{1}{x^3}}{\frac{3x^3}{x^3} - \frac{5}{x^3}}$$

Cancelling out similar items, we get

$$\lim_{x\to\infty} \frac{4 - \frac{2}{x} + \frac{1}{x^3}}{3 - \frac{5}{x^3}}$$

We know that any number divided by a very very large number is equal (almost) to zero. So, as x approaches infinity, all the numbers divided by x to any power will approach zero. Plugging this we have:

$$\lim_{x\to\infty} \frac{4 - \frac{2}{x} + \frac{1}{x^3}}{3 - \frac{5}{x^3}} = \frac{4 - 0 + 0}{3 - 0} = \frac{4}{3}$$

8.4.5.4 Other Techniques for Solving Limits at Infinity

In the following examples we won't be using the basic technique of dividing by the greatest power of x. We'll be using something even more basic.

Suppose we have this limit:

$$\lim_{x \to \infty} \frac{x+1}{x}$$

We can "separate" this fraction as:

$$\lim_{x \to \infty} \left(\frac{x}{x} + \frac{1}{x} \right) = \lim_{x \to \infty} \left(1 + \frac{1}{x} \right)$$

Whenever we have two or more terms in the numerator, and only one term in the denominator, we may try to do this.

We use Limits theorem here. We know that 1/x approaches zero when x approaches infinity. So, we have:

$$\lim_{x \to \infty} \left(1 + \frac{1}{x} \right) = \lim_{x \to \infty} 1 + \lim_{x \to \infty} \frac{1}{x} = 1$$

Now, we will look at a really interesting problem. Let's consider the limit:

$$\lim_{n \to \infty} \frac{1+2+3+\dots\dots\dots+n}{n^2}$$

In the numerator we have the sum of all numbers from 1 to "n"; where "n" can be any natural number.

Now, as n approaches infinity, the number of terms in the numerator also approaches infinity, because there are n terms. So, the numerator approaches an infinite sum. To solve this limit, let's try to remember basic facts about arithmetic progressions.

We know the formula that gives us the sum of "n" terms of an arithmetic progression:

$$S_n = \frac{(a_1 + a_n) \times n}{2}$$

a_1 = first term of the arithmetic progression

a_n = last term of the arithmetic progression

n = number of terms

So, when we apply the formula for sum of AP to the given expression, we get

$$1+2+3+\dots\dots\dots+n = \frac{(a_1 + a_n)n}{2} = \frac{(1+n)n}{2}$$

Thereby reducing our limit to:

$$\lim_{n\to\infty}\frac{1+2+3+.........+n}{n^2}=\lim_{n\to\infty}\frac{\frac{(1+n)n}{2}}{n^2}$$

This equals:

$$\lim_{n\to\infty}\frac{(1+n)n}{2n^2}=\lim_{n\to\infty}\frac{1+n}{2n}$$

Now, we can use the technique we used in the previous example. We have only one term in the denominator, so we will "separate" the fraction. We have:

$$\lim_{n\to\infty}\frac{1}{2n}+\lim_{n\to\infty}\frac{n}{2n}=\lim_{n\to\infty}\frac{1}{2n}+\lim_{n\to\infty}\frac{1}{2}$$

And finally we have value of the above limit as :

$$0+\frac{1}{2}=\frac{1}{2}$$

A surprising result that the sum of all natural numbers up to a large n, divided by n^2, approach a finite value - one half.

8.4.5.5 Special case - Limits at Infinity not Existing

Some limits at infinity may not exist. For example, let's try to calculate this limit:

$$\lim_{x\to\infty}\frac{x^2+x-1}{2x+5}$$

We will use the basic technique of dividing by the greatest power of x. Let's divide all terms by x^2 leading to

$$\lim_{x\to\infty}\frac{\frac{x^2}{x^2}+\frac{x}{x^2}-\frac{1}{x^2}}{\frac{2x}{x^2}+\frac{5}{x^2}}=\lim_{x\to\infty}\frac{1+\frac{1}{x}-\frac{1}{x^2}}{\frac{2}{x}+\frac{5}{x^2}}$$

All numbers divided by any power of x will approach 0 as x approaches infinity. So, we have:

$$\frac{1+0-0}{0+0}=\frac{1}{0}$$

Division by zero is undefined, so this limit does not exist.

8.4.5.6 Limits at Infinity with radicals

Now let's turn our attention to limits at infinity of functions involving radicals. In this case we can also use the basic technique of dividing by x to the greatest exponent.

Let's consider the limit:

$$\lim_{x\to\infty} \frac{\sqrt{x^2+1}}{x+1}$$

We have a radical in the numerator. We use the basic technique of dividing both the numerator and denominator. In this case we divide by x, leading to:

$$\lim_{x\to\infty} \frac{\dfrac{\sqrt{x^2+1}}{x}}{\dfrac{x+1}{x}}$$

By inserting the x in the numerator inside the radical, we get:

$$\lim_{x\to\infty} \frac{\sqrt{\dfrac{x^2+1}{x^2}}}{\dfrac{x+1}{x}}$$

On further simplifying, we get:

$$\lim_{x\to\infty} \frac{\sqrt{\dfrac{x^2}{x^2}+\dfrac{1}{x^2}}}{\dfrac{x}{x}+\dfrac{1}{x}} = \lim_{x\to\infty} \frac{\sqrt{1+\dfrac{1}{x^2}}}{1+\dfrac{1}{x}}$$

Once again, all the terms divided by x will approach zero. So we have:

$$\frac{\sqrt{1+0}}{1+0} = \pm 1$$

Since square of ± 1 is 1, we can have either a positive or negative sign in the answer. This depends on whether x approaches positive or negative infinity. We can see this in the graph below:

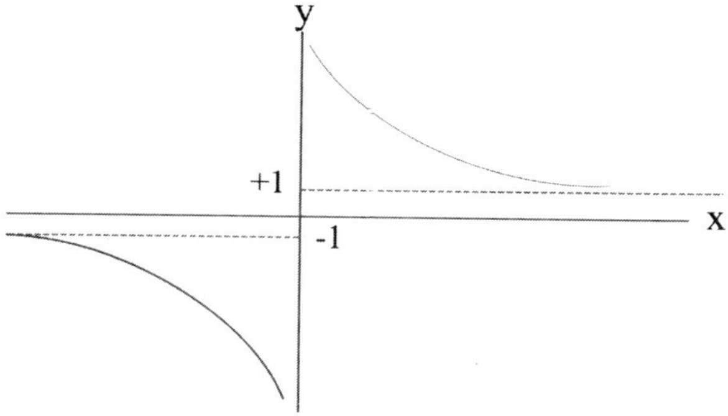

Figure 8.6

When x approaches positive infinity, the function approaches positive 1. And when x approaches negative infinity, the function approaches negative 1.

8.4.6 Type 5: Trigonometric Limits

In most limits that involve trigonometric functions we need to apply the fundamental limit of Sin(x) over x: With anything involving trigonometric functions, one always needs practice, because there are many trigonometric identities to choose from.

8.4.6.1 The Limit of Sin(x) over x

This limit is very useful, because it can be used to solve other limits that involve Trigonometric functions. We will calculate this limit using Squeeze theorem discussed above:

$$\lim_{x \to 0} \frac{\sin x}{x} = 1$$

To begin, let's draw a quarter circle of **radius 1** and a triangle, as in the figure below:

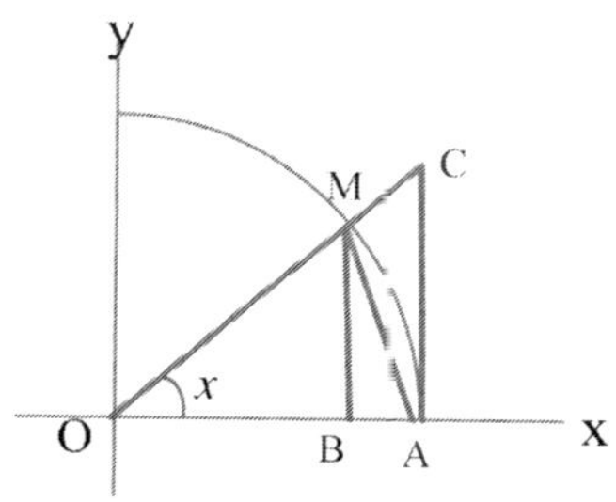

Figure 8.7

O is center of the circle intersecting x-axis at point A. Draw angle x in radians to x-axis meeting the circle at point M. Connect M to A and draw a perpendicular from M to x-axis meeting x-axis at B. Now draw a tangent to the circle at A meeting extended line OM at C. Note the following.

Area of triangle MOA (ΔMOA) is smaller than the area of circular sector MOA which in turn is smaller than area of triangle COA. That is:

$\Delta MOA < circular\ sector\ MOA < \Delta COA$

Area of triangle MOA = $\dfrac{OA \times MB}{2}$

Since OA = OM = 1 and $\sin x = \dfrac{MB}{OM} = \dfrac{MB}{1} = MB$

We find that Area of triangle MOA = $\dfrac{\sin x}{2}$

Now, let's find the area of circular sector MOA of radius 1 and an internal angle x:

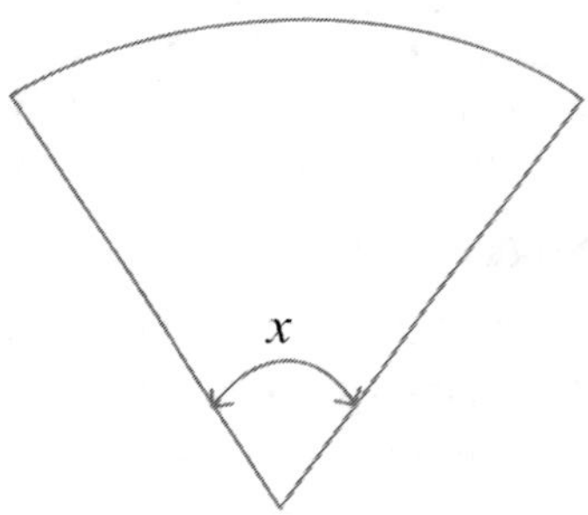

Figure 8.8

We know that area of full circle with center angle 2π radians is πr^2. Hence area of a sector with angle x radians $= \dfrac{xr^2}{2} = \dfrac{x}{2}$ in this case, because radius = 1.

Let's find the area of triangle COA:

Area of triangle COA = $\dfrac{OA \times AC}{2}$

Since OA = 1 and $\tan x = \dfrac{AC}{OA} = \dfrac{AC}{1} = AC$

Hence we see that Area of triangle COA = $\dfrac{\tan x}{2}$

We now have the three areas! Let's replace them in our inequality:

$$\frac{\text{Sin}\,x}{2} < \frac{x}{2} < \frac{\text{Tan}\,x}{2}$$

We can multiply the three members by 2 and the inequality holds. Let's do that, let's multiply by 2 leading to

$$\text{Sin}\,x < x < \frac{\text{Sin}\,x}{\text{Cos}\,x}$$

Now, let's divide the three members by sin(x). The inequality holds:

$$1 < \frac{x}{\text{Sin}\,x} < \frac{1}{\text{Cos}\,x}$$

Now, we will use a property of inequalities. This property says "if A is less than B, the inverse of A is greater than B". This would mean we have the inequality:

$$1 > \frac{\text{Sin}\,x}{x} > \text{Cos}\,x$$

Now, we're ready to use the squeeze theorem! We know that:

$$\underset{x \to 0}{Lim} \cos x = 1$$

So, the middle function $\frac{\text{Sin}\,x}{x}$ is squeezed between two 1s. By the squeeze theorem, we can conclude that:

$$\underset{x \to 0}{Lim} \frac{\sin x}{x} = 1$$

Here's the graph of the function sin(x) over x:

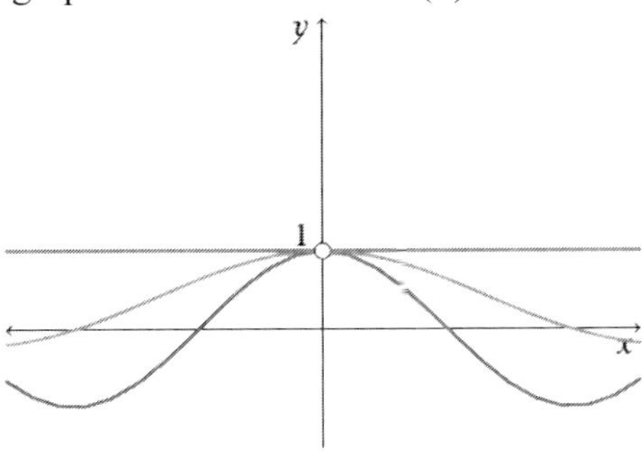

Figure 8.9

The above figure shows the three functions to which we apply the squeeze theorem.

The green function is sinx/x, the red function is cosx and the brown function is the constant function 1.We can see how sinx/x is squeezed by the other two when x approaches 0.

As we progress further, we will see that this is really a very useful limit. We need to use it often to solve other limits involving trigonometric functions.

8.4.6.2 More Examples of Trigonometric Limits

Example 1:

$$\underset{x\to 0}{Lim}\frac{\tan x}{x}$$

By rearrangement

$$\underset{x\to 0}{Lim}\frac{\tan x}{x}=\underset{x\to 0}{Lim}\frac{\sin x}{\cos x}\times\frac{1}{x}=\underset{x\to 0}{Lim}\frac{\sin x}{x}\times\frac{1}{\cos x}$$

The rearranged limit looks quite like the limit we just proved.

The limit of sin(x) over x is 1. And the limit of cos(x) is also 1. So, we can write the limit as:

$$\underset{x\to 0}{Lim}\frac{\tan x}{x}=\underset{x\to 0}{Lim}\frac{\sin x}{x}\times\underset{x\to 0}{Lim}\frac{1}{\cos x}=1\times 1=1$$

Example 2

Here's another example. Let's find this limit:

$$\underset{x\to 0}{Lim}\frac{1-\cos x}{x}$$

If we just replace the x's by zero, we get 0/0! We need to look at our trigonometric formulae and try to find something that may be useful.

Remember the formula 1 - cos 2x = 2 Sin^2x,

Substituting appropriately in the above Limit function and rearranging, we arrive at

$$\underset{x\to 0}{Lim}\frac{1-\cos x}{x}=\underset{x\to 0}{Lim}\frac{2\sin^2\frac{x}{2}}{x}=\underset{x\to 0}{Lim}\frac{\sin^2\frac{x}{2}}{\frac{x}{2}}$$

Using the formula for limit of $\frac{\text{Sin}\,x}{x}$ as x→ 0 and substituting x = 0 in $\text{Sin}\frac{x}{2}$, we get

$$\underset{x\to 0}{Lim}\frac{\sin^2\frac{x}{2}}{\frac{x}{2}}=\underset{x\to 0}{Lim}\frac{\sin\frac{x}{2}}{\frac{x}{2}}\times\frac{\sin\frac{x}{2}}{1}=1\times 0=0$$

Example 3

Now, let's find this limit:

$$\underset{x\to 0}{Lim}\frac{\sin^2\frac{x}{3}}{x^2}=\underset{x\to 0}{Lim}\frac{\left(\sin\frac{x}{3}\right)^2}{\left(\frac{x}{3}\right)^2\times 3^2}=\underset{x\to 0}{Lim}\left(\frac{\sin\frac{x}{3}}{\frac{x}{3}}\right)^2\times\frac{1}{9}=1\times\frac{1}{9}=\frac{1}{9}$$

Note that we just simplify the expression and use the result we already know.

The rule of thumb is: whenever you have a limit with trigonometric functions, and x approaching 0, you may try to use the limit of sin(x) over x to find the limit.

8.4.6.3 Type 6: Limits Involving Mathematical constant 'e'

The number ***e*** is a mathematical constant, approximately equal to **2.71828**, which appears in many different topics throughout mathematics. It was discovered by the Swiss mathematician Jacob Bernoulli while studying compound interest (reference https://en.wikipedia.org/wiki/Compound interest), where *e* arises as

$$\underset{n\to\infty}{Lim}\left(1+\frac{1}{n}\right)^n \;\ldots\ldots..(1)$$

Remember that compounded amount of a sum P with annual rate of interest say **a%**

$$\left(\textit{in numerical value that is}\frac{a}{100},\textit{for example } 10\%=\frac{1}{10}\right)$$ and

number of compounding durations per year as n will equal to $P\times\left(1+\frac{a}{n\times 100}\right)^n$.

If rate of interest per interval, that could be monthly, quarterly or annually, is such that over a period of **n** intervals, the amount doubles with simple interest. Hence rate of interest per interval will be equal to $\frac{1}{n}$. Now if number of compounding intervals **n** is taken as infinite (very very large), with rate of interest per interval as $\frac{1}{n}$, the compounded amount will be achieved as per Limit shown at (1) above which will approach a constant e (=2.71828). Readers are encouraged to try calculating values of $\left(1+\frac{1}{n}\right)^n$ for values of n = 10, 100, 1000, 10000 etc. It will be found that value of this expression approaches value of "e" as value of n keeps increasing.

The number *e* can also be calculated as the sum of the infinite series

$$\sum_{n=0}^{\infty}\frac{1}{n!}=\frac{1}{1}+\frac{1}{1}+\frac{1}{1\times 2}+\frac{1}{1\times 2\times 3}\ldots\ldots$$

'**e**' is also known as **Euler's number** after the Swiss mathematician Leonhard Euler,

Like the constant π, ***e*** is irrational. It is not a ratio of integers. Also like π, *e* is not a root of any non-zero polynomial with rational coefficients.

There are some limits that can be solved using this fundamental limit. This is similar to what we do with trigonometric limits. We try to manipulate the function algebraically to apply the limit we already know.

Now let us solve the following limit that at first sight has no connect to number **e**.

$$\lim_{x\to\infty}\left(\frac{x+4}{x-1}\right)^{x+4}$$

However, if we notice carefully it is in some way similar to the definition of e involving limit. Let us use a technique that involves change of variables to make the question look like the limit as definition of e:

Let us define a variable y such that $\frac{x+4}{x-1}=1+\frac{1}{y}$

If we solve for x in the above equation we get:

$$5y+1=x \text{ and } x+4=5y+5$$

This implies that if x approaches infinity, y also approaches infinity. So, replacing in the original limit we get:

$$\lim_{x\to\infty}\left(\frac{x+4}{x-1}\right)^{x+4} = \lim_{y\to\infty}\left(1+\frac{1}{y}\right)^{5y+5}$$

Now, using the properties of exponents and limits we have:

$$\lim_{y\to\infty}\left(1+\frac{1}{y}\right)^{5y+5} = \lim_{y\to\infty}\left[\left(1+\frac{1}{y}\right)^{y+1}\right]^5$$

Using the properties of exponents once again to simplify the terms:

$$\lim_{y\to\infty}\left[\left(1+\frac{1}{y}\right)^{y+1}\right]^5 = \lim_{y\to\infty}\left[\left(1+\frac{1}{y}\right)\times\left(1+\frac{1}{y}\right)^{y}\right]^5$$

Now, applying the limit, first factor becomes one, and second factor is definition of e! So:

$$\lim_{y\to\infty}\left[\left(1+\frac{1}{y}\right)\times\left(1+\frac{1}{y}\right)^{y}\right]^5 = e^5$$

8.5 The Idea of Continuity Functions

Limits and continuity are quite related. The basic idea of continuity is very simple, and the "formal" definition uses limits.

Basically, we say a function is continuous when you can graph it without lifting your pencil from the paper. It also implies that small changes in the input of the function result in small changes in the output. Here is an example of what a continuous function looks like:

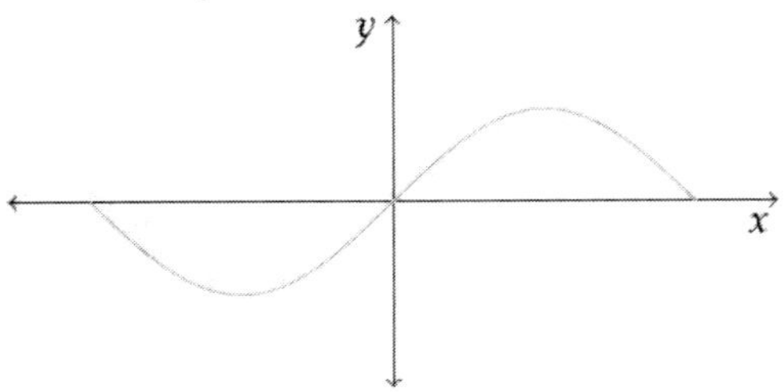

Figure 8.10

There is a precise mathematical definition of continuity that uses limits. ***This definition says that if 'a' be a point in the domain of the function f(x), then f is continuous at***

x=a if and only if

$$\lim_{x\to a} f(x) = f(a)$$

and $\lim_{x\to a^-} f(x)$ as well as $\lim_{x\to a^+} f(x)$ are both $= f(a)$

A function essentially is discontinuous when it has any "gap" in its graph. A discontinuous function looks like the example below:

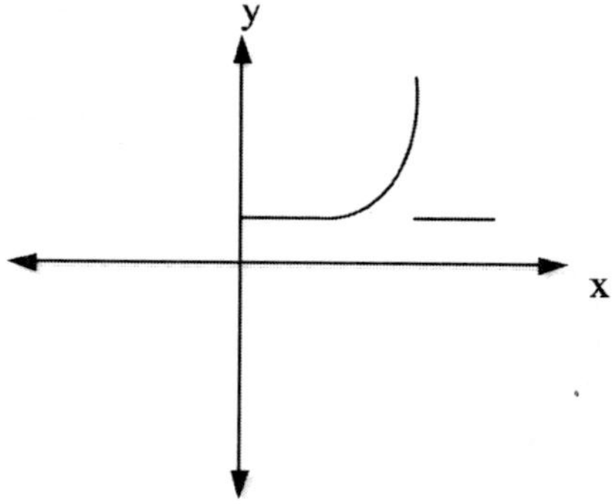

Figure 8.11

8.6 Derivatives

If you study different graphs shown above carefully, you will see that value of y changes with value of x by a defined equation. For example, equation y=x implies that value of y equals value of x and value of y changes by exact same value as changes in value of x. That is if value of x is doubled, value of y also doubles implying 2y= 2x.

Similarly $y = x^2$ implies that value of y equals square of value of x. And if x is doubled, y goes 4 times its original value. Value of y changes continuously as value of x varies.

Consider the straight line segment AB in the following graph (Figure 8.12). As seen in Chapter 7 on Co-ordinate geometry, slope of the line in the graph = ratio of change in y divided by change in x. With given co-ordinates of points A and B, slope $= \frac{y_2 - y_1}{x_2 - x_1}$.

For functions y where value of y keeps changing continuously, we denote small change in x at any point as Δx resulting in corresponding change in y as Δy, the slope of the line at x will then be $=\frac{\Delta y}{\Delta x}$. Derivatives are all about slope. It seems quite easy for a straight line as y changes by a uniform proportion with reference to x. thus $\frac{\Delta y}{\Delta x}$ is a constant. ***And we say derivative of a straight line with generic equation*** $y = mx + c$ ***where m is slope and c is offset of the line on y-axisis a constant = m (value of the slope).***

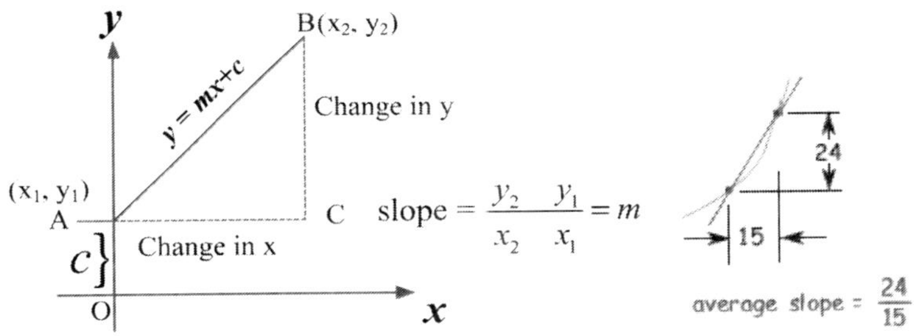

Figure 8.12

So far, so good. Now if the function is such that value of y continuously varies by a varying degree with reference to x-axis, then slope will also vary at different points. An easy example of such a function is distance and speed of an object forcefully thrown up with reference to time. If we draw the vertical distance with reference to time, it will result in a graph below (Figure 8.13).

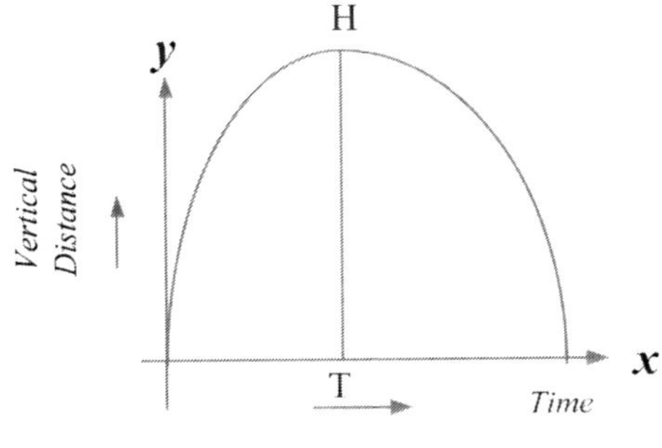

Figure 8.13

Vertical distance keeps increasing till the object reaches a peak and then starts decreasing as it starts falling under gravity and comes to the ground level. Here we notice that upward as well as downwards distance and speed keeps changing with time.In such instances, we take a point and find **small change in y (Δ y)**corresponding to a **small change in x (Δ x)** depicted below and then try to shrink the difference Δ **x** towards zero and find corresponding ratio $\frac{\Delta y}{\Delta x}$.

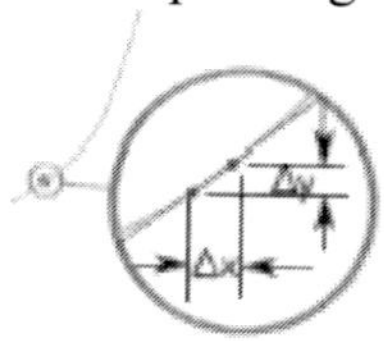

Figure 8.14

So derivative of a function at a point is nothing but $\lim_{\Delta x \to 0} \frac{\Delta y}{\Delta x}$ which is denoted as $\frac{dy}{dx}$. Here y being a function of x, we can also denote y as $f(x)$. In such a case, derivative $\frac{dy}{dx}$ is also denoted as $f'(x)$.

For better understanding, let us once again consider figure 8.13 above. We notice that distance changes rapidly to start with, rate of change reduces after some time till it reaches zero at point H,at time T. We can see that rate of change of distance with time at any point is the velocity of the object at that point. Mathematically velocity at any time t is written as

$$v(t)=\frac{ds}{dt}$$

where s is distance from the origin. Velocity is positive when the object is going up, goes to zero when object reaches peak (point H) and starts increasing negatively till it reaches its peak on reaching the ground. Velocity at any point is also seen as slope of the curve at that point. Similarly we can define acceleration or deceleration (negative acceleration) as rate of change of velocity with time. We write acceleration as

$$a(t)=\frac{dv}{dt}$$

Example of derivative of a generic function f(x) taking f(x) = x^n

Let us try and find derivative of a function $f(x)$ at point (x,y). It effectively means, we need to find slope of the function at point (x,y). To do this, we would assume a small change Δx in value of x and find corresponding change in function $f(x)$ at x+Δx. Thereafter we find difference in value of $f(x)$ at x and x+Δx compared to Δx and find its limiting value as Δx shrinks to 0. Mathematically we state this process as:

Derivative of $f(x)$ at x (written as $f'(x)$ is

$$f'(x) = \lim_{\Delta x \to 0} \frac{f(x+\Delta x) - f(x)}{\Delta x}$$

Let us find derivative of function f(x) = x^2 by the above definition as below:

$$f(x) = x^2$$

$$f(x+\Delta x) = (x+\Delta x)^2 = x^2 + 2x\Delta x + \Delta x^2$$

$$f'(x) = \lim_{\Delta x \to 0} \frac{f(x+\Delta x) - f(x)}{\Delta x} = \lim_{\Delta x \to 0} \frac{x^2 + 2x\Delta x + \Delta x^2 - x^2}{\Delta x} =$$

$$\lim_{\Delta x \to 0} 2x + \Delta x = 2x$$

Thus derivative is also written as $\frac{d}{dx}(x^2) = 2x$

Let us try derivative of x^3 by the same process.

$$f(x) = x^3$$

$$f(x+\Delta x) = (x+\Delta x)^3 = x^3 + 3x^2\Delta x + 3x\Delta x^2 + \Delta x^3$$

$$f'(x) = \lim_{\Delta x \to 0} \frac{f(x+\Delta x) - f(x)}{\Delta x} = \lim_{\Delta x \to 0} \frac{x^3 + 3x^2\Delta x + 3x\Delta x^2 + \Delta x^3 - x^3}{\Delta x} =$$

$$\lim_{\Delta x \to 0} 3x^2 + 3x\Delta x = 3x^2$$

Thus derivative of x^3 is $\frac{d}{dx}(x^3) = 3x^2$

From the above two examples, we can generalise this result as $\frac{d}{dx}x^n = nx^{n-1}$

8.6.1 Chain Rule of differentiation:

You may also encounter functions of ***y*** in a different variable ***say u which in turn is a variable of x***. In such a situation we apply the following Chain Rule of differentiation

$$\frac{dy}{dx} = \frac{dy}{du} \times \frac{du}{dx}$$

8.6.2 Double differentiation:

Refer to the example given above of acceleration at a given time as a derivative of velocity with reference to time at that point. Since we know that velocity in turn is derivative of distance with reference to time, we can also write acceleration as

$$a(t) = \frac{dv}{dt} = \frac{d}{dt}\left(\frac{ds}{dt}\right) = \frac{d^2s}{dt^2}$$

Clear understanding of these concepts helps a great deal in solving problems on derivatives as well as in practical situations.

8.6.3 Uniqueness of derivative of exponential function e^x

The function $f(x) = e^x$ is called the (natural) exponential function and is the unique function **of type a^x which is equal to its own derivative $\{f(x) = f'(x) = e^x\}$,** so ***e*** can be spotted as the unique positive number ***'a'*** such that the graph of the function $y = a^x$ has unit slope at $x = 0$.

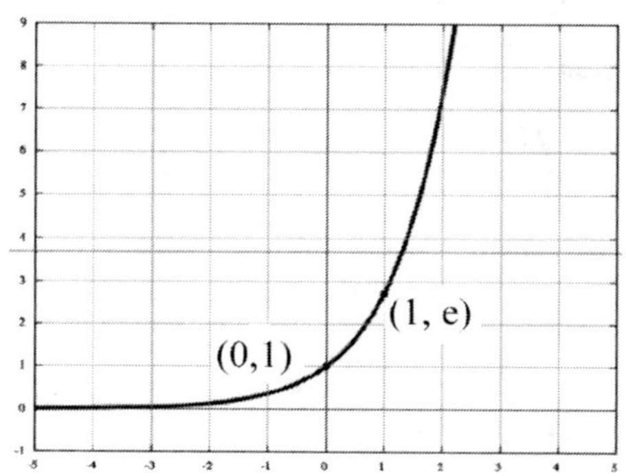

Figure 8.15: Natural exponential function $y = e^x$*(courtesy: Wikipedia)*

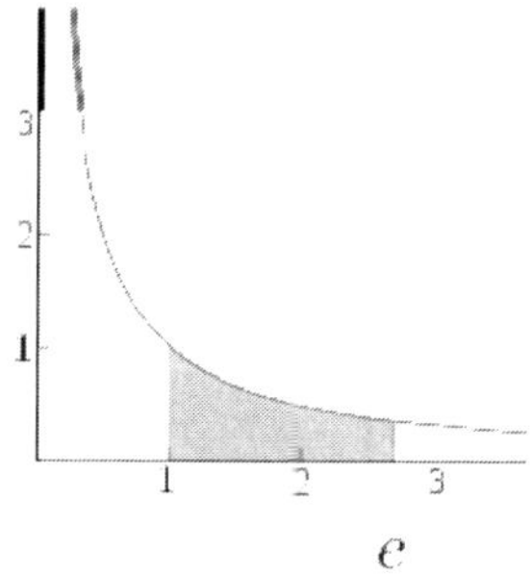

Figure 8.16

Reference figure 8.16. This is graph of the function f(x) = 1/x

We have learnt about the concept of logarithm and natural number ***e*** in Chapter 1. ***Logarithm to the base e is called natural logarithm, also written as ln. The natural logarithm of a positive number k*** can also be defined directly as the ***area under*** the curve y = 1/x between x = ***1 and x = k, in which case e*** is the ***unique value of k for which this area equals unity. This also implies that* e *is the unique number whose natural logarithm is equal to one.***

We will learn this concept as we study the topic of Integration which is just inverse process of differentiation, in the next chapter.

8.6.4 Limit of exponential functions

Let the **exponential function** be ***f(x) = b^x,*** where the input variable *x* occurs as an exponent. A function of the form ***f(x)=b^{x+c},*** where c is a constant, is also considered an exponential function and can be rewritten as ***f(x)=ab^x***, with ***a = b^c***

As functions of a real variable, exponential functions are uniquely characterized by the fact that the growth rate of such a function (i.e., its derivative) is directly proportional to the value of the function itself. The constant of proportionality of this relationship is the natural logarithm of the base b

$$\frac{d}{dx}(b^x) = b^x \log_e(b)$$

The constant $e \approx 2.71828...$ is the unique base for which the constant of proportionality is 1, so that the function's derivative is itself:

$$\frac{d}{dx}(e^x) = e^x \log_e(e) = e^x$$

Since changing the base of the exponential function merely results in the appearance of an additional constant factor, it is computationally convenient to reduce the study of exponential functions to the study of this particular function, called the "natural exponential function", or simply, "the exponential function" and denoted by e^x or exp (x)

The exponential function satisfies the fundamental multiplicative identity

$$e^{x+y} = e^x \times e^y \text{ for all real x and y}$$

The exponential function is considered "the most important function in mathematics". Exponential functions occur widely in the natural and social sciences; thus, the exponential function also appears in a variety of contexts within physics, chemistry, engineering, mathematical biology, and economics.

In the figure 8.17 below, graph of $y=e^x$ is shown. It is upward-sloping, and increases faster as x increases. The graph always lies above the x-axis but can get arbitrarily close to it for negative x; thus, the x-axis is an asymptote to the graph curve.

From any point P on the curve, let a tangent line, and a vertical line be drawn, forming a right triangle with height h and base *b* on the *x*-axis. It will be noticed that base b in both cases equals 1. Since the slope of the tangent line (the derivative) at *P* is equal to the ratio of the triangle's height to the triangle's base (rise over run), the derivative equals height h which is nothing but the exponential function y itself. Thus the derivative is equal to the value of the function y itself. Its inverse function is the natural logarithm, denoted $\log_e$ or ln; because of this, some texts refer to the exponential function as the natural **antilogarithm.**

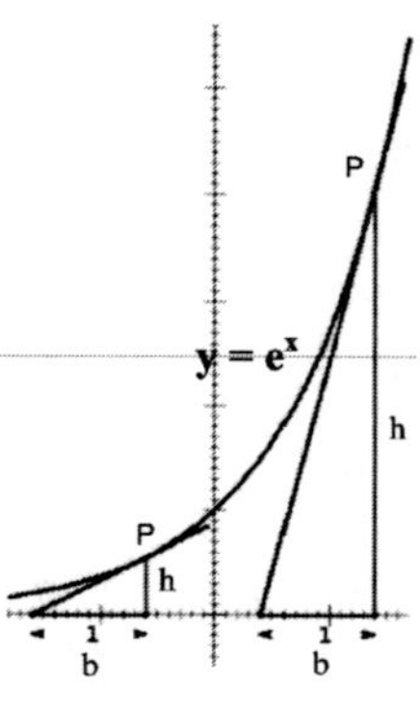

Figure 8.17

8.6.5 Derivative Rules

We need to find derivatives of many different combinations of functions. Rules for finding derivatives of different combinations are as below:

Rules	Function	Derivative
Multiplication by constant	cf	cf'
Power Rule	x^n	nx^{n-1}
Sum Rule	$f+g$	$f'+g'$
Difference Rule	$f-g$	$f'-g'$
Product Rule	fg	$f'g+gf'$
Quotient Rule	$\frac{f}{g}$	$\frac{gf'-g'f}{g^2}$
Reciprocal (same as power rule with quotient –1)	$\frac{1}{f}$	$\frac{-f'}{f^2}$
Chain Rule using ′	$f(g(x))$	$f'(g)\times g'(x)$
Chain Rule using $\frac{d}{dx}$	$\frac{d}{dx}f(g(x))$	$\frac{d}{dg}f(g)\times\frac{d}{dx}g(x)$

8.6.5.1 Proof for product rule

We need to find $\frac{d}{dx}u(x)v(x)$

Using the fundamental definition of derivative, we will get

$$\frac{d}{dx}\{u(x)v(x)\} = \lim_{\Delta x\to 0}\frac{u(x+\Delta x)v(x+\Delta x)-u(x)v(x)}{\Delta x}$$

Subtracting and adding a common term, $u(x+\Delta x)\,v(x)$, we get

$$\frac{d}{dx}u(x)v(x) =$$

$$\lim_{x\to 0}\frac{u(x+\Delta x)v(x+\Delta x)-u(x+\Delta x)v(x)+u(x+\Delta x)v(x)-u(x)v(x)}{\Delta x}\Rightarrow$$

$$=\lim_{\Delta x\to 0}\frac{u(x+\Delta x)\big(v(x+\Delta x)-v(x)\big)+v(x)\big(u(x+\Delta x)-u(x)\big)}{\Delta x}\Rightarrow$$

$$= \lim_{\Delta x \to 0} u(x+\Delta x) \times \lim_{\Delta x \to 0} \frac{(v(x+\Delta x) - v(x))}{\Delta x} +$$

$$\lim_{\Delta x \to 0} v(x) \times \lim_{\Delta x \to 0} \frac{(u(x+\Delta x) - u(x))}{\Delta x} \Rightarrow$$

$$\frac{d}{dx} u(x)v(x) = u(x) \times v'(x) + v(x) \times u'(x) \text{ ...... } \textbf{\textit{Hence proved}}$$

8.6.5.2 Proof for quotient rule:

Here we need to find derivative of $\frac{u(x)}{v(x)}$.

While it is possible to have proof of quotient rule by fundamental definition of derivative, it is easier to prove it by the product rule by considering $\frac{1}{v}$ as v^{-1}

Hence $\frac{d}{dx} \frac{u(x)}{v(x)} = \frac{d}{dx} u(x) v^{-1}(x) = u(x) \times \frac{-1}{v^2(x)} v'(x) + \frac{1}{v(x)} \times u'(x) \Rightarrow$

$$\frac{d}{dx} \frac{u(x)}{v(x)} = \frac{v(x)u'(x) - u(x)v'(x)}{v^2(x)} \text{ ...... } \textbf{\textit{Hence proved}}$$

8.6.5.3 Derivatives of trigonometric functions

Normal definition of derivatives is used to arrive at derivative of say **sin x.**

$$\frac{d}{dx} \sin x = \lim_{\Delta x \to 0} \frac{\sin(x+\Delta x) - \sin x}{\Delta x} \Rightarrow$$

$$\frac{d}{dx} \sin x = \lim_{\Delta x \to 0} \frac{\sin x \cos \Delta x + \cos x \sin \Delta x - \sin x}{\Delta x}$$

$$= \lim_{\Delta x \to 0} \left(\frac{\sin x \cos \Delta x - \sin x}{\Delta x} + \frac{\cos x \sin \Delta x}{\Delta x} \right)$$

Since $\lim_{\Delta x \to 0} \cos \Delta x = 1$ leading the first term to vanish. Thus we get

$$\frac{d}{dx}\sin x = \lim_{\Delta x \to 0}\left(\frac{\cos x \sin \Delta x}{\Delta x}\right).$$

Since $\lim_{\Delta x \to 0} \frac{\sin \Delta x}{\Delta x} = 1$, we get $\frac{d}{dx}\sin x = \cos x$

Similarly **derivative of cosx** can be proved as $\frac{d}{dx}(\cos x) = -\sin x$

Let us try another derivative (of tan x) and discover an interesting lesson. We will try two approaches to derive this derivative.

Approach 1: By definition of derivative as in the case of sin x

$$\frac{d}{dx}(\tan x) = \lim_{\Delta x \to 0} \frac{\tan(x+\Delta x) - \tan x}{\Delta x} = \lim_{\Delta x \to 0} \frac{\frac{\tan x + \tan \Delta x}{1 - \tan x \tan \Delta x} - \tan x}{\Delta x}$$

$$= \lim_{\Delta x \to 0} \frac{\tan \Delta x + \tan^2 x \tan \Delta x}{\Delta x} \times \frac{1}{1 - \tan x \tan \Delta x}$$

We will notice that limit of the second term $\lim_{\Delta x \to 0}\left(\frac{1}{1 - \tan x \tan \Delta x}\right)$ reduces to 1 as $\lim_{\Delta x \to 0} \tan \Delta x = 0$

Thus

$$\frac{d}{dx}(\tan x) = \lim_{\Delta x \to 0} \frac{\tan x}{\Delta x} \times \left(1 + \tan^2 x\right) = \left(1 + \tan^2 x\right) \times \lim_{\Delta x \to 0} \frac{\tan x}{\Delta x} \times 1$$

Since $\lim_{\Delta x \to 0} \frac{\tan x}{\Delta x} = 1$ we get,

$$\boxed{\frac{d}{dx}(\tan x) = \left(1 + \tan^2 x\right) = \sec^2 x}$$

Approach 2: By Quotient Rule:

$$\frac{d}{dx}(\tan x) = \frac{d}{dx}\frac{\sin x}{\cos x} = \frac{\cos x \times \frac{d}{dx}\sin x - \sin x \times \frac{d}{dx}\cos x}{\cos^2 x}$$

$$= \frac{\cos^2 x - \sin x \times (-\sin x)}{\cos^2 x} = \frac{\cos^2 x + \sin^2 x}{\cos^2 x} = \frac{1}{\cos^2 x} = \sec^2 x$$

It may be noticed that the second approach is much simpler. You may carefully select the approach appropriate to solve different problems on derivatives.

8.6.5.4 Derivatives of common functions

Common functions	Function	Derivative
Constant	c	0
Line	$mx + c$	m
Square	x^2	$2x$
Square root	$\sqrt{x}$	$\frac{1}{2\sqrt{x}}$
Exponential	e^x a^x	e^x $a^x \log_e a$
Logarithms	$\log_e x$ $\log_a x$	$\frac{1}{x}$ $\frac{1}{x} \log_e a$
Trigonometry- angle x in radians	$Sin\ x$	$Cos\ x$
	Cos x	-Sin x
	Tan x	Sec^2 x
	Cosec x	-Cot x Cosec x
	Sec x	Tan x Sec x
	$Cot\ x$	$-Cosec^2\ x$
Inverse Trigonometry	$Sin^{-1}x$	$\frac{1}{\sqrt{1-x^2}}$
	$Cos^{-1}x$	$-\frac{1}{\sqrt{1-x^2}}$
	$Tan^{-1}x$	$\frac{1}{1+x^2}$

8.6.5.5 Proof for derivative of e^x

Let h be a small variant of x like Δx

By first principles,

$$\frac{d}{dx}e^x = \lim_{h\to 0}\frac{e^{x+h} - e^x}{h} = \lim_{h\to 0}\frac{e^x \times e^h - e^x}{h} = e^x \lim_{h\to 0}\frac{\left(e^h - 1\right)}{h}$$

We have seen that $e^x = 1 + x + \frac{x^2}{2!} + \frac{x^3}{3!} + ...$

Hence $\frac{(e^h - 1)}{h} = 1 + \frac{h}{2!} + \frac{h^2}{3!} + ... \Rightarrow \lim_{h\to 0}\frac{(e^h - 1)}{h} = 1$

So $\frac{d}{dx}e^x = e^x \lim_{h\to 0}\frac{(e^h - 1)}{h} = e^x$

8.6.5.6 Proof for Derivative of ln x

Let $\ln x = p$.

As per definition of natural logarithm, $\ln x = p$ means $e^p = x$

Differentiating both sides, $\frac{d}{dx}e^p = \frac{d}{dx}(x) \Rightarrow e^p\frac{dp}{dx} = 1$

Or $\frac{dp}{dx} = \frac{1}{e^p} = \frac{1}{x} \Rightarrow \frac{d}{dx}(\ln x) = \frac{1}{x}$.

8.6.5.7 Proof for derivative of a^x

Let $\ln a = p \Rightarrow e^p = a$

Hence $a^x = e^{px}$ .....(1)

We have seen above that $\frac{d}{dx}e^x = e^x$

Differentiating both sides of eq. (1),

$\frac{d}{dx}a^x = \frac{d}{dx}e^{px} = e^{px}\frac{d}{dx}px = a^x \times p$ (p being a constant)

Hence $\frac{d}{dx}a^x = a^x \times \ln a$ or $\frac{d}{dx}a^x = a^x \times \log_e a$

8.6.6 Limits by L'Hopital's Rule using derivatives

With L'Hopital's Rule, we can solve limits using our skills for finding derivatives.

This rule says that to find the limit of a quotient, you only need to find the derivatives of both the numerator and denominator and apply the limit again. This works only if the quotient is an indeterminate form 0/0 or infinity over infinity. For example:

$$\lim_{x \to \infty} \frac{\log x}{x}$$

Both the numerator and denominator approach infinity. So, this is an indetermination of the form infinity over infinity. As per L'Hopital's Rule, we derive the derivative of numerator and denominator and apply the limit again and arrive at the following result:

$$\lim_{x \to \infty} \frac{\log x}{x} = \lim_{x \to \infty} \frac{\frac{1}{x}}{1} = 0$$

Here is a challenging limit problem. This is a problem where it is not initially an indetermination of the type 0/0 or infinity/infinity.

$$\lim_{x \to 0} \left(\frac{a^x - x\ln a}{b^x - x\ln b} \right)^{\frac{1}{x^2}}$$

To transform this into 0/0 or infinity/infinity we apply natural logarithm. That is, suppose that we define the function:

$$f(x) = \left(\frac{a^x - x\ln a}{b^x - x\ln b} \right)^{\frac{1}{x^2}}$$

If we apply natural logarithms to both sides:

$$\ln f(x) = \frac{\ln\left(\frac{a^x - x\ln a}{b^x - x\ln b} \right)}{x^2}$$

Now, as the argument of the logarithm approaches 1, the natural logarithm approaches zero leading to the numerator going to zero as x approaches 0. The denominator obviously approaches 0 when x approaches zero. So, if we take the limit as x approaches zero of this new function, we have 0/0. Therefore, we can apply L'Hopital's rule to this limit.

To solve this we need to apply L'Hopital's Rule twice. It is not an easy problem even from here. To get the original limit back, apply limit to the log of original function and get back the original limit thereafter by evaluating the function:

$$\lim \ln f(x) = \ln \lim f(x)$$

If we solve the limit to the left, we will have the natural logarithm of the limit we originally were trying to solve. Then, we just need to solve the logarithmic equation.The final answer is:

$$\lim_{x \to 0} f(x) = e^{\frac{(\ln a)^2 - (\ln b)^2}{2}}$$

Readers are encouraged to try and solve this problem on their own. However, the solution is provided in the Annexure if it needs to be referred to.

8.7 Unsolved problems

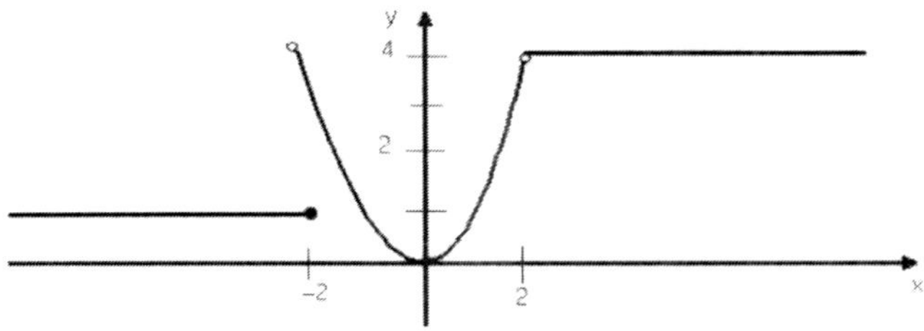

1. With reference to the above graph, find the following limits:

$\lim_{x \to 2^+} f(x)$ $\lim_{x \to 2^-} f(x)$ $\lim_{x \to 2} f(x)$

$\lim_{x \to -2^+} f(x)$ $\lim_{x \to -2^-} f(x)$ $\lim_{x \to -2} f(x)$

$\lim_{x \to \infty} f(x)$ $\lim_{x \to -\infty} f(x)$ $\lim_{x \to 0} f(x)$

2. Find the following limits

$\lim_{x \to -1} \frac{x^2 + 3x + 2}{x+1}$ $\lim_{x \to 2} \frac{x^2 + 3x - 10}{x - 2}$ $\lim_{x \to \infty} \frac{73 + x^3 - x^4}{-5 + 3x^2 - 7x^3}$

$\lim_{x \to 0} \frac{x}{1 - e^x}$ $\lim_{x \to 0} \frac{x^2}{1 - \sec x}$ $\lim_{x \to 0} x \cot x$

$\lim_{x \to 0} \csc x - \frac{1}{x}$ $\lim_{x \to \infty} \frac{\ln(1+x)}{x}$ $\lim_{x \to \infty} (1+x)^{\frac{1}{x}}$

3. Consider the function f(x) = $|x|$ -x where $|x| = \begin{cases} x \text{ for } x > 0 \\ 0 \text{ for } x = 0 \\ -x \text{ for } x < 0 \end{cases}$

(a) What is f(x) when x>0

(b) What is f(x) when x<0

(c) Find the following limits

$\lim_{x \to 0+} f(x) \qquad \lim_{x \to 0-} f(x)$

Is f(x) continuous at x=0

(d) Find the following limits

$$\lim_{h \to 0+} \frac{f(0+h)-f(0)}{h}$$

$$\lim_{h \to 0-} \frac{f(0+h)-f(0)}{h}$$

Is f(x) differentiable at x=0

4. Find derivative $\frac{dy}{dx}$ of the following functions

(i) $y = e^{(x^2+4)}$

(ii) $y = e^{\cos x}$

(iii) $y = \ln(\cos^2 x)$

(iv) $y = \ln(\sin x)$

(v) $y = \sqrt{u}$ where $u = 1 + x^2$

(vi) $y = \frac{u+1}{u-1}$ where $u = e^{2x}$

5. Find derivative $\frac{dy}{dx}$ given the following equations in terms of x and y:

(i) $xy = 1$

(ii) $e^x y = 1$

(iii) $y \cos x = x$

(iv) $x \cos y = y$

(v) $x^2 + 2xy + y^2 = 1$

(vi) $y^3 + 4xy + x^3 = 7$

(vii) $\cos(x+y) = x$

(viii) Find slope of the tangent line at any point (x,y) for the curve with following equation:

$y^2 = x^3 + 3x^2$

At what points other than the origin, does this curve has

(a) Horizontal slope

(b) Vertical slope

(ix) The equation $x^2 - xy + y^2 = 3$ represents an ellipse.

(a) Find slope of the tangent line at any point (x,y)

(b) Find the points where the ellipse cuts the x-axis

(c) Show that tangent lines at points in (b) above are parallel.

Answers to Unsolved Problems:

1. 4, 4, Does not exist

   . 4, 1, Does not exist

   4, 1, 0

2. 1, 7, ∞

   -1, -2, 1

   0, 0, 1

3. 0, -2x , 0

   0, 0, yes=0

   0, -2, No

4.

   (i) $2x \times e^{(x^2+4)}$

   (ii) $-\sin x \times e^{\cos x}$

   (iii) $-2\tan x$

   (iv) $\cot x$

   (v) $\dfrac{x}{\sqrt{1+x^2}}$

   (vi) $\dfrac{-4e^{2x}}{(e^{2x}-1)^2}$

5.

   (i) $\dfrac{-y}{x}$

   (ii) $-y$

   (iii) $\dfrac{1+y\sin x}{\cos x}$

   (iv) $\dfrac{\cos y}{1+x\sin y}$

   (v) -1

   (vi) $-\dfrac{3x^2+4y}{3y^2+4x}$

(vii) $-\dfrac{1+\sin(x+y)}{\sin(x+y)}$..

(viii) Slope = $\dfrac{3x(x+2)}{2y}$, (a) (-2,2) and (-2,-2) (b) (-3,0)

(ix) Slope = $\dfrac{y-2x}{2y-x}$, (a) ($\sqrt{3}$,0) and ($-\sqrt{3}$,0) (b) slope of tangent at $y = 0$ is same and equals 2

Annexure

Solution for Type 3 Limits- double rationalisation

First rationalizing $\sqrt{x^2+p^2}-\sqrt{p}$,

$$\lim_{x\to 0}\frac{\sqrt{x^2+p^2}-\sqrt{p}}{\sqrt{x^2+q^2}-\sqrt{q}} = \lim_{x\to 0}\frac{(\sqrt{x^2+p^2}-p)\times(\sqrt{x^2+p^2}+p)}{(\sqrt{x^2+q^2}-q)\times(\sqrt{x^2+q^2}+p)}$$

$$= \lim_{x\to 0}\frac{x^2}{(\sqrt{x^2+q^2}-q)\times(\sqrt{x^2+q^2}+p)}$$

Now rationalizing $\sqrt{x^2+q^2}-\sqrt{q}$,

$$\lim_{x\to 0} = \frac{x^2}{(\sqrt{x^2+q^2}-q)\times(\sqrt{x^2+p^2}+p)} \Rightarrow$$

$$\lim_{x\to 0} = \frac{x^2\times(\sqrt{x^2+q^2}+q)}{(\sqrt{x^2+q^2}-q)\times(\sqrt{x^2+q^2}+q)\times(\sqrt{x^2+p^2}+p)} \Rightarrow$$

$$\lim_{x\to 0} = \frac{x^2\times(\sqrt{x^2+q^2}+q)}{x^2\times(\sqrt{x^2+p^2}+p)} = \lim_{x\to 0}\frac{(\sqrt{x^2+q^2}+q)}{(\sqrt{x^2+p^2}+p)} = \frac{2q}{2p} = \frac{q}{p}$$

Solution for Limits- double L'Hopital's Rule

We need to find the following limit:

$$\lim_{x\to 0}\left(\frac{a^x - x\ln a}{b^x - x\ln b}\right)^{\frac{1}{x^2}}$$

Let $f(x) = \left(\frac{a^x - x\ln a}{b^x - x\ln b}\right)^{\frac{1}{x^2}}$

$$\lim_{x\to 0} Ln\, f(x) = \lim_{x\to 0}\frac{1}{x^2}\ln\frac{a^x - x\ln a}{b^x - x\ln b} =$$

$$= \lim_{x\to 0}\frac{1}{x^2}\left[\ln(a^x - x\ln a) - \ln(b^x - x\ln b)\right]$$

$$= \lim_{x\to 0}\left(\frac{\ln(a^x - x\ln a)}{x^2} - \frac{\ln(b^x - x\ln b)}{x^2}\right)$$

If we take the limit as x approaches zero of this new function, we have 0/0. Therefore, we can apply L'Hopital's rule to this limit. Applying L'Hopital's Rule to each of the two components

$$= \lim_{x\to 0}\frac{\frac{d}{dx}\ln(a^x - x\ln a)}{\frac{d}{dx}x^2} - \lim_{x\to 0}\frac{\frac{d}{dx}\ln(b^x - x\ln b)}{\frac{d}{dx}x^2}$$

$$= \lim_{x\to 0}\frac{\frac{1}{(a^x - x\ln a)}\times(a^x\ln a - \ln a)}{2x} - \lim_{x\to 0}\frac{\frac{1}{(b^x - x\ln b)}\times(b^x\ln b - \ln b)}{2x}$$

Applying L'Hopital's Rule a second time, the above equation transforms to

$$= \lim_{x\to 0}\frac{\frac{(a^x - x\ln a)\times a^x(\ln a)^2 - (a^x\ln a - \ln a)^2}{(a^x - x\ln a)^2}}{2} - \lim_{x\to 0}\frac{\frac{(b^x - x\ln b)\times b^x(\ln b)^2 - (b^x\ln b - \ln b)^2}{(b^x - x\ln b)^2}}{2}$$

Applying limit as x approaches zero, the above expression further reduces to

$$\frac{\frac{1\times(\ln a)^2 - 0}{1} - \frac{1\times(\ln b)^2 - 0}{1}}{2} = \frac{(\ln a)^2 - (\ln b)^2}{2}$$

Thus $\lim_{x\to 0} Ln\ f(x) = \frac{(\ln a)^2 - (\ln b)^2}{2} = Ln \lim_{x\to 0} f(x)$ leading to

$$\lim_{x\to 0} f(x) = e^{\frac{(\ln a)^2 - (\ln b)^2}{2}}$$

9

INTEGRAL CALCULUS

9.1 What is integral Calculus

Integral calculus also called integration is the reverse of differentiation. Recall the example of time vs distance curve discussed in chapter 8. There, we found rate of change of distance with time to get derivative $\frac{ds}{dt}$ which meant velocity v(t) of the particle at time t. On the contrary, if we have a graph of velocity over time and we want to find distance covered by the moving object over time period from time t_1 to time t_2, we need to perform the reverse process to find product of instantaneous velocity v(t) and very small time interval dt (that is close to 0) and find sum of all these areas from time t_1 to t_2 yielding area ABCD in figure 9.1 below:

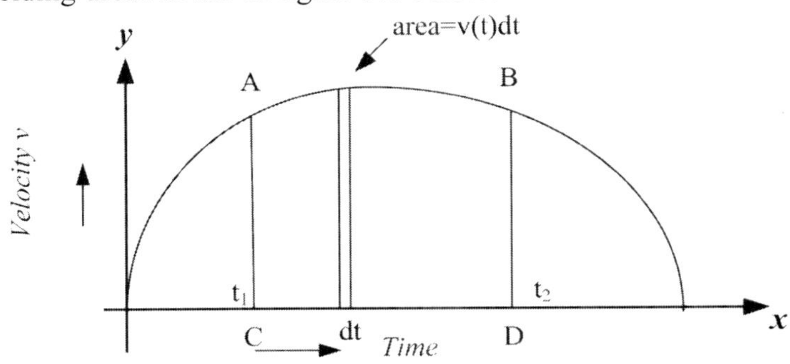

Figure 9.1

This process is symbolically expressed as $\int_{t_1}^{t_2} v(t)dt$. So integral of a function is primarily a product of two variables, an independent variable (time in above case) and dependent variable (continuously varying velocity with independent variable time in the above case). The product of the two variables velocity and time is nothing but distance covered over given time interval. This concept remains the same irrespective of the variables. As another example if we integrate acceleration with reference to time over a time interval, we get average velocity in the given interval.

9.2 Comparison of integration and differentiation

Let us consider a simple function $y=x$, shown in figure 9.2 below

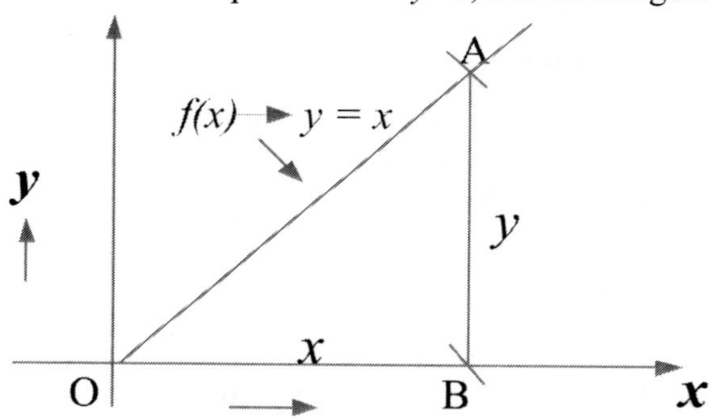

Figure 9.2

As per definition of integration explained above, integral of the above function $y = x$ over interval 0 to x represented as $\int_0^x ydx$ equals area of the right angled triangle OAB where base OA = x and altitude AB = y = x (due to the function y = x). Thus

Integral value of function y = x over period 0 to x equals area of $\Delta OAB = \frac{x^2}{2}$. We can thus state that

$$\int xdx = \frac{x^2}{2} = \frac{x^{1+1}}{(1+1)}$$

The above result is true even for the generic function $y = x^n$ as below

$$\int x^n dx = \frac{x^{n+1}}{(n+1)}.$$

Recall that derivative $\frac{dy}{dx}$ of function $x^n = nx^{n-1}$. We can also state that derivative $\frac{dy}{dx}$ of function $\frac{x^n}{n} = x^{n-1}$. If we consider integral of Right hand side of this equation, the result is same as the Left hand side of the equation. This is exact same result mentioned above for $\int_0^x x^n dx$. Since differential of a constant is 0, we normally add a constant C to the generic integration value. Thus complete value of $\int x^n dx = \frac{x^{n+1}}{(n+1)} + C$

Thus we can conclude that integration is simply reverse of derivative of any function. ***Integral is also sometimes called anti-derivative***. This will be evident from the examples given in the table in section 9.5 later in the chapter.

9.3 Definite and Indefinite integrals

We have defined in section 9.1. $\int_{t_1}^{t_2} v(t)dt$ as area of the curve v(t) with reference to time depicted on the x-axis from t_1 to t_2. This is a case of definite integral in the defined time range (t_1 to t_2). On the contrary, indefinite integral defines an integral without any defined limits. Thus indefinite integral defines an unknown extent of the curve $v(t)$. This definition leads us to fundamental theorems of Integral Calculus defined below. Definite integrals have more practical applications and will be dealt later.

9.3.1 First fundamental theorem of Integral Calculus

Let an area function A(x) be defined as $\int_a^b u(x)dx$ for all values of x such that $a \geq x \geq b$ and u(x) is continuous over the range $[a,b]$.

Then for all $x \in [a,b]$

9.3.2 Second fundamental theorem of Integral Calculus

Let u be a continuous function of x over a closed interval $[a,b]$ and let U be another function such that $\frac{d}{dx}U(x)=u(x)$ for all x in the domain of u(x), then $\int_a^b u(x)dx=[U(x)+C]_a^b=U(b)-U(a)$ (Constant C gets cancelled here)

This is known as definite integral of u(x) over range $[a,b]$ where ***a***&**b** are called limits of integration, **a** being the lower limit.

9.3.3 Properties of indefinite Integrals:

9.3.3.1 Two indefinite integrals with the same derivative are equivalent to each other, separated only by a constant.

If $\frac{d}{dx}\int f(x)dx=\frac{d}{dx}\int g(x)dx$,

$$\int f(x)dx=\int g(x)dx+C$$

9.3.3.2 For any real number k,

$$\int kf(x)dx=k\int f(x)dx$$

9.3.3.3 Integral of sum of two functions equals integral of the two functions separately.

$$\int[f(x)+g(x)]dx=\int f(x)dx+\int g(x)dx$$

Example:

Find $\int(2x^2+3x)dx$

Here we notice that the expression to be integrated is sum of two functions of x say g(x) and f(x) in the above formula. Accordingly,

$$\int(2x^2+3x)dx=\int 2x^2dx+\int 3xdx$$

$$=\frac{2x^3}{3}+\frac{3x^2}{2}+C$$

Other similar problems can be solved knowing integral of individual functions.

9.4 Different methods of Integration

9.4.1 Integration by method of inspection and substitution

Some of the integration problems of a function in x having product of two functions can be reduced to a single function by suitable substitution of the function. For example if f(x) is a product of two functions in x that cannot be integrated straight away and if the product can be split such that second function is derivative of the first function, we can easily solve that problem. We can express this statement as below:

If $\int f(x)dx = \int g(t(x))t'(x)dx$, then

$\int f(x)dx = \int g(t)dt$. This can be solved by known integration formulae. Value of t is then substituted in terms of x. Let us consider a couple of examples:

Example 1:

$$\int 2x\sin(x^2+1)dx$$

Here we notice that 2xdx is derivative of (x^2+1). So we can substitute $t(x)=(x^2+1)$, leading us to dt =2xdx, thereby

$$\int 2x\sin(x^2+1)dx = \int \sin t dt = -\cos t + C = -cos\left(x^2+1\right) + C$$

Example 2:

$$\int \frac{\tan^4\sqrt{x}\sec^2\sqrt{x}}{\sqrt{x}}dx$$

We know that derivative of $\sqrt{x} = \frac{1}{2\sqrt{x}}dx$ or $\frac{1}{\sqrt{x}}dx = 2\sqrt{x}$

We also know that derivative of tan x is $\sec^2 x$. With this knowledge, we can try and substitute $\tan\sqrt{x} = t$ in the given problem. This will lead us to $dt = \frac{d}{dx}(\tan\sqrt{x}) = \sec^2\sqrt{x} \times \frac{1}{2\sqrt{x}}dx$

By substituting in the given question

$$\int \frac{\tan^4\sqrt{x}\sec^2\sqrt{x}}{\sqrt{x}}dx = \int 2t^4 dt = \frac{2t^5}{5} + C = \frac{2\tan^5\sqrt{x}}{5} + C$$

Example 3:

$\int \tan x dx$

$\tan x = \dfrac{\sin x}{\cos x}$ and $\dfrac{d}{dx}\cos x = -\sin x$

If we substitute t = cos x,dt=$-\sin x dx$. the original problem reduces to

$$\int \tan x dx = \int \frac{\sin x}{\cos x}dx = \int -\frac{1}{t}dt = -\log|t| + C = \log\frac{1}{|t|} + C$$

Since t = $\cos x$, it implies $\int \tan x dx =$ $\log\dfrac{1}{|t|} + C = \log|\sec x| + C$

9.4.2 Integration using trigonometric identities

Some problems involving trigonometric functions can be solved by using knowledge of trigonometric identities. Consider following examples:

Example 1:

Find $\int \sin 3x \cos 4x dx$

We know the identity $\sin x \cos y = \dfrac{1}{2}[\sin(x+y) + \sin(x-y)]$

Using this identity,

$$\int \sin 3x \cos 4x dx = \int \frac{1}{2}[\sin(3x+4x) + \sin(3x-4x)]dx$$

$$= \int \frac{1}{2}[\sin 7x) + \sin(-x)]dx$$

$$= \frac{1}{2}\left[\int \sin 7x dx + \int -\sin x dx\right]$$

$$= \frac{1}{2}\left[-\frac{\cos 7x}{7} + \cos x\right] + C = \frac{7\cos x - \cos 7x}{14} + C$$

Let us verify this answer by differentiation

$$\frac{d}{dx}\left[\frac{7\cos x - \cos 7x}{14} + C\right] = \frac{1}{14}(-7\sin x + 7\sin 7x)$$

$$= \frac{1}{2}(\sin 7x - \sin x) = \frac{2}{2}\left(\cos\frac{7x+x}{2}\sin\frac{7x-x}{2}\right)$$

$= \cos 4x \sin 3x$, same as the original expression.

Example 2:

Find $\int \frac{\cos x - \sin x}{1 + \sin 2x} dx$

Using the identity, sin2x = 2sinx cosx, the problem reduces to

$$\int \frac{\cos x - \sin x}{1 + 2\sin x \cos x} dx$$.

Now since $\sin^2 x + \cos^2 x = 1$

$$\int \frac{\cos x - \sin x}{1 + 2\sin x \cos x} dx = \int \frac{\cos x - \sin x}{(\sin x + \cos x)^2} dx$$

Substituting (sinx + cosx) =t, dt= (cosx-sinx)dx

The above problem then reduces to

$$\int \frac{\cos x - \sin x}{(\sin x + \cos x)^2} dx = \int \frac{1}{t^2} dt = -\frac{1}{t} + C = -\frac{1}{\sin x + \cos x} + C$$

Readers may try and verify this result by differentiating it. Best way of mastering this technique is to try and solve unknown problems available online in plenty.

9.4.3 Integration using partial fractions

Recall that a rational fraction is the one where numerator is of lesser degree than the denominator and the denominator is $\neq 0$. We can integrate rational fractions where the denominator can be factored into two or more factors. Suppose we want to integrate a rational fraction $\frac{P(x)}{Q(x)}$ we can split it into a sum of rational fractions using the method of partial fractions that we will discuss here. After partial fractions, integration is possible using standard formulae.There are a few standard forms of partial fractions related to different type of rational fractions. These are as below:

Sl. No.	Type of rational function	Form of Partial fraction
1	$\frac{ax+b}{(x-p)^2}$	$\frac{A}{x-p}+\frac{B}{(x-p)^2}$
2	$\frac{ax+b}{(x-p)(x-q)} p \neq q$	$\frac{A}{x-p}+\frac{B}{(x-q)}$
3	$\frac{ax^2+bx+c}{(x-p)(x-q)(x-r)}$	$\frac{A}{x-p}+\frac{B}{(x-q)}+\frac{C}{x-r}$
4	$\frac{ax^2+bx+c}{(x-p)^2(x-q)}$	$\frac{A}{x-p}+\frac{B}{(x-p)^2}+\frac{C}{x-q}$
5	$\frac{ax^2+bx+c}{(x-p)(x^2+qx+r)}$ (x^2+qx+r) cannot be factorised	$\frac{A}{x-p}+\frac{Bx+C}{(x^2+qx+r)}$

Here A, B and C are real numbers that can be determined by algebraic methods.

Let us consider a few examples for further understanding of the concept.

Example 1:

Find $\int \frac{x}{x^2-3x+2}dx$

We notice that the denominator can be easily factorised into $(x-1)(x-2)$. We can carry out partial fractions of the above expression as below:

$$\frac{x}{x^2-3x+2} = \frac{A}{x-1}+\frac{B}{x-2} = \frac{(A+B)x-(2A+B)}{x^2-3x+2}$$

Equating the two sides

A + B = 1 and 2A + B = 0

So A=- 1 and B = 2, Thus

$$\int \frac{x}{x^2 - 3x + 2} dx = \int \frac{-1}{x-1} dx + \int \frac{2}{x-2} dx = -\log|x-1| + 2\log|x-2| + C$$

$$= \log\left|\frac{(x-2)^2}{x-1}\right| + C$$

Example 2:

Find $\int \frac{x}{(x-1)^2(x+2)} dx$

Proceeding as per standard form 4 above,

$$\frac{x}{(x-1)^2(x+2)} = \frac{A}{x-1} + \frac{B}{(x-1)^2} + \frac{C}{x+2}$$

$$\frac{x}{(x-1)^2(x+2)} = \frac{A(x-1)(x+2) + B(x+2) + C(x-1)^2}{(x-1)^2(x+2)}$$

Expanding the numerator and sorting, we get

$$\frac{x}{(x-1)^2(x+2)} = \frac{x^2(A+C) + x(A+B-2C) - (2A-2B-C)}{(x-1)^2(x+2)}$$

Comparing the two sides.

A+C=0, yielding A = - C and A+B-2C=1.

By substituting the above value of A, we get 3A+B=1 ...(1)

Similarly, 2A-2B – C=0, leads to 3A-2B = 0 ...(2)

Subtracting (2) from (1) above,

3B = 1, or B=$\frac{1}{3}$. Substituting this value of B, A= $\frac{2}{9}$ and C= - $\frac{2}{9}$

$\therefore \frac{x}{(x-1)^2(x+2)} = \frac{2}{9(x-1)} + \frac{1}{3(x-1)^2} - \frac{2}{9(x+2)}$; leading us to

$$\int \frac{x}{(x-1)^2(x+2)} dx = \int \frac{2}{9(x-1)} dx + \int \frac{1}{3(x-1)^2} dx - \int \frac{2}{9(x+2)} dx,$$

On solving using standard formulae,

$$\int \frac{x}{(x-1)^2(x+2)} dx = \frac{2}{9}\log|x-1| - \frac{1}{3(x-1)} - \frac{2}{9}\log|x+2| + C$$

$$= \frac{2}{9}\log\left|\frac{x-1}{x+2}\right| - \frac{1}{3(x-1)} + C$$

Example 3:

Find $\int \frac{1}{x(x^n+1)}dx$

This appears to be a little tricky but gets simplified if we multiply numerator and denominator by x^{n-1}, yielding

$$\int \frac{1}{x(x^n+1)}dx = \int \frac{x^{n-1}}{x^n(x^n+1)}dx$$

If we substitute t = x^n, dt = nx^{n-1} and we get

$$\int \frac{1}{x(x^n+1)}dx = \int \frac{x^{n-1}}{x^n(x^n+1)}dx = \frac{1}{n}\int \frac{dt}{t(t+1)}.$$

Now this can be solved by partial fractions as below

$\frac{1}{t(t+1)} = \frac{A}{t} + \frac{B}{t+1}$, on simplifying we get

A+B = 0 and A = 1, thereby, B = -1

$$\therefore \frac{1}{n}\int \frac{dt}{t(t+1)} = \frac{1}{n}\left[\int \frac{1}{t}dt - \frac{1}{t+1}dt\right] = \frac{1}{n}\left[\log|t| - \log|t+1|\right]$$

$$= \frac{1}{n}\log\left|\frac{t}{t+1}\right|$$

Or $\int \frac{1}{x(x^n+1)}dx = \frac{1}{n}\log\left|\frac{x^n}{x^n+1}\right| + C$

9.4.4 Integration by parts

Integration by Parts is a special method of integration that is often useful when integrating two functions that are multiplied together. If the two functions are denoted as u(x) and v(x), integral of their product as per formula of integration by parts is: $\int uvdx = u\int vdx - \int u' \times \int vdx\,dx$ where

- u is the function u(x)
- v is the function v(x)
- u′ = $\frac{du}{dx}$

Explanation: This rule comes from the Product Rule of differentiation (given below) that we have learnt earlier.

$$\frac{d}{dx}(u(x)v(x)) = v(x)\frac{d}{dx}u(x) + u(x)\frac{d}{dx}v(x)$$

We can deduce the formula for Integration by parts by integrating both sides of the above equation leading to

$$\int\frac{d}{dx}(u(x)v(x)) = \int u'(x)v(x)dx + \int u(x)v'(x)dx$$

By transposing $\int u'vdx$ ***from right to left of the equation, we get***

or $\int uv'dx = uv - u'vdx$

We know that $v' = \frac{dv}{dx} \Rightarrow v'dx = dv$ ***and*** $u' = \frac{du}{dx} \Rightarrow u'dx = du$

Substituting above values in equation (1) will yield

$\int udv = uv - \int du \times v$, ***where u is the first part and dv is the second part of the product***

Result of the Product Rule of integration can be described as " first term as it is multiplied by integral of the second term minus integral of product of derivative of the first part and integral of the second part "

Let us consider a couple of examples of integration by parts to get a hang of this complex process:

Example 1: What is $\int x\cos x\,dx$?

This is in the format $\int udv$ where $u = x$ and $dv = cos\ x\ dx$

Using the above formula, we get

$$\begin{aligned}\int x\cos x\,dx &= x\int\cos x\,dx - \int(1\times\int\cos x)dx\\ &= x\sin x - \int\sin x dx\\ &= x\sin x - (-\cos x)\\ &= x\sin x + \cos x + C\end{aligned}$$

Readers should verify the result by differentiating the term on RHS to arrive at the original function being integrated.

Example 2: Find $\int e^x x dx$

If we assume $u = e^x$ and $dv = x\mathrm{d}x$ leading us to

$$\int e^x x dx = e^x \int x dx - \int \frac{d}{dx} e^x \left(\int x dx\right) dx$$
$$= e^x \frac{x^2}{2} - \int e^x \frac{x^2}{2} dx$$

It will be noticed that the result gets complicated as we find $\int e^x \frac{x^2}{2} dx$ is as or even more complicated than the original function.

Let us now assume $u=x$ and $v=e^x$ and see what happens. Using the same formula, we get

$$\int e^x x dx = x \int e^x dx - \int \frac{d}{dx} x \left(\int e^x dx\right) dx$$
$$= xe^x - \int 1 \times e^x dx = xe^x - e^x = e^x (x-1) + C$$

This leads us to an important conclusion that we should choose functions u and v carefully. Thumb rule is to choose **u** such that it simplifies on differentiation and **v** that does not get any more complicated on integration.

A helpful rule of thumb to correctly choose u and v is **"I LATE"** (Courtesy: mathsisfun.com), meaning thereby to choose **u** based on which of these comes first:

I: Inverse trigonometric functions such as $\sin^{-1}(x)$, $\cos^{-1}(x)$, $\tan^{-1}(x)$

L: Logarithmic functions such as ln(x), log(x)

A: Algebraic functions such as x^2, x^3

T: Trigonometric functions such as sin(x), cos(x), tan (x)

E: Exponential functions such as e^x, 3^x

Now let us try a tricky example below:

Example 3: Find $\int e^x \sin x dx$

We assume $u = e^x$ *and* $dv = \sin x dx$

$$\int e^x \sin x dx = e^x \int \sin x dx - \int \frac{d}{dx} e^x \left(\int \sin x dx\right) dx$$

$$= e^x(-\cos x) - \int e^x(-\cos x) dx$$

$$= -e^x \cos x + \int e^x \cos x dx$$

$$= -e^x \cos x + e^x \int \cos x dx - \int \frac{d}{dx} e^x (\int \cos x dx) dx$$

$$= -e^x \cos x + e^x \sin x - \int e^x \sin x dx$$

We notice that the last term on RHS is same as the original term. Hence on transposing, we get

$2\int e^x \sin x dx = -e^x \cos x + e^x \sin x$, or

$$\int e^x \sin x dx = \frac{1}{2} e^x (\sin x - \cos x) + C$$

An important note: We can try this example by assuming u and v, the other way round i.e. $u = \sin x$ *and* $dv = e^x dx$. The result will be same. In fact last two types (trigonometric and Exponent) for the ILATE rule can be selected interchangeably as first term without any change in complexity of the solution.

Here is an interesting type of problem. Let us say we need to find integration of $e^x f(x)$, where $f(x)$ can be split into two parts such that the second part is differential of the first part i.e. $f(x) = f_1(x) + f_1'(x)$

Now let us integrate the given product by parts as below:

As per ILATE Rule, we take *f(x)* as the first part and e^x as the second part. By substituting assumed expansion of f(x) we get

$$\int e^x f(x) dx = \int f_1(x) e^x dx + \int f_1'(x) e^x dx \qquad ...(1)$$

Now integrating first part on the RHS, we get

$$\int e^x f_1(x) dx = f_1(x) \int e^x dx - \int f_1'(x) \int e^x dx = f_1(x) e^x - \int f_1'(x) e^x dx \quad ..(2)$$

Substituting result obtained in equation (2) above in equation (1), we get

$$\int e^x f(x) dx = f_1(x) e^x - \int f_1'(x) e^x dx + \int f_1'(x) e^x dx = e^x f_1(x) + C$$

Let us consider a couple of real life examples below:

$\int e^x \frac{x-1}{(x+1)^3} dx$.

We can split $\frac{x-1}{(x+1)^3}$ as $\frac{x+1-2}{(x+1)^3} = \frac{1}{(x+1)^2} - \frac{2}{(x+1)^3}$.

If we consider $\frac{1}{(x+1)^2}$ as $f_1(x)$, we will see that $-\frac{2}{(x+1)^3}$ is $f_1'(x)$ reducing the given problem as $\int e^x f(x)dx = \int f_1(x)e^x dx + \int f_1'(x)e^x dx$

Using the concept explained above, we get

$$\int e^x \frac{x-1}{(x+1)^3}dx = e^x \frac{1}{(x+1)^2} + C$$

$$\int e^x \frac{\sin x - 1}{(1-\cos x)}dx$$

Simplifying the trigonometric function we get,

$$\frac{\sin x - 1}{(1-\cos x)} = \frac{2\sin\frac{x}{2}\cos\frac{x}{2} - 1}{2\sin^2\frac{x}{2}} = \cot\frac{x}{2} - \frac{1}{2}\operatorname{cosec}^2\frac{x}{2}$$

Now if we assume $\cot\frac{x}{2}$ as $f_1(x)$, $f_1'(x) = -\frac{1}{2}\operatorname{cosec}^2\frac{x}{2}$ leading to the problem reducing to $\int e^x f(x)dx = \int f_1(x)e^x dx + \int f_1'(x)e^x dx$ yielding the solution as

$$\int e^x \frac{\sin x - 1}{(1-\cos x)}dx = e^x \cot\frac{x}{2} + C$$

By practicing more similar examples and carefully choosing u and v, readers can easily master Integration of product of two functions.

9.5 Some formulae with their differential equivalent

Here are examples of some standard integration formulae with their equivalent derivative value

Derivative	***Integral (anti-derivative)***
$\frac{d}{dx}n = 0$ (where n is a constant)	$\int 0dx = C$, a constant
$\frac{d}{dx}x = 1$	$\int 1dx = x + C$

$\frac{d}{dx}x^{n+1} = (n+1)x^n$	$\int x^n dx = \frac{x^{n+1}}{n+1} + C$
$\frac{d}{dx}(ax+b)^{n+1} = (n+1)(ax+b)^n \times a$	$\int (ax+b)^n dx = \frac{(ax+b)^{n+1}}{a(n+1)} + C$, n≠-1
$\frac{d}{dx}e^x = e^x$	$\int e^x dx = e^x + C$
$\frac{d}{dx}\ln x = \frac{1}{x}$ (***ln,*** read as natural logarithm is same as $\log_e$)	$\int \frac{1}{x} dx = \ln x + C$
$\frac{d}{dx}a^x = a^x \times \ln a$	$\int a^x dx = \frac{a^x}{\ln a} + C$
$\frac{d}{dx}(\sin x) = \cos x$	$\int \cos x dx = \sin x + C$
$\frac{d}{dx}(-\cos x) = \sin x$	$\int \sin x dx = -\cos x + C$
$\frac{d}{dx}(\tan x) = \sec^2 x$	$\int \sec^2 x dx = \tan x + C$
$\frac{d}{dx}(-\cot x) = \cos ec^2 x$	$\int -co\sec^2 x dx = \cot x + C$
$\frac{d}{dx}(\sec x) = \sec x \tan x$	$\int \sec x \tan x dx = \sec x + C$
$\frac{d}{dx}(-co\sec x) = co\sec x \cot x$	$\int co\sec x \cot x dx = -co\sec x + C$
$\frac{d}{dx}(\sin^{-1} x) = \frac{1}{\sqrt{1-x^2}}$	$\int \frac{dx}{\sqrt{1-x^2}} = \sin^{-1} x + C$
$\frac{d}{dx}(-co\text{s}^{-1} x) = \frac{1}{\sqrt{1-x^2}}$	$\int \frac{dx}{\sqrt{1-x^2}} = -co\text{s}^{-1} x + C$
$\frac{d}{dx}(\tan^{-1} x) = \frac{1}{1+x^2}$	$\int \frac{dx}{1+x^2} = \tan^{-1} x + C$

$\frac{d}{dx}(-\cot^{-1} x) = \frac{1}{1+x^2}$	$\int \frac{dx}{1+x^2} = -\cot^{-1} x + C$
$\frac{d}{dx}(\sec^{-1} x) = \frac{1}{x \times \sqrt{1-x^2}}$	$\int \frac{dx}{x \times \sqrt{1-x^2}} = \sec^{-1} x + C$
$\frac{d}{dx}(-co\sec^{-1} x) = \frac{1}{x \times \sqrt{1-x^2}}$	$\int \frac{dx}{x \times \sqrt{1-x^2}} = -co\sec^{-1} x + C$

9.6 Formulae for integration of inverse trigonometric functions:

Formulae for integration of common inverse trigonometric functions that are used frequently in solving real life problems are given below.

Reverse trignometric functions	**Corresponding Integral**
$\int \sin^{-1} x dx$	$x \sin^{-1} x + \sqrt{1-x^2} + C$
$\int \cos^{-1} x dx$	$x \cos^{-1} x - \sqrt{1-x^2} + C$
$\int \tan^{-1} x dx$	$x \tan^{-1} x - \ln\sqrt{1+x^2} + C$
$\int -\cos ec^{-1} x dx$	$x \cos ec^{-1} x + x \times \sqrt{1-x^2} + C$
$\int sec^{-1} x dx$	$x sec^{-1} x - x \times \sqrt{1-x^2} + C$
$\int -cot^{-1} x dx$	$-x cot^{-1} x - \ln\sqrt{1+x^2} + C$

Readers are encouraged to verify these formulae by differentiating the formula for integration by using method for differentiating product of two variables discussed in chapter 8.

9.7 Integration results of a few special functions:

Here are formulae for integration of a few very useful special functions:

1. $\int \frac{dx}{x^2 - a^2} = \frac{1}{2a} \log\left|\frac{x-a}{x+a}\right| + C$

2. $\int \frac{dx}{a^2 - x^2} = \frac{1}{2a} \log\left|\frac{a+x}{a-x}\right| + C$

3. $\int \frac{dx}{a^2+x^2} = \frac{1}{a}\tan^{-1}\frac{x}{a} + C$

4. $\int \frac{dx}{\sqrt{x^2-a^2}} = \log\left|(x+\sqrt{x^2-a^2})\right| + C$

5. $\int \frac{dx}{\sqrt{a^2-x^2}} = \sin^{-1}\frac{x}{a} + C$

6. $\int \frac{dx}{\sqrt{x^2+a^2}} = \log\left|(x+\sqrt{x^2+a^2})\right| + C$

Let us prove a few of these results

Here is proof for result 2 above.

To find $\int \frac{dx}{a^2-x^2}$.

$$\int \frac{dx}{a^2-x^2} = \int \frac{dx}{(a-x)(a+x)} = \int \frac{(a+x)^2}{(a-x)(a+x)} \times \frac{1}{(a+x)^2}dx$$

$$= \int \frac{(a+x)}{(a-x)} \times \frac{1}{(a+x)^2}dx \qquad ...(1)$$

Let us substitute,

$$\frac{a-x}{a+x} = t \text{ , dt} = \frac{(a+x)(-1)-(a-x)(1)}{(a+x)^2}dx = \frac{-2a}{(a+x)^2}dx$$

Substituting in the original expression,

$$\int \frac{dx}{a^2-x^2} = -\frac{1}{2a}\int \frac{1}{t}dt$$

$$= -\frac{1}{2a}\log|t| + C = \frac{1}{2a}\log\left|\frac{1}{t}\right| + C = \frac{1}{2a}\log\left|\frac{a+x}{a-x}\right| + C$$

Proof for result 4:

To find $\int \frac{dx}{\sqrt{x^2-a^2}}$.

Substitute $x = a\sec t$ leading to $dx = a\tan t\sec t$

$$\int \frac{dx}{\sqrt{x^2-a^2}} = \int \frac{a\tan t\sec t}{a(\sqrt{\sec^2 t - 1})}dt = \int \frac{a\tan t\sec t}{a\tan t}dt = \int \sec t dt$$

$(\because 1+\tan^2 t = \sec^2 t\ )$

Multiplying numerator and denominator by (tan t + sec t)

$$\int \sec t dt = \int \frac{\sec t(\tan t + \sec t)}{(\sec t + \tan t)} dt = \int \frac{(\tan t \sec t + \sec^2 t)}{(\sec t + \tan t)} dt$$

Let us make a second substitution

$(\sec t + \tan t) =$ p, implying dp $=$ $(\tan t \sec t + \sec^2 t)$ dt. This leads us to

$$\int \frac{(\tan t \sec t + \sec^2 t)}{(\sec t + \tan t)} dt = \int \frac{1}{p} dp = \log|p| + C$$

Substituting back, the result is

$$\log|p| + C = \log|\tan t + \sec t| + C = \log\left|\sqrt{\sec^2 t - 1} + \sec t\right| + C$$

$$= \log\left|\left[\sqrt{\frac{x}{a^2}^2 - 1}\right] + \frac{x}{a}\right| + C$$

$$= \log\left|\frac{1}{a}\left[\sqrt{x^2 - a^2} + x\right]\right| + C$$

$$= \log\left|x + \sqrt{x^2 - a^2}\right| - \log|a| + C$$

$$= \log\left|x + \sqrt{x^2 - a^2}\right| + C', \text{ because}$$

$\log|a|$ is a constant and can be combined with constant C

Proof for result 5:

To find $\int \frac{dx}{\sqrt{a^2 - x^2}}$

Let $x = a \sin t$

dx $=$ $a \cos t dt$; substituting in the original expression,

$$\int \frac{dx}{\sqrt{a^2 - x^2}} = \int \frac{a \cos t dt}{\sqrt{a^2 - a^2 \sin^2 t}} = \int \frac{a \cos t dt}{a \cos t} = \int dt = t + C$$

$$= \sin^{-1} \frac{x}{a} + C$$

Similarly, by suitable substitution, it should be easy to prove rest of the above results.

These standard formulae are useful in solving many complicated integration problems by suitably reducing them to the standard form. Let us see a couple of examples below:

1. Find integral $\int \frac{dx}{ax^2 + bx + c}$

We can reduce the denominator to the form $t^2 \pm p^2$ and use corresponding standard formula to get the result. Here are the necessary steps to achieve desired result:

$$\int \frac{dx}{ax^2 + bx + c} = \int \frac{dx}{a\left(x^2 + \frac{b}{a}x + \frac{c}{a}\right)} = \int \frac{dx}{a\left(\left(x + \frac{b}{2a}\right)^2 + \left(\frac{c}{a} - \frac{b^2}{4a^2}\right)\right)}$$

Assuming t = $\left(x + \frac{b}{2a}\right)$ and $\left(\frac{c}{a} - \frac{b^2}{4a^2}\right) = \pm$ p^2, depending upon value of the expression $\left(\frac{c}{a} - \frac{b^2}{4a^2}\right)$. We can see that dt = dx. The problem now reduces to

$\int \frac{dx}{ax^2 + bx + c} = \frac{1}{a} \int \frac{dt}{t^2 \pm p^2}$, which will give the result using standard formulae.

2. Find integral $\int \frac{px + q}{ax^2 + bx + c} dx$

Here px+q is similar to the derivative $\frac{d}{dx}(ax^2 + bx + c)$. We can equate quotient p and constant q to those corresponding to the quotients of x and the constant in the above derivative as below:

$\frac{d}{dx}(ax^2 + bx + c) = 2ax + b$: thus we equate p to 2a with a suitable multiplier and q to b with a suitable addition/subtraction of a constant.

Now if we assume $(ax^2 + bx + c)$ = t, the given expression will reduce to

$\int \frac{px + q}{ax^2 + bx + c} dx = l \times \int \frac{dt}{t} + \int \frac{k}{ax^2 + bx + c} dx$ where l and k

are constants. Integration result will be $l \times \log\left|ax^2+bx+c\right|$ + $k\int \frac{1}{ax^2+bx+c}dx$.

The second part is solved by converting denominator to the form $t^2 \pm n^2$ as shown above.

3. Similar is the case for $\int \frac{px+q}{\sqrt{ax^2+bx+c}}dx$.

   Let us consider a couple of examples for clarity of these concepts.

Example 1:

Find $\int \frac{x+3}{x^2-2x-5}dx$

$$\frac{d}{dx}(x^2-2x-5) = 2x-2$$

x + 3 can be altered as $\frac{1}{2}(2x-2)+4$ leading us to

$$\int \frac{x+3}{x^2-2x-5}dx = \int \frac{\frac{1}{2}(2x-2)}{x^2-2x-5}dx + \int \frac{4}{x^2-2x-5}dx$$

If we assume x^2-2x-5=t, the problem reduces to

$$\int \frac{x+3}{x^2-2x-5}dx = \frac{1}{2}\int \frac{dt}{t} + \int \frac{4}{x^2-2x-5}dx$$

$$= \frac{1}{2}\log\left|x^2-2x-5\right| + \int \frac{4}{(x-1)^2-(\sqrt{6})^2}dx$$

$$= \frac{1}{2}\log\left|x^2-2x-5\right| + 4\int \frac{1}{(x-1)^2-(\sqrt{6})^2}dx$$

$$= \frac{1}{2}\log\left|x^2-2x-5\right| + \frac{4}{2\sqrt{6}}\log\left|\frac{x-1-\sqrt{6}}{x-1+\sqrt{6}}\right| + C$$

$$= \frac{1}{2}\log\left|x^2-2x-5\right| + \frac{2}{\sqrt{6}}\log\left|\frac{x-(\sqrt{6}+1)}{x+(\sqrt{6}-1)}\right| + C$$

Example 2

Find $\int \frac{1}{\sqrt{9x-4x^2}}dx$.

This expression can be transformed to the form $\int \frac{1}{\sqrt{a^2-x^2}}dx$ yielding the solution $\int \frac{1}{\sqrt{a^2-x^2}}dx = \sin^{-1}\frac{x}{a} + C$

Let us transform the given expression to the above standard form as below:

$$\sqrt{9x-4x^2} = \sqrt{-\left(2x-\frac{9}{4}\right)^2 + \left(\frac{9}{4}\right)^2} = \sqrt{\left(\frac{9}{4}\right)^2 - \left(2x-\frac{9}{4}\right)^2}$$

Let us assume $\left(2x-\frac{9}{4}\right)$=t, thereby, 2dx = dt. And the problem reduces to

$$\int \frac{1}{\sqrt{9x-4x^2}}dx = \frac{1}{2}\int \frac{dt}{\sqrt{\left(\frac{9}{4}\right)^2 - t^2}} = \frac{1}{2}\sin^{-1}\frac{4t}{9} + C$$

$$= \frac{1}{2}\sin^{-1}\frac{8x-9}{9} + C$$

9.8 Definite integrals

Recall section 9.1 and concept of definite and indefinite integral given in section 9.3. We have already learnt indefinite integrals in detail. All properties of indefinite integrals apply to definite integrals.

Definite integral gives area under the curve defined by function say y= f(x) over a range of x between two given limits say [a,b]. If F(x) be integral of function f(x), Then,

$$\int_a^b f(x)dx = F(b) - F(a)$$

Now we will look at a few properties unique to the definite integrals.

9.8.1 Properties of definite integrals

9.8.1.1 $\int_a^b f(x)dx = -\int_b^a f(x)dx$

9.8.1.2 $\int_a^b f(x)dx = \int_a^c f(x)dx + \int_c^b f(x)dx$ for a<c<b

9.8.1.3 $\int_a^b f(x)dx = \int_b^a f(a+b-x)dx$

9.8.1.4 $\int_0^a f(x)dx = \int_0^a f(a-x)dx$

9.8.1.5 $\int_a^{2a} f(x)dx = \int_0^a f(x)dx + \int_0^a f(2a-x)dx = 2\int_0^a f(x)dx$

if $f(2a-x) = f(x)$

$= 0$ if $f(2a-x) = -f(x)$

9.8.1.6 $\int_{-a}^a f(x)dx = 2\int_0^a f(x)dx$ if $f(x)$ *is an even function*

$= 0$ if $f(x)$ *is an odd function*

9.8.1.7 $\int_a^b [f(x)dx + g(x)dx] = \int_a^b f(x)dx + \int_a^b g(x)dx$

These properties of Definite integral are quite important. Many problems of definite integral cannot be solved by normal methods/ formulae of indefinite integrals but need use of some of these properties. It will be evident from some of the Practice problems for Definite Integrals listed below.

9.9 Application of integrals

In Section 9.1, we saw that we can find area under a constantly changing curve using technique of integration. We shall now study how we use this technique to find the area under simple curves, area between lines and arcs of circles, parabolas and ellipses etc.

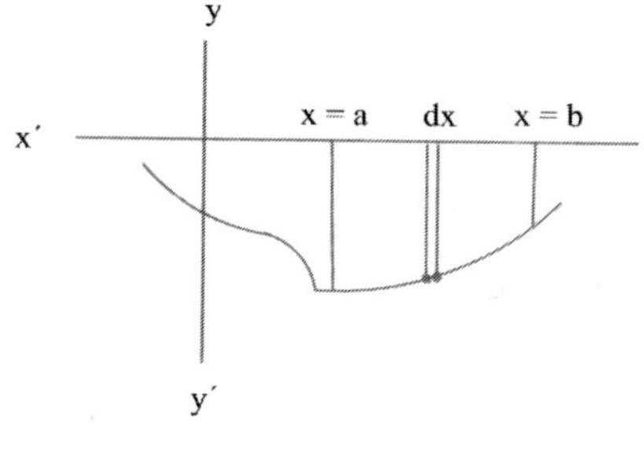

Figure 9.3

If the position of the curve under consideration is below the x-axis as shown in the above figure 9.3, then since f (x) < 0 from x = a to x = b, the area bounded by the curve, x-axis and the ordinates x = a, x = b comes out to be negative. But, it is only the numerical value of the area which is taken into consideration. Thus, if the area is negative, wetake its absolute value, i.e., $\left|\int_a^b f(x)dx\right|$.

9.9.1 Area of a circle with radius r

Let us find area of a circle with radius r. We know that formula for the circle centered at the origin is $x^2 + y^2 = r^2$. Refer to Figure 9.4 below. Let us assume an arbitrary point C with coordinates x,y. We now consider a strip in quadrant I of width dx at this point. Area of this strip will be $y \times dx = ydx$. If we integrate this expression over value of x from 0 to r, we will get area of quadrant 1 as $\int_0^r ydx$

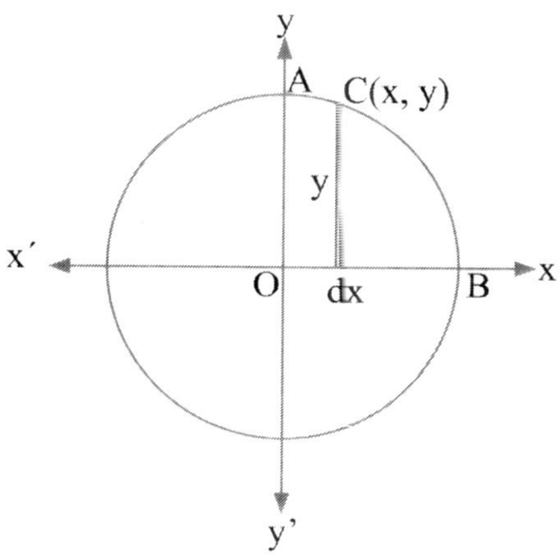

Figure 9.4

Accordingly, area of circle will be $4 \times \int_0^r ydx$. From equation of the circle we get

$y = \sqrt{r^2 - x^2}$; substituting we get,

$$4\times\int_0^r ydx$$

$$=4\times\int_0^r\sqrt{r^2-x^2}\,dx=4\times\int_0^r\frac{r^2-x^2}{\sqrt{r^2-x^2}}dx=4\left(\int_0^r\frac{r^2}{\sqrt{r^2-x^2}}dx-\int_0^r\frac{x^2}{\sqrt{r^2-x^2}}dx\right)$$

$$=4\left\{\left|r^2\sin^{-1}\frac{x}{r}\right|_0^r-\int_0^r\frac{x^2}{\sqrt{r^2-x^2}}dx)\right\}$$

For integral 2, let us assume $x=r\sin\theta$ leading to $dx=r\cos\theta d\theta$; substituting

$$\int_0^r\frac{x^2}{\sqrt{r^2-x^2}}dx=\int_0^{\pi/2}\frac{r^2\sin^2\theta\times r\cos\theta}{\sqrt{r^2-r^2\sin^2\theta}}d\theta=\int_0^{\pi/2}\frac{r^2\sin^2\theta\times r\cos\theta}{r\cos\theta}d\theta=\int_0^{\pi/2}r^2\sin^2\theta d\theta$$

Now $\int_0^{\pi/2}r^2\sin^2\theta d\theta=\int_0^{\pi/2}r^2\frac{(1-\cos2\theta)}{2}d\theta=\left|\frac{r^2\theta}{2}-\frac{\sin2\theta}{4}\right|_0^{\pi/2}$

$$\therefore 4\times\int_0^r\sqrt{r^2-x^2}\,dx=4\left(\left|r^2\sin^{-1}\frac{x}{r}\right|_0^r-\left|\frac{r^2\theta}{2}-\frac{\sin2\theta}{4}\right|_0^{\pi/2}\right)$$

$$=4\times\left(\left|r^2\times\frac{\pi}{2}-0\right|-\left|\frac{r^2\pi}{4}-0\right|\right)=4\times\frac{r^2\pi}{4}=\pi r^2,$$

well-known formula for area of a circle.

The same way, one can derive formula for area of an ellipse using equation of the ellipse as $\frac{x^2}{a^2}+\frac{y^2}{b^2}=1$ where a and b are the semi-major and semi-minor axis.

We will try another example of area of the region bounded by a parabolic curve and a lineat 9.9.3 below.

9.9.2 Volume of a right circular cone

To derive the formula for volume of a cone, the principle is to slice a small disc, shaded in, of thickness dy, and radius x. If we were to slice many discs of the same thickness and summate their volume then we should get volume of the cone.

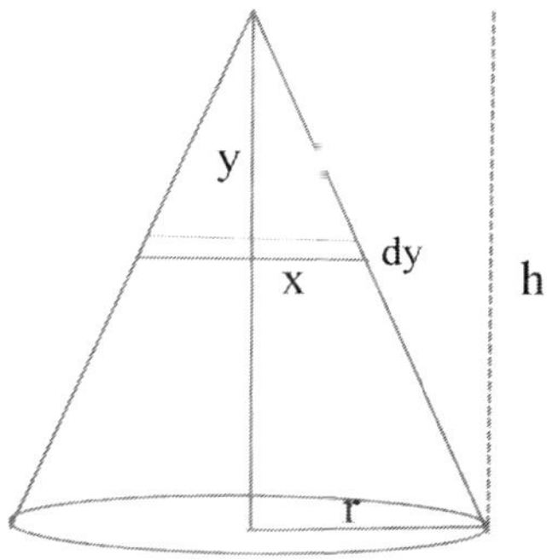

Figure 9.5

To derive the formula for volume of a cone, the principle is to slice a small disc of thickness dy, and radius x. If we were to slice many discs of the same thickness and summate their volume then we should get volume of the cone.

As we can see, there are two right-angled triangles one with base x and the other with base r with heights as y and h respectively. Sides of these two triangles are proportional giving us the equality $\frac{r}{h} = \frac{x}{y}$ or $x = \frac{ry}{h}$

Volume of the small disc = $\pi x^2 dy$;

Substituting value of x, we get Volume of the small disc $= \pi \frac{r^2 y^2}{h^2} dy$

At this stage, the integration principle involves adding the volume of all the discs between y = zero, and y = h to give us the volume of the cone.

Thus volume of the cone = $\int_0^h \pi \frac{r^2}{h^2} y^2 dy$

As r and h are constants,

$$\int_0^h \pi \frac{r^2}{h^2} y^2 dy = \pi \frac{r^2}{h^2} \int_0^h y^2 dy = \pi \frac{r^2}{h^2} \left| \frac{y^3}{3} \right|_0^h = \pi \frac{r^2}{h^2} \times \frac{h^3}{3} = \frac{1}{3} \pi r^2 h$$

9.9.3 Area of the region bounded by a parabolic curve and a line

Let the parabolic curve be $y=x^2$ and the line be $y= 4$ as in the figure 9-6 below

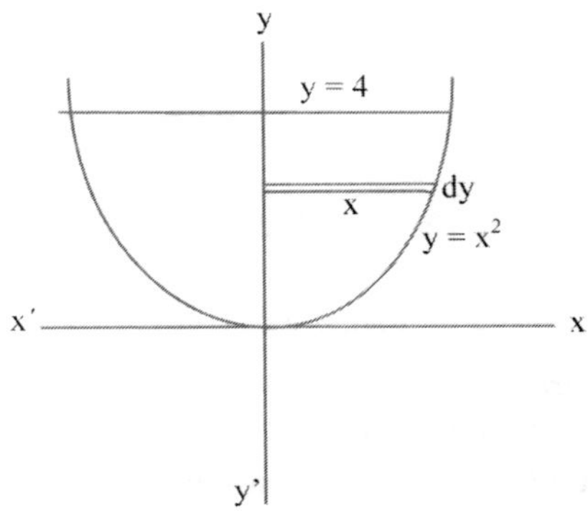

Figure 9.6

From the given equation of the parabola, $x=\pm\sqrt{y}$. And area of the curve bound by the straight line will be $\int_0^4 x\,dy$. Positive values of x will yield area in quadrant 1 and negative values of x will yield area in quadrant 2. Both these areas will be equal. Hence,

Area of the curve bound by the straight line $= 2\times\int_0^4 x\,dy = 2\times\int_0^4 \sqrt{y}\,dy$

$$= 2\times\left|\frac{y^{3/2}}{3/2}\right|_0^4 = 2\times\left|\frac{2}{3}\times y^{3/2}\right|_0^4 = \frac{4}{3}\times(8-0) = \frac{32}{3}$$

Readers will notice that the process of finding area in different situations is the same. One has to be careful in selecting the small strip parameters and end points of applicable integration. A few questions are listed for practice at end of the chapter.

9.9.4 Sum of Harmonic Progression of n numbers:

Let us refer to graph of function $f(x)=\frac{1}{x}$ in figure 9.7 below

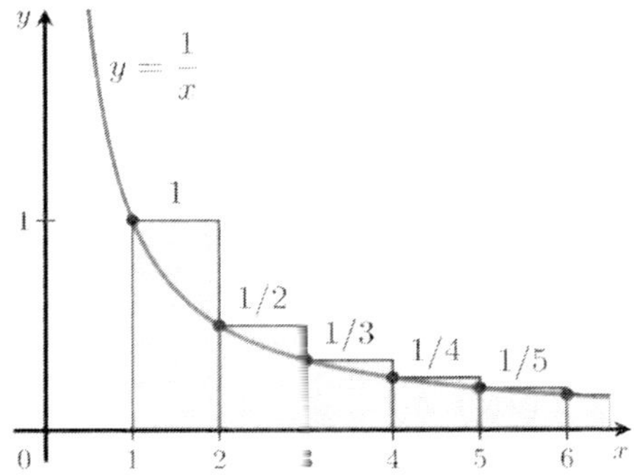

Figure 9.7

The shaded rectangles shown on the graph have fixed width (1 here) and varying length equal to the numbers in harmonic series with $a = 1$ and $d = 1$. It would be noticed that area of rectangles is quite close to the area under the graph as n goes beyond 4 or so. Hence area under the curve is quite close to the sum of area of all rectangles.

If we assume the constant interval d as width of a rectangle and $\frac{1}{a}$ as length of the first rectangle, the next rectangle has same width d but $\frac{1}{a+d}$ as its length. Thus length of every succeeding rectangle constantly keeps .on decreasing/increasing depending upon d being positive or negative. Since d is fixed, area of the rectangle also keeps on decreasing/increasing. Length of the ith rectangle will accordingly be $\frac{1}{a+(i-1)d}$ and the rectangles will be sitting next to each other being equally spaced.

Sum of the Harmonic Progression with $\frac{1}{a}$ as the first term and $\frac{1}{a+(i-1)d}$ as the ith term will virtually mean that we need to find sum of length of the rectangles each of width d sitting next to each other.

Since d is small, each rectangle is very thin and length of every adjoining rectangle are close to each other and top of the graph representing length of the rectangles is more or less a smooth curve. Sum of area of all rectangles will approximately equal to area of the curve defined by the function $\frac{1}{x}$ that can be found by integration process.

Since width of the rectangles is fixed *(d)*, area under the curve yields close to *d* times the sum of the HP.

If numbers in the HP of n terms are considered as mid-point of the respective rectangles with $\frac{1}{a}$ as mid-point of first rectangle and $\frac{1}{a+(n-1)d}$ as mid-point of the nth rectangle, edges of the graph

of $1/x$ on x-axis will be $d/2$ away from the mid points towards respective outer edge of the graph. Hence boundaries of the graph shall be (first term $-d/2$) and (the nth term $+d/2$) i.e. $\left(a-\frac{d}{2}\right)$ and

$\left(a+(n-1)d+\frac{d}{2}\right)=\left(a+\left(n-\frac{1}{2}\right)d\right)$ respectively.

With the above explanation we can see that Sum of an HP of n terms equals $\frac{1}{d}\times$ area of the curve represented by $\frac{1}{x}$ within the boundaries defined above.

Hence sum of the HP $=\frac{1}{d}\times\int_{a-\frac{d}{2}}^{a+\left(n-\frac{1}{2}\right)d}\frac{1}{x}dx=\frac{1}{d}\times\left|\ln x\right|_{a-\frac{d}{2}}^{a+\left(n-\frac{1}{2}\right)d}$, where x are the numbers in the typical HP explained above.

Representing x by i[th] term of the HP, we get $x=a+(i-1)d$ and the boundaries in terms of i will be

$$a-\frac{d}{2}=a+(i-1)d\Rightarrow(i-1)=-\frac{1}{2}\Rightarrow i=\frac{1}{2}\text{ and}$$

$$a+\frac{(n-1)}{2}d=a+(i-1)d\Rightarrow(i-1)=\frac{n-1}{2}\Rightarrow i=\frac{n+1}{2}$$

By substituting the value of x and boundaries of the graph in terms of ***i***, we get approximate sum of the HP S_n as

$$S_n = \frac{1}{d} \times \left| \ln\left(a + (i-1)d\right) \right|_{\frac{1}{2}}^{n+\frac{1}{2}} = \frac{1}{d} \times \left| \ln\left(a + \left(n + \frac{1}{2} - 1\right)d\right) - \ln\left(a + \left(\frac{1}{2} - 1\right)d\right) \right|$$

$$S_n = \frac{1}{d} \times \ln \frac{\left(a + \frac{(2n-1)d}{2}\right)}{a - \frac{1}{2}d} = \frac{1}{d} \times \ln \frac{2a + (2n-1)d}{2a - d}$$

It should be remembered that this formula gives approximate value of the sum, with the condition that where $2a \neq d$, $d \neq 0$ and n is large compared to d. We can verify the formula for say a=1, d = $1/2$ and n = 10

As per above data, HP is $1, \frac{2}{3}, \frac{1}{2}, \frac{2}{5}, \frac{1}{3}, \frac{2}{7}, \frac{1}{4}, \frac{2}{9}, \frac{1}{5}, \frac{2}{11}$

Hence Sum by arithmetical calculation

$$= 1 + \frac{2}{3} + \frac{1}{2} + \frac{2}{5} + \frac{1}{3} + \frac{2}{7} + \frac{1}{4} + \frac{2}{9} + \frac{1}{5} + \frac{2}{11} = 1 + \left(\frac{2}{3} + \frac{1}{3} + \frac{2}{9}\right) + \left(\frac{1}{2} + \frac{1}{4}\right) + \left(\frac{2}{5} + \frac{1}{5}\right) + \frac{1}{7} + \frac{2}{11}$$

$$= 1 + \frac{11}{9} + \frac{3}{4} + \frac{3}{5} + \frac{2}{7} + \frac{2}{11} = 1 + 1.222 + 0.75 + 0.6 + 0.2857 + 0.1818 = 4.0395$$

As per the Sum formula, S_n of the above HP

$$= \frac{1}{1/2} \times \ln \frac{2 \times 1 + (2 \times 10 - 1) \times 1/2}{2 \times 1 - 1/2} = 2 \ln \frac{11.5}{1.5} = 2 \times 2.0369 = 4.0738$$

It may be noticed that its very easy to compute sum by the above formula compared to the manual process and the result is quite close.

We have learnt in Chapter 4 that Mean of a Harmonic series is "Inverse of the Mean of Inverse of terms of the HP (that, by very definition of Harmonic Progression (HP) are actually in AP)". This further implies that each term in an HP is the harmonic mean of its neighboring terms. This is quite evident from a typical HP of n terms

$\frac{1}{a}, \frac{1}{a+d}, \frac{1}{a+2d}, \frac{1}{a+3d}, \ldots\ldots \frac{1}{a+(n-1)d}$ as $\frac{1}{a+d}$ is inverse of Arithmetic Mean of a and $(a + 2d)$ and $\frac{1}{a+2d}$ is inverse of Arithmetic Mean of $(a + d)$ and $(a + 3d)$.

Thus sum of an HP of n terms can also be understood as sum of a series of Harmonic Means.

9.10 Concluding remarks:

Subject of integration is somewhat complex. Readers must grasp and master

1. Fundamental theorems of integration
2. Integration formulae for standard functions of x and
3. All different methods of integration.

While attempting a problem, first job is to observe and identify the type of function. Next job is to alter the same to bring it in the form of known integral formula by

(a) Suitable substitution

(b) Converting into product of two functions with known integration formulae

(c) Using trigonometric identities

(d) Partial fractions etc.

And finally we arrive at the result using known standard formulae to the transformed function. It is a good practice to differentiate the result and verify that you reach the original function given in the problem. This gives enormous confidence to tackle difficult problems and master your knowledge of differentiation.

9.11 Practice problems:

9.11.1 Indefinite integrals

Evaluate each of the following indefinite integrals:

1. $\int \frac{x^2+4}{x+3}dx$
2. $\int \frac{x^2+3x+4}{\sqrt{x}}dx$
3. $\int (\sin 2x - 4e^{3x})dx$
4. $\int \frac{\sin 4x}{\sin x}dx$
5. $\int 7^x e^x dx$
6. $\int \frac{\cos x}{\sqrt{1+\sin x}}dx$
7. $\int \frac{(x+1)(x+\log x)^2}{x}dx$
8. $\int \cos^5 x dx$
9. $\int \frac{dx}{a^2\sin^2 x + b^2\cos^2 x}$
10. $\int \frac{x}{(x-1)(x-2)(x-3)}dx$
11. $\int \frac{4x^2-16x-3}{(3x+1)(2x-1)^2}dx$
12. $\int \frac{\sin\theta\cos\theta}{\cos^2\theta - \cos\theta - 2}d\theta$
13. $\int x^2 \log x dx$
14. $\int e^{ax}\cos bx dx$
15. $\int e^x \frac{x^2+1}{(x+1)^2}dx$
16. $\int \sqrt{a^2+x^2}dx$
17. $\int \tan^{-1}\left(\frac{2x}{1+x^2}\right)dx$
18. $\int \frac{x+\sin x}{1+\cos x}dx$
19. $\int \frac{dx}{x+\sqrt{x}}$
20. $\int \sqrt{e^x - 1}dx$
21. $\int \sec^4 x dx$
22. $\int \frac{1}{x}\sqrt{\frac{x+m}{x-m}}dx$
23. $\int \frac{dx}{x^{\frac{1}{2}} + x^{\frac{1}{3}}}$
24. $\int \left[\frac{1}{\log x} - \frac{1}{(\log x)^2}\right]dx$
25. $\int \frac{\sin x\cos x}{\sin^4 x + \cos^4 x}dx$
26. $\int \frac{\cos^9 x}{\sin x}dx$
27. $\int \frac{5x+3}{\sqrt{x^2+4x+10}}dx$
28. $\int (\sqrt{\tan x} + \sqrt{\cot x})dx$
29. $\int \frac{dx}{\sqrt{\sin^3 x \sin(x+\alpha)}}$
30. Prove that

$$\int \sec x dx = \log\left|\tan\left(\frac{\pi}{4}+\frac{x}{2}\right)\right| + C$$

9.11.2 Definite integrals

Evaluate each of the following definite integrals:

1. $\int_0^1 \frac{2x+3}{5x^2+1}dx$

2. $\int_1^2 \frac{5x^2}{x^2+4x+3}dx$

3. $\int_2^4 \frac{x^3}{\sqrt{x^4-1}}dx$

4. $\int_0^1 \frac{x}{1+\sqrt{x}}dx$

5. $\int_0^{1/2} \frac{1}{2+8x^2}dx$

6. $\int_0^{\pi/2} \frac{\sin x}{1+\cos^2 x}dx$

7. $\int_0^{\pi/4} \sqrt{1+\sin 2x}dx$

8. $\int_0^{\pi/2} \sqrt{\sin\theta}\cos^5\theta d\theta$

9. $\int_0^{2\pi} \frac{\cos x}{\sqrt{4+3\sin x}}dx$

10. $\int_0^{\pi/2} \frac{1}{a^2\sin^2 x+b^2\cos^2 x}dx$

11. $\int_0^{\pi} \frac{\theta}{a^2\cos^2\theta+b^2\sin^2\theta}d\theta$

12. $\int_0^{\pi/2} \log\tan x dx$

13. $\int_0^1 \frac{\log(1+x)}{x^2}dx$

14. $\int_0^{\pi} \frac{x}{a^2-\cos^2 x}dx$

15. $\int_0^{\infty} \frac{x\tan^{-1}x}{(1+x^2)^2}dx$

16. $\int_0^1 x(\tan^{-1}x)^2 dx$

17. $\int_0^{\infty} \frac{x^2}{(x^2+1)(x^2+2)(x^2+3)}dx$

18. $\int_0^1 \frac{1}{(x-1)\sqrt{x-x^2}}dx$

19. where

$$f(x)=\begin{cases} \sin x, 0\le x<\pi/2 \\ 1, \pi/2\le x\le 3 \\ e^{x-3}, 3<x\le 9 \end{cases}$$

20. $\int_0^{\pi/2} (2\log\sin x-\log\sin 2x)dx$

9.11.3 Application of Integrals

1. Find the area of the region in the first quadrant enclosed by the x-axis, the line y=x and the circle $x^2 + y^2 = 32$
2. Find the area between the curve $\frac{x^2}{a^2} + \frac{y^2}{b^2} = 1$ and the x-axis between x = 0 and x = a
3. Find the area included between the curves $y^2 = 4ax$ and $x^2 = 4ay$
4. Find area of smaller part of the circle $x^2 + y^2 = a^2$ cutoff by the line $x = \frac{a}{\sqrt{2}}$
5. Find the area of the region enclosed between the circles $x^2 + y^2 = 1$ and $(x-1)^2 + y^2 = 1$

Answers to Practice Problems:

Indefinite integrals

1. $x^2 - 3x + 13\log|x+3| + C$

2. $\frac{2}{5}x^{\frac{5}{2}} + 2x^{\frac{3}{2}} + 8\sqrt{x}) + C$

3. $-\frac{\cos 2x}{2} - \frac{4}{3}e^{3x} + C$

4. $4\sin x - \frac{8}{3}\sin^3 x + C$

5. $\frac{7^x \times e^x}{(1+\log_e 7)}$

6. $2\sqrt{1+\sin x}$ +C

7. $\frac{(x+\log x)^3}{3} + C$

8. $\sin x + \frac{\sin^5 x}{5} - \frac{2\sin^3 x}{3} + C$

9. $\frac{1}{ab}\tan^{-1}\frac{a\tan x}{b} + C$

10. $\frac{1}{2}\log|x-1| - 2\log|x-2| + \frac{3}{2}\log|x-3| + C$

11. $\frac{1}{3}\log|3x+1| + \frac{2}{(2x-1)} + C$

12. $\frac{2}{3}\log|\cos x - 2| - \frac{1}{3}\log|\cos x + 1| + C$

13. $\frac{x^3}{3}|\log x| - \frac{x^3}{9} + C$

14. $\frac{e^{ax}}{a^2+b^2}(b\sin bx + a\cos bx) + C$

15. $e^x - \frac{2}{(x+1)}e^x + C$

16. $\frac{x}{2}\sqrt{a^2+x^2}-\frac{a^2}{2}\log\left|x+\sqrt{a^2+x^2}\right|+C$

17. $2x\tan^{-1}x-\log\left|1+x^2\right|+C$

18. $x\tan\frac{x}{2}+C$

19. $2\log\left|1+\sqrt{x}\right|+C$

20. $2\sqrt{e^x-1}-2\tan^{-1}\sqrt{e^x-1}+C$

21. $\tan x+\frac{\tan^3 x}{3}+C$

22. $\log\left|x+\sqrt{x^2-m^2}\right|+\sec^{-1}\frac{x}{m}+C$

23. $2\sqrt{x}-3x^{\frac{1}{3}}+6x^{\frac{1}{6}}-6\log\left|(x^{\frac{1}{6}}+1)\right|+C$

24. $\frac{x}{\log|x|}+C$

25. $\frac{1}{2}\tan^{-1}(\tan^2 x)+C$

26. $\log|\sin x|+\frac{\sin^8 x}{8}-\frac{2\sin^6 x}{3}+\frac{3\sin^4 x}{2}-2\sin^2 x+C$

27. $5\sqrt{x^2+4x+10}-7\log\left|(x+2)+\sqrt{x^2+4x+10}\right|+C$

28. $\sqrt{2}\tan^{-1}\log\left|\frac{\tan x-1}{\sqrt{2\tan x}}\right|+C$

29. $-\frac{2}{\sin\alpha}\sqrt{\frac{\sin(x+\alpha)}{\sin x}}+C$

Definite integral problems

1. $\frac{1}{5}\log 6+\frac{3}{\sqrt{5}}\tan^{-1}\sqrt{5}$

2. $5-\frac{5}{2}\left[9\log\frac{5}{4}-\log\frac{3}{2}\right]$

3. $\frac{1}{2}\left(\sqrt{255}-\sqrt{15}\right)$

4. $\frac{5}{3}-2\log 2$

5. $\frac{\pi}{16}$

6. $\frac{\pi}{4}$

7. 1

8. $\frac{64}{231}$

9. $\frac{4}{3}$

10. $\frac{\pi}{2ab}$

11. $\frac{\pi^2}{ab}$

12. 0

13. $\frac{\pi^2}{2a\sqrt{a^2-1}}$

14. $\frac{\pi}{8}\log 2$

15. $\frac{\pi}{8}$

16. $\frac{\pi^2}{16}-\frac{\pi}{4}+\frac{1}{2}\log 2$

17. $\frac{\pi}{2}\left(\frac{2\sqrt{2}-\sqrt{3}-1}{2}\right)$

18. 2

19. $3-\frac{\pi}{2}+e^6$

20. $-\frac{\pi}{2}\log 2$

Application of integrals

1. 2π

2. $\frac{\pi ab}{2}$

3. $\frac{16a^2}{3}$

4. $\frac{a^2}{2}\left(\frac{\pi}{2}-1\right)$

5. $\frac{2\pi}{3}-\frac{\sqrt{3}}{2}$

10

COMPLEX NUMBERS

10.1 Introduction

We have seen that square of a positive or negative number is always positive. Hence square root of a positive number can be positive or negative. For example $\sqrt{9}=\pm 3$ How about square root of a negative number? It is clear that square root of a negative number cannot be a real number. Hence a need was felt to extend the real number system to a system where we can find square root of negative numbers. Leonhard Euler in the 18th century introduced the symbol ***i*** (iota) for positive square root of -1 i,e, $i=\sqrt{-1}$

Let us consider equation $x^2+4=0$

Solution of this equation will be $x=\pm\sqrt{-4}$

Knowing the concept of ***i***, we can see that

$$x=\pm\sqrt{-4}=\pm\sqrt{(-1)\times 4}=\pm i\sqrt{4}=\pm 2i$$

i (iota) has interesting exponential values. $i^2=-1$, $i^3=-i$ and $i^4=1$. In generic terms we can say that for any integer k, $i^{4k}=1$, $i^{4k+1}=i$, $i^{4k+2}=-1$ and $i^{4k+3}=-i$

It will be noticed that for generic quadratic equation $ax^2+bx+c=0$ where $D^2=(b^2-4ac)$ is negative, the solution to this quadratic equation is $x=\dfrac{-b\pm iD}{2a}$ resulting in complex numbers $\left(\dfrac{-b}{2a}+i\dfrac{D}{2a}\right)$ and

$\left(\frac{-b}{2a} - i\frac{D}{2a}\right)$. Here $\frac{-b}{2a}$ is the real part and $\pm\frac{D}{2a}$ is the imaginary part. The two complex numbers are equal if their real and imaginary parts are equal respectively. We will learn to deal with complex numbers in this chapter.

10.2 Arithmetical operations on complex numbers

10.2.1 Addition of two complex numbers

Let us assume two complex numbers $x_1 = a + ib$ and $x_2 = c + id$ sum of these two complex numbers is $x_1 + x_2 = (a+c) + i(b+d)$. That is real and imaginary components are added respectively.

Addition of complex numbers satisfies the following properties (similar to those of real numbers)

(a) Closure law: Sum of two complex numbers is always a complex number.

(b) The Commutative law: For any two complex numbers x_1 and x_2, $x_1 + x_2 = x_2 + x_1$

(c) The Associative law: For any three complex numbers $x_{1,}$ x_2 and x_3, $(x_1 + x_2) + x_3 = x_1 + (x_2 + x_3)$

(d) Additive identity: Complex number 0+i0, denoted as 0, also known as zero complex number is called Additive Identity such that for every complex number x, x+0 = x

(e) Additive inverse: For every complex number x = a+ib, there is a complex number –a+i(-b), known as –x, which is called the additive inverse of x yielding x+(-x) = 0

10.2.2 Difference of two complex numbers

Similar to the addition of complex numbers, difference of two complex numbers is achieved by subtracting real part and imaginary part of the two numbers respectively. Thus difference of two complex number $x_1 = a + ib$ and $x_2 = c + id$ is $x_1 - x_2 = (a-c) + i(b-d)$

10.2.3 Multiplication of two complex numbers

Product of two complex numbers is achieved by familiar principles of algebraic multiplication. Let us multiply two complex numbers, $x_1 = a + ib$ and $x_2 = c + id$

By algebraic multiplication we get, $x_1 x_2 = ac + iad + icd + i^2 bd$

Since $i^2 = -1$, we get

$$x_1x_2 = ac - bd + iad + icd = (ac - bd) + i(ad + cd)$$

Multiplication of complex numbers satisfies the following properties (similar to those of real numbers)

(a) Closure law: Product of two complex numbers is always a complex number.

(b) The Commutative law: For any two complex numbers x_1 and x_2, $x_1 \times x_2 = x_2 \times x_1$

(c) The Associative law: For any three complex numbers x_1, x_2 and x_3, $(x_1x_2)x_3 = x_1(x_2x_3)$

(d) Multiplicative identity: Complex number 1+i0, denoted as 1, is called the multiplicative Identity such that for every complex number x, $x \times 1 = x$

(e) Multiplicative inverse: For every non zero complex number x = a+ib where $a \neq 0, b \neq 0$, there is a complex number $\frac{a}{a^2+b^2} + i\frac{-b}{a^2+b^2}$, denoted by $\frac{1}{x}$ or x^{-1}, is called the multiplicative inverse of x such that $x \times \frac{1}{x} = 1$

(f) The Distributive law: For any three complex numbers x_1, x_2 and x_3, $(x_1 + x_2)x_3 = x_1x_3 + x_2x_3$ and $x_1(x_2 + x_3) = x_1x_2 + x_1x_3$

10.2.4 Division of two complex numbers

Division of two complex numbers say x_1= a+ib andx_2=c+id is a little complex

$\frac{x_1}{x_2}$ is equivalentto $x_1 \times \frac{1}{x_2}$. We have already learnt that $\frac{1}{x_2}$ is multiplicative inverse of x_2 and is $= \frac{c}{c^2+d^2} + i\frac{-d}{c^2+d^2}$

$$\therefore \frac{x_1}{x_2} = (a+ib) \times \left(\frac{c}{c^2+d^2} + i\frac{-d}{c^2+d^2}\right) = \left(\frac{ac+bd}{c^2+d^2} + i\frac{bc-ad}{c^2+d^2}\right)$$

The same result can be obtained by following the familiar rationalizing process:

$\frac{x_1}{x_2} = \frac{a+ib}{c+id}$; rationalizing by multiplying with $\frac{c-id}{c-id}$ we get

$$\frac{x_1}{x_2}=\frac{a+ib}{c+id}\times\frac{c-id}{c-id}=\frac{(ac+bd)+i(bc-ad)}{c^2+d^2}$$

10.2.5 Algebraic identities

Familiar algebraic identities applicable to real numbers are equally applicable to complex numbers. Let us look at an example with two familiar complex numbers x_1=a+ib and x_2=c+id.

$$(x_1+x_2)^2=((a+c)+i(b+d))^2$$

$$=\left[(a^2+c^2+2ac)-(b^2+d^2+2bd)+2i(a+c)(b+d)\right]$$

$$=\left[(a^2+c^2+2ac)-(b^2+d^2+2bd)+2i(ab+bc+ad+cd)\right]$$

By rearranging we get

$$=(x_1+x_2)^2=\left[(a^2-b^2+2iab)+2\left[(ac-bd)+i(ad+bc)\right]+(c^2-d^2+2icd)\right]$$

$$=(a+ib)^2+2((a+ib)(c+id))+(c+id)^2$$

$$=x_1{}^2+2x_1x_2+x_2{}^2 \qquad .....(1)$$

Similarly we can prove that all other identities true for real numbers are also valid for complex numbers. Some of the examples are

$$(x_1-x_2)^2=x_1{}^2-2x_1x_2+x_2{}^2 \qquad ...(2)$$

$$(x_1+x_2)^3=x_1{}^3+3x_1x_2(x_1+x_2)+x_2{}^3 \qquad ...(3)$$

$$(x_1-x_2)^3=x_1{}^3-3x_1x_2(x_1-x_2)-x_2{}^3 \qquad ...(4)$$

$$x_1{}^2-x_2{}^2=(x_1+x_2)\times(x_1-x_2) \qquad ...(5)$$

It will be noticed that identities (2) and (4) are quite similar to the identities (1) and (3). Easy way to recall these identities is to substitute ($-x_2$) for x_2 in the original identities. Thus there is no need to remember identities (2) and (4).

10.2.6 Modulus and Conjugate of a complex numbers

Modulus of a complex number z = $a+ib$ is denoted as $|z|$ and is defined to be the non negative real number $\sqrt{a^2+b^2}$ and Conjugate of the complex number z = $a-ib$ and is denoted as $\bar{z}$

Recall that Multiplicative inverse of complex number

$$z^{-1} = \frac{a}{a^2+b^2} + i\frac{-b}{a^2+b^2} = \frac{a-ib}{a^2+b^2} = \frac{\bar{z}}{|z|^2}$$

We also know that $z^{-1} = \frac{1}{z}$

$$\therefore |z|^2 = z\bar{z}$$

Similarly the following results can easily be proved/ verified for any two complex numbers x_1 and x_2

$$|x_1 x_2| = |x_1| \times |x_2|$$

$$\left|\frac{x_1}{x_2}\right| = \frac{|x_1|}{|x_2|}$$, provided $|x_2| \neq 0$

$$\overline{x_1 . x_2} = \overline{x_1} \times \overline{x_2}$$

$$\overline{x_1 \pm x_2} = \overline{x_1} \pm \overline{x_2}$$

$$\overline{\left[\frac{x_1}{x_2}\right]} = \frac{\overline{x_1}}{\overline{x_2}}$$, provided $|x_2| \neq 0$

10.3 Representation of complex numbers on x-y and polar plane

Complex numbers can be represented graphically on x-y coordinates with the real part represented on x-axis and imaginary part on y axis. Thus complex number $z = a + ib$ is represented as below in figure 10.1

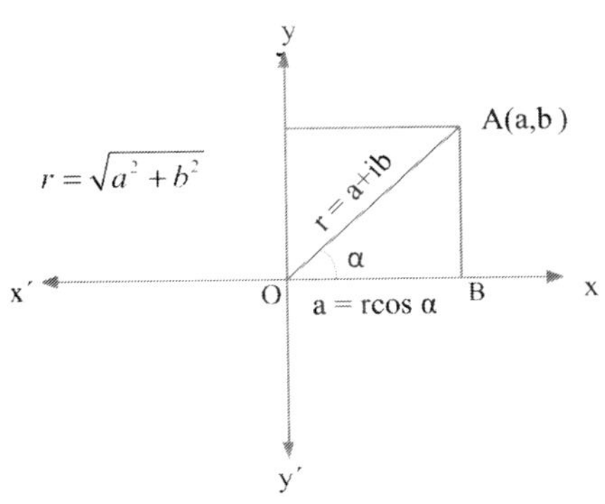

Figure 10.1

In order to understand the concept of complex numbers, it is important to know about different type of quantities. In our day to day life, we come across two types of quantities like speed of a car, wind velocity, hitting of ball by cricket batsman etc. Observe that speed of a car is stated by only a magnitude (kms/hour) without any reference to direction in which car is travelling whereas wind velocity involves both - its magnitude as well as direction of wind flow. Similarly, when a cricketer hits the ball, it involves force with which the ball is hit as well as direction of the hit.

Quantities like speed of a car that have only magnitude are called ***scalar quantities*** *whereas the quantities like wind velocity are* ***called vectors.*** *Other examples of scalar quantities are mass, volume, temperature etc. and of vector quantities like velocity, momentum, weight etc.*

Let us say that OA (with length say $r = \sqrt{a^2 + b^2}$) is referred to as a Vector (say z). $\sqrt{a^2 + b^2}$ is clearly the modulus of the vector z and is written as $|z| = r = \sqrt{a^2 + b^2}$. Since it is length of a line segment, modulus $|z|$ is always positive.

In vector form, angle subtended by the vector OA to the x-axis, considered as α here is known as Argument of the complex number z written as arg z, leading to the real part of the complex number z (a) = $r\cos\alpha$ and imaginary part (b) = $r\sin\alpha$. Value of α is $-\pi < \alpha \leq \pi$ and is known as Principal argument of z.

The complex number z is written in polar form as $z = r\cos\alpha + ir\sin\alpha$

10.4 Euler's formula for complex numbers

We have already seen expansion of Euler's number ***e*** in Chapter 8 as below:

$$e = \sum_{n=0}^{\infty} \frac{1}{n!} = \frac{1}{1} + \frac{1}{1} + \frac{1}{1\times 2} + \frac{1}{1\times 2\times 3} \ldots\ldots$$

This expansion derives from the famous Taylor series defined below for a function f(x) around an arbitrary point x=a

$f(x) = \sum_{0}^{\infty} \frac{f^n(a)}{n!} \times (x-a)^n$, where $f^n(a)$ is nth derivative of $f(x)$ at x = a

$$= f(a) + f'(a)(x-a) + \frac{f''(a)}{2!}(x-a)^2 + \frac{f'''(a)}{3!}(x-a)^3 + ...$$

Special case of this Taylor series is at x=0, which is also known as Maclaurin Series as below (credit: study.com):

$$f(x) = \sum_{0}^{\infty} \frac{f^n(0)}{n!} \times x^n = f(0) + \frac{f'(0)}{1!}x + \frac{f''(0)}{2!}x^2 + \frac{f'''(0)}{3!}x^3 + ...$$

Let us consider $f(x) = e^x$, and since derivative of $e^x = e^x$ itself, f(0), f′(0), f″(0) etc. yield value 1. The Maclaurin Series will thus yield

$$\therefore e^x = 1 + \frac{x}{1!} + \frac{x^2}{2!} + \frac{x^3}{3!} + \frac{x^4}{4!} + \frac{x^5}{5!} + ...$$

Substituting x = 1, we get $e = 1 + 1 + \frac{1}{2!} + \frac{1}{3!} + ...$, the same result as stated above

Now if we substitute x by ix, we get

$$e^{ix} = 1 + ix + \frac{(ix)^2}{2!} + \frac{(ix)^3}{3!} + \frac{(ix)^4}{4!} + \frac{(ix)^5}{5!} + ...$$

$$= 1 + ix - \frac{x^2}{2!} - \frac{ix^3}{3!} + \frac{x^4}{4!} + \frac{ix^5}{5!} + ...$$

Separating the real and imaginary parts, we get

$$e^{ix} = 1 - \frac{x^2}{2!} + \frac{x^4}{4!} + ... + i\left(x - \frac{x^3}{3!} + \frac{x^5}{5!} .....\right) \quad ....(6)$$

Let us now consider f(x) to be sin x.f′(x) = cos x, f″(x) = -sin x, $f^3(x)$ = -cos x, $f^4(x)$=sinx again. The sequence repeats itself for further series of derivatives. The Maclaurin series for sin x thus yields

$$f(x) = f\ (Sin\ x) = f(0) + f'(0)x + \frac{f''(0)}{2!}x^2 + \frac{f'''(0)}{3!}x^3 + ...$$

$$= x - \frac{x^3}{3} + \frac{x^5}{5} - .....$$

Similarly $f(x) = f(Cos x) = f(0) + f'(0)x + \frac{f''(0)}{2!}x^2 + \frac{f'''(0)}{3!}x^3 + ...$

$$= 1 - \frac{x^2}{2!} + \frac{x^4}{4!} - .....$$

Substituting in Equation (6) above we get

$e^{ix} = cos\, x + isin\, x$ where value of x is in radians and modulus = 1.

From the above definition, a Complex number z = a+ib can be defined as $|z|e^{ix}$, where $|z| = \sqrt{a^2 + b^2}$ is modulus of the complex number. This is known as Euler's formula for defining a complex number and is quite the same as vector representation of the complex number depicted in section 10.3 above. Euler's definition is depicted in the figure 10.2 below. Euler's definition results in a circle with modulus as the radius and variable x as the angle in radians.

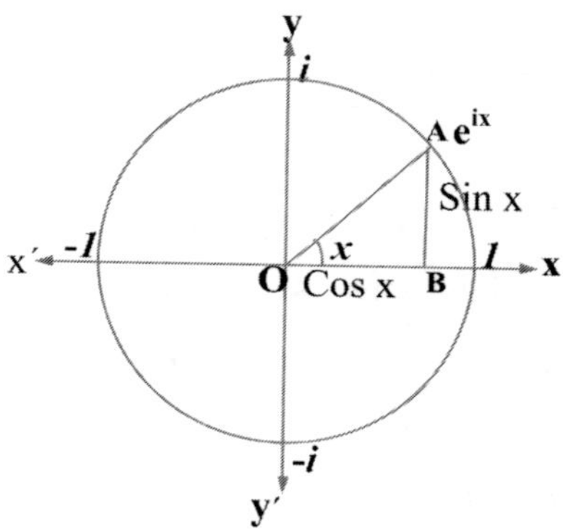

Figure 10.2

Let us depict a few more complex numbers in Euler's form. Refer to figure 10.3 below

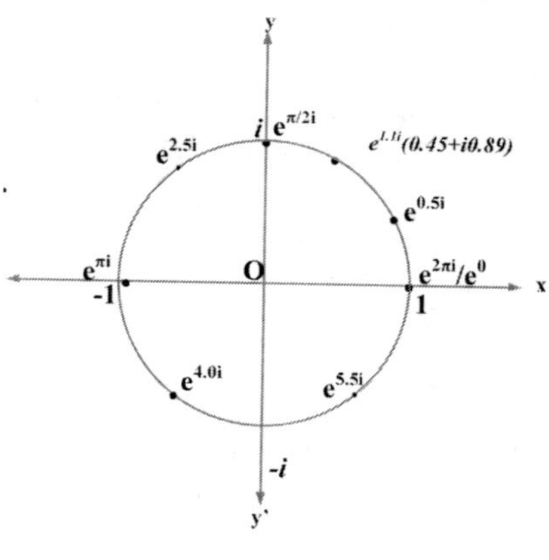

Figure 10.3

Value of x is shown to lie between 0 and 2π. However conventionally value of x is taken to lie between –π to +π. We can easily convert values larger than π to its equivalent value > -π. For example $e^{4.0} = e^{-(2\pi-4)} =\sim e^{-2.28}$ and $e^{5.5} = e^{-(2\pi-5.5)} =\sim e^{-0.78}$. Euler's form is quite convenient in algebraical operations like multiplication, division and exponents etc. on complex numbers

10.5 Solution of Quadratic equation with imaginary roots

We have already studied that roots of quadratic equation $ax^2 + bx + c = 0$ are

$$\frac{-b \pm \sqrt{b^2 - 4ac}}{2a}$$

In case where value of $\sqrt{b^2 - 4ac}$ is negative, we have imaginary roots. In such a case we write the roots as $\frac{-b \pm i\sqrt{4ac - b^2}}{2a}$

10.6 Triangle Inequality of Modulus of two Complex numbers

Triangle inequality theorem for modulus of complex numbers is defined as below:

If there are any two complex numbers say x_1 and x_2, then

$$|x_1| + |x_2| \geq |x_1 + x_2| \qquad ...(7)$$

and

$$|x_1| - |x_2| \leq |x_1 - x_2| \qquad ...(8)$$

Proof:

$$|x_1| + |x_2| \geq |x_1 + x_2|$$

Let $x_1 = a + ib$ and $x_2 = c + id$ leading to $x_1 + x_2 = (a + c) + i(b + d)$ and

$$|x_1| = \sqrt{a^2 + b^2}\,, |x_2| = \sqrt{c^2 + d^2}\ \&\ |x_1 + x_2| = \sqrt{(a + c)^2 + (b + d)^2}$$

$$|x_1| + |x_2| = \sqrt{a^2 + b^2} + \sqrt{c^2 + d^2}\ ;\ \text{squaring}$$

$$(|x_1| + |x_2|)^2 = a^2 + b^2 + 2\sqrt{a^2 + b^2}\sqrt{c^2 + d^2} + c^2 + d^2$$

$$= (a + c)^2 - 2ac + (b + d)^2 - 2bd + 2\sqrt{a^2 + b^2}\sqrt{c^2 + d^2}$$

$$= \left|(a + c)^2 + (b + d)^2\right| - 2(ac + bd) + 2\sqrt{a^2 + b^2}\sqrt{c^2 + d^2}$$

$$= |x_1 + x_2|^2 - 2(ac + bd) + 2\sqrt{a^2 + b^2}\sqrt{c^2 + d^2} \quad ....(9)$$

Remember that we are dealing with modulus which is always a positive number.

Now let us consider $(ac + bd)$ and $\sqrt{a^2 + b^2}\sqrt{c^2 + d^2}$

Squaring both terms we get

$$(ac + bd)^2 = a^2c^2 + b^2d^2 + 2abcd \quad ...(10)$$

and $\left(\sqrt{a^2 + b^2} \times \sqrt{c^2 + d^2}\right)^2 =$

$$\left(a^2 + b^2\right)\left(c^2 + d^2\right) = a^2c^2 + b^2d^2 + b^2c^2 + a^2d^2 \quad ....(11)$$

Subtracting (10) from (11), we get $b^2c^2 + a^2d^2 - 2abcd = (bc - ad)^2$

Since square of a number is always positive or in worst case 0, it implies that term (11) is $\geq$ term (10)

$$\therefore \sqrt{a^2 + b^2}\sqrt{c^2 + d^2} \geq (ac + bd)$$

Now going back to equation (9), we get that

$$(|x_1| + |x_2|)^2 \geq |x_1 + x_2|^2$$

Since we are dealing with positive numbers, $(|x_1| + |x_2|) \geq |x_1 + x_2|$

2. $|x_1| - |x_2| \leq |x_1 - x_2|$

As in the above proof, $|x_1| - |x_2| = \sqrt{a^2 + b^2} - \sqrt{c^2 + d^2}$; squaring

$$(|x_1| - |x_2|)^2 = a^2 + b^2 - 2\sqrt{a^2 + b^2}\sqrt{c^2 + d^2} + c^2 + d^2$$

$$= (a - c)^2 + 2ac + (b - d)^2 + 2bd - 2\sqrt{a^2 + b^2}\sqrt{c^2 + d^2}$$

$$= \left|(a - c)^2 + (b - d)^2\right| + 2(ac + bd) - 2\sqrt{a^2 + b^2}\sqrt{c^2 + d^2}$$

$$= |x_1 - x_2|^2 + 2(ac + bd) - 2\sqrt{a^2 + b^2}\sqrt{c^2 + d^2} \quad ...(12)$$

As already proved above, $\sqrt{a^2 + b^2}\sqrt{c^2 + d^2} \geq (ac + bd)$ yielding LHS $\geq$ RHS in eq 12

$$\therefore (|x_1| - |x_2|)^2 \leq |x_1 - x_2|^2$$

Since we are dealing with positive numbers, $(|x_1| - |x_2|) \leq |x_1 - x_2|$

These are important results that are handy in solving some of the problems.

10.7 Concluding remarks

Managing complex numbers is quite handy in solving many engineering problems. Complex numbers look a bit complicated but follow normal rules learnt for dealing with real numbers. Concepts of vector algebra and coordinate geometry come handy in dealing with the complex numbers. Most important part is to get comfortable in dealing with the term iota (i) and visualizing complex numbers on coordinate plain. Solving practice problems will go a long way in mastering the complex numbers.

10.8 Practice problems:

1. If $\frac{z_1}{z_2}$ is imaginary, then find value of $\left|\frac{2z_1+3z_2}{2z_1-3z_2}\right|$
2. If $z_k = \cos\left[\frac{k\pi}{10}\right] + i\sin\left[\frac{k\pi}{10}\right]$, find value of $z_1z_2z_3z_4$
3. If ω is a complex cube root of unity $(\omega \neq 1)$, then find the value of $\frac{a+b\omega+c\omega^2}{c+a\omega+b\omega^2}+\frac{a+b\omega+c\omega^2}{b+c\omega+a\omega^2}$
4. If $x = 1 + i\tan\alpha$ where $\pi < \alpha < \frac{3\pi}{2}$, find the value of $|x|$
5. If in polar form $x_1 = \cos\alpha + i\sin\alpha$, $x_2 = \cos\beta + i\sin\beta$ and $x_3 = \cos\Phi + i\sin\Phi$ and $x_1 + x_2 + x_3 = 0$, then find value of $x_1^{-1} + x_2^{-1} + x_3^{-1}$
6. Find the value of $\sum_{n=1}^{n=100} i^n$
7. If α is the nth root of unity, find the value of $1+2\alpha+3\alpha^2+4\alpha^3+.......$ to n terms
8. Find the value of $\sum_{p=1}^{6} 2\left[\sin\frac{2p\pi}{7} - i\cos\frac{2p\pi}{7}\right]$
9. If x and y are different complex numbers with $|x|=1$, find the value of $\left|\frac{y-x}{1-\bar{x}y}\right|$
10. If $|x| = 5$, find the least value of $\left|x+\frac{2}{x}\right|$

11. For any complex number z, find the maximum value of $|z| - |z - 1|$

12. If α, β are the roots of $x^2 + px + q = 0$, and ω is a cube root of unity, then find the value of $(\omega\alpha + \omega^2\beta)(\omega^2\alpha + \omega\beta)$

13. If $\omega(\neq 1)$ is a cube root of unity and $(1+\omega^2)^{11} = a + b\omega + c\omega^2$, find the value of (a, b, c)

14. If $z = \dfrac{1+\cos\theta + i\sin\theta}{\sin\theta + i(1+\cos\theta)}\left(0 < \theta < \dfrac{\pi}{2}\right)$, find the value of $|z|$

15. If $x^2 + x + 1 = 0$, where x is a complex number, then find the value of

$$S = \left(x+\frac{1}{x}\right)^2 + \left(x^2+\frac{1}{x^2}\right)^2 + \left(x^3+\frac{1}{x^3}\right)^2 \ldots\ldots \left(x^6+\frac{1}{x^6}\right)^2$$

Answers to Practice Problems:

1. 1.

2. $z_1z_2z_3z_4 = \cos[\pi] + i\sin[\pi] = -$

3. $\dfrac{a+b\omega+c\omega^2}{c+a\omega+b\omega^2} + \dfrac{a+b\omega+c\omega^2}{b+c\omega+a\omega^2} = -1$

4. $|x| = -\sec\alpha$

5. $x_1^{-1} + x_2^{-1} + x_3^{-1} = 0$

6. $\sum_{n=1}^{n=100} i^n = 0$

7. $Sum = \dfrac{-n}{(1-\alpha)}$

8. $\sum_{p=1}^{6} 2\left[\sin\dfrac{2p\pi}{7} - i\cos\dfrac{2p\pi}{7}\right] = 2i$

9. $\left|\dfrac{y-x}{1-\bar{x}y}\right| = 1$

10. $\dfrac{23}{5}$

11. $|z| - |z-1| = 1$

12. If α, β are the roots of $x^2 + px + q = 0$, and ω is a cube root of unity, then find the value cf$(\omega\alpha + \omega^2\beta)(\omega^2\alpha + \omega\beta) = p^2 - 3q$

13. a=1, b= 1 and c = 0

14. $|z| = 1$

15. S = 12

11

PERMUTATIONS, COMBINATIONS AND PROBABILITY

11.1 Introduction

In this Chapter we will deal with three important but slightly tricky topics of Permutations, Combinations and Probability. In simple language, number of Permutations tell us about the possible different arrangements of a group of items where sequence of items is important and is considered distinct. Number of combinations tells us the number of ways of selecting one or more items out of a group of items. Here the sequence is not important. Probability on the other hand quantifies chance of getting desired results out of numerous possible outcomes. Let us go ahead and enjoy the topics.

11.1.1 Permutations

Let us assume that you have shirts of 3 different colours. As you hang these shirts in a cupboard, there will be many possible arrangements (sequence in this case) in which shirts can be hanged. For example let us denote shirts as abc. We can hang these shirts in the following sequences- **abc, acb, bac, bca, cab, cba**. Thus there are 6 possible arrangements of hanging these three shirts. There will similarly be different combinations if we decide that shirt **a** will always be hanged first. Then there will be only two possible arrangements as abc and acb. ***Formally, each of the different arrangements that can***

be made out of a given number of things by taking some or all of them at a time is called a Permutation.

11.1.2 Combinations

In the above example, let us see how many different ways we can select two out of three shirts. We will see that three different selections can be made by taking two shirts at a time example: - ab, bc and ca; each of these selections is a combination. ***Formally, each of the different selections (or groups) that can be made out of a given number of items by taking some or all of them at a time is called a combination.***

Note that, order of items being important in permutations (i.e., **ab** and **ba** denote different permutations); but the order is immaterial in forming combinations (i.e., **ab** and **ba** denote the same combination).

11.1.3 Probability

Let us again consider the above example of three shirts a, b, c. Now let us assume that the shirts are in a bag. We take out one shirt from the bag without looking at the contents of the bag. We want to know the measure of possibility of drawing shirt **a**. Since there are 3 shirts, there is an equal possibility of drawing **shirt a, shirt b or shirt c**. Thus possibility of drawing shirt **a** is $\frac{1}{3}$. **This measure is known as Probability**.

Similarly, consider throw of a dice. If we want to know possibility of getting a specific value say **2** on the dice, it will be 1/6. And if we want to find possibility of getting an odd number (1, 3, 5), the obvious result will be 3 out of 6 possible outcomes yielding probability of $\frac{1}{2}$ ***Formally,***

measure of likelihood of an event's occurrence over total number of possibilities is called Probability of the said event's occurrence.

11.2 Calculating Permutations and Combinations for simple cases

We saw in section 11.1.1 that there are 6 possible ways/sequences (**abc, acb, bac, bca, cab, cba**) of arranging three shirts in a cupboard. It can be easily understood this way that for each first shirt (a, b or c), there are two possible ways of arranging the remaining two shirts. We thus know that there are $3 \times 2 = 6$ permutations of a Group of 3 shirts. Similarly for 4 items there will be $4 \times 3 \times 2 = 24$ permutations with 4 items in each group.

Now let us assume that we have 4 books **(a,b,c,d)** and we need to arrange these on different shelves with 2 books per shelf. As discussed already, sequence of books on the shelf is not important for finding out number of Combinations but is important for finding number of permutations. Here we can see that there are 6 possible groups of two books each as below:

ab, ac, ad, bc, bd, cd.

By the same logic as above, we can see that:

(a) If we choose first book as **a**, there are **three** other possibilities (**b, c or d** yielding three groups of 2 books as **ab, ac and ad**).

(b) Now if we choose book **b** as the first book, there are two new possibilities **bc** and **bd** (as **ab** is already covered).

(c) Finally if we choose **c** as the first book, there is one additional possibility **cd** as all other possibilities with **c or d** as one book are already covered,

Thus there are Six (3+2+1) possibilities of choosing 2 out of 4 books. These are thus all possible combinations of 2 out of 4 books where sequence is not important.

Since sequence of each pair of books yields two possibilities, there are a total of 6×2=12 permutations of arranging two out of 4 books on shelves.

As number of events/items and consequently number of possibilities go up, this will obviously be a tedious way of finding values of Permutations and Combinations. There are arithmetical ways of solving problems of Permutation and Combination that we will study shortly.

11.2.1 Concept of Factorial

We introduce the concept of factorial with symbol '!' ' in Chapter 1. Let us recall that Factorial of a number n is product of all integers starting with n upto 1. Thus

$$n! = n \times (n-1) \times (n-2) .... \times 2 \times 1$$

It can be seen that 3! = 3×2×1 = 6, 2! =2×1 = 2 and 1! = 1. ***As a special case, 0! is also considered as 1.***

The concept of factorial is used extensively in Permutation and Combination calculations.

11.2.2 Formula for Permutations

We have seen that there are 6 permutations of 3 distinct items, in groups of 3, 24 permutations of 4 distinct items, in groups of 4 and there are 12 permutations of 4 distinct items, in groups of 2. **We denote the formula for Permutations of n items in groups of r as ^nP_r.**

Arithmetical formula for $^nP_r = \dfrac{n!}{(n-r)!}$

With n = 4 and r = 4, $^4P_4 = \dfrac{4!}{(4-4)!} = \dfrac{4!}{0!} = \dfrac{4\times3\times2\times1}{1} = 24$

With 4 and r= 2, $^4P_2 = \dfrac{4!}{(4-2)!} = \dfrac{4!}{2!} = \dfrac{4\times3\times2\times1}{2\times1} = 4\times3 = 12$

In generic terms we can see that

$$^nP_r = \frac{n\times(n-1)\times(n-2)\times.......(n-r+1)\times(n-r)!}{(n-r)!} = n\times(n-1)\times(n-2)\times.......(n-r+1)$$

11.2.3 Formula for Combinations

In Combinations, sequence of items is irrelevant. As we have seen above, number of sequences for n items = n! so number of sequences for r items = r! Number of combinations will thus be lower than the corresponding permutations by a factor r!

Formula for Combinations of n items in groups of r is denoted as ^nC_r and as mentioned above is an extension of formula for permutations with

$$^nC_r = \frac{n!}{(n-r)!\times r!}$$

11.2.3.1 Extensions of formula for combinations:

Let us consider formula for $^nC_{n-r}$

$$^nC_{n-r} = \frac{n!}{(n-(n-r)!\times(n-r)!} = \frac{n!}{r!\times(n-r)!}$$

Thus $^nC_r = {}^nC_{n-r}$

This is an important formula used frequently in solving Binomial theorem related problems.

A couple of other important formulae are as below:

1. $^{n}C_{r} + {}^{n}C_{r-1} = {}^{n+1}C_{r}$

Proof:

Expanding left hand side we find

$$^{n}C_{r} + {}^{n}C_{r-1} = \frac{n!}{(n-r)!\times r!} + \frac{n!}{(n-r+1)!\times(r-1)!} = \frac{n!}{(n-r)!\times r!}\left(1+\frac{r}{n-r+1}\right)$$

$$= \frac{n!}{(n-r)!\times r!}\left(\frac{n+1}{(n+1)-r}\right) = \frac{(n+1)!}{(n+1-r)!\times r!} = {}^{n+1}C_{r}$$

2. $^{n}C_{r} = \left({}^{n-1}C_{r-1} + {}^{n-2}C_{r-1} + {}^{n-3}C_{r-1} + \ldots\ldots\ldots\ldots + {}^{r}C_{r-1} + {}^{r-1}C_{r-1}\right)$

Proof:

In this case, proof can be understood by a different process of finding number of possible combinations. Let us assume an example of finding number of ways of choosing 3 out of 6 given colours say

1. Blue
2. Green
3. Yellow
4. Orange
5. Red
6. Pink

The straight away answer to the given question is $^{6}C_{3}$

Alternate way of finding the solution is to consider possible ways of selecting 3 out of 6 colours as below:

If we select 4th colour (Orange), there is only 1 way of selecting both the colours below it that can be defined as $^{2}C_{2}$. Now if we select colour 3 (yellow), we can select 2 out of 3 colours below yellow by $^{3}C_{2}$ ways. Similarly by selecting colour 2 (Green) and colour 1 (Blue) we will have $^{4}C_{2}$ and $^{5}C_{2}$ ways respectively.

Thus the total number of combinations are $^{2}C_{2} + {}^{3}C_{2} + {}^{4}C_{2} + {}^{5}C_{2}$

Thus $^{6}C_{3} = {}^{5}C_{2} + {}^{4}C_{2} + {}^{3}C_{2} + {}^{2}C_{2}$

This relationship can be expressed in general terms as

$$^{n}C_{r} = \left({}^{n-1}C_{r-1} + {}^{n-2}C_{r-1} + {}^{n-3}C_{r-1} + \ldots\ldots\ldots\ldots + {}^{r}C_{r-1} + {}^{r-1}C_{r-1}\right)$$

The above formula derived in case of colours can be easily verified as below:

$$\text{And } {}^6C_3 = \frac{6\times5\times4}{3\times2\times1} = 20,\ {}^5C_2 = \frac{5\times4}{2\times1} = 10,\ {}^4C_2 = \frac{4\times3}{2\times1} = 6,$$

$${}^3C_2 = \frac{3\times2}{2\times1} = 3,\ {}^2C_2 = 1$$

We can now see that ${}^6C_3 = {}^5C_2 + {}^4C_2 + {}^3C_2 + {}^2C_2$

This is a very important formula again. The readers can try and verify the formula by assuming any other values of n and r.

11.3 Probability theories

In general, to find the probability of an event, we find the ratio of the number of outcomes favourable to the event, to the total number of ***equally likely*** outcomes. ***This theory of probability is known as classical theory of probability.***Here ***equally likely*** part is very important,

Another approach is to find the probability on the basis of observations and collected data. ***This is called statistical approach to probability***. This latter approach has many practical applications. We will consider this approach later in the Chapter.

11.4 A few important concepts for Classical approach to probability

In our day to day life, we carry out many experiments/events where the result is not the same even when the event is performed under identical conditions. Examples are tossing of a coin, throw of a dice, pulling a card from deck of playing cards etc. For probability, we deal with Random Experiments. An experiment is called random experiment if it satisfies the following two conditions:

1. It has more than one possible outcome.
2. It is not possible to predict the outcome in advance.

Let us understand a few concepts related to this approach to Probability.

11.4.1 Events, Sample space and sample point

All possible outcomes of an event are termed as Sample space. Let us denote it by symbol S. For example: when a dice is rolled, 1, 2, 3, 4, 5, 6 is the sample space for the number on top of the dice. Each element of the Sample space is called the Sample point.

We will normally be interested in finding probability of a specific or a set of outcome(s) from out of the Sample space. A specific outcome is a single point on the Sample space and is referred to as a **Simple Event**. Let us denote it as ω. A set of desired outcomes is termed as a **Compound Event (E).** Both **ω** and **E** are subset of the Sample space S. This is represented as $\omega \in E \in S$.

11.4.2 Algebra of Events:

(i) For every event A, there is a Compliment of that event denoted as $\overline{A}$.

Obviously $\overline{A}$ = 'not A' and A+$\overline{A}$ = S

(ii) If the desired event is event A OR event B, it is represented as A+B or $A \cup B$ (A Union B as in the Sets)

(iii) If the desired event is Event A AND Event B, it is represented as A×B or $A \cap B$ (A intersection B as in Sets)

(iv) If the desired event is A but not B, it will be represented as A-B or $A \cap B'$

(v) If we are interested in finding a set of events that excludes all events of another set,we term it as Mutually Exclusive Event(s). For example if we are looking for an odd number on throw of a dice (say Event A), it is mutually exclusive of the event of getting an even number on the dice (say Event B). Then the set $A \cap B$ will be a Null set denoted as φ.

If we are able to list all possible outcomes of a Sample space S, we will get a set of events that may or may not be mutually exclusive. However, assuming there are **n** such events, $E_1+E_2+\dots\dots E_n = S$. Set of these events is called Exhaustive Events. Further, if $E_i \cap E_j = \varphi$ for $i \neq j$ i.e, events E_i and E_j are pairwise disjoint and $\sum_{i=1}^{n} E_i = S$, then ***events*** ***E_1, E_2, ..., E_n are called mutually exclusive and exhaustive events.***

Classical probability can be used for very basic events, like rolling a dice and tossing a coin. It can also be used for other events like guessing on a multiple choice test or pulling a card from a deck of cards etc., where occurrence of all events is equally likely. Choosing a card from a standard deck of cards gives you a $\frac{1}{52}$ chance of getting a particular card, no matter what card you choose. On the other hand, figuring out will it rain tomorrow or not isn't something you can figure out with this basic type of probability.

11.5 Axiomatic approach to probability

Axiomatic approach is another way of describing probability of an event. In this approach some axioms or rules are depicted to assign probabilities. Let us think of a random experiment. All possibilities of this experiment put together is called its Sample space or its Domain say S. The probability P is a real valued function whose domain is the probability set of the entire sample space S and range is the interval [0,1] satisfying the following axioms

(i) For any event E, Probability $P(E) \geq 0$

(ii) $P(S) = 1$

(iii) If E and F are mutually exclusive events, then $P(E \cup F) = P(E) + P(F)$.

To prove that probability of NULL set $P(\varphi) = 0$, we take $F = \varphi$ and note that E and φ are disjoint events. Therefore, from axiom (iii) above, we get $P(E \cup \varphi) = P(E) + P(\varphi)$ or $P(E) = P(E) + P(\varphi)$ i.e. $P(\varphi) = 0$

(iv) In case the events E and F are not mutually exclusive, there will be another set of events say G= E F. Thus if we look at $E \cup F$, the common events would be counted twice. Therefore, $E \cup F = E + F - (E \cap F)$

Hence Probability $P(E \cup F) = P(E) + P(F) - P(E \cap F)$. This result is also evident from the following Venn diagram

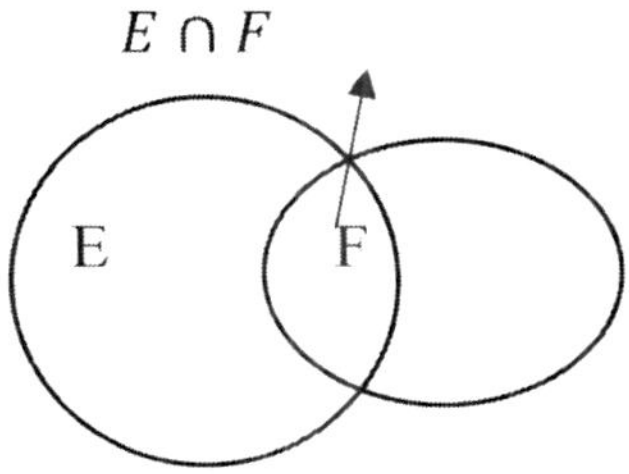

Figure 11.1

Let S be a sample space containing disjoint outcomes, $\omega_1, \omega_2, \ldots, \omega_n$ i.e., $S = \{\omega_1, \omega_2, \ldots, \omega_n\}$ It follows from the axiomatic definition of probability that

(i) $0 \leq P(\omega_i) \leq 1$ for each $\omega_i \in S$

(ii) $P(\omega_1) + P(\omega_2) + \ldots + P(\omega_n) = 1$

(iii) For any event A, $P(A) = \sum P(\omega_i), \omega_i \in S$.

Let us consider a few **examples (1-5)** below for clarity of concepts:

1. Two dice are thrown together. Find the probability that the product of the numbers on the top of the dice is:

   (i) 8

   (ii) 12

   Solution: We know that there are total 6 possibilities on one dice and 6 possibilities on the other dice for each of the 6 possibilities on the first dice i.e a total of $6 \times 6 = 36$ possibilities with 2 dice.

   For part (i), the only possibilities are (2,4) or (4,2) on the first and second dice respectively. Thus the first dice has two possibilities (2 or 4) yielding a probability of $\frac{2}{6}$ Correspondingly, the second dice can have either 2 or 4 depending upon the number on the first dice yielding a probability of $\frac{1}{6}$. Thus probability of the product 8 is $\frac{2}{36} = \frac{1}{18}$

   For part (ii), possibilities are (2,6), (3,4), (4,3), (6,2), yielding probability of $\frac{4}{36} = \frac{1}{9}$.

2. At a gala, cards bearing numbers 1 to 500, one number on one card are put in a box. Each player selects one card at random and that card is not replaced. If the selected card has a perfect square greater than 250, the player wins a prize. What is the probability that:

   (i) The first player wins a prize?

   (ii) The second player wins a prize if the first has won?

   Solution: We know that perfect squares greater than 250 but less than 500 are 256, 289, 324, 371, 400, 441 and 484 - a total of 7.

   There being total 500 possibilities, clearly probability of first player winning the prize is $\frac{7}{500}$ Since the card drawn by Player 1 is not replaced, probability of second player winning the prize is $\frac{6}{499}$.

3. A box contains 10 red marbles, 5 blue marbles and 7 green marbles. One marble is taken out of at random. What is the probability that the marble taken out will be:

   (i) Neither green nor blue

   (ii) Blue

   (iii) Not green

   Solution: There are total 22 marbles.

   For question, (i), the probability will be $= \frac{10}{22} = \frac{5}{11}$

   Similarly, for question, (ii), the probability will be $= \frac{5}{22}$ and

   for question, (iii), the probability will be $= \frac{15}{22}$

4. A lot consists of 50 mobile phones of which 43 are good, 3 have only minor defects and 4 have major defects. Mani will buy a phone only if it is good. And the trader will buy a mobile if it has no major defect. One phone is selected at random from the lot. What is the probability that it is:

   (i) Acceptable to Mani?

   (ii) Acceptable to Trader?

Solution: There are total 50 phones. For question, (i), the probability will be $= \frac{43}{50}$ and for question (ii), the probability will be $= \frac{46}{50} = \frac{23}{25}$

5. In a game, the entry fee is Rs 10. The game consists of tossing a coin 3 times. If two heads show, Shiva gets his entry fee back. If he throws 3 heads, he receives double the entry fee. Otherwise, he will lose. For tossing a coin three times, find the probability that he:

   (i) Loses the entry fee

   (ii) Gets double the entry fee

   (iii) Just gets his entry fee

   (iv) What is the probability of the Game host making money

   (v) If 200 customers play the Game, what is likely profit/loss

   Solution: There are a total of 8 possibilities in tossing 3 coins. Possible outcomes are TTT, TTH, THT, THH, HTT, HTH, HHT, HHH. It may be noticed that there are 2 Heads in 3 possible outcomes and 3 Heads in 1 outcome. That leaves us with 4 outcomes in which the customer loses entry fee.

We can therefore conclude that

(i) There is $\frac{4}{8} = \frac{1}{2}$ or 50% probability of the customer losing entry fee

(ii) Probability of getting double the entry fee is $\frac{1}{8}$

(iii) Probability of getting back the entry fee is $\frac{3}{8}$

(iv) We can thus see that the Game host returns entry fee for 3 out of 8 customers pays double the entry fee to 1 customer (amounting to refunding 2 more customers. As a result he gains entry fee equivalent to 3 out of 8 customers.

(v) From (iv), it is evident that if 200 customers play the Game, his likely profit $= \frac{3}{8} \times 200 \times 10$=Rs.750

11.6 Conditional probability

Until now we have discussed the methods of finding probability of random events. If we have multiple events from the same sample space, does the information about the occurrence of one of the events affect the probability of the other event? To answer this question, let us consider the following example:

A family has three children. What is the probability that atleast 2 children are boys given that the first child is a boy?

Sample space S of the total possibilities is {(g,g,g) (g,g,b), (g,b,g), (g,b,b), (b,g,g), (b.g.b), (b,b,g), (b,b,b)}

Thus there are total 8 possibilities. Let us assume event of at least two children being boys as E and event of first child being a boy as F. So we need to find probability of event E given that F has occurred, denoted as P (E|F).

When we examine Sample space S, we notice that the sample space E= {(g,b,b), (b.g.b), (b,b,g), (b,b,b)} and Sample space F = { (b,g,g), (b.g.b), (b,b,g), (b,b,b)}. There are 4 elements in each of these sample spaces.

Now if we consider that the first child is a boy i.e. event F has occurred, we should not consider cases in sample space E where the first child is a girl. Sample points F that are favorable to E reduces to {(b.g.b), (b,b,g), (b,b,b)}

Note that the elements of F which favor the event E are the common elements of E and F, i.e. the sample points of E ∩ F, which is effectively three events out of 4 possible events depicted in F.

This gives us probability of E given that event F has happened $P(E|F) = \frac{3}{4}$

Thus, we can also write the conditional probability of E given that F has occurred P (E|F) as $\frac{Number\ of\ event\ favorable\ to\ E \cap F}{Number\ of\ events\ favourable\ to\ F}$

$$= \frac{n(E \cap F)}{n(F)} = \frac{\frac{n(E \cap F)}{n(S)}}{\frac{n(F)}{n(S)}} = \frac{P(E \cap F)}{P(F)}$$

This is also considered definition of Conditional Probability P(E|F). Note that this expression will be valid only if P(F) ≠ 0 i.e. F is not a NULL set.

We can see from the above discussion that $P(F) = \frac{4}{8}$ and $P(E \cap F) = \frac{3}{8}$, giving us

$$\mathbf{P(E|F)} = \frac{3/8}{4/8} = \frac{3}{4}$$

11.6.1 Properties of conditional probability

Property 1:

If E and F are events of a sample space S, P (S|F) = P(F|F) =1

Proof:

As per formula for Conditional probability $P(S|F) = \frac{P(S \cap F)}{(P(F))}$

F being a subset of S, $(S \cap F) = F$

$$\therefore P(S|F) = \frac{P(S \cap F)}{P(F)} = \frac{P(F)}{P(F)} = 1,$$

,Similarly $(F \cap F) = F, \therefore P(F|F) = \frac{P(F)}{P(F)} = 1$

Property 2:

If Aand B are any two events of a sample space S and F is an event of S such that P (F) ≠ 0, then $P((A \cup B)|F) = P(A|F) + P(B|F) - P((A \cap B)|F)$.

In particular, if A and B are disjoint events, then

$P((A \cup B)|F) = P(A|F) + P(B|F)$

Proof:

From the formula for Conditional probability,

$P((A \cup B)|F) =$

$$\frac{P[(A \cup B) \cap F}{P(F)} = \frac{P\{(A \cap F) \cup (B \cap F)\}}{P(F)} = \frac{(P[A \cap F) + P[B \cap F) - P[(A \cap B) \cap F)}{P(F)}$$

$$= \frac{P(A \cap F)}{P(F)} + \frac{P(B \cap F)}{P(F)} - \frac{P(A \cap B) \cap F}{P(F)}$$

$$= P(A|F) + P(B|F) - P((A \cap B)|F)$$

When A and B are disjoint events, then $P((A \cap B)|F) = 0$ leading to $P((A \cup B)|F) = P(A|F) + P(B|F)$

Note: It may be noted that this property of Conditional property is similar to the normal probability with overlapped/ disjoint events.

Property 3:

P(E|F) = 1 – P($\overline{E}$ |F)

Proof:

We know that event sets E and $\overline{E}$ form the complete event set S and Event set $\overline{E}$ is exclusive of the event set E. Hence E∩ $\overline{E}$ = 0 and E∪ $\overline{E}$ =S, leading us to

P (E∩ $\overline{E}$) = 0 and P (E∪ $\overline{E}$) = 1

Now from the formula for Conditional probability,

P((E∪ $\overline{E}$)|F) = P(E|F) + P($\overline{E}$ |F) - P((E ∩ $\overline{E}$)|F).

Since P (E∪ $\overline{E}$) = 1 and P (E∩ $\overline{E}$) = 0, we get 1 = P(E|F) + P($\overline{E}$ |F)

Or P (E|F) = 1 – P ($\overline{E}$ |F)

Let us consider a few examples for clarity of this concept.

Examples: 6 to 8

A coin is tossed three times, determine P(E|F) for the following:

6. E: head on third toss, F: heads on first two tosses
7. E: at least two heads, F: at most two heads
8. E: at most two tails, F: at least one tail

Solution:

Here the entire sample space is (t,t,t), (t,t,h) (t,h,t) (t,h,h), (h,t,t), (h,t,h), (h,h,t), (h,h,h). Hence probability of each event is $\frac{1}{8}$. Let us consider the questions now

6. Outcome of last toss being fixed, E will have $2 \times {}^2P_1 = 2 \times 2 = 4$ possibilities namely {(t,t,h), (t,h,h), (h,t,h), (h,h,h)},similarly for F, outcome of first two tosses being fixed,F will have ${}^2P_1$ = 2 possibilities {(h,h,t), (h,h,h)}

   ∴E ∩ F will be (h,h,h)

   Using the formula for conditional probability, P(E|F) = $\frac{P(E \cup F)}{P(F)} = \frac{1/8}{2/8} = \frac{1}{2}$

7. As in question (6), one can easily see that E will have 4 possibilities (t,h,h), (h,t,h), (h,h,t), (h,h,h) and F will have all possibilities except all 3 heads, i.e. F will have 7 possibilities viz (t,t,t), (t,t,h) (t,h,t) (t,h,h), (h,t,t), (h,t,h), (h,h,t). Accordingly E ∩ F will be (t,h,h), (h,t,h), (h,h,t),

   $$P(E|F) = \frac{3/8}{7/8} = \frac{3}{7}$$

8. In this case E will have all possibilities except all 3 tails i.e. (t,h,h),(h,t,h), (h,h,t), (h,h,h) (t,t,h) (t,h,t) (t,h,h) and F will have all possibilities except all 3 heads, i.e(t,h,h), (h,t,h), (h,h,t), (t,t,h) (t,h,t) (t,h,h), (t,t,t)

   E ∩ F will be (t,h,h), (h,t,h), (h,h,t), (t,t,h) (t,h,t) (t,h,h),

   $$\therefore P(E|F) = \frac{6/8}{7/8} = \frac{6}{7}$$

Example 9:

An instructor has a question bank consisting of 300 easy True / False questions, 200 difficult True / False questions, 500 easy multiple choice questions and 400 difficult multiple choice questions. If a question is selected at random from the question bank, what is the probability that it will be an easy question given that it is a multiple choice question?

Solution:

Let E be the Sample space for Easy questions. There are total 1400 questionsand there are 800 easy questions. So the total sample space S is 1400 and sample space E is 800.

Let P (E) be the probability of selecting an easy question at random from the entire set. It will clearly be $\frac{800}{1400}$. Now let P (F) be the probability that the selected question is a multiple choice question. P (F) will be $= \frac{900}{1400}$.

Since there are 500 easy multiple choice questions out of total 900 multiple choice questions, P (E∩F) will equal $\frac{500}{1400}$

Using formula for conditional probability,

$$P(E|F) = \frac{P(E \cap F)}{P(F)} = \frac{500/1400}{900/1300} = \frac{500}{900} = \frac{5}{9}$$

11.6.2 Multiplication theorem on probability

Let us consider the case where we want to know probability of occurrence of two events E and F simultaneously. This will imply that we are interested in finding probability of overlap of event space E and F, which clearly is P (E∩F).

While discussing Conditional probability, we have seen that Conditional probability of occurrence of an event E when event F has already taken place is given as

$$P(E|F) \quad \frac{P(E \cap F)}{P(F)}$$

Thus P (E∩F) = P (E|F) × P (F) **...(1)**

It is worth noting that sequence of events is not important. We could as well find probability of occurrence of event F assuming that even E has already occurred, resulting in the following expression:

P (F∩E) = P (F|E) × P (E) **...(2)**

The above two expressions represent Multiplication theorem or **Multiplication Rule** of Probability denoted as P (E∩F) or P($E \times F$)

Multiplication rule of probability can be extended for more than two events. If E, F and G are three events of sample space S, we have P (E ∩ F ∩ G) = P(E) × P(F|E) × P(G|(E ∩ F)) = P(E) P(F|E) P(G| ($E \times F$)). Similarly, the multiplication rule of probability can be extended for four or more events.

Let us consider a couple of examples:

Example 10

Two cards are drawn from a pack of Playing Cards. What is the probability that the cards drawn are a Red and a Black King?

Solution:

There being two Black kings and two Red kings in pack of cards, probability of drawing a Black king P (E) =. $\frac{2}{52} = \frac{1}{26}$

Now the pack still has two red kings. Hence $P(F|E) = \frac{2}{51}$

$$\therefore P(E \cap F) = \frac{2}{51} \times \frac{1}{26} = \frac{1}{663}$$

Example 11

A box contains 12 blue and 8 yellow balls. Two balls are drawn from the box one after the other without replacement. What is the probability that both drawn balls are yellow?

Solution

In this case, if we denote event of drawing first ball as A and the second ball as B, we will see that

$P(A) = \frac{8}{20}$ and $P(B|A) = \frac{7}{19}$ leading us to

$$P(A \cap B) = \frac{8}{20} \times \frac{7}{19} = \frac{14}{95}$$

11.6.3 Independent and Mutually Exclusive events

Two events say A and B are said to be independent if occurrence of one event say A does not affect occurrence of the other event (B here) and vice versa, In other words,

$P(A|B) = P(A)$ and $P(B|A) = P(B)$

From the Multiplication theorem we know that

$P(A \cap B) = P(A|B)\, P(B)$

Since $P(A|B) = P(A)$,

$P(A \cap B) = P(A) \times P(B)$

Similarly, $P(B \cap A) = P(B) \times P(A)$

The above expressions are valid provided P (A) and P (B) are $\neq 0$

Two events A and B are said to be dependent if they are not independent, i.e. if $P(A \cap B) \neq P(A) \times P(B)$

Similarly, three events A, B and C are said to be independent if

$P(A \cap B) = P(A) \times P(B)$

$P(A \cap C) = P(A) \times P(C)$

$P(B \cap C) = P(B) \times P(C)$ and

$P(A \cap B \cap C) = P(A) \times P(B) \times P(C)$

If at least one of the above is not true for three given events, we say that the events are not independent.

Two events A and B are said to be **Mutually Exclusive** if they have no common occurrence. This implies that $P(A \cap B) = 0$. From the above expression of Independent events, it will be evident that two event with non-zero probability cannot be mutually exclusive.

Additional Examples on Independent events

Example 12

A circlular wheel labelled with integers from 1 to 10 is spun twice. If the rolls are independent, what is the probability that we will get two 6's?

Solution

Probability of getting a 6 on spinning the wheel $= \frac{1}{10}$

Since the two events of spinning the wheel are independent,

Probability of getting two 6's $= \frac{1}{10} \times \frac{1}{10} = \frac{1}{100}$

Example 13:

Asha, Sonu and Vivek each have a coin and are about to flip it at the same time. What is the probability that the coins of all three of them will land showing heads?

Solution:

Probability of getting head on a toss $= \frac{1}{2}$

Since the three tosses are independent, Probability of getting all 3 heads will be $= \frac{1}{2} \times \frac{1}{2} \times \frac{1}{2} = \frac{1}{8}$

Example 14:

Three girls each roll a fair standard dice only once for each. What is the probability that at least one of the three girls gets a prime number?

Solution:

Probability of getting a Prime number P (2, 3 or 5) on rolling a die $= \frac{1}{2}$. So we have two possibilities i.e. getting a Prime number or a Non-Prime (NP) number. When the three girls roll the die, total possibilities are 8 as in tossing three coins simultaneously.

Probability of at least one girl getting Prime number is same as (1- all three girls getting non-prime numbers).

Probability of all 3 numbers being non-prime = $\frac{1}{8}$

Probability of at least one girl getting Prime number is = $1 - \frac{1}{8} = \frac{7}{8}$

Example 15

A die marked 1, 2, 3 in red and 4, 5, 6 in green is tossed. Let A be the event, 'the number is even,' and B be the event, 'the number is red'. Are A and B independent?

We notice that probability of even number P (A) = $\frac{3}{6} = \frac{1}{2}$ and probability of red number P (B) = $\frac{3}{6} = \frac{1}{2}$. We can also see that probability of even number and in red P(A|B) = $\frac{1}{3}$.

Hence, probability of the number being even as well as red $P(A \cap B) = P(A|B) \times P(B) = \frac{1}{3} \times \frac{1}{2} = \frac{1}{6}$

Since in this case, $P(A \cap B) \neq P(A) \times P(B)$, the events A and B are not independent.

Example 16:

If A and B are two independent events, then the probability of occurrence of at least one of A and B is given by $1 - P(\overline{A})\, P(\overline{B})$

P (at least one of A and B) = $P(A \cup B) = P(A) + P(B) - P(A \cap B)$

Since A and B are independent events, $P(A \cap B) = P(A) \times P(B)$

Hence $P(A \cup B) = P(A) + P(B) - P(A) \times P(B)$

$$= P(A) + P(B)\,[1 - P(A)] = P(A) + P(B).\,P(\overline{A})$$

$$= 1 - P(\overline{A}) + P(B)\,P(\overline{A}) = 1 - P(\overline{A})\,[1 - P(B)]$$

$$= 1 - P(\overline{A})\,P(\overline{B})$$

Example 17:

For a loaded die, the probabilities of outcome are given as under:

P (1) = P (2) = 0.2, P (3) = P (5) = P (6) = 0.1 and P (4) = 0.3

The die is thrown two times. Let A and B be the events "Same number each time" and "a total score of 10 or more" respectively. Determine whether or not A and B are independent

Sample space A is (1,1), (2,2)..............(6,6)

Sample space B is (4,6), (5,5), (5,6), (6,4), (6,5) , (6,6)

And sample space A ∩ B is (5,5) and (6,6)

From the above data,

$P(A) = 2 \times (0.2)^2 + 3 \times (0.1)^2 + (0.3)^2 = 0.2$

$P(B) = 2 \times (0.3) \times (0.1) + 4 \times (0.1)^2 = 0.1$

And $P(A \cap B) = 2 \times (0.1)^2 = 0.02$

Now $P(A) \times P(B) = 0.2 \times 0.1 = 0.02$

Since value of P (A ∩ B) = P (A) × P (B), the events A and B are Independent

Example 18:

Consider the experiment of throwing a die, if a multiple of 3 comes up, throw the die again and if any other number comes, toss a coin. Find the conditional probability of the event 'the coin shows a tail', given that 'at least one die shows a 3'.

Solution:

Probability of getting a multiple of 3 on the die say $P(E) = \frac{2}{6} = \frac{1}{3}$ Accordingly event of not getting multiple of 3 on the die P) P (F) $= \frac{2}{3}$. Let P (F|E) be the probability of getting a non- multiple of 3 knowing that at least one toss of die yields a multiple of 3. Using formula for conditional probability and knowing that E and F are two independent events,

$$P(F|E) = \frac{P(E \cap F)}{P(E)} = \frac{(P(E) \times P(F)}{P(E)} = P(F) = \frac{2}{3}$$

Let G be the event of getting tail on toss of the coin. P (G) = $\frac{1}{2}$.

We need to find probability of getting a Tail on the coin when atleast one toss of die gets multiple of 3. These being again two independent events,

$$P\,(G|F|E) = \frac{2}{3} \times \frac{1}{2} = \frac{1}{3}$$

11.6.4 Bayes' theorem (credit Mario F. Triola)

Bayes' Theorem is an extension of Conditional Probability. We noted that the conditional probability of an event is a probability obtained with the additional information that some other event has already occurred. We used P(E|F) to denote the conditional probability of event E occurring, given that event F has already occurred. The following formula was derived for finding $P(E|F) = \frac{P(E \cap F)}{P(F)}$.

In addition to the above formal rule, the concept also outlined an "intuitive approach for finding a conditional probability": The conditional probability of F given E has already occurred can be found by assuming that event E has occurred and, working under that assumption, calculating the probability that event F will occur.

In this section we extend the discussion of conditional probability to include applications of Bayes' theorem (or Bayes' rule), which we use for revising a probability value based on additional information that is obtained later. One key to understanding the essence of Bayes' theorem is to recognize that we are dealing with sequential events, whereby new additional information is obtained for a subsequent event, and that new information is used to revise the probability of the initial event. In this context, the terms Prior Probability and Posterior Probability are commonly used.

Definitions

A **prior probability** is an initial probability value obtained before any additional information is obtained.

A **posterior probability** is a probability value that has been revised by using additional information that is obtained later.

Example 19

An organization randomly selects a citizen for a survey about credit card usage. Use subjective probabilities to estimate the following.

a. What is the probability that the selected subject is a male?

b. After selecting a subject, it is later learnt that this person was smoking a cigarette during the interview. What is the probability that the selected subject is a male?

c. Which of the preceding two results is a prior probability? Which is a posterior probability?

Solution:

(a) Roughly half of all citizens are males, so we estimate the probability of selecting a male subject to be 0.5. Denoting a male by M, we can express this probability as: P(M) = 0.5.

(b) Although some women smoke cigarettes, the vast majority of cigarette smokers are males. A reasonable guess is that 85% of cigarette smokers are males. Based on this additional subsequent information that the survey respondent was smoking a cigarette, we estimate the probability of this person being a male as 0.85. Denoting a male by M and denoting a cigarette smoker by C, we can express this result as P(M|C)= 0.85

(c) In part (a), the value of 0.5 is the initial probability, so we refer to it as the prior probability. Because the probability of 0.85 in part (b) is a revised probability based on the additional information that the survey subject was smoking a cigarette, this value of 0.85 is referred to a posterior probability. Thomas Bayes included the following theorem (or rule) that he developed for determining probabilities of events by incorporating information about subsequent events.

The probability of event E, given that event F has subsequently occurred, is P(E|F) = $\dfrac{P(F \mid E)\times P(E)}{P(F \mid E)\times P(E)+P(F \mid \overline{E})\times P(\overline{E})}$

Proof:

Using the Venn diagram in figure 11.2 below, we can see that $P(F)=P(F\cap E)+P(F\cap \overline{E})$

Since $E \cap F$ and $E \cap \bar{F}$ are disjoint,

$$P(F) = (F \mid E) \times P(E) + (F \mid \bar{E}) \times P(\bar{E})$$

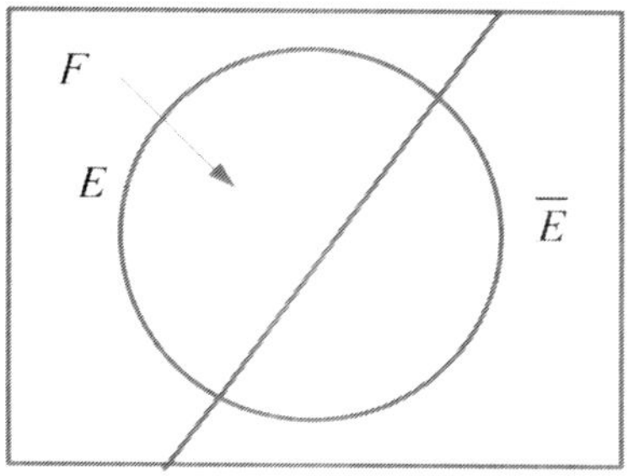

Figure 11.2

This leads us to

$$P(E \mid F) = \frac{P(E \cap F)}{P(F)} = \frac{(F \mid E) \times P(E)}{(F \mid E) \times P(E) + (F \mid \bar{E}) \times P(\bar{E})}$$

That looks complicated but we will simplify its calculation. See the following example, which illustrates use of the above expression, but also see the alternative method based on a more intuitive application of Bayes' theorem.

Example 20:

In a state, 51% of the adults are males. One adult is randomly selected for a survey involving credit card usage.

a. Find the prior probability that the selected person is a male.

b. It is later learnt that the selected survey subject was smoking a cigarette. Also, 9.5% of all males smoke cigarette, whereas 1.7% of females smoke cigarette. Use this additional information to find the probability that the selected subject is a male.

Solution:

Let's use the following notation:

M = male $\overline{M}$ = female (or not male) C = cigarette smoker $\overline{C}$ = not a cigarette smoker.

a. Before using the information given in part b, we know only that 51% of the adults in the given state are males, so the probability of randomly selecting an adult and getting a male is given by P(M) = 0.51.

b. Based on the additional given information, we have the following: P(M) = 0.51 and P($\bar{M}$) = 0.49 because 49% of the adults are (not males) P(C|M) = 0.095 because 9.5% of the males smoke cigarettes P(C|$\bar{M}$) = 0.017, because 1.7% of the females smoke cigars Let's now apply Bayes' theorem by using the preceding formula with M in place of E, and C in place of F.

$$P(M|C) = \frac{P(M) \times P(C|M)}{P(M) \times P(C|M) + P(\bar{M}) \times P(C|\bar{M})}$$

$$= \frac{0.51 \times 0.095}{0.51 \times 0.095 + 0.49 \times 0.017} = \frac{0.04845}{0.04845 + 0.00833} = \frac{0.04845}{0.05678}$$

$$= 0.853 \text{ or } 85.3\%$$

Thus we initially knew that there was a 51% probability that the survey subject was a male. However, after learning that the subject smoked a cigarette, we revised the probability to 85.3%. This makes sense, because the likelihood of a male increases dramatically with the additional information that the subject smokes cigarette (because so many more males smoke cigarette than females).

11.6.4.1 Intuitive Bayes Theorem

The preceding solution illustrates the application of Bayes' theorem with its calculation using the formula. Unfortunately, that calculation is complicated enough to create opportunities for errors and/or incorrect substitution of the involved probability values. Fortunately, here is another approach that is much more intuitive and easier:

Assume some convenient value for the total of all items involved, then construct a table of rows and columns with the individual cell frequencies based on the known probabilities. For the preceding example, simply assume some value for the adult population of the state, such as 10 Million, then use the given information to construct a table, such as the one shown below. Finding the number of males who smoke cigarettes: If 51% of the 10 Million adults are males, then there are 5.1 Million males. If 9.5% of the males smoke cigarettes, then the number of cigarettesmoking males is 9.5% of 5.1 Million , or 0.095 × 5.1 = 0.4845 Million.

The other males who do not smoke cigarettes must be 5.1–0.4845 = 4.6155 Million.Using similar reasoning, 49% of the 10 Million adults are females, so the number of females is 4.9 Million. Given that 1.7% of the females smoke cigarettes, the number of cigarette–smoking females is 0.017 × 4.9 = 0.0833. The number of females who do not smoke cigars is 4.9 –0.0833 = 4.8167 Million.

	Cigarette Smokers (Millions)	**Non Cigarette Smokers (Millions)**	**Total (Millions)**
Males M	0.4845	4.6155	5.1
Females	0.0833	4.8167	4.9
Total	0.5678	9.4322	10

The above table involves relatively simple arithmetic. Now we can easily address the key question as follows:

To find the probability of getting a male given that the subject smokes, restricts the table to the column of cigarette smokers, then find the probability of getting a male in that column. Among the 0.5678 Million cigarette smokers, there are 0.4845 Million males, so the probability we seek is 0.4845/0.5678 = 0.85329341 = 0.853 (rounded)

An important point to note is that we have to account for all subjects in the required category (cigarette smokers, whether they are males or females in this case) to arrive at the right result. Expression in the denominator of Bayes' theorem achieves this while the numerator defines the required category meeting the defined criteria (a Cigarette smoking Male in this case).

11.6.4.2 Bayes' Theorem Generalized

The preceding formula for Bayes' theorem and the preceding example uses exactly two categories for event A (male and female), but the formula can be extended to include more than two categories. The following example illustrates this extension and it also illustrates a practical application of Bayes' theorem to quality control in industry.

When dealing with more than the two events of A and , we must be sure that the multiple events satisfy two important conditions:

1. The events must be disjoint (with no overlapping).
2. The events must be exhaustive, which means that they combine to include allpossibilities.

Example 21:

An air-conditioner unit (A/C) is a device designed to bring down room temperature. Manufacturing Companies A, B and C make 80%, 15% and 5% of the air-conditioner units A/Cs respectively. The A/Cs manufactured by company A have a 4% rate of defects, the company B A/Cs have a 6% rate of defects, and the company C A/Cs have a 9% rate of defects

a. If an A/C is randomly selected from the general population of all A/Cs, find the probability that it was made by the Manufacturing Company A.

b. If a randomly selected A/C is then tested and is found to be defective, find the probability that it was made by the Manufacturing Company A.

Solution

We use the following notation: A = A/C manufactured by A

B = A/C manufactured by B and C = A/C manufactured by C

D = A/C is defective $\overline{D}$= A/C is not defective

a. If an A/C is randomly selected from the general population of all A/Cs, the probability that it was made by A is 0.8 (because A manufactures 80% of them).

b. If we now have the additional information that the A/C was tested and was found to be defective, we want to revise the probability from part (a) so that the new information can be used. We want to find the value of P(A|D), which is the probability that the A/C was made by the company A given that it is defective.

Based on the given information, we know these probabilities:

$P(A) = 0.80$, $P(B) = 0.15$ and $P(C) = 0.05$

$P(D|A) = 0.04$ because 4% of the company A A/Cs are defective

$P(D|B) = 0.06$ because 6% of the company B A/Cs are defective

$P(D|C) = 0.09$ because 9% of the company C A/Cs are defective

Here is Bayes' theorem extended to include three events corresponding to the selection of A/Cs from the three manufacturers (A, B, C):

$$P(A|D) = \frac{P(A).P(D|A)}{P(A).P(D|A)+P(B).P(D|B)+P(C).P(D|C)}$$

$$= \frac{0.8\times0.04}{0.8\times0.04+0.15\times0.06+0.05\times0.09} = 0.703 \text{ (rounded)}$$

Now let's find P(A|D) by using the table below. Let's arbitrarily assume that 10,000 A/Cs were manufactured. Because 80% of the A/Cs are made by A, we have 8000 A/Cs made by A, and 4% of them (or 320) are defective. Also, if 320 of the company A A/C s are defective, the other 7680 are not defective. See the values of 320 and 7680 in the table below. The other values are found using the same reasoning.

Manufacturer	Defective (D)	Non -Defective ($\bar{D}$)	Total
A	320	7680	8000
B	90	1410	1500
C	45	455	500
Total	455	9545	10000

We want to find the probability that an A/C was made by company A, given that it is known to be defective. Because we know the condition that the A/C is defective, we can refer to the first column of values where we see that among the 455 total defective A/Cs, 320 were manufactured by A, so that the probability is 320/455 = 0.703 (rounded). This is the same result obtained with the formula from Bayes' theorem.

The preceding example involves an extension of Bayes' theorem to three events denoted by A, B, C. Based on the format of the formula used in the solution, it is easy to extend Bayes' theorem so that it can be used with four or more events.

Example 22:

Pregnancy Test Results.

Refer to the results summarized in the table below for questions (a) and (b)

Subject	Positive test result (Pregnancy indicated) (P)	Negative Test Result (Pregnancynot indicated ($\bar{P}$)
Pregnant	80	5
Not pregnant	3	11

(a) If one of the 99 test subjects is randomly selected, what is the probability of getting a subject who is pregnant?

Solution:

Data in the table shows that there are 85 pregnant subjects out of total 99. Hence probability of the a randomly selected subject being pregnant P (P) = $\frac{85}{99}$

(b) A test subject is randomly selected and is given a pregnancy test. What is theprobability of getting a subject who is pregnant, given that the test result is positive?

Solution:

Success rate of Pregnancy test on subjects who are tested positive is $\frac{80}{83}$

Since the Test result is positive, probability of the randomly selected subject being actually pregnant will $= \frac{80}{83}$

Example 23:

Pleas and Sentences

In a study of pleas and prison sentences, it is found that 45% of the subjects studied were sent to prison. Among those sent to prison, 40% chose to plead guilty. Among those not sent to prison, 55% chose to plead guilty.

If one of the study subjects is randomly selected, find the probability of getting someone who was not sent to prison.

If a study subject is randomly selected and it is then found that the subject entered a guilty plea, find the probability that this person was not sent to prison.

Solution: Let us assume total number of subjects as 100 for sake of simplicity.

Results as per the given data are summarized below:

Subject	Total	Pleaded Guilty (G)	Not pleaded Guilty ($\bar{G}$)
Sent to Prison (P)	45	18	27
Not sent to prison (NP)	55	30.25	24.75
	100	48.25	51.75

a. Clearly, probability of a randomly selected subject not sent to prison $= \frac{55}{100} = 0.55$

(b) In this case, we need to find the probability of a person selected at random that this person was not sent to prison knowing that he filed a guilty plea. Here we notice that 48.25 persons filed Guilty plea out of which 30.25 were not sent to prison.

Hence the required probability $= \frac{30.25}{48.25} = \frac{121}{193} = 0.6269$

Using Bayes' theorem, this results will translate to

$$P(G|NP) = \frac{P(NP|G) \times P(NP)}{P(NP|G) \times P(NP) + P(P|G) \times P(P)} = \frac{0.55 \times 0.55}{0.55 \times 0.55 + 0.4 \times 0.45}$$

$$= \frac{0.3025}{0.4825} = 0.6269$$

Example 24

HIV: A State Health Department reports a 10% rate of the HIV virus for the "at-risk" population. Under certain conditions, a preliminary screening test for the HIV virus is correct 95% of the time. If someone is randomly selected from the at-risk population, what is the probability that they have the HIV virus if it is known that they have tested positive in the initial screening?

Let us define the following:

P (H) – Probability of HIV infection for the at Risk population

P($\bar{H}$) – Probability of no HIV infection for the at Risk population

P (T|H) – Probability that the HIV test is positive when the subject is actually HIV infected

P($\overline{H}$) - Probability that the HIV test is positive when the subject is actually not HIV infected

We need to find P (H|T)

Using Bayes' theorem, this will translate to

$$P(H \mid T)=\frac{P(T \mid H)\times P(H)}{P(T \mid H)\times P(H)+P\left(T \mid \overline{H}\right)\times P\left(\overline{H}\right)}=$$

$$\frac{0.1\times 0.95}{0.1\times 0.95+0.9\times 0.05}=\frac{0.095}{0.14}=0.6786$$

11.6.4.3 Extending Bayes' Theorem to more than 3 events

Example 21 in this section included an extension of Bayes' theorem to include three events, denoted by A, B, C. We can now write an expression that extends Bayes' theorem so that it can be used to find P(A|Z), assuming that the initial event can occur in one of four ways: A, B, C, D as below:

$$P\,(A|Z) = \frac{P(A).P(Z|A)}{P(A).P(Z|A)+P(B).P(Z|B)+P(C).P(Z|C)+ P(D).P(Z|D)}$$

11.6.4.4 Partition of a sample space

Reference figure 11.3 below:

A set of events $E_1, E_2, ..., E_n$ is said to represent a partition of the sample space S if

(a) $E_i \cap E_j = \varphi,\ i \neq j,\ i, j = 1, 2, 3, ..., n$

(b) $E_1 \cup E_2 \cup ... \cup E_n = S$ and

(c) $P(E_i) > 0$ for all i = 1, 2, ..., n.

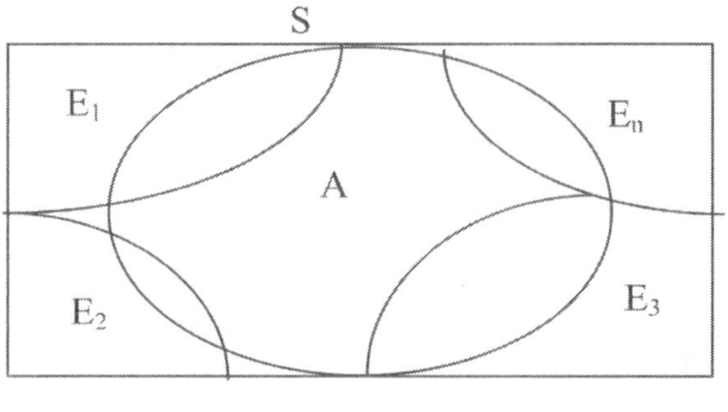

Figure 11.3

In other words, the events $E_1, E_2, ..., E_n$ represent a partition of the sample space S if they are pairwise disjoint, exhaustive and have non-zero probabilities.

As an example, we see that any nonempty event E and its complement E′ form a partition of the sample space S since they satisfy $E \cap E' = \varphi$ and $E \cup E' = S$.

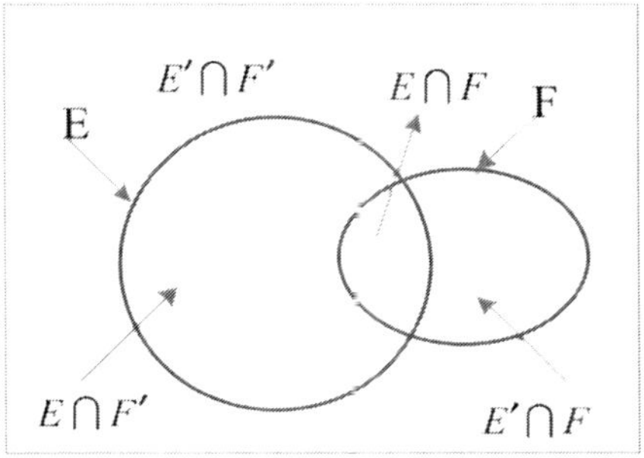

Figure 11.4

From the Venn diagram in Fig 11.4, one can easily observe that if E and F are any two events associated with a sample space S, then the set $\{E \cap F', E \cap F, E' \cap F, E' \cap F'\}$ is a partition of the sample space S such that

$$(E \cap F') + (E \cap F) + (E' \cap F) + (E' \cap F') = S$$

It may be mentioned that the partition of a sample space is not unique. There can be several partitions of the same sample space. We shall now prove a theorem known as Theorem of Total probability.

11.6.4.5 Theorem of total probability

Let a set $\{E_1, E_2, ..., E_n\}$ be a partition of the sample space S, and suppose that each of the events $E_1, E_2, ..., E_n$ has non-zero probability of occurrence.

Let A be any event associated with S, then

$$P(A) = P(E_1)\,P(A|E_1) + P(E_2)\,P(A|E_2) + ... + P(E_n)\,P(A|E_n)$$

$$= \sum_{j=n}^{j=1} P(Ej)P(A \mid Ej)$$

Proof

Given that $(E_1, E_2, ..., E_n)$ is a partition of the sample space S (See Fig 11.3). Therefore, $S = E_1 \cup E_2 \cup ... \cup E_n$ ... (1) and $Ei \cap Ej = \varphi$, $i \neq j$, $i, j = 1, 2, ..., n$

Now, we know that for any event A,

$A = A \cap S = A \cap (E_1 \cup E_2 \cup \ldots \cup E_n)$

$=(A \cap E_1) \cup (A \cap E_2) \cup \ldots \cup (A \cap E_n)$

Also A ∩ Ei and A ∩ Ej are respectively the subsets of Ei and Ej. We know that Ei and Ej are disjoint, for $i \neq j$, therefore, A ∩ Ei and A ∩ Ej are also disjoint for all $i \neq j$, i, j = 1, 2, ..., n.

Thus, $P(A) = P[(A \cap E_1) \cup (A \cap E_2) \cup \ldots \cup (A \cap E_n)]$

$=P(A \cap E_1) + P(A \cap E_2) + \ldots + P(A \cap E_n)$

Now, by multiplication rule of probability, we have

$P(A \cap Ei) = P(Ei)\,P(A|Ei)$ as $P(Ei) \neq 0$ for all i = 1,2,..., n

Therefore, $P(A) = P(E_1)\,P(A|E_1) + P(E_2)\,P(A|E_2) + \ldots + P(E_n)\,P(A|E_n)$

$$\text{or } P(A) = \sum_{j=n}^{j=1} P(Ej)\,P(A \mid Ej)$$

Let us consider a few examples for sake of further clarity:

Example 25:

We draw two cards from a deck of shuffled cards without replacement. Find the probability of getting the second card a queen.

Solution: Let 'A' represent the event of getting the first card a queen.

Let 'B' represent the event that the first card is not a queen.

Let 'E' represent the event that the second card is a queen.

Then the probability that the second card will be a queen or not will be represented by the law of total probability as:

$P(E) = P(A)P(E|A)+P(B)P(E|B)$

Where P(E) is the probability that second card is a queen

P(A) is the probability that the first card is a queen

P(E|A) is the probability that the second card is a queen given that first card is a queen.

P(B) is the probability that the first card is not a queen

P(E|B) is the probability that the second card is a queen but the first card drawn is not a queen.

According to question:

$$P(A)= \frac{4}{52} \quad P(B) = \frac{48}{52} \quad P(E|A) = \frac{3}{51} \quad P(E|B) = \frac{4}{51}$$

On substituting in the above equation we get, $P(E) = \frac{4}{52}\times\frac{3}{51}+\frac{48}{52}\times\frac{4}{51}$

$$= \frac{12}{52\times51} + \frac{192}{52\times51} = \frac{204}{52\times51} = \frac{4}{52} = \frac{1}{13}$$

Example 26:

Patients are recruited onto the two arms, (0- Control, 1-Treatment) of a clinical trial. The probability that an adverse outcome occurs on the control arm is p_0, and is p_1 for the treatment arm. Patients are allocated alternately onto the two arms, and their outcomes are independent. What is the probability that the first adverse event occurs on the control arm?

Solution: Define:

- Sample space S to consist of all possible infinite sequences of patient outcomes;
- Event E_1 - first patient (allocated onto the control arm) suffers an adverse outcome;
- Event E_2 - first patient (allocated onto the control arm) does not suffer an adverse outcome, but the second patient (allocated onto the treatment arm) does suffer an adverse outcome;
- Event E_0 - neither of the first two patients suffer adverse outcomes;
- Event F - first adverse event occurs on the control arm.

We want P(F). Now the events E_1, E_2 and E_0 are partition of S, so, by the Theorem of Total Probability,

$P(F) = P(F|E_1)P(E_1) + P(F|E_2)P(E_2) + P(F|E_0)P(E_0)$.

Now, $P(E_1) = p_0$, $P(E_2) = (1 - p_0)\, p_1$, $P(E_0) = (1 - p_0)\,(1 - p_1)$ and $P(F|E_1) = 1$, $P(F|E_2) = 0$.

Finally, as after two non-adverse outcomes the allocation process effectively re-starts, we have $P(F|E_0) = P(F)$.

Hence $P(F) = (1 \times p_0) + (0 \times (1 - p_0)\, p_1) + (P(F) \times (1 - p_0)\,(1 - p_1))$

$= p_0 + (1 - p_0)\,(1 - p_1)\, P(F)$, which can be re-arranged to give`

First adverse event occurs on the control arm $P(F) = \dfrac{p_0}{p_0 + p_1 - p_0 p_1}$

Example 27

In a tennis match, with the score at deuce, the game is won by the first player who gets a clear lead of two points. If the probability that given player wins a particular point is θ, and all points are played independently, what is the probability that player eventually wins the game?

Solution: Define:

- Sample space S to consist of all possible infinite sequences of points;
- event Wi - given player wins the ith point;
- event Vi - given player wins the game on the ith point;
- event V - given player wins the game.

We want P (V). The events $\{W_1, W'_1\}$ partition S, and thus, by the Theorem of Total Probability

$$P(V) = P(V|W_1)P(W_1) + P(V|W'_1)P(W'_1). \quad ...(3)$$

Now $P(W_1)$ (probability of winning first point) = θ and $P(W'_1)$ (probability of losing first point) = 1 − θ.

To get $P(V|W_1)$ and $P(V|W'_1)$, we need to apply further condition on the result of the second point and again use the Theorem: For example

$P(V|W_1)$ $= P(V|W_1 \cap W_2)P(W_2|W_1) + P(V|W_1 \cap W'_2)P(W'_2|W_1)$... }&

$P(V|W'_1)$ $= P(V|W'_1 \cap W_2)P(W_2|W'_1) + P(V|W_1 \cap W'_2)P(W'_2|W'_1)$, }...(4) where

$P(V|W_1 \cap W_2)$ = 1, implying that the given player wins first and second points consecutively and wins

$P(W_2|W_1)$ $= P(W_2) = \theta$, shows that probability of winning second point having won the first point is independent and equals θ

$P(V|W_1 \cap W'_2)$ = P(V), shows that given player's probability of win is back to P(V) if he loses second point having won the first point

$P(W'_2|W_1)$ $= P(W'_2) = 1 - \theta$, shows probability of losing second point having won the first point.

P(V |W′$_1$ ∩ W$_2$) = P(V), shows that given player's probability of win is back to P(V) if he wins second point having lost the first point

P(W$_2$|W′$_1$) = P(W$_2$) = θ,

P(W′$_2$ |W′$_1$) = P(W'$_2$) = 1 − θ, shows probability of losing second point having lost the first point.

P(V |W′$_1$ ∩ W′$_2$) = 0, implies that the given player loses first and second points consecutively and loses the match

Substituting in equation 4, we get

$P(V \mid W_1) = (1 \times \theta) + (P(V) \times (1 - \theta)) = \theta + (1 - \theta) P(V)$

$P(V \mid W'_1) = (P(V) \times \theta) + 0 \times (1 - \theta) = \theta \times P(V)$.

Hence, combining (3) and (4) we have

$$P(V) = (\theta + (1 - \theta) P(V))\, \theta + \theta \times P(V)(1 - \theta)$$
$$= \theta^2 + 2\theta(1 - \theta)P(V)$$

$$\Rightarrow P(V) = \frac{\theta \times \theta}{1 - 2\theta(1 - \theta)} = \frac{\theta^2}{1 - 2\theta(1 - \theta)}$$

Alternatively, $\{V_i, i = 1, 2, ...\}$ partitions V .

Hence $P(V) = \sum_{i=1}^{\infty} P(V_i)$

Now, $P(V_i) = 0$ if i is odd, as the game can never be completed after an odd number of points.

Since the given player has to win two successive points (odd and even in that order), probability of victory for the given player will be $P(V_2) = \theta^2$

If 1stpoint is won by the given player and the 2nd point is won by the rival player the game will be back to deuce. Similarly if 1st point is won by the rival player and the 2nd point is won by the given player the game will be back to deuce again. Probability of two such pairs of points is same and can be denoted as $2\theta(1-\theta)$. And for k such pairs of points probability $=(2\theta(1-\theta))^k$.

The score must stand at deuce after 2k pairs of points and the game must not have been completed prior to this, indicating that there must have been 2k successive drawn pairs of points, each of which could be arranged win/lose or lose/win for the nominated player. Then the given player must win the final two points. Thus if we assume there are k pairs of points leading to k numbers of deuce, before the given player wins, the victory will come after 2k+2 points

For last two points i=2k+1 and i=2k+2, $P(V_{2k+1/2k+2}) = \theta^2$,

Thus for i = 2k + 2 points (k = 1, 2, 3, ...) we have

$P(V_i) = P(V_{2k+2}) = \{2\theta(1 - \theta)\}^k \times \theta^2$

Hence $P(V) = \sum_{k=0}^{\infty} P(V_{2k+2}) = \theta^2 \sum_{k=0}^{\infty} (2\theta(1 - \theta))^k = \frac{\theta^2}{1 - 2\theta(1 - \theta)}$

Because it can be seen by equating $\frac{d}{d\theta}\{2\theta(1-\theta)\}$ to 0 that the term in the geometric series $|2\theta\ (1 - \theta)|$ has maximum value of 0.5 for all values of θ in the range $0<\theta<1$ and satisfies the condition $|2\theta\ (1 - \theta)| < 1$.

11.6.5 Random variables and its probability distribution

We have already learnt about random experiments and formation of sample spaces. In most of these experiments, we were not only interested in the particular outcome that occurs but rather in some number associated with those outcomes as shown in following examples/experiments.

i. In tossing two dice, we may be interested in the sum of the numbers obtained on the two dice.

ii. In tossing a coin 50 times, we may want the number of heads obtained.

iii. In the experiment of taking out four articles (one after the other) at random from a lot of 20 articles in which 6 are defective, we want to know the number of defectives in the sample of four and not in the particular sequence of defective and non-defective articles.

In all of the above experiments, we define a rule which assigns to each outcome of the experiment a single real number. This single real number may vary with different outcomes of the experiment. Hence, it is a variable. Also its value depends upon the outcome of a random experiment and, hence, is called **random variable**. A random variable is usually denoted by a capital letter such as X or Y.

If you recall the definition of a function, you will realize that the random variable X is really speaking a function whose domain is the set of outcomes (or sample space) of a random experiment. A random variable can take any real value, therefore, its co-domain is the set of real numbers. Hence, a random variable can be defined as follows:

Definition: Suppose that to each point of a sample space we assign a number. We then have a function defined on the sample space. This function is called a **Random Variable** or more precisely a random function. In general, a random variable has some specified physical, geometrical, or other significance.

Example 28:

Suppose that a coin is tossed twice so that the sample space is S = {HH, HT, TH, TT}. If X denotes the number of heads obtained, then X is a random variable and for each outcome, its value is as given below: X(HH) = 2, X (HT) = 1, X (TH) = 1, X (TT) = 0. More than one random variable can be defined on the same sample space. For example, let Y denote the number of heads minus the number of tails for each outcome of the above sample space S. Then Y(HH) = 2, Y (HT) = 0, Y (TH) = 0, Y (TT) = – 2. Thus, X and Y are two different random variables defined on the same sample space S as depicted in the Table below.

Sample point	**HH**	**HT**	**TH**	**TT**
X	2	1	1	0
Y	2	0	0	-2

Note: One outcome can generate more than one random variable depending on the requirement.

It should be noted that many other random variables could also be defined on this sample space like, square of the number of heads etc. A random variable that takes on a finite or countably infinite number of values is called a discrete random variable while one which takes on a non-countably infinite number of values is called a non-discrete random variable.

Example 29:

A person plays a game of tossing a coin thrice. For each head, he is given Rs 2 by the organizer of the game and for each tail, he has to give Rs 1.50 to the organizer. Let X denote the amount gained or lost by the person. Show that X is a random variable and exhibit it as a function on the sample space of the experiment.

Solution X is a number whose values are defined on the outcomes of a random experiment. Therefore, X is a random variable. Now, sample space of the experiment is S = {HHH, HHT, HTH, THH, HTT, THT, TTH, TTT}

Then X(HHH) = Rs (2 × 3) = Rs 6

X(HHT) = X(HTH) = X(THH) = Rs (2 × 2 – 1 × 1.50) = Rs 2.50

X(HTT) = X(THT) = (TTH) = Rs(1 × 2) – (2 × 1.50) = – Re 1 and

X(TTT) = – Rs (3 × 1.50) = – Rs 4.50 where, (-) sign shows loss to the player.

Thus, for each element of the sample space, X takes a unique value, hence, X is a function on the sample space whose range is {– 4.50, –1, 2.50, 6}

Example 30:

A bag contains 2 white and 1 red balls. One ball is drawn at random and then put back in the box after noting its colour. The process is repeated again. If X denotes the number of red balls recorded in the two draws, describe X.

Solution

Let the balls in the bag be denoted by w_1, w_2, r. Then the sample space is

$S = \{w_1w_1; w_1\ w_2; w_2w_2; w_2w_1; w_1\ r; w_2\ r;\ r\ w_1;\ r\ w_2;\ r\ r\}$

Now, for $\omega \in S, X(\omega)$ = number of red balls

Therefore $X(\{w_1w_1\}) = X(\{w_1w_2\}) = X(\{w_2w_2\}) = X(\{w_2w_1\}) = 0$

$X(\{w_1\ r\}) = X(\{w_2\ r\}) = X(\{r\ w_1\}) = X(\{r\ w_2\}) = 1$ and $X(\{r\ r\}) = 2$

Thus, X is a random variable which can take values 0, 1 or 2.

11.6.5.1 Probability distribution of a random variable

Let us look at the experiment of selecting one family out of ten families f_1, f_2 ,..., f_{10} in such a manner that each family is equally likely to be selected.

Let the families f_1, f_2, ..., f_{10} have 3, 4, 3, 2, 5, 4, 3, 6, 4, 5 members, respectively. Let us select a family and note down the number of members in the family as X. Clearly, X is a random variable defined as below : $X(f_1) = 3$, $X(f_2) = 4$, $X(f_3) = 3$, $X(f_4) = 2$, $X(f_5) = 5$, $X(f_6) = 4$, $X(f_7) = 3$, $X(f_8) = 6$, $X(f_9) = 4$, $X(f_{10}) = 5$

Thus, X can take any value 2,3,4,5 or 6 depending upon which family is selected.

Since we had assumed that each family is equally likely to be selected, the probability that family f_4 is selected is 1/10. Thus, the probability that X can take the value 2 is 1/10. We write P(X = 2) = 1/10

Also, the probability that any one of the families f_1, f_3 or f_7 is selected is $P(\{f_1, f_3, f_7\}) = 3/10$. Thus, the probability that X can take the value 3 = $\frac{3}{10}$. We write $P(X = 3) = \frac{3}{10}$

Similarly, we obtain $P(X = 4) = P(\{f_2, f_6 \text{ or } f_9\}) = \frac{3}{10}$

$P(X = 5) = P(\{f_5 \text{ or } f_{10}\}) = \frac{2}{10}$ and

$P(X = 6) = P(\{f_8\}) = \frac{1}{10}$

Such a description giving the values of the random variable along with the corresponding probabilities is called the probability distribution of the random variable X. In general, the probability distribution of a random variable X is defined as follows:

Definition: The probability distribution of a random variable X is the system of numbers

$X: x_1 x_2 \ldots\ldots\ldots x_n$

$P(X): p_1 p_2 \ldots\ldots\ldots p_n$

Where, $p_i > 0$ and $\sum_{i=0}^{n} p_i = 1$, I=1, 2,….n

The real numbers $x_1, x_2, \ldots, x_n$ are the possible values of the random variable X and p(i) (i = 1,2,..., n) is the probability of the random variable X taking the value x_i i.e., $P(X = x_i) = p_i$

Note: If x_i is one of the possible values of a random variable X, the statement $X = x_i$ is true only at some point (s) of the sample space. Hence, the probability that X takes value x_i is always nonzero, i.e. $P(X = x_i) \neq 0$. Also for all possible values of the random variable X, all elements of the sample space are covered. Hence, the sum of all the probabilities in a probability distribution must be one.

11.6.5.2 Mean of a Random variable

In many problems, it is desirable to describe some feature of the random variable by means of a single number that can be computed from its probability distribution. Few such numbers are mean, median and mode. In this section, we shall discuss mean only. Mean is a measure of location or central tendency in the sense that it roughly locates a middle or average value of the random variable.

Definition: Let X be a random variable whose possible values $x_1, x_2, x_3, \ldots, x_n$ occur with probabilities $p_1, p_2, p_3, \ldots, p_n$, respectively. The mean of X, denoted by μ, is thenumber $\sum_{i=1}^{n} x_i \times p_i$ i.e. the mean of X is the weighted average of the possible values of X, each value being weighted by its probability with which it occurs. The mean of a random variable X is also called the expectation of X, denoted by E(X).

$$\text{Thus, } E(X) = \mu = \sum_{i=1}^{n} x_i \times p_i = x_1 p_1 + x_2 p_2 + \ldots + x_n p_n.$$

In other words, the mean or expectation of a random variable X is the sum of the products of all possible values of X by their respective probabilities.

Example 31:

Let a pair of dice be thrown and the random variable X be the sum of the numbers that appear on the two dice. Find the mean or expectation of X.

Solution The sample space of the experiment consists of 36 elementary events in the form of ordered pairs (xi, yi), where xi = 1, 2, 3, 4, 5, 6 and yi = 1, 2, 3, 4, 5, 6. The random variable X i.e. the sum of the numbers on the two dice takes the values 2, 3, 4, 5, 6, 7, 8, 9, 10, 11 or 12.

$$\text{Now } P(X = 2) = P(\{(1,1)\}) = \frac{1}{36}$$

$$P(X = 3) = P(\{(1,2), (2,1)\}) = \frac{2}{36}$$

$$P(X = 4) = P(\{(1,3), (2,2), (3,1)\}) = \frac{3}{36}$$

$$P(X = 5) = P(\{(1,4), (2,3), (3,2), (4,1)\}) = \frac{4}{36}$$

$$P(X = 6) = P(\{(1,5), (2,4), (3,3), (4,2), (5,1)\}) = \frac{5}{36}$$

$P(X = 7) = P(\{(1,6), (2,5), (3,4), (4,3), (5,2), (6,1)\}) = \dfrac{6}{36}$

$P(X = 8) = P(\{(2,6), (3,5), (4,4), (5,3), (6,2)\}) = \dfrac{5}{36}$

$P(X = 9) = P(\{(3,6), (4,5), (5,4), (6,3)\}) = \dfrac{4}{36}$

$P(X = 10) = P(\{(4,6), (5,5), (6,4)\}) = \dfrac{3}{36}$

$P(X = 11) = P(\{(5,6), (6,5)\}) = \dfrac{2}{36}$

$P(X = 12) = P(\{(6,6)\}) = \dfrac{1}{36}$

The probability distribution of X is

X	2	3	4	5	6	7	8	9	10	11	12
P(X)	$\frac{1}{36}$	$\frac{2}{36}$	$\frac{3}{36}$	$\frac{4}{36}$	$\frac{5}{36}$	$\frac{6}{36}$	$\frac{5}{36}$	$\frac{4}{36}$	$\frac{3}{36}$	$\frac{2}{36}$	$\frac{1}{36}$

Therefore, $\mu = E(X) =$

$$2\times\frac{1}{36}+3\times\frac{2}{36}+4\times\frac{3}{36}+5\times\frac{4}{36}+6\times\frac{5}{36}+7\times\frac{6}{36}+8\times\frac{5}{36}+9\times\frac{4}{36}$$

$$+10\times\frac{3}{36}+11\times\frac{2}{36}+12\times\frac{1}{36}$$

$$= \frac{2+6+12+20+30+42+40+36+30+22+12}{36} = \frac{252}{36} = 7$$

Thus, mean of the sum of the numbers that appear on throwing two fair dice is 7.

11.6.5.3 Variance of a Random variable

The mean of a random variable does not give us information about the variability in the values of the random variable. In fact, if the variability is small, then the values of the random variable are close to the mean. Also random variables with different probability distributions can have equal means

11.6.5.4 The Variance and Standard Deviation

We have already noted that the expectation of a random variable X is often called the mean and is denoted by μ. Another quantity of great importance in probability and statistics is called the **variance** and is defined by

$$\text{Var}(X) = E[(X - \mu)^2]$$

The variance is a non-negative number. The positive square root of the variance is called the standard deviation and is given by

$$\sigma_x = \sqrt{Var\ (X)} = \sqrt{E\left[(X - \mu)^2\right]}$$

Where no confusion can result, the standard deviation is often denoted by σ instead of σ_x, and the variance in such case is σ^2.

If variable X is a discrete random variable taking the values x_1, x_2, x_3, ..., x_n and has probability function f(x), then the variance is given by

$$\sigma_x^2 = \sum_{j=1}^{n} (x_j - \mu)^2 f(x_j)$$

In the special case where the probabilities are all equal, we have

$$\sigma^2 = \frac{\left[(x_1 - \mu)^2 + (x_2 - \mu)^2 \ldots\ldots\ldots\ldots + (x_n - \mu)^2\right]}{n}$$, which is the variance for a set of n variables x_1, x_2, x_3, ..., x_n.

If X takes on an infinite number of values x_1, x_2, . . . , then

$\sigma_x^2 = \sum_{j=1}^{\infty} (x_j - \mu)^2 f(x_j)$, provided that the series converges.

If X rather being discrete is a continuous random variable having density function f(x), then the variance is given by

$\sigma_x^2 = \int^{\infty} (x - \mu)^2 f(x) dx$, provided that the integral converges.

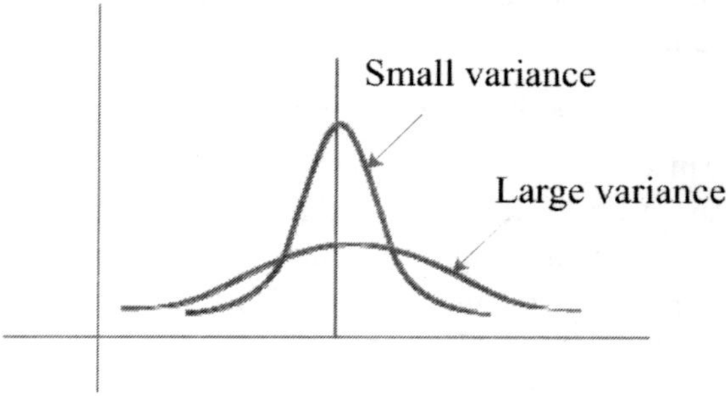

Figure 11.5

The variance (or the standard deviation) is a measure of the dispersion, or scatter of the values of the random variable about the mean . If the values tend to be concentrated near the mean, the variance is small; while if the values tend to be distributed far from the mean,the variance is large. The situation is indicated graphically in Fig. 11.5 above for the case of two continuous distributions having the same mean μ .

Example 32:

Two cards are drawn simultaneously (or successively without replacement) from a well shuffled pack of 52 cards. Find the mean, variance and standard deviation of the number of kings.

Solution

Let X denote the number of kings in a draw of two cards. X is a random variable which can assume the values 0, 1 or 2.

Now P(X = 0) = P (no king) = $\frac{^{48}C_2}{^{52}C_2} = \frac{48 \times 47}{52 \times 51} = \frac{188}{221}$

P(X = 1) = P (one king and one non-king)

$= \frac{^{4}C_1 \times {^{48}C_1}}{^{52}C_2} = \frac{4 \times 48 \times 2}{52 \times 51} = \frac{32}{221}$

And P(X = 2) = P (two kings) $= \frac{^{4}C_2}{^{52}C_2} = \frac{4 \times 3}{52 \times 51} = \frac{1}{221}$

Thus, the probability distribution of X is

X	0	1	2
P(X)	$\frac{188}{221}$	$\frac{32}{221}$	$\frac{1}{221}$

Now, Mean of X = E(X)

$$= \sum_{i=1}^{n} x_i P(x_i) = 0 \times \frac{188}{221} + 1 \times \frac{32}{221} + 2 \times \frac{1}{221} = \frac{34}{221}$$

Also E(X²) = $\sum_{i=1}^{n} x_i^2 P(x_i) = 0 \times \frac{188}{221} + 1 \times \frac{32}{221} + 4 \times \frac{1}{221} = \frac{36}{221}$

Now Var(X) = E(X²) – [E(X)]²= $\frac{36}{221}$ - $\left[\frac{34}{221}\right]^2$

$$= \frac{36 \times 221 - 1156}{(221)^2} = \frac{7956 - 1156}{(221)^2} = \frac{6800}{(221)^2}$$

Therefore $\sigma_x = \sqrt{Var(X)} = \dfrac{\sqrt{6800}}{221} = 0.37$

Example 33

A random variable X has the following probability distribution:

X	0	1	2	3	4	5	6	7
(X)	0	k	2k	2k	3k	k^2	$2k^2$	$7k^2+k$

Determine

(i) k (ii) P(X < 3) (iii) P(X > 6) (iv) P(0 < X < 3)

Solution:

Sum of all probabilities = $0+k+2k+2k+3k+ k^2 +2 k^2 +7 k^2 +k$ $=10k^2 +9k$

Since the total probability = 1

$10k^2 +9k =1$ or $10k^2 +9k-1 =0$

Factorising this quadratic equation leads us to ((10k-1) × (k+1) = 0, yielding values of k as $\dfrac{1}{10} or -1$. Since probability cannot be negative, $k = \dfrac{1}{10}$

$P(X < 3) = P(X=0)+P(X=1)+P(X=2) = 0+k+2k = 3k = \dfrac{3}{10}$

$P(X > 6) = P(X=7) = 7k^2 + k = 7\times\left(\dfrac{1}{10}\right)^2 + \dfrac{1}{10} = \dfrac{7}{100} + \dfrac{1}{10} = \dfrac{17}{100}$

$P(0 < X < 3) = P(X=1)+P(X=2) = k+2k = 3k = \dfrac{3}{10}$

11.6.6 Bernoulli trials and Binomial distribution

11.6.6.1 Bernoulli trials

Many experiments are dichotomous (variables that have only two variables) in nature. For example, a tossed coin shows a 'head' or 'tail', a manufactured item can be 'defective' or 'non-defective', the response to a question might be 'yes' or 'no', an egg has 'hatched' or 'not hatched', the decision is 'yes' or 'no' etc. In such cases, it is customary to call one of the outcomes a 'success' and the other 'not success' or 'failure'. For example, in tossing a coin, if the occurrence of the head is considered a success, then occurrence of tail is a failure.

Each time we toss a coin or roll a die or perform any other experiment, we call it a trial. If a coin is tossed, say, 4 times, the number of trials is 4, each having exactly two outcomes, namely, success or failure. The outcome of any trial is independent of the outcome of any other trial. In each of such trials, the probability of success or failure remains constant. Such independent trials which have only two outcomes usually referred as 'success'or'failure' are called Bernoulli trials.

Definition: Trials of a random experiment are called Bernoulli trials, if they satisfy the following conditions:

(i) There should be a finite number of trials.

(ii) The trials should be independent.

(iii) Each trial has exactly two outcomes: success or failure.

(iv) The probability of success remains the same in each trial

For example, throwing a die 50 times is a case of 50 Bernoulli trials, in which each trial results in success (say an even number) or failure (an odd number) and the probability of success (p) is same for all 50 throws. Obviously, the successive throws of the die are independent experiments. If the die is fair and have six numbers 1 to 6 written on six faces, then $p = \frac{1}{2}$ (probability of success) and $q = 1 - p = \frac{1}{2}$ = probability of failure.

11.6.6.2 Binomial distribution

Consider the experiment of tossing a coin in which each trial results in success (say, heads) or failure (tails). Let S and F denote respectively success and failure in each trial. Suppose we are interested in finding the ways in which we have one success in six trials. Clearly, six different cases are there as listed below: SFFFFF, FSFFFF, FFSFFF, FFFSFF, FFFFSF, FFFFFS.

Similarly, two successes and four failures can have ${}^6C_2 = {}^6C_4 = \frac{6!}{4! \times 2!}$ combinations. It will be lengthy job to list all of these ways. Therefore, calculation of probabilities of 0, 1, 2, ..., n number of successes may be lengthy and time consuming. To avoid the lengthy calculations and listing of all the possible cases, for the probabilities of number of successes in n-Bernoulli trials, a formula is derived.

For this purpose, let us take the experiment made up of three Bernoulli trials with probabilities p and q = 1 – p for success and failure respectively in each trial. The sample space of the experiment is the set S = {SSS, SSF, SFS, FSS, SFF, FSF, FFS, FFF} The number of successes is a random variable X and can take values 0, 1, 2, or 3. The probability distribution of the number of successes is as below :

P(X = 0) = P(no success) =P ({FFF}) = P (F) P (F) P (F) = $q \times q \times q = q^3$

Since the trials are independent

P(X = 1) = P(one success) =P({SFF, FSF, FFS})

=P({SFF}) + P({FSF}) + P({FFS})

=P(S) P(F) P(F) + P(F) P(S) P(F) + P(F) P(F) P(S)

$= \ p \times q \times q + q \times p \times q + q \times q \times p = 3pq^2$

P(X = 2) = P (two successes) =P({SSF, SFS, FSS})

=P({SSF}) + P ({SFS}) + P({FSS})

=P(S) P(S) P(F) + P(S) P(F) P(S) + P(F) P(S) P(S)

$= \ p \times p \times q + p \times q \times p + q \times p \times p = 3p^2 q$ and

P(X = 3) = P(three success) = P ({SSS}) =P(S). P(S). P(S) = p^3

Thus, the probability distribution of X is

X	0	1	2	3
P(X)	q^3	$3q^2p$	$3qp^2$	p^3

Also, the binominal expansion of $(q + p)^3$ is $q^3+3q^2 p+3q p^2+p^3$

Note that the probabilities of 0, 1, 2 or 3 successes are respectively the 1st, 2nd, 3rd and 4th term in the expansion of $(q + p)^3$.

Also, since q + p = 1, it follows that the sum of these probabilities, as expected, is 1. Thus, we may conclude that in an experiment of n-Bernoulli trials, the probabilities of 0, 1, 2,..., n successes can be obtained as 1st, 2nd,...,$(n + 1)^{th}$ terms in the expansion of $(q + p)^n$.

To prove this result, let us find the probability of x-successes in an experiment of n-Bernoulli trials. Clearly, in case of x successes (S), there will be (n – x) failures (F).

Now, x successes (S) and (n – x) failures (F) can be obtained in $\frac{n!}{x!(n-x)!}$ ways.

In each of these ways, the probability of x successes and (n − x) failures is

P(x successes)×P{ (n–x) failures} =P(S)×P(S)×...P(S) (x times) × P(F)×P(F) ×... ×P(F) (n-x times) = $p^x q^{n-x}$

Thus, the probability of x successes in n-Bernoulli trials is

$$\frac{n!}{x!(n-x)!} \times p^x q^{n-x} \text{or } {}^nC_x p^x q^{n-x}$$

Thus P(x successes) ${}^nC_x\ p^x q^{n-x}$ x = 0, 1, 2,...,n. (q = 1 – p)

Clearly, P(x successes), i.e. ${}^nC_x\ p^x q^{n-x}$ is the $(x + 1)^{th}$ term in the binomial expansion of $(q + p)^n$.

Thus, the probability distribution of number of successes in an experiment consisting of n Bernoulli trials may be obtained by the binomial expansion of $(q + p)^n$. Hence, this distribution of number of successes X can be written as

X	0	1	2	....	x	...	n
P(X)	${}^nC_0\ q^n p^0$	${}^nC_1\ q^{n-1} p$	${}^nC_2\ q^{n-2} p^2$	....	${}^nC_x\ q^{n-x} p^x$		${}^nC_n\ p^n$

The above probability distribution is known as binomial distribution with parameters n and p because for given values of n and p, we can find the complete probability distribution. The probability of x successes P(X = x) is also denoted by P(x) and is given by P(x) = ${}^nC_x\ q^{n-x} p^x$, x=0, 1,.., n. and (q =1 – p)

This P(x) is called the probability function of the binomial distribution.

A binomial distribution with n-Bernoulli trials and probability of success in each trial as p, is denoted by B(n, p). Let us now take up some examples.

Example 34:

A die is thrown 6 times. If 'getting an odd number' is a success, what is the probability of

(i) 5 successes?

(ii) at least 5 successes?

(iii) at most 5 successes?

Solution:

When we throw a dice, possible outcomes are 1, 2, 3, 4, 5, 6. Thus there are 3 out of 6 possibilities of getting an odd number on each throw. This gives us probability of getting an odd number as $\frac{3}{6}=\frac{1}{2}$

The repeated tosses of a dice are Bernoulli trials. Let X denote the number of odd numbers in an experiment of 6 trials. Clearly, X has the binomial distribution with n = 6 and $p = \frac{1}{2}$

Therefore $P(X = x) = {}^nC_x\, q^{n-x}p^x$, x = 0, 1, 2,..., n where n = 6, $p = \frac{1}{2}\ \&\ q = 1 - \frac{1}{2} = \frac{1}{2}$

Therefore $P(X = x) = {}^6C_x \times \frac{1}{2^{6-x}} \times = {}^6C_x \times \frac{1}{2^6}$

Now

(i) $P(X = 5) = {}^6C_5 \times \frac{1}{2^6} = \frac{6}{64} = \frac{3}{32}$

(ii) P(at least 5 odd numbers)

$= P(X \geq 5) = P(X = 5) + P(X = 6)$

$= {}^6C_5 \times \frac{1}{2^6} + {}^6C_6 \times \frac{1}{2^6} = (6+1) \times \frac{1}{2^6} = \frac{7}{64}$

(iii) P(at most 5 odd numbers)= $P(X \leq 5)$

$= P(X= 0) + P(X= 1) + P(X= 2) + P(X= 3) + P(X= 4) + P(X= 5)$

$= {}^6C_0 \times \frac{1}{2^6} + {}^6C_1 \times \frac{1}{2^6} + {}^6C_2 \times \frac{1}{2^6} + {}^6C_3 \times \frac{1}{2^6} + {}^6C_4 \times \frac{1}{2^6} + {}^6C_5 \times \frac{1}{2^6}$

$= (1+6+15+20+15+6) \times \frac{1}{2^6} = \frac{63}{64}$

Another way to look at this problem is that we have to find probability of getting odd numbers on every toss except one. Since sum of all probabilities of the Bernaulli trials = 1 and probability of getting odd number on all tosses except one = ${}^6C_6 \times \frac{1}{2^6} = \frac{1}{64}$, we get probability of getting at most 5 odd numbers = $1 - \frac{1}{64} = \frac{63}{64}$

Example 35:

Five cards are drawn successively with replacement from a well-shuffled deck of 52 cards. What is the probability that

(i) all the five cards are spades?

(ii) only 3 cards are spades?

(iii) none is a spade?

Solution:

Let X denote the number of spades in the 5 cards drawn. Since the drawing is done with replacement, the trials are Bernoulli trials. Clearly, X has the binomial distribution with n = 5 and $p = \frac{13}{52} = \frac{1}{4}$ & $q = 1 - p = 1 - \frac{1}{4} = \frac{3}{4}$

Now P(all the 5 cards are spade) = P(X= 5) = $^5C_5\, q^0 p^5$

$$= 1 \times 1 \times \left[\frac{1}{4}\right]^5 = \frac{1}{4^5} = \frac{1}{1024}$$

P (only 3 cards are spade) = P(X= 3) = $^5C_3\, q^2 p^3$

$$= \frac{5 \times 4}{2 \times 1} \times \left(\frac{3}{4}\right)^2 \times \left(\frac{1}{4}\right)^3 = 10 \times \frac{9 \times 1}{16 \times 64} = \frac{45}{512}$$

P (none is a spade) = P(X= 0) = $^5C_0\, q^5 p^0$

$$= 1 \times \left(\frac{3}{4}\right)^5 \times \left(\frac{1}{4}\right)^0 = 1 \times \frac{243}{1024} \times 1 = \frac{243}{1024}$$

Example 36:

A bag consists of 10 balls each marked with one of the digits 0 to 9. If four balls are drawn successively with replacement from the bag, what is the probability that none is marked with the digit 0?

Solution:

Let x denote the number of times ball with number 0 is drawn. Since the drawing is done with replacement, the trials are Bernoulli trials. Clearly, x has the binomial distribution with n = 4 and $p = \frac{1}{10}$ & $q = 1 - p = 1 - \frac{1}{10} = \frac{9}{10}$

We need to find P (x=0) = $^4C_0\, q^4 p^0 = 1 \times \left(\frac{9}{10}\right)^4 \times \left(\frac{1}{10}\right)^0 = 1 \times \frac{6561}{10000}$

Example 37:

The probability of a shooter hitting a target is $\frac{3}{4}$. How many minimum number of times must he/she fire so that the probability of hitting the target at least once is more than 0.99?

Solution:

Let the shooter fire n times. Obviously, n fires are n Bernoulli trials. In each trial, p = probability of hitting the target = $\frac{3}{4}$

And q = probability of not hitting the target = $1-\frac{3}{4}=\frac{1}{4}$

Then P(X = x) $= {}^nC_x q^{n-x} p^x = {}^nC_x\left(\frac{1}{4}\right)^{n-x} \times \left(\frac{3}{4}\right)^x = {}^nC_x \times \left(\frac{3^x}{4^n}\right)$

Now, given that, P(hitting the target at least once) > 0.99 i.e. P(x ≥ 1) > 0.99

Therefore, 1 – P (x = 0) > 0.99 or $1-{}^nC_0 \times \left(\frac{3^0}{4^n}\right) > 0.99$

i.e. ${}^nC_0 \times \left(\frac{1}{4^n}\right) > 0.01$ or $\frac{1}{4^n} > 0.01 \Rightarrow 4^n > 100$

The minimum value of n to satisfy the above inequality is 4. Thus, the shooter must fire 4 times.

Example 38:

On a multiple choice examination with three possible answers for each of the five questions, what is the probability that a candidate would get four or more correct answers just by guessing?

Solution:

By guessing the answer,

Probabiity of getting a question right p = $\frac{1}{3}$ and getting it wrong q = $\frac{2}{3}$

Let X be the number of correct answers. We need to find $P(X \geq 4)$

$P(X \geq 4)$ = P(X=4) + P(X=5) = ${}^5C_4 q^1p^4 + {}^5C_5q^0p^5$

$$= 5\times\frac{2}{3}\times\left(\frac{1}{3}\right)^4 + \left(\frac{1}{3}\right)^5 = \frac{10}{243}+\frac{1}{243}=\frac{11}{243}$$

11.7 Statistical approach to probability

Dividing the number of events by the number of all possible events is very simplistic, and it isn't suited to finding probabilities for a lot of situations. For example, natural events like weights, heights, and test scores need normal distribution probability charts to calculate probabilities. In fact, most "real life" situations aren't simple events like coins, cards, or dice. You'll need something more complicated than classical probability theory to solve them.

For example, the risk associated with an asset can be assessed more accurately by the use of probability distribution. For example, if the expectation is that a given outcome will occur six out of ten times, it can be said to have sixty percent probability (chance of happening); if it is certain to happen, the probability of happening is 100%. An outcome which has a probability of zero will never occur. So, on the basis of the probability distributed or assigned to different outcomes (return), the expected value of the return can be computed. The expected rate of return is the weighted average of possible returns multiplied by their respective probabilities and those probabilities of the various outcomes are used as weights.

The expected return $(\mathrm{R}) = \sum_{i=1}^{n} R_i \times P_{ri}$

Where:

Ri = Return for the ith possible outcome

P_{ri} = Probability of its outcome or return

n= Number of outcomes considered

Example 39:

Let us calculate the expected return of two assets X & Y, whose probability of generating pessimistic returns of 13% and 7% is 30%, most likely returns of 15% and 14% is 30% and optimistic returns of 17% and 21% is 40%, respectively.

Let us tabulate the given information as below:

Possible outcomes	Probability	Returns	Expected Returns
Asset X	(P)	(R)	(P×R)
Pessimistic	0.30	13%	3.9%
Most likely	0.30	15%	4.5%

Possible outcomes	Probability	Returns	Expected Returns
Optimistic	0.40	17%	6.8%
Expected return			15.2%
Asset Y			
Pessimistic	0.30	7%	2.1%
Most likely	0.30	14%	4.2%
Optimistic	0.40	21%	8.4%
Expected return			14.7%

From the above risk analysis, Asset X seems to have a higher expected return and would be preferred over Asset Y.

11.8 Practice problems:

1. A die is rolled, find the probability that an even number is obtained.
2. Two dice are rolled, find the probability that the sum is
   (a) equal to 1 , (b) equal to 4 (c) less than 13
3. A card is drawn at random from a deck of cards. Find the probability of getting a queen.
4. The blood groups of 200 people is distributed as follows: 50 have type A blood, 65 have B blood type, 70 have O blood type and 15 have type AB blood. If a person from this group is selected at random, what is the probability that this person has O blood type?
5. Rolling three dices, evaluate the probability of having k equal faces, with $k \in [0;2;3]$.
6. Assuming women and men exist in equal number, and assuming that 5% of the men are colour blind and that 0.25% of the women are colour blind, evaluate the probability that a person drawn at random is colour blind. Then evaluate the probability that, having drawn a colour-blind person, this is a male.
7. You spin a spinner with equal probability of landing on each of the numbers 1 through 10. Your friend covers up the number it lands on, but you can tell that it only has one digit (and therefore can't be 10). What is the probability it landed on a 5?
8. Suppose that Paul is a compulsive liar. Every statement he says has a 75% probability of being a lie. He flips a fair coin and

tells you it's a head. Is the coin flip more likely to have been a head or a tail?

9. Paul is still a compulsive liar whose every statement has a 75% probability of being a lie. However, we know that he is aware of the winning number in a lottery that consists of choosing a single integer from 1 to a million. He says the winning number is 123. If you were to enter the lottery, which number should you pick to maximize the probability of winning?

   Note: if Paul decides to lie about the winning number, he will pick any incorrect but plausible number with equal probability.

10. Suppose that in a given population of people everyone is either good or bad at jumping, and good or bad at running. These are independent of each other (so knowing if someone is good at jumping tells you nothing about how they are at running). An athletic training camp only accepts people who are either good at running or good at jumping (possibly good at both). You meet a random person who went to the camp, and you find out they are good at running. Are they more, less, or equally likely to be good at jumping than someone random at the camp whose running ability you don't know?

    Note: Assume that there is at least one person good at only running and one person good at only jumping at the camp.

11. You flip two coins and your friend tells you that both landed on the same side. Conditional on this information, what is the probability that at least one of the coins landed on heads?

12. A dice A has four red faces and two white faces. A dice B, vice-versa, has two red faces and four white faces. You flip a coin once, if heads the game continues with dice A, otherwise it continues with dice B.

    (a) On rolling the dice, what is the probability that a red face appears on the dice?

    (b) and at the second rolling of the same dice?

    (c) If the first two rollings show a red face, what is the probability that also on the third rolling is red?

    (d) If the first n rollings show a red face, what is the probability that you are using dice A?

13. The probabilities that three different archers, A, B, C hit the

mark, independently of one another, are respectively 1/6, 1/4 and 1/3. Everyone shoots an arrow.

(a) Find the probability that only one hits the mark.

(b) If only one hits the mark, what is the probability he is archer A?

14. From a deck of 52 cards we draw two cards. Find the probabilities of the following events A ={ first card is a King; the second figure}={K1;F2} B ={at least one figure}

15. An installation consists of two boilers and one engine. Let A be the event that the engine is in good condition, Bi the event that the ith boiler (i = 1, 2) is in good condition, and the event that the installation can operate. If the installation requires that at least one boiler and the engine are in good condition before it can operate, express C and $\overline{C}$ in terms of A, B_1 and B_2.

16. A, B and C are events such that

    P(A) = 0.4, P(B) = 0.7, P(C) = 0.3

    $P[C \cap (A \cup B)] = 0.2$, $P[B \cap (A \cup C)] = 0.4$

    (a) Find the probability that exactly two of A, B and C occur.

    (b) Find the probability that none of A, B, C occur.

17. A cube, all sides of which are painted, is sawn into a thousand small cubes of the same dimensions. The small cubes obtained are carefully mixed. Determine the probability that a small cube selected at random will have two painted sides.

18. According to the weather report, there is a 50% chance of rain on Saturday and a 70% chance of rain on Sunday. If these events are independent, what is the probability that it would rain on both days?

19. Sonia flips a coin and tosses a 6-sided dice. What is the probability that either the coin lands on Heads, or the dice lands on 4?

20. 100 high school students are asked if they speak English. 70% of male students answer that they do not speak English, while 40% of female students answer they speak English. It is known that 40% of the responding students are female. If a randomly chosen student answers that he/she speaks English what is the probability that this student is a female?

Answers to Practice Problems

1. $\frac{1}{2}$
2. (a) 0 , (b) $\frac{1}{12}$ (c) 1
3. $\frac{1}{13}$
4. $\frac{7}{20}$
5. Probability of having 0 equal faces $= \frac{5}{9}$, Probability of having 2 equal faces $= \frac{5}{36}$ and probability of having 3 equal faces $= \frac{1}{36}$
6. $\frac{20}{21}$
7. $\frac{1}{9}$
8. It is thus more likely to be a Tail.
9. Number 123
10. Group of persons goodatrunning are less likely to be good at jumping.
11. $\frac{1}{2}$
12. (a) $\frac{1}{3}$ (b) $\frac{1}{5}$ (c) $\frac{1}{9}$ (d) 5
13. (a) $\frac{31}{72}$ (b) $\frac{1}{12}$
14. A: $\frac{11}{663}$ B: $\frac{3}{13}$
15. $C = \left(A \cap B_1 \cap \overline{B_2}\right) + \left(A \cap B_2 \cap \overline{B_1}\right) + \left(A \cap B_1 \cap B_2\right)$ and

$$\overline{C} = \left(\overline{A} \cap B_1 \cap \overline{B_2}\right) + \left(\overline{A} \cap B_2 \cap \overline{B_1}\right) + \left(\overline{A} \cap B_1 \cap B_2\right) + \left(\overline{A} \cap \overline{B_2} \cap \overline{B_1}\right) + \left(A \cap \overline{B_2} \cap \overline{B_1}\right)$$

16. (a) 36% (b) 20%
17. $\frac{12}{125}$
18. 35%
19. $\frac{2}{3}$
20. $\frac{8}{17}$

12

STATISTICS

12.1 Introduction

Statistics is a branch of Mathematics dealing with collection, organisation, analysis, interpretation and presentation of data. We need to understand methods of representing data graphically and in tabular form. This representation reveals certain salient features or characteristics of the data. We also need to study methods of finding a representative value for the given data. This value is called the measure of central tendency. Mean (Arithmetic mean), Median and Mode are three measures of central tendency that we will understand shortly. A measure of central tendency gives us a rough idea where data points are centred. But, in order to make better interpretation of the data, we should also have an idea how the data are scattered or how much they are bunched around the measure of a central tendency.

While applying statistics to a practical problem say in the field of science, industry, or social engineering, it is conventional to begin with a statistical population or a statistical model process to be studied. Statistical populations can be for diverse topics such as "all people living in a country" or "data of all cricket players" etc. Statistics deals with all aspects of data including planning of data collection in terms of the design of surveys and experiments. For example, if census data cannot be collected, statisticians collect data by developing specific experimental designs and survey samples. Representative sampling is designed such that inferences and conclusions can reasonably extend from the sample to the population as a whole.

Statistics is one of the most relevant fields in the area of business. Today, most business decisions are taken after looking at varied data points from multiple sources. This makes the skill of statistical analyses very valuable.

12.2 Models of Data sets

Data sets are primarily of three types:

1. Distinct numbers in the set. For example runs scored by different batsmen in a cricket match innings or height of different students in a class etc.
2. Grouped distinct numbers where there is repetition of data numbers. For example, age in years of students in a class where multiple students may have same age (in years)
3. Grouped numbers in defined group size. An example could be scores out of 100 of different students in an examination. Here students could be clubbed in score range say 31-35, 36-40, 41-45 and so on in groups of 5 marks each, again getting single or multiple students in different groups. This is also called **Continuous frequency distribution.** *The groups here seem discontinuous. This is because the marks are whole numbers. In case the data points are in decimal numbers, the groups are like 30-35, 35-40, 40-45 etc. where data points between 30.5 and 35.5 are counted in the group 30-35 and >35.5 and <40.5 are counted in the group 35-40 etc.*

We will deal with all these types of data as we go along

12.3 Central Value Parameters

Mean, Median, and Mode are three different measures of centre in a numerical data set. They each try to summarize a dataset with a single number to represent a "typical" central data point from the dataset.

Mean is "The average number" found by adding all data points and dividing by the number of data points.

Example: The mean of (4, 1, and 7) is (4+1+7)/3 = 12/3 = 4

Median is "The middle number" found by ordering all data points and picking out the one in the middle (or if there are two middle numbers, taking the mean of those two numbers). Example: The median of (4, 1, and 7) is 4 because when the numbers are put in ascending or descending order (1, 4, 7), the number 4 is in the middle.

Mode is "The most frequent number"—that is, the number that occurs the highest number of times in the data set

Example: The mode of (4, 2, 4, 3, 2, 2) is 2 because it occurs three times, which is more than any other number.

12.3.1 Calculating the Mean

There can be many different types of mean, but usually when we say Mean, we are talking about the Arithmetic Mean.We have already studied **Arithmetic Mean** of a set of numbers in Chapter 4. In Statistics we denote Arithmetic mean of n numbers $x_1, x_2, x_3, \ldots\ldots,$ x_n denoted as x_i where 'i' varies from 1 to n, as **Mean ($\bar{x}$)** that gives us value of average of all the numbers given in the Data set. Using notation sigma $\left(\sum_{i=1}^{n} x_i\right)$ for sum of the n numbers, we get, $\bar{x} = \frac{1}{n}\sum_{i=1}^{n} x_i$

In case we do not have discrete data, as already mentioned, data may be grouped into two ways:

(a) Discrete frequency distribution,

(b) Continuous frequency distribution.

Let us discuss the method of finding mean for both these types of data. The fundamental concept for evaluating Mean value remains unchanged. Let us assume the following data set with discrete frequency distribution:

Table 12.1

Data value x_i	2	5	6	8	10	12	15
No. of data points (frequency) f_i	2	8	10	7	8	5	2
Total data values $f_i x_i$	4	40	60	56	80	60	30

Here total number of data points (say N) is arrived at by adding all data points (frequency) as $N = \sum_{i=1}^{n} f_i$ where n is the number of data values. From the given data it can be seen that N= 2+8+10+7+8+5+2 = 42

Similarly total data value X is computed as $X = \sum_{i=1}^{n} f_i x_i$, where $\sum_{i=1}^{n} f_i x_i$ represents the sum of the products of observations x_i with their respective frequencies fi.

So, from the Table above, X= sum of $(f_i \times x_i)$ = 4+40+60+56+80+60+30 = 330

$$\text{Thus } \bar{x} = \frac{1}{N}\sum_{i=1}^{n} f_i x_i = \frac{330}{42} = 7.86$$

Let us now take an example of third data type **(continuous frequency distribution)** with following table showing marks obtained by different students in a paper

Table 12.2

Marks range x (mid point x_i)	**21-30 (25)**	**31-40 (35)**	**41-50 (45)**	**51-60 (55)**	**61-70 (65)**	**71-80 (75)**	**81-90 (85)**
No. of students (frequency) f_i	10	18	27	20	17	6	2
Total data values (Mid-point × No. of students	250	630	1215	1100	1105	450	170

Total number of students (say N) is again arrived at by adding all Grouped data points (frequency) as $N = \sum_{i=1}^{n} f_i$ where n is the number of data groups. From the given data it can be seen that N= 10+18+27+20+17+6+2 = 100

Similarly total data value X is computed as $X = \sum_{i=1}^{n} f_i x_i$, where x_i is mid-point of different marks groups. In the given example, mid-point of different groups is taken as x_i. Hence

X= 250+630+1215+1100+1105+450+170 = 4920

$$\text{Thus } \bar{x} = \frac{1}{N}\sum_{i=1}^{n} f_i x_i = \frac{4920}{100} = 49.20$$

12.3.2 Finding the Median

As already defined, the median is the middle point in a dataset-half of the data points are smaller than the median and the other half of the data points are larger.

To find the median:

- Arrange the data points from smallest to largest.
- If the number of data points is odd, the median is the middle data point in the list.
- If the number of data points is even, the median is the average of the two middle data points in the list.

Example 1

Find the median of this data: 1, 4, 2, 5, 0

Put the data in order first: 0, 1, 2, 4, 5

There is an odd number of data points, so the median is the middle data point, i.e.

The median is 2.

Example 2

Find the median of this data: 10, 40, 20, 50

Put the data in order first: 10, 20, 40, 50

There is an even number of data points, so the median is the average of the middle two data points, i.e. **The Median** = $\frac{20+40}{2}=30$

12.3.3 Finding the Mode

The mode is the most commonly occurring data point in a dataset. The mode is useful when there are a lot of repeated values in a dataset. There can be no mode, one mode, or multiple modes in a dataset.

Example 3

A teacher asked students in her class how many siblings they each had. They responded with their answers listed below:

0, 0, 1, 1, 1, 1, 1, 1, 1, 2, 2, 2, 3, 5

Let us find the mode of this data by looking for the value that occurs the most: **In this data set, the Mode is clearly 1.**

Example 4

In another class the teacher asked students in her class the same question. In this case data was 0, 1, 1, 1, 1, 2, 2, 2, 2, 4

To find the mode of this data, we look for the value that occurs the most: There is a tie here for the value that occurs the most often.

Thus the modes are 1 and 2.

12.4 Adequacy of Central value parameters

Let us consider two data sets (Runs scored by two batsmen in their last ten matches) as follows:

Batsman A: 20, 91, 10, 58, 48, 90, 30, 5, 87, 81

Batsman B: 53, 43, 43, 50, 58, 58, 53, 53, 47, 62

If we calculate Mean of these two sets, it will work out as 52 for both batsmen. If we arrange the scores in ascending order, Median for Batsmen A = $\frac{48+58}{2} = 53$ and Median for Batsman B is also 53.

However, it may be noticed that scores of batsman A are quite spread out while those of Batsman B are quite consistent (close in value to each other). ***Thus the measures of Central values Mean and Median (being equal) while being a single number to represent a data set, do not give complete information, especially when it comes to consistency.*** Thus **Variability** is another factor which is required to be studied under statistics, which is used to gauge how spread out is the data.

Like 'measures of central values' we want to have a single number to describe variability. This single number is called a 'measure of dispersion'. We shall now learn some of the important measures of dispersion and their methods of calculation for ungrouped and grouped data.

12.5 Measures of Dispersion

The dispersion or scatter in a data is measured on the basis of the observations and the types of the measure of central tendency used there. There are following common measures of dispersion:

(i) Range,

(ii) Mean deviation,

(iii) Standard deviation.

We shall now study all of these measures of dispersion.

12.5.1 Range

Range is the simplest of the measures of dispersion to calculate. The Range is the difference between the highest and lowest values within a set of numbers. To calculate range, subtract the smallest number from the largest number in the set. For example, if height of students in a class of say 50 students varies from 145 cms to 165 cms, the Range of this data set is 165 cms - 145 cms = 20 cms.

In the example given in section 12.4, Range of Batsman A is **86** (91-5) and the same for Batsman B is **19** (62-43).

The range of data gives us a rough idea of variability or scatter but does not tell about the dispersion of the data from the measure of a central tendency (Mean or Median). For this purpose, we need other measures of variability. Clearly, such measures must depend upon the difference (or deviation) of the values from the central tendency. Important measures of dispersion, which depend upon the deviations of the observations from a central tendency are **Mean deviation** and **Standard deviation**. Let us discuss these in detail.

12.5.2 Mean Deviation

Deviation of an observation x from a fixed value 'a' is the difference x – a. In order to find the dispersion of values of x from the central value Mean ($\bar{x}$), we need to find the deviations of x about $\bar{x}$. An absolute measure of dispersion is the average of these deviations. Since Mean is a central value, it lies between the maximum and the minimum values of the set of observations. Therefore, some of the deviations will be negative and some positive resulting in the sum of deviations about $\bar{x}$ to be zero.

Also Mean of deviations = Sum of deviations/ Number of observations n= $\frac{0}{n} = 0$

Thus, finding the Mean of deviations about the Mean is not of any use for us, as far as the measure of dispersion is concerned. Since we are primarily interested in knowing spread of data with reference to the Mean value, we take the mean of the absolute values of the deviations from the central value. This Mean is called the 'Mean deviation'. Thus Mean deviation about a central value '$\bar{x}$' is the mean of the absolute values of deviations of the observations from '$\bar{x}$'. The mean deviation from '$\bar{x}$'is denoted as M.D. ($\bar{x}$). Therefore,

$$\text{M.D}(\bar{x}) = \frac{\textit{Sum of absolute values of deviations from } \bar{x}}{\textit{Number of observations n}}$$

Mean deviation, however, may be obtained from any measure of central tendency, be it Mean or Median. Mean deviation from Mean and Median are commonly used in statistical studies. Let us now learn how to calculate mean deviation about Mean and mean deviation about Median for various types of data

12.5.2.1 Mean deviation for ungrouped data

Let n observations be $x_1, x_2, ..., x_n$. Then mean deviation about Mean $(\overline{x})$ and Median (M) are:

M.D. $(\bar{x}) = \frac{1}{n}\sum_{i=1}^{n}\left|(x-\overline{x})\right|$ where $\bar{x}$ = Mean and

M.D. (M) $= \frac{1}{n}\sum_{i=1}^{n}\left|x-M\right|$ where M = Median

12.5.2.2 Mean deviation for grouped data

We know that data can be grouped in two ways; discrete frequency distribution and Continuous frequency distribution.

Let us discuss the method of finding mean deviation for both types of the data.

12.5.2.2.1 Discrete frequency distribution

Let the given data consist of n distinct values $x_1, x_2, x_3, ...,$ x_n occurring with frequencies $f_1, f_2, f_3, ...,$ fn respectively. This data can be represented in the tabular form as given below:

Table 12.3

x	x_1	x_2	x_3	....	x_n
f	f_1	f_2	f_3	....	f_n

(i) Mean deviation about mean

First of all we find the mean $\bar{x}$ of the given data by using the formula

$$\bar{x} = \frac{1}{N}\sum_{i=1}^{n} f_i x_i \text{ where}$$

N = Total number of data points or sum of the frequencies $= \sum_{i=1}^{n} f_i$

Then, we find deviations of observations x_i from the mean $\bar{x}$ and take their absolute values, i.e., $|x_i - \bar{x}|$ for all i =1, 2,..., n. After this, we find mean of the absolute values of the above deviations, which is the required mean deviation about the mean. Thus

$$\text{M.D. } (\bar{x}) = \frac{1}{N}\sum_{i=1}^{n} f_i |x_i - \bar{x}|$$

(ii) Mean deviation about median

Similarly to find mean deviation about median, we find the median of the given discrete frequency distribution. For this the observations are first arranged in ascending order. Then the cumulative frequencies are obtained. Further, we identify the observation whose cumulative frequency is equal to or just greater than $\frac{N}{2}$ where N is the sum of frequencies. This value of the observation lies in the middle of the data, therefore, it is the required median. After finding median, we obtain the mean of the absolute values of the deviations from median.Thus,

$$\text{M.D.(M)} = \frac{1}{N}\sum_{i=1}^{n} f_i |x_i - M|$$

Example 5 Find the mean deviation about the median for the following data:

Table 12.4

x_i	2	4	8	11	13	15	18	22
fi	3	4	6	3	4	6	4	3

Solution The given observations are already in ascending order. Adding a row corresponding to cumulative frequencies to the given data, we get

Table 12.5

x_i	2	4	8	11	13	15	18	22
fi	3	4	6	3	4	6	4	3
cf	3	7	13	16	20	26	30	33

Now, N=33 which is odd.Median is thus the 17th observation. This observations lies in the cumulative frequency 20, for which the corresponding observation is 13.

∴ Median M = 13

Now, absolute values of the deviations from median, i.e., $|x - M|$ are shown in Table below

Table 12.6

$\lvert x_i - M\rvert$	11	9	5	2	0	2	5	9
fi	3	4	6	3	4	6	4	3
fi $\lvert x_i - M\rvert$	33	36	30	6	0	12	20	27

We have N= $\sum_{i=1}^{8} f_i = 33$ and $\sum_{i=1}^{8} f_i |x_i - M| = 164$

Therefore, M.D. (M) = $\frac{164}{33} = 4.97$

12.5.2.2.2 Continuous frequency distribution

We have already discussed the continuous frequency distribution data in section 12.3.1 and worked out Mean of one such data. We will now work out Mean Deviation for the same Data.

(i) Mean deviation about mean

While calculating the mean of a continuous frequency distribution in the above section, we had made the assumption that the frequency in each class is centred at its mid-point and proceeded further as for the case of discrete frequency distribution. Here also, we will do the same to find the mean deviation. Let us consider the same example for simplicity (Table reproduced below) and find the mean deviation about the mean. Mean $\bar{x}$ was calculated as 49.2. We add two more rows to show $|x - \bar{x}|$ and $f_i |x - \bar{x}|$

Table 12.7

x x_i	21-30 (25)	31-40 (35)	41-50 (45)	51-60 (55)	61-70 (65)	71-80 (75)	81-90 (85)	Total
No. of students (frequency f_i)	10	18	27	20	17	6	2	

Total data values (Mid-point x_i × f_i)	250	630	1215	1100	1105	450	170	
$\left\|x-\overline{x}\right\|$	24.2	14.2	4.2	5.8	15.8	25.8	35.8	125.8
$f_i\left\|x-\overline{x}\right\|$	242	255.6	113.4	116	268.6	154.8	71.6	1222

$\sum_{i=1}^{7} f_i\left|x_i-\overline{x}\right| = 1222$ and we already know that $N=\sum_{i=1}^{n} f_i = 100$

Therefore M.D. (M) = $\frac{1}{N}\times\sum_{i=1}^{n} f_i\left|x_i-M\right| = \frac{1222}{100} = 12.22$

We noticed in section 12.3.1 that calculations for arriving at the Mean value can get quite difficult. We can avoid the tedious calculations by following a **step-deviation method**.In this method, we take an assumed mean (a round figure) which is in the middle or just close to it in the data. Then deviations of the observations (or mid-points of classes) are taken from theassumed mean. This is like shifting of origin from zero to the assumed mean on the number line, as depicted below:

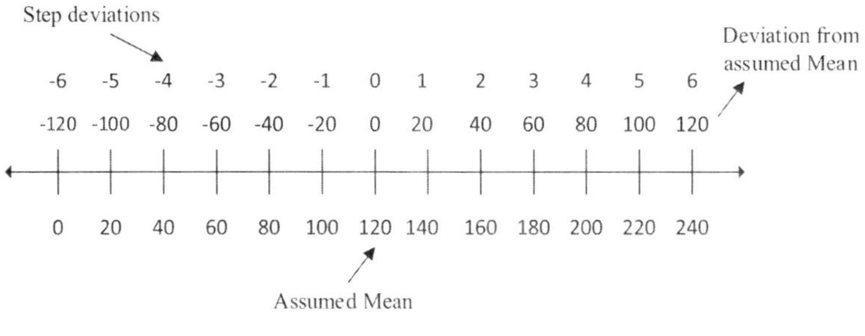

Let us re-workout the example of section 12.3.1 for clarity of the concept. Let us assume the mean $\overline{a}$ to be 50. Further simplification is possible in case there is a common factor in the deviations of the observations with respect to the assumed mean. We divide the deviations by this common factor (h). These are known as step-deviations. The deviations and step-deviations reduce the size of the observations, so that the computations viz. multiplication, etc., become simpler. Let, the step deviations arrived at by using common factor h be denoted by $d_i = \dfrac{x_i - \overline{a}}{h}$

The Table for calculations changes as below, taking assumed mean $\overline{a}$ = 50 and h=5:

Table 12.8

x **x_i**	**21-30 (25)**	**31-40 (35)**	**41-50 (45)**	**51-60 (55)**	**61-70 (65)**	**71-80 (75)**	**81-90 (85)**
No. of students (frequency f_i)	10	18	27	20	17	6	2
Total data values (Mid-point $x_i \times f_i$)	250	630	1215	1100	1105	450	170
Deviations $x_i - a$	-25	-15	-5	5	15	25	35
Step deviations $d_i = \dfrac{x_i - a}{5}$ assuming h= 5	-5	-3	-1	1	3	5	7
$f_i d_i$	-50	-54	-27	20	51	30	14
$\lvert x - \overline{x} \rvert$	24.2	14.2	4.2	5.8	15.8	25.8	35.8
$f_i \lvert x - \overline{x} \rvert$	242	255.6	113.4	116	268.6	154.8	71.6

Now $\overline{x} = a + \dfrac{\sum_{i=1}^{n} f_i d_i}{\sum_{i=1}^{n} f_i} \times h$.

From the Table above, $\sum_{i=1}^{7} f_i d_i = -16$ and $\sum_{i=1}^{7} f_i = 100$

$\therefore \overline{x} = 50 - \dfrac{16}{100} \times 5 = 49.2$

Step deviation method thus helps in simplifying calculation of Mean value. Rest of the calculations for Mean Deviation about Mean remains same.

Therefore M.D. (M) = $\frac{1}{N} \times \sum_{i=1}^{n} f \left| x_i - \bar{x} \right| = \frac{1222}{100} = 12.22$

(ii) Mean deviation about median:

The process of finding the mean deviation about median for a continuous frequency distribution is similar to the procedure we performed for mean deviation about the mean. The only difference lies in the replacement of the mean by median while taking deviations.

We have already understood the process of finding median for discreet frequency distribution in Example 5 above. We will apply the same concept for Continuous Frequency distribution. In the above example, considered for Mean deviation about mean, N being 100, it is observed that cumulative frequency 50 and 51 falls in the Continuous Frequency band of 40-50. Value of Median is calculated using the formula

Median M = $l + \frac{\frac{N+1}{2} - C}{f} \times h$ where $\frac{N+1}{2}$ is the central data point

l = *lower end of the class interval*

C = *Cumulative frequency before the class interval under consideration*

f = *frequency of the class interval*

h = *size of the class interval*

Since the class intervals are 31-40, 41-50 etc. We need to assume that value of x from 30.5–40.5 will be considered in the class interval 31-40 and above 40.5 upto 50.5 in the class interval 41-50. The Median

M, therefore = $40.5 + \frac{50.5 - 28}{27} \times 10 = 48.83$

Now we can calculate Mean Deviation about Median by using the following Table 12.9

Table 12.9

x $\mathbf{x_i}$	**21-30 (25)**	**31-40 (35)**	**41-50 (45)**	**51-60 (55)**	**61-70 (65)**	**71-80 (75)**	**81-90 (85)**
No. of students (frequency f_i)	10	18	27	20	17	6	2

Total data values (Mid-point $x_i \times f_i$)	250	630	1215	1100	1105	450	170
$\lvert x_i - M \rvert$	23.83	13.83	3.83	6.83	16.83	26.83	36.83
$f_i \lvert x_i - M \rvert$	238.3	248.94	103.14	136.6	286.11	160.98	73.66

Here $\sum_{i=1}^{7} f_i \lvert x_i - M \rvert = 1247.73$ and we already know that $N = \sum_{i=1}^{n} f_i = 100$

$$\text{Therefore M.D. (Median)} = \frac{1}{N} \times \sum_{i=1}^{n} f_i \lvert x_i - M \rvert = \frac{1247.73}{100} = 12.477$$

12.5.3 Limitations of mean deviation

It will be noticed that where the degree of variability is very high, the median is not a representative central tendency. Therefore, the mean deviation about median calculated for such series cannot be fully relied upon. The sum of the deviations from the mean (minus signs ignored) is more than the sum of the deviations from median. Therefore, the mean deviation about the mean is also not very scientific. In many cases, mean deviation may give unsatisfactory results. This led to another measure of dispersion known as **Standard deviation.**

12.5.4 Standard Deviation and Variance

Recall that while calculating mean deviation about mean or median, the absolute values of the deviations were taken in order to give meaning to the mean deviation, otherwise the deviations may cancel among themselves. Another way to overcome this difficulty is to take squares of all the deviations. Obviously all these squares of deviations are positive. Let $a_1, a_2, a_3, .., a_n$ be n observations and $\bar{a}$ be their mean. Then

$$(a_1 - \bar{a})^2 + (a_2 - \bar{a})^2 + (a_3 - \bar{a})^2 + \ldots (a_n - \bar{a})^2 = \sum_{i=1}^{n} (a_i - \bar{a})^2$$

Value of the square of the deviations represents dispersion from the Mean. If this value is zero, it implies that there is no dispersion meaning thereby that all values are equal to the Mean. However, this value is not adequate to represent dispersion as will be evident from the following example. Let there be two sets of data as below:

Set 1 (9 data points): 25, 30, 35, 40, 45,50, 55, 60, 65 – Mean is 45

Set 2 (29 data points): 31, 32, 33, 34, 35, 36, 37, 38, 39, 40, 41, 42, 43, 44, 45, 46, 47, 48, 49, 50, 51, 52, 53, 54, 55, 56, 57, 58, 59 – Mean is again 45

(Deviations)^2 set 1 = $2\times(20^2+15^2+10^2+5^2)=1500$

(Deviations)^{2}set 2 =

$$2\times(14^2+13^2+12^2+11^2+10^2+9^2+8^2+7^2+6^2+5^2+4^2+3^2$$
$$+2^2+1^2+0^2)=2030$$

It will be noticed that even though data spread in Set 1 is larger (25-65) than in set 2 (31-59), the (Deviations)^2 is more in set 2 than set 1 which is contradictory. In order to remove this contradiction, we take mean of the (Deviations)^2 leading us to the Term **Variance** denoted as $\sigma^2 = \frac{1}{N}\sum_{i=1}^{n}(a_i-\bar{a})^2$.

Since units of Variance are not the same as Mean or Median, positive square root of Variance is denoted as **Standard Deviation** and is the most popular measure of data Dispersion. In real life, it is used very frequently in various fields like Weather forecasting, Health Insurance, performance of advertisements in Marketing, assessing Test scores in universities, salary structure by Human Resource departments in corporates etc.

Standard Deviation $\sigma = \sqrt{\frac{1}{N}\sum_{i=1}^{n}(a_i-\bar{a})^2}$

12.5.4.1 Standard deviation of a discrete frequency distribution:

Let the given discrete frequency distribution be data a: a_1, a_2, a_3 ,. . . , $a_n$and corresponding frequencies f : f_1, f_2,f_3 ,. . . , f_n

In this case standard deviation $\sigma = \sqrt{\frac{1}{N}\sum_{i=1}^{n}f_i(a_i-\bar{a})^2}$ where $N=\sum_{i=1}^{n}f_i$

12.5.4.2 Standard deviation of a continuous frequency distribution

The given continuous frequency distribution can be represented as a discrete frequency distribution by replacing each class by its mid-point. Then, the standard deviation is calculated by the same technique adopted in the case of a discrete frequency distribution with the same formula

$\sigma = \sqrt{\frac{1}{N}\sum_{i=1}^{n} f_i(a_i - \bar{a})^2}$, where there are n classes each class defined by its mid-point a_i with frequency f_i and $\bar{a}$ is mean of the dataset.

12.5.4.3 Alternate method of calculating Standard Deviation

We can rework the formula for Variance of frequency distribution data as below:

$$\sigma^2 = \frac{1}{N}\sum_{i=1}^{n} f_i(a_i - \bar{a})^2 = \frac{1}{N}\sum_{i=1}^{n} f_i\left\{a_i^2 + \left(\bar{a}\right)^2 - 2a_i\bar{a}\right\}$$

$$= \frac{1}{N}\left(\sum_{i=1}^{n} f_i a_i^2 + \sum_{i=1}^{n} f_i \bar{a}^2 - \sum_{i=1}^{n} 2 f_i a_i \bar{a}\right)$$

$$= \frac{1}{N}\left(\sum_{i=1}^{n} f_i a_i^2 + \bar{a}^2 \sum_{i=1}^{n} f_i - 2\bar{a}\sum_{i=1}^{n} f_i a_i\right)$$

Since $\sum_{i=1}^{n} f_i = \text{N}$ and $\sum_{i=1}^{n} f_i a_i = N \times \bar{a}$, the above expression simplifies to

$$= \frac{1}{N}\left(\sum_{i=1}^{n} f_i a_i^2 + \bar{a}^2 N - 2\bar{a} \times N\bar{a}\right) = \frac{1}{N}\sum_{i=1}^{n} f_i a_i^2 + \bar{a}^2 - 2\bar{a}^2$$

$$\therefore \sigma^2 = \frac{1}{N}\sum_{i=1}^{n} f_i a_i^2 - \bar{a}^2 = \frac{1}{N}\sum_{i=1}^{n} f_i a_i^2 - \left(\frac{\sum_{i=1}^{n} f_i a_i}{N}\right)^2$$

$$= \frac{1}{N^2}\left[N\sum_{i=1}^{n} f_i a_i^2 - \left(\sum_{i=1}^{n} f_i a_i\right)^2\right]$$

Thus alternative expression for Standard Deviation

$$\sigma = \frac{1}{N}\left[\sqrt{N\sum_{i=1}^{n} f_i a_i^2 - \left(\sum_{i=1}^{n} f_i a_i\right)^2}\right]$$

12.5.4.4 Shortcut method to find variance and standard deviation:

As in the case of Mean and Mean deviations, it is possible to simplify the procedure for calculating Variance and Standard Deviations by using step-deviation method.

Let the assumed mean be 'A' and the scale used is 's' for reducing size of deviations from assumed mean, then the step deviation 'b_i' for the deviation a_i is given by

$$b_i = \frac{a_i - A}{s} \text{ or } a_i = A + sb_i \qquad ...(1)$$

We know that $\bar{a} = \frac{1}{N}\sum_{i=1}^{n} f_i a_i$, substituting value of a_i from above we get

$$\bar{a} = \frac{1}{N}\sum_{i=1}^{n} f_i(A + sb_i) = \frac{1}{N}\left(\sum_{i=1}^{n} f_i A + \sum_{i=1}^{n} sf_i b_i\right)$$

$$\bar{a} = A \times \frac{N}{N} + \frac{1}{N} \times N \times s\bar{b} = A + s\bar{b} \qquad ...(2)$$

(as $\sum_{i=1}^{n} f_i = N$ and $\frac{1}{N}\sum_{i=1}^{n} f_i b_i = \bar{b}$)

Now Variance of the variable a $\sigma_a^2 = \frac{1}{N}\sum_{i=1}^{n} f_i(a_i - \bar{a})^2$

Substituting values of a_i and $\bar{a}$, $\sigma_a^2 = \frac{1}{N}\sum_{i=1}^{n} f_i(A + sb_i - A - s\bar{b})^2$

$$= \frac{1}{N}\sum_{i=1}^{n} f_i s^2 (b_i - \bar{b})^2$$

$$= \frac{s^2}{N}\sum_{i=1}^{n} f_i (b_i - \bar{b})^2 =$$

$s^2 \times$ *Variance of* b_i

Or $\sigma_a^2 = s^2 \times \sigma_b^2$ and $\sigma_a = s \times \sigma_b$

The Step Deviation method is an easier way of calculating Mean, Variance and Standard Deviation of the given data.

To illustrate this aspect let us work out Mean and Variance by both methods for the following data:

Table 12.10

Classes	0-30	30-60	60-90	90-120	120-150	90-120	180-210
Frequencies	2	3	5	10	3	5	2

Conventional method:

Table 12.11

Class	Mid point (a_i)	Frequency f_i	$f_i a_i$	$f_i a_i^2$	$f_i(a_i-\bar{a})^2$
0-30	15	2	30	450	16928
30-60	45	3	135	6075	11532
60-90	75	5	375	28125	5120
90-120	105	10	1050	110250	40
120-150	135	3	405	54675	2352
150-180	165	5	825	136125	16820
180-210	195	2	390	76050	15488
		N=30	3210	411750	68280

Mean $\bar{a} = \frac{1}{N}\sum_{i=1}^{n} f_i a_i = \frac{1}{30} \times 3210 = 107$

There are two alternate formulae for calculating Variance σ^2 as below:Taking values from Table 12.11

Table 12.12

	Approach 1	Approach 2
	$\frac{1}{N^2}\left[N\sum_{i=1}^{n} f_i a_i^2 - \left(\sum_{i=1}^{n} f_i a_i\right)^2\right]$	$\frac{1}{N}\sum_{i=1}^{n} f_i(a_i-\bar{a})^2$
Variance	$\frac{1}{30^2}\left(30 \times 411750 - (3210)^2\right)$ $=\left(\frac{411750}{30} - \left(\frac{3210}{30}\right)^2\right)$ $=\left(13725 - (107)^2\right)$ $= 13725 - 11449 = 2276$	$\frac{1}{30} \times 68280 = 2276$

Here we see Approach 2 to be easier provided the value of Mean is known or calculated before hand.

Standard Deviation $\sigma = \sqrt{2276} = 47.71$

Step deviation method

Let the assumed Mean '**A**' be 105 and scale factor '**s**' be 10. Required details are tabulated below:

Table 12.13

Class	Mid point (a_i)	Freq-uency f_i	$b_i = \frac{a_i - 105}{10}$	b_i^2	$f_i b_i$	$f_i b_i^2$
0-30	15	2	-9	81	-18	162
30-60	45	3	-6	36	-18	108
60-90	75	5	-3	9	-15	45
90-120	105	10	0	0	0	0
120-150	135	3	3	9	9	27
150-180	165	5	6	36	30	180
180-210	195	2	9	81	18	162
		N=30			6	684

Now $\bar{b} = \frac{1}{N}\sum_{i=1}^{n} f_i b_i = \frac{1}{30} \times 6 = 0.2$

And Mean $\bar{a} = A + s\bar{b} = 105 + 10 \times 0.2 = 107$

Variance of b_i; $\sigma_b^2 = \frac{1}{N^2}\left[N\sum_{i=1}^{n} f_i b_i^2 - \left(\sum_{i=1}^{n} f_i b_i\right)^2\right]$

$$= \frac{1}{900}\left(30 \times 684 - 6^2\right) = 22.8 - 0.04 = 22.76$$

Variance of a_i; $\sigma_a^2 = s^2 \times \sigma_b^2 = 100 \times 22.76 = 2276$

It will be noticed that calculations in the Step deviation method can be carried out orally and are thus much simpler than the conventional method.

A couple of interesting corollaries follow:

(i) If each observation of a data set is multiplied by a constant say 'k', how will the variance change?

Since each data point say x_i is multiplied by k, the mean value will also be $k \times \overline{x}$ where $\overline{x}$ is the original mean. Thus the new

Variance $\sigma^2{}_{new} = \frac{1}{N}\sum_{i=1}^{n}(kx_i - k\overline{x})^2 = k^2 \times \frac{1}{N}\sum_{i=1}^{n}(x_i - \overline{x})^2 = k^2\sigma^2$,

which is k^2 times the original Variance.

(ii) If a constant 'k' is added to each observation of a data set, how will the variance change?

Here we will notice that the new mean value will also be $\overline{x}+k$, where $\overline{x}$ is the original mean leading to the new Variance being $\sigma^2{}_{new} = \frac{1}{N}\sum_{i=1}^{n}(x_i + k - (\overline{x}+k))^2 = \frac{1}{N}\sum_{i=1}^{n}(x_i - \overline{x})^2 = \sigma^2$,

indicating that the variance remains unchanged.

12.5.5 Analysis of Data Dispersion

We have studied different types of measures of dispersion. These measures, like mean deviation and standard deviation, have same units in which data are given. If we want to compare the variability of two series, which are measured in different units, we require a measure which is independent of the units of data. Such a measure of variability is called **coefficient of variation** (denoted as C.V.) The coefficient of variation is defined as

$CV = \frac{\sigma}{\overline{x}} \times 100$, where σ and $\overline{x}$ are the standard deviation and mean of the data and $\overline{x} \neq 0$.

The series having greater C.V. is said to be more variable than the other. The series having lesser C.V. is said to be more consistent than the other.

If we have two series with the same Mean, it will be evident that the series with higher Standard deviation will be more variable than the other and the series having lesser Standard deviation will be more consistent than the other.

12.6 Solved examples:

1. Coefficients of variation of 2 distributions are 70% & 75%, & S.D. are 28 & 27 respectively. Find their arithmetic means?

Solution:

Coefficient of variationin % $CV = \frac{\sigma}{\overline{x}} \times 100$, where σ and $\overline{x}$ are the standard deviation and mean of the data and $\overline{x} \neq 0$.

Substituting the given values,

Mean of distribution#1 $= \frac{28}{70} \times 100 = 40$

And Mean of distribution#2 $= \frac{27}{75} \times 100 = \frac{27 \times 4}{3} = 36$

2. For a group of 200 candidates, the mean and S.D. were found to be 40 and 15 respectively. Later on it was found that the score 43 was misread as 34. Find the correct mean and correct S.D.

Solution:

(i) Mean $= \frac{\textit{Total score}}{\textit{Total number of candidates}}$

Substituting given values, total score with assumed value of 34 = $40 \times 200 = 8000$

Now using the correct value of 43, Total score = 8000+(43-34)=8009

Hence correct Mean $= \frac{8009}{200} = 40.045$

(ii) Standard deviation $\sigma = \sqrt{\frac{1}{n}\sum_{i=1}^{200}(a_i - \overline{a})^2}$

There are two options for the solution.

(a) With assumed value of 34, $\sum_{i=1}^{n}(a_i - \overline{a})^2$

$= \sigma^2 \times n = 15^2 \times 200 = 45000$

Since each value of $\mathbf{a}_i$ remains same except for one data point and value of new mean is 40.045, the difference $(a_i - \overline{a})$ will be ± 0.045 for 199 data points.

When value is taken as 43 instead of 34, the difference $(a_i - \bar{a})$ will be 2.955 in place of original value of 6.

Hence $\sum_{i=1}^{200}(a_i - \bar{a})^2 =$

$$45000 + 0.045^2 \times 199 + 2.955^2 - 6^2 = 44973.135$$

Accordingly Standard deviation

$$\sigma = \sqrt{\frac{1}{n}\sum_{i=1}^{200}(a_i - \bar{a})^2} = \sqrt{\frac{44973.135}{200}} = \sqrt{224.8656} = 14.995$$

With assumed value of 34, $\sum_{i=1}^{n}(a_i - \bar{a})^2 =$

$$\sigma^2 \times n = 15^2 \times 200 = 45000$$

Or $\sum_{1=1}^{200}(a_i^2 + \bar{a}^2 - 2a_i\bar{a}) =$

$$\sum_{i=1}^{200} a_i^2 + 200 \times 40^2 - 2 \times 40 \times \sum_{i=1}^{200} a_i = 45000$$

$$\sum_{i=1}^{200} a_i^2 = 45000 - 200 \times 40^2 + 2 \times 40 \times 200 \times 40 = 365000$$

With correct value of 43 for the data point in place of assumed value of 34, actual value of $\sum_{i=1}^{200} a_i^2$ will be $365000 + 43^2 - 34^2 = 365693$

$\therefore$ New value of $\sum_{i=1}^{200}(a_i - \bar{a})^2 = \sum_{1=1}^{200}(a_i^2 + \bar{a}^2 - 2a_i\bar{a})$

$$= \sum_{i=1}^{200} a_i^2 + 200 \times 40.045^2 - 2\bar{a} \times \sum_{i=1}^{200} a_i$$

Since $\sum_{i=1}^{200} a_i = 200 \times \bar{a} = 200 \times 40.045 = 8009$

New value of $\sum_{i=1}^{200}(a_i - \bar{a})^2$

$$= 365693 + (40.045)^2 \times 200 - 2 \times (40.045 \times 8009)$$

$$= 365693 + (40.045) \times 8009 - 2 \times (40.045 \times 8009)$$

$=365693-(40.045\times 8009)$

$=365693-320720.40=44972.6$

Thus new Standard deviation

$$\sigma_{new}=\sqrt{\frac{1}{200}\times 44972.6}=\sqrt{224.863}=14.995$$

3. Find the mean deviation about the median of the following frequency distribution:- Class 0-6 6-12 12-18 18-24 24-30

   Frequency 8 10 12 9 5

   Solution:

   Mid point of classes is respectively 3, 9, 15, 21 and 27

   Median M $=l+\dfrac{\frac{N}{2}-C}{f}\times h$ where

   l = *lower end of the class interval*

   C = *Cumulative frequency before the class interval under consideration*

   f = *frequency of the class interval*

   h = *size of the class interval*

   Median being part of the group 12-18.

   Hence Median M $=12+\dfrac{22-18}{12}\times 6=14$

   Now we can calculate mean deviation about the median as below:

Table 12.14

x **x_i**	**0-6** **(3)**	**6-12** **(9)**	**12-18** **(15)**	**18-24** **(21)**	**24-30** **(27)**	**Total**
Frequency f_i	8	10	12	9	5	44
$\lvert x_i-M\rvert$	11	5	1	7	13	
$f_i\lvert x_i-M\rvert$	88	50	12	63	65	278

$$N=\sum_{i=1}^{n}f_i=44$$

$$\text{Mean deviation about Median} = \frac{1}{N}\sum_{i=1}^{n} f_i |x_i - M| = \frac{278}{44} = 6.318$$

4. The following table gives the distribution of income of 100 families in a village. Calculate mean and variance.

Income (in Rs.) 0-1000 1000-2000 2000-3000 3000-4000 4000-5000 5000-6000

No. of families 18 26 30 12 10 4

Let us tabulate the data as below:

Income	**0-1000**	**1000-2000**	**2000-3000**	**3000-4000**	**4000-5000**	**5000-6000**	**Total**
No. of families f	18	26	30	12	10	4	100

Let the assumed mean be 2500

Table 12.15

Class	**Mid point (a_i)**	**Frequency f_i**	$b_i = \frac{a_i - 2500}{1000}$	b_i^2	$f_i b_i$	$f_i b_i^2$
0-1000	500	18	-2	4	-36	72
1000-2000	1500	26	-1	1	-26	26
2000-3000	2500	30	0	0	0	0
3000-4000	3500	12	1	1	12	12
4000-5000	4500	10	2	4	20	40
5000-6000	5500	4	3	9	12	36
		N=100			-18	186

$$\text{Now } \bar{b} = \frac{1}{N}\sum_{i=1}^{n} f_i b_i = \frac{1}{100} \times -18 = -0.18$$

And Mean $\bar{a} = A + s\,\bar{b}$ =2500 – 1000 × 0.18 = 2320

$$\text{Variance of } b_{i\,;}\ \sigma_b^{\,2} = \frac{1}{N^2}\left[N\sum_{i=1}^{n} f_i b_i^2 - \left(\sum_{i=1}^{n} f_i b_i\right)^2\right]$$

$$= \frac{1}{10000}\left(100 \times 186 - (-18)^2\right) = 1.86 - 0.0324 = 1.8276$$

$$\text{Variance of } a_{i\,;}\ \sigma_a^{\,2} = s^2 \times \sigma_b^{\,2} = 1000000 \times 1.8276 = 1827600$$

5. Calculate the mean, median and standard deviation of the following distribution:-

C.I. 31-35 36-40 41-45 46-50 51-55 56-60 61-65 66-70

F 2 3 8 12 16 5 2 3

Tabulating the data as below:

Class	Mid point (a_i)	Frequency f_i	a_i^2	$f_i a_i$	$f_i a_i^2$	fi $\lvert ai - M \rvert$
31-35	33	2	1089	66	2178	35.73
36-40	38	3	1444	114	4332	38.595
41-45	43	8	1849	344	14792	62.92
46-50	48	12	2304	576	27648	34.38
51-55	53	16	2809	848	44944	34.16
56-60	58	5	3364	280	16820	35.675
61-65	63	2	3969	126	7938	24.27
66-70	68	3	4624	204	13872	51.405
		N=51		2558	132524	317.135

$$\text{Mean } \overline{a} = \frac{2558}{51} = 50.16$$

$$\text{Median M} = l + \frac{\frac{N+1}{2} - C}{f} \times h = 50.5 + \frac{26-25}{16} \times 5 = 50.865$$

$$\text{Variance } \sigma_a^2 = \frac{1}{N}\sum_{i=1}^{n} f_i(a_i - \overline{a})^2 = \frac{132524}{51} = 2598.51$$

$$\text{Standard deviation } \sigma = \sqrt{\frac{1}{N}\sum_{i=1}^{n}(a_i - \overline{a})^2} = \sqrt{2598.51} = 50.96$$

6. The mean of 5 observations is 4.4 & their variance is 8.24. If three of observations are 1, 2 & 6, find other two observations.

Let the missing observations be a and b. Since mean of 5 observations is 4.4, sum of the observations = $4.4 \times 5 = 22$

$\therefore a + b = 22 - (1 + 2 + 6) = 13$ or $b = 13 - a$

Since Variance $\sigma_a^2 = \frac{1}{N}\sum_{i=1}^{n} f_i(a_i - \overline{a})^2 = 8.24$ and N =5

$\sum_{i=1}^{n} f_i(a_i - \overline{a})^2 = 8.24 \times 5 = 41.2$

Or

$$(4.4-1)^2+(4.4-2)^2+(4.4-6)^2+(4.4-a)^2+(4.4-b)^2=41.2$$

leading to

$$(11.56+5.76+2.56+38.72)+a^2+b^2-8.8(a+b)=41.2$$

$$a^2+b^2=41.2+8.8\times13-58.6$$

$$a^2+b^2=41.2+114.4-58.6=97$$

Substituting value of b, $a^2+(13-a)^2=97$

Or $2a^2-26a+72=0$; $a^2-13a+36=0$

Factorising, we get $(a-9)\times(a-4)=0$ - giving us two possible values of a=4 or a=9.

Since a+b =13, If a= 4, b =9 and vice versa.

Hence two missing observations are 4 and 9

7. An analysis of monthly wages paid to workers of two firms A & B belonging to the same industry gives following result:

	Firm A	Firm B
No. of workers	1000	1200
Average monthly wages	Rs. 2800	Rs. 2800
Variance of distribution of wages	100	169

in which firm, A or B is there greater variability in individual wages ?

Since Variance

$$\sigma_a{}^2=\frac{1}{N}\sum_{i=1}^{n}(a_i-\overline{a})^2 \text{ or } \sum_{i=1}^{n}(a_i-\overline{a})^2=\sigma_a{}^2\times N$$

Now because average (Mean $\overline{a}$) wage being same in the two companies, we get

For Firm A $\sum_{i=1}^{n}(a_{i1}-\overline{a})^2=100\times1000$ and

For Firm B $\sum_{i=1}^{n}(a_{i2}-\overline{a})^2=169\times1200$

We can conclude from the above comparison that since $\overline{a}$ is same for the two firms, variability $(a_{i2} - \bar{a})$ of wages in Firm 2 is larger than the same for Firm 1

12.7 Practice sums:

1. There are 60 students in a class. The following is the frequency distribution of the marks obtained by students in a test, where x is a positive integer.

Marks	0	1	2	3	4	5
Frequency	x-2	x	x^2	$(x+1)^2$	2x	x+1

Find Mean and Standard deviation of the marks.

2. Calculate Mean deviation about mean of first n natural numbers where n is an odd number.

3. Find the mean deviation about the median for the data.

   13, 17, 16, 14, 11, 13, 10, 16, 11, 18, 12, 17

4. Calculate the mean deviation about the mean for following frequency distribution

Class interval	0-4	4-8	8-12	12-16	16-20
Frequency	4	6	8	5	2

5. Calculate the mean deviation from the median for following frequency distribution

Class interval	0-6	6-12	12-18	18-24	24-30
Frequency	4	5	3	6	2

6. The mean and standard deviation of some data for the time taken to complete a test are calculated with the following results:

   Number of observations = 25, Mean = 18.2 seconds and standard deviation = 3.25 seconds

   Further, another set of 15 observations, $x_1, x_2, x_3, \ldots\ldots\ldots x_{15}$, also in seconds becomes available. With the results $\sum_{i=1}^{15} x_i = 279$ and $\sum_{i=1}^{15} x_i^2 = 5524$. calculate the Standard deviation based on all 40 observations.

7. The mean and standard deviation of a set of n_1 observations are respectively $\bar{x}_1$ and s_1 while the mean and standard deviation

of a set of another n_2 observations are $\overline{x_2}$ and s_2 respectively. Show that Standard deviation of the combined set of $(n_1 + n_2)$ observations is given by

$$S.D. = \sqrt{\frac{n_1(s_1)^2 + n_2(s_2)^2}{n_1 + n_2} + \frac{n_1 n_2(\overline{x_1} - \overline{x_2})^2}{(n_1 + n_2)^2}}$$

8. The mean and standard deviation of 100 observations were found to be 40 and 10 respectively. If at the time of calculation, two observations were wrongly taken as 30 and 70 instead of 3 and 27 respectively, find the correct standard deviation.

9. The following information relates to a sample of size 60. $\sum_{n=1}^{60} x_n^2 = 18000$ and $\sum_{n=1}^{60} x_n = 960$. Find the variance.

10. Coefficient of variation of two distributions are 50 and 60 and their arithmetic mean are 30 and 25 respectively. Find difference of their standard deviations.

11. Standard deviation of some temperature data in ^oC is 5. If data are converted to ^oF, find the variance.

12. An analysis of monthly wages paid to workers in two firms A and B, belonging to the same industry, gives the following results:

	Firm A	**Firm B**
No. of wage earners	586	648
Mean of monthly wages	Rs 5253	Rs 5253
Variance of the distribution of wages	100	121

Find

(i) Which firm A or B pays larger amount as monthly wages?

(ii) Which firm, A or B, shows greater variability in individual wages?

13. The mean and standard deviation of six observations are 8 and 4, respectively. If each observation is multiplied by 3, find the new mean and new standard deviation of the resulting observations.

14. The mean and standard deviation of a group of 100 observations were found to be 20 and 3, respectively. Later on it was found

that three observations were incorrect, which were recorded as 21, 21 and 18. Find the mean and standard deviation if the incorrect observations are omitted.

15. The mean and standard deviation of marks obtained by 50 students of a class in three subjects, Mathematics, Physics and Chemistry are given below:

Subject	**Mathematics**	**Physics**	**Chemistry**
Mean	42	32	40.9
Standard deviation	12	15	20

Which of the three subjects shows the highest variability in marks and which shows the lowest?

Answers to the Practice Sums:

1. Mean =2.8 SD = 1.125
2. $\frac{n^2-1}{4n}$
3. 2.33
4. 3.873 sec
5. 3.76
6. 5.52
7. –
8. 10.24
9. 44
10. 0
11. 81
12. Firm B, Firm B
13. 24, 12
14. 20, 3.036
15. Chemistry, Mathematics

13

MATRICES AND DETERMINANTS

13.1 Introduction:

A matrix is a rectangular arrangement of numbers (real or complex) which may be represented as

$$\begin{pmatrix} a_{11} & a_{12} & a_{13}\ldots\ldots & a_{1n} \\ a_{21} & a_{22} & a_{23}\ \ldots\ldots & a_{2n} \\ \ldots & \ldots & \ldots & \ldots\ \ldots \\ a_{m1} & a_{m2} & a_{m3}\ \ldots\ldots & a_{mn} \end{pmatrix}$$

The numbers a_{11}, a_{12} ... etc., in the above matrix are known as the **elements of the matrix**, generally represented as a_{ij}, which denotes element in ith row and jth column. *Since the above matrix has m rows and n columns,* ***it is said to be of the order*** m×n. The above matrix can be represented in compact form as $A = [a_{ij}]_{m \times n}$. If number of rows and columns is equal, its called a **Square Matrix**.

Associated with each square matrix $A = [a_{ij}]_{n\times n}$, is a function of the matrix called its **Determinant** denoted as det(A), det A, or |A|. The **determinant** is a scalar value that can be computed from the elements of the associated square matrix. This value helps encode certain properties of the linear transformation described for the matrix.

13.2 Matrices

13.2.1 Types of Matrices

1. **Row Matrix:** A matrix having only one row and any number of columns is called a row matrix.

Example of a row matrix is $\begin{bmatrix} 2 & 9 & -12 \end{bmatrix}$

2. **Column Matrix:** A matrix having only one column and any number of rows is called column matrix.

Example of a column matrix is $\begin{bmatrix} -1 \\ 5 \\ -7 \end{bmatrix}$

3. **Rectangular Matrix:** A matrix of order m x n, such that m ≠ n, is called Rectangular matrix. The rectangular matrix has sub-categories as Horizontal or Vertical matrix.
   (a) **Horizontal Matrix:** A matrix in which the number of rows is less than the numberof columns, is called a horizontal matrix.
   (b) **Vertical Matrix:** A matrix in which the number of rows is greater than the numberof columns, is called a vertical matrix.

Examples of Rectangular matrix are as below:

$\begin{pmatrix} 2 & 16 & -7 \\ 15 & -8 & 12 \end{pmatrix}$ Horizontal Matrix $\begin{pmatrix} 2 & 9 \\ 15 & 8 \\ 11 & -7 \end{pmatrix}$ Vertical Matrix

4. **Null/Zero Matrix:** A matrix of any order having all its elements as zero, is called a null/zero matrix. i.e., $a_{ij} = 0$, for all i,j
5. **Square Matrix:** As already mentioned, a matrix of order m x n, such that m = n, is called square matrix.
6. **Principal Diagonal of a Matrix**: In a square matrix, the diagonal from the first element of the first row to the last element of the last row is called the principal diagonal of a matrix. These elements are $a_{11} \quad a_{22} \quad a_{33} \ldots\ldots\ldots a_{nn}$
7. **Diagonal Matrix:** A square matrix $A = [a_{ij}]_{m\times n}$, is called a diagonal matrix, if all the elements except those in the principal diagonal are zero, i.e., $a_{ij} = 0$ for $i \neq j$. Itcan berepresentedasA = diag[a_{11} a_{22}... a_{nn}]. A Diagonal matrix is depicted below:

$$\begin{pmatrix} a_{11} & 0 & 0 & \ldots\ldots & 0 \\ 0 & a_{22} & 0 & \ldots\ldots & 0 \\ 0 & 0 & a_{33} & \ldots\ldots & 0 \\ .. & .. & .. & .. & .. \\ 0 & 0 & 0 & \ldots\ldots & a_{nn} \end{pmatrix}$$

8. Scalar Matrix: A square matrix in which every non-diagonal element is zero and all diagonal elements are equal, is called scalar matrix. i.e., in scalar matrix $a_{ij} = 0$, for $i \neq j$ and $a_{ij} = k$, for $i = j$. Example of Scalar matrix is

$$\begin{pmatrix} k & 0 & 0 & \ldots\ldots & 0 \\ 0 & k & 0 & \ldots\ldots & 0 \\ 0 & 0 & k & \ldots\ldots & 0 \\ .. & .. & .. & .... & .. \\ 0 & 0 & 0 & \ldots\ldots & k \end{pmatrix}$$

There is a special case of Scalar matrix where every diagonal element is 1. It is called **Unit or Identity Matrix, i.e.**$a_{ij} = 1$, for $i = j$

9. Upper and Lower Triangular Matrix:A **triangular matrix** is a special kind of **squarematrix.** A triangular matrix is one that is either lower triangular or upper triangular.A square matrix is called **lower triangular** if all the entries ***above*** the main diagonal are zero. Similarly, a square matrix is called **upper triangular** if all the entries ***below*** the main diagonal are zero.

In other words, $A = a[_{ij}]_{n \times n}$ is called an Upper triangular matrix if $a[_{ij}] = 0$, for all $i > j$. Similarly, a square matrix $A = a[_{ij}]_{n \times n}$ is called a lowertriangular matrix, if $a[_{ij}] = 0$, for all $i < j$.

It will also be noticed that a matrix that is both upper and lower triangular is a **Diagonal Matrix**

Example of Lower and Upper Triangular matrix is given below:

$$\begin{pmatrix} a_{11} & 0 & 0 & \ldots\ldots & 0 \\ a_{21} & a_{22} & 0 & \ldots\ldots & 0 \\ a_{31} & a_{32} & a_{33} & \ldots\ldots & 0 \\ .. & .. & .. & .. & .. \\ a_{n1} & a_{n2} & a_{n3} & \ldots\ldots & a_{nn} \end{pmatrix} \quad \begin{pmatrix} a_{11} & a_{12} & a_{13} & a_{44} & \ldots a_{1n} \\ 0 & a_{22} & a_{23} & a_{24} & \ldots a_{2n} \\ 0 & 0 & a_{33} & a_{34} & \ldots a_{3n} \\ 0 & 0 & 0 & a_{44} & \ldots a_{4n} \\ \ldots & \ldots & \ldots & \ldots & \ldots. \\ 0 & 0 & 0 & 0 \ldots\ldots & a_{nn} \end{pmatrix}$$

10. Submatrix: A matrix which is obtained from a given matrix by deleting any number of rows or columns or both is called a submatrix of the given matrix.

11. **Equal Matrices:** Two matrices A and B are said to be equal, if both having the sameorder and corresponding elements of the matrices are equal.
12. **Singular Matrix** A square matrix A is said to be singular matrix, if determinant of A denoted by det (A) or |A| is zero, i.e., |A|= 0, otherwise it is a non-singular matrix.

    (***Note:We will learn about determinant value of a matrix in the next section***)

13.3 Determinants

The **determinant** of a matrix is already defined in the introductory para 13.1. The determinant value can be positive/ negative or zero.

In the case of a 2 × 2 matrix $A=\begin{bmatrix} a_{11} & a_{12} \\ a_{21} & a_{22} \end{bmatrix}$, its determinant is defined as $|A|=\begin{vmatrix} a_{11} & a_{12} \\ a_{21} & a_{22} \end{vmatrix}=a_{11}a_{22}-a_{21}a_{12}$, implying thereby that we get value of determinant for a 2×2 matrix by cross multiplication left to right with a subtraction sign in between.

For a 3 × 3 matrix, procedure to find determinant value is as below:

Let the determinant for a 3 × 3 matrix A be the following

$$|A|=\begin{vmatrix} a_{11} & a_{12} & a_{13} \\ a_{21} & a_{22} & a_{23} \\ a_{31} & a_{32} & a_{33} \end{vmatrix}$$

We take each element a_{1j} of the first row of the matrix (j = 1, 2, 3) starting with first element of the first row (a_{11}) and multiply it by the 2 × 2 determinant, known as ***cofactor of a_{ij}*** arrived at by ignoring i^{th} row as well as j^{th}column. The cofactor has sign determined by $(-1)^{i+j}$. We go on to take all other elements of the first row of the matrix and multiply it with its respective cofactor and add it to the value of the product arrived at of the first element. Same procedure is repeated for rest of the elements of the first row. Adopting this procedure, we arrive at the following value of the determinant A given above.

$$(-1)^{1+1}\times a_{11}\times\begin{vmatrix} a_{22} & a_{23} \\ a_{32} & a_{33} \end{vmatrix}+(-1)^{1+2}\times a_{12}\times\begin{vmatrix} a_{21} & a_{23} \\ a_{31} & a_{33} \end{vmatrix}+(-1)^{1+3}\times a_{13}\times\begin{vmatrix} a_{21} & a_{22} \\ a_{31} & a_{32} \end{vmatrix}$$

$$= a_{11}\left(a_{22}a_{33} - a_{32}a_{23}\right) - a_{12}\left(a_{21}a_{33} - a_{31}a_{23}\right) + a_{13}\left(a_{21}a_{32} - a_{31}a_{22}\right) \quad ....(1)$$

$$= a_{11}a_{22}a_{33} + a_{12}a_{23}a_{31} + a_{13}a_{21}a_{32} - a_{31}a_{22}a_{13} - a_{32}a_{23}a_{11} - a_{33}a_{21}a_{12} \quad ...(2)$$

(by re-arrangement)

Another interesting way of finding value of 3×3 determinant is as follows:

Column 1 and column 2 are placed next to column 3 as shown below.

$$\begin{vmatrix} a_{11} & a_{12} & a_{13} & a_{11} & a_{12} \\ a_{21} & a_{22} & a_{23} & a_{21} & a_{22} \\ a_{31} & a_{32} & a_{33} & a_{31} & a_{32} \end{vmatrix}$$

Now one set of three downwards diagonal products are taken starting with a_{11} and continuing with a_{12} & a_{13} and another set of three upwards diagonal products are taken starting with a_{31} (column 1 and row3) a_{11} continuing with a_{32} & a_{33}. The downwards set of products are positive and the upwards set is negative. This way we arrive at the same result for value of the determinant as shown at (2) above but with a little ease. This method has been explained as we may come across ***3×3*** determinants often. The student should try to find value of ***3×3*** determinants with real numbers by both methods to get a hang of these.

Note: This Method of diagonal multiplication for evaluation of determinants of matrices of order 3×3 IS NOT valid for determinants of matrices of higher order.

Let us look at an example and find value of the determinant by both methods as below:

$$\text{Evaluate A} = \begin{vmatrix} 3 & -1 & 2 \\ 0 & 4 & 3 \\ -1 & -5 & 3 \end{vmatrix}$$

Conventional method:

$$|A| = (-1)^{1+1} \times 3 \times \begin{vmatrix} 4 & 3 \\ -5 & 3 \end{vmatrix} + (-1)^{1+2} \times (-1) \times \begin{vmatrix} 0 & 3 \\ -1 & 3 \end{vmatrix} + (-1)^{1+3} \times 0 \times \begin{vmatrix} 0 & 4 \\ -1 & -5 \end{vmatrix}$$

$$= 3\left(4 \times 3 - (-5) \times 3\right) - (-1)\left(-(-1) \times 3\right) + 1 \times 2 \times \left(0 - (-1) \times 4\right)$$

$$= 3(12 + 15) + 3 + 2 \times (0 + 4) = 81 + 3 + 8 = 92$$

Alternative method:

Determinant once Column 1 and column 2 are placed next to column 3 is

$$\begin{vmatrix} 3 & -1 & 2 & 3 & -1 \\ 0 & 4 & 3 & 0 & 4 \\ -1 & -5 & 3 & -1 & -5 \end{vmatrix}$$

Evaluating by diagonal multiplication, we get .

$$|A| = (3\times 4\times 3) + \{(-1)\times 3\times(-1)\} + (2\times 0\times(-5))$$
$$-\{(-1)\times 4\times 2\} - \{(-5)\times 3\times 3\} - \{3\times 0\times(-1)\}$$

$$|A| = 36 + 3 + 0 - (-8) - (-45) - 0 = 92$$

We can see that the alternative method is somewhat simpler.

In general, a Determinant of square matrix $A(a_{ij})$ of order n is defined as

$$|A| = \begin{vmatrix} a_{11} & a_{12} & \cdots & \cdots & a_{1n} \\ a_{21} & a_{22} & \cdots & \cdots & a_{2n} \\ \cdots & \cdots & \cdots & \cdots & \cdots \\ a_{n1} & a_{n2} & \cdots & \cdots & a_{nn} \end{vmatrix}$$

The value of the determinant of a matrix of order n is defined as sum of n! terms of the form $(-1)^k a_{1i_1} a_{2i_2} \ldots\ldots.. a_{ni_n}$

Each term contains one and only one element from each row and one and only one element from each column; that is the second subscript $i_1, i_2, \ldots, i_n$ are equal to 1, 2, .,n taken in some order. The exponent k represents the number of interchanges of two elements necessary for the second subscripts to be in the order 1, 2, 3,….,n. For example consider the term $a_{13}a_{21}a_{34}a_{42}$ in the evaluation of determinant of a matrix of order 4. The value of k is 3 since three interchanges of two elements are necessary for the second subscripts in the order 1, 2, 3, 4 as

$$a_{13}a_{21}a_{34}a_{42} = a_{21}a_{13}a_{34}a_{42} = a_{21}a_{13}a_{42}a_{34} = a_{21}a_{42}a_{13}a_{34}.$$

Hence the term containing the factor $a_{13}a_{21}a_{34}a_{42}$ has the additional factor $(-1)^3 = -1$

13.4 Algebra of Matrices

13.4.1 Addition of Matrices

Provided they have the same size; two matrices can be added or subtracted element by element. Let A and B be two matrices each of order m×n, then, the sum of matrices A + B is defined as below:

If $A = [a_{ij}]_{m\times n}$, $B = [b_{ij}]_{m\times n}$

Then, $A + B = [a_{ij} + b_{ij}]_{m\times n}$

Let us see an example of addition of following two matrices. As mentioned above, the corresponding elements are added together in each row to get the following result. Subtraction is exactly a similar process.

$$\begin{pmatrix} 2 & 9 & -12 \\ 15 & 8 & -15 \end{pmatrix} + \begin{pmatrix} 6 & 4 & 10 \\ 5 & 80 & 18 \end{pmatrix} = \begin{pmatrix} 2+6 & 9+4 & -12+10 \\ 15+5 & 8+80 & -15+18 \end{pmatrix} = \begin{pmatrix} 8 & 13 & -2 \\ 20 & 88 & 3 \end{pmatrix}$$

13.4.1.1 Properties of Addition of Matrices:

If A, B and C are three matrices of order m×n, then

1. **Commutative Law** $A + B = B + A$
2. **Associative Law** $(A + B) + C = A + (B + C)$
3. **Existence of Additive Identity** A zero matrix (0) of order m×n (same order as of A), is additive identity, if $A + 0 = A = 0 + A$
4. **Existence of Additive Inverse** If A is a square matrix, then the matrix (–A) is called additive inverse, if $A + (-A) = 0 = (-A) + A$
5. **Cancellation Law**

   $A + B = A + C \Rightarrow B = C$ (left cancellation law)
   $B + A = C + A \Rightarrow B = C$ (right cancellation law)

13.4.2 Subtraction of Matrices

Let A and B be two matrices of the same order, then subtraction of matrices, A – B, is defined as

$A - B = [a_{ij} - b_{ij}]_{m\times n}$,

where $A = [a_{ij}]_{m\times n}$, $B = [b_{ij}]_{m\times n}$

13.4.3 Multiplication of a Matrix by a Scalar

Let $A = [a_{ij}]_{m\times n}$ be a matrix and k be any scalar. Then, the matrix obtained by multiplying each element of A by k is called the scalar multiple of A by k and is denoted by kA, given as $kA= [ka_{ij}]_{m\times n}$

13.4.3.1 Properties of Scalar Multiplication

If A and B are matrices of order m×n, then

1. $k(A + B) = kA + kB$
2. $(k_1 + k_2) A = k_1A + k_2A$
3. $k_1k_2A = k_1(k_2A) = k_2(k_1A)$
4. $(-k) A = -(kA) = k(-A)$

13.4.4 Multiplication of Matrices

The rule for matrix multiplication, however, is that two matrices can be multiplied only when the number of columns in the first matrix equals the number of rows in the second i.e. an (m×n) matrix can only be multiplied with an (n×p) matrix, resulting in an (m×p) matrix.

Symbolically multiplication of two matrices $A=[a_{ij}]_{m\times n}$ known as Pre-multiplier or Pre-factor and $B=[b_{ij}]_{n\times p}$ known as Post-multiplier or post factor is given by $c_{ij} = \sum_{k=1}^{n} a_{ik} b_{kj}$ where c_{ij} is the element of product matrix C = AB

The rule of multiplication of matrices is row column wise (or → ↓ wise). In simple language, process of multiplication of two matrices envisages that each element in each row of the multiplier matrix (A here) multiplies respectively each corresponding element of all columns of the matrix to be multiplied (matrix B) and added. For example, if matrix A_{3x2} multiplies Matrix B_{2x2}, each of the two elements in row 1 of pre-multiplier matrix A multiplies respective element in column 1 of Matrix B and are added to get element 1 of row 1 of the product matrix C. Similarly, each of the two elements in row 1 of matrix A multiplies respective elements in column 2 of Matrix B and are added to get element 2 of row 1 of the product matrix. This process continues for the remaining two rows of the multiplier matrix till we get all elements of the three rows of the product matrix. The result, as mentioned above is a matrix $AB_{3\times 2}$.

Example of such a Matrix multiplication is as follows:

$$\begin{pmatrix} 3 & 7 \\ -6 & 8 \\ 2 & 5 \end{pmatrix} \times \begin{pmatrix} 6 & 4 \\ 5 & 8 \end{pmatrix} = \begin{pmatrix} 3\times6 + 7\times5 & 3\times4 + 7\times8 \\ -6\times6 + 8\times5 & -6\times4 + 8\times8 \\ 2\times6 + 5\times5 & 2\times4 + 5\times8 \end{pmatrix} = \begin{pmatrix} 53 & 68 \\ 4 & 40 \\ 37 & 48 \end{pmatrix}$$

In this case, there is no product the other way round as a 2×2 matrix cannot be multiplied with a 3×2 matrix, implying thereby that matrix multiplication is not commutative. This is also true for a square matrix. We can try this for

$$A=\begin{pmatrix} 3 & 7 \\ -6 & 8 \end{pmatrix} \text{ and } B=\begin{pmatrix} 6 & 4 \\ 5 & 8 \end{pmatrix}$$

$$\text{We get } AB = \begin{pmatrix} 3\times6+7\times5 & 3\times4+7\times8 \\ -6\times6+8\times5 & -6\times4+8\times8 \end{pmatrix} = \begin{pmatrix} 53 & 68 \\ 4 & 40 \end{pmatrix} \text{ and}$$

$$BA = \begin{pmatrix} 6\times3+4\times-6 & 6\times7+4\times8 \\ 5\times3+8\times-6 & 5\times7+8\times8 \end{pmatrix} = \begin{pmatrix} -6 & 74 \\ -33 & 99 \end{pmatrix} \Rightarrow AB \neq BA$$

13.4.4.1 Properties of Multiplication of Matrices

1. *Commutative Law:* Generally AB ≠ BA (as seen above)
2. *Associative Law*: (AB)C = A(BC)

Once again we can take matrices A and B as above and $C=\begin{pmatrix} 1 & 2 \\ 3 & 4 \end{pmatrix}$

$$\text{We know } AB = \begin{pmatrix} 53 & 68 \\ 4 & 40 \end{pmatrix} \Rightarrow$$

$$AB(C) = \begin{pmatrix} 53 & 68 \\ 4 & 40 \end{pmatrix} \times \begin{pmatrix} 1 & 2 \\ 3 & 4 \end{pmatrix} =$$

$$\begin{pmatrix} 53\times1+68\times3 & 53\times2+68\times4 \\ 4\times1+40\times3 & 4\times2+40\times4 \end{pmatrix} = \begin{pmatrix} 257 & 378 \\ 124 & 168 \end{pmatrix}$$

And

$$BC = \begin{pmatrix} 6 & 4 \\ 5 & 8 \end{pmatrix} \times \begin{pmatrix} 1 & 2 \\ 3 & 4 \end{pmatrix} = \begin{pmatrix} 6\times1+4\times3 & 6\times2+4\times4 \\ 5\times1+8\times3 & 5\times2+8\times4 \end{pmatrix} = \begin{pmatrix} 18 & 28 \\ 29 & 42 \end{pmatrix}$$

leading to

$$A(BC)=\begin{pmatrix}3 & 7\\ -6 & 8\end{pmatrix}\times\begin{pmatrix}18 & 28\\ 29 & 42\end{pmatrix}=$$

$$\begin{pmatrix}3\times18+7\times29 & 3\times28+7\times42\\ -6\times18+8\times29 & -6\times28+8\times42\end{pmatrix}=\begin{pmatrix}257 & 378\\ 124 & 168\end{pmatrix}$$

Thus we can see that(AB)C = A(BC)

3. *Existence of multiplicative Identity:* A.I = A = I.A, where I is a unit/identity matrix defined above and is also called multiplicative Identity.

   To elaborate on the concept, let us multiply a matrix A_{3x2} by a Unit matrix B_{2x2} as below:

$$\begin{pmatrix}2 & 5\\ 5 & 7\\ 8 & 4\end{pmatrix}\times\begin{pmatrix}1 & 0\\ 0 & 1\end{pmatrix}=\begin{pmatrix}2\times1\ +\ 5\times0 & 2\times0\ +\ 5\times1\\ 5\times1\ +\ 7\times0 & 5\times0\ +\ 7\times1\\ 8\times1\ +\ 4\times0 & 8\times0\ +\ 4\times1\end{pmatrix}=\begin{pmatrix}2 & 5\\ 5 & 7\\ 8 & 4\end{pmatrix}$$

4. *Distributive Law* A(B+ C) = AB + AC

   We have seen in section 13.4.1 that addition of matrices is straight forward.

$$Let\ A\ =\begin{pmatrix}a_{11} & a_{12}\\ a_{21} & a_{22}\end{pmatrix},\ B=\begin{pmatrix}b_{11} & b_{12}\\ b_{21} & b_{22}\end{pmatrix}\ \&\ C=\begin{pmatrix}c_{11} & c_{12}\\ c_{21} & c_{22}\end{pmatrix}$$

$$\text{Then B + C}=\begin{pmatrix}b_{11}+c_{11} & b_{12}+c_{12}\\ b_{21}+c_{21} & b_{22}+c_{21}\end{pmatrix},$$

$$\text{AB}=\begin{pmatrix}a_{11}b_{11}+a_{12}b_{21} & a_{11}b_{12}+a_{12}b_{22}\\ a_{21}b_{11}+a_{22}b_{21} & a_{21}b_{12}+a_{22}b_{22}\end{pmatrix}\text{ and}$$

$$\text{AC}=\begin{pmatrix}a_{11}c_{11}+a_{12}c_{21} & a_{11}c_{12}+a_{12}c_{22}\\ a_{21}c_{11}+a_{22}c_{21} & a_{21}c_{12}+a_{22}c_{22}\end{pmatrix}\Rightarrow$$

AB + AC =

$$\begin{pmatrix}a_{11}b_{11}+a_{12}b_{21}+a_{11}c_{11}+a_{12}c_{21} & a_{11}b_{12}+a_{12}b_{22}+a_{11}c_{12}+a_{12}c_{22}\\ a_{21}b_{11}+a_{22}b_{21}+a_{21}c_{11}+a_{22}c_{21} & a_{21}b_{12}+a_{22}b_{22}+a_{21}c_{12}+a_{22}c_{22}\end{pmatrix}\Rightarrow$$

AB + AC =

$$\begin{pmatrix} a_{11}(b_{11}+c_{11})+a_{12}(b_{21}+c_{21}) & a_{11}(b_{12}+c_{12})+a_{12}(b_{22}+c_{22}) \\ a_{21}(b_{11}+c_{11})+a_{22}(b_{21}+c_{21}) & a_{21}(b_{12}+c_{12})+a_{21}(b_{22}+c_{22}) \end{pmatrix}$$

= A(B+C)

5. *Cancellation Law* If A is non-singular matrix, then

   AB = AC ⇒ B = C (left cancellation law)

   BA = CA ⇒B = C (right cancellation law)

   Proof for this property will be clear later in the chapter when we discuss singular and non-singular matrices.

6. AB = 0, does not necessarily imply that A = 0 or B = 0 or both A and B = 0

Important Points to be remembered

(i). If A and B are square matrices of the same order, say n, then both the product AB and BA are defined and each is a square matrix of order n.

(ii) In the matrix product AB, the matrix A is called pre-multiplier (pre-factor) and B is called post-multiplier (post-factor).

(iii) The rule of multiplication of matrices is row column wise (or → ↓ wise) the first row of AB is obtained by multiplying the first row of A with each element of first, second, third,... columns of B respectively; similarly second row of A with first, second, third, ... columns of B, respectively and so on.

13.4.4.2 Positive Integral Powers of a Square Matrix

Rules here are similar to the normal algebraic multiplication rules.

Let A be a square matrix. Then, we can define

1. $A^{n+1} = A^n . A$, where n is a natural number.
2. $A^m . A^n = A^{m+n}$
3. $(A^m)^n = A^{mn}$, for all m, n to be natural numbers

Let us try to prove Rule 3 with n = m =2.

Let $A = \begin{pmatrix} a & b \\ c & d \end{pmatrix} \Rightarrow A^2 = \begin{pmatrix} a^2+bc & ab+bd \\ ac+cd & bc+d^2 \end{pmatrix}$ and

$$A^3 = \begin{pmatrix} a(a^2+bc)+b(ac+cd) & a(ab+bd)+b(bc+d^2) \\ c(a^2+bc)+d(ac+cd) & c(ab+bd)+d(bc+d^2) \end{pmatrix}$$

$$= \begin{pmatrix} a^3+2abc+bcd & a^2b+abd+b^2c+bd^2 \\ a^2c+bc^2+acd+cd^2 & abc+2bcd+d^3 \end{pmatrix}$$

Multiplying by A again, we get

$$A^4 = \begin{pmatrix} a(a^3+2abc+bcd)+b(a^2c+bc^2+acd+cd^2) & a\left(a^2b+abd+b^2c+bd^2\right)+b\left(abc+2bcd+d^3\right) \\ c(a^3+2abc+bcd)+d(a^2c+bc^2+acd+cd^2) & c(a^2b+abd+b^2c+bd^2)+d(abc+2bcd+d^3) \end{pmatrix} \Rightarrow$$

$$A^4 = \begin{pmatrix} a^4+2a^2bc+abcd+a^2bc+b^2c^2+abcd+bcd^2 & a^3b+a^2bd+ab^2c+abd^2+ab^2c+2b^2cd+bd^3 \\ (a^3c+2abc^2+bc^2d)+(a^2cd+bc^2d+acd^2+cd^3) & a^2bc+abcd+b^2c^2+bcd^2+abcd+2bcd^2+d^4 \end{pmatrix} \Rightarrow$$

By rearrangement, value of A⁴

$$A^4 = \begin{pmatrix} a^4+b^2c^2+2a^2bc+abcd+a^2bc+abcd+bcd^2 & a^3b+a^2bd+ab^2c+b^2cd+ab^2c+b^2cd+abd^2+bd^3 \\ (a^3c+abc^2+a^2cd+bc^2d+abc^2+acd^2+bc^2d+cd^3) & a^2bc+abcd+abcd+bcd^2+b^2c^2+2bcd^2+d^4 \end{pmatrix} \Rightarrow$$

$$A^4 = \begin{pmatrix} \left(a^4+b^2c^2+2a^2bc\right)+ab(cd+ac)+bd(ac+cd) & a^2(ab+bd)+bc(ab+bd)+bc(ab+bd)+d^2(ab+bd) \\ ac(a^2+bc)+cd(a^2+bc)+ac(bc+d^2)+cd(bc+d^2) & ac(ab+bd)+cd(ab+bd)+(b^2c^2+d^4+2bcd^2) \end{pmatrix}$$

$$A^4 = \begin{pmatrix} \left(a^2+bc\right)^2+(ab+bd)(ac+cd) & (a^2+bc)(ab+bd)+(bc+d^2)(ab+bd) \\ \left(a^2+bc\right)(ac+cd)+\left(bc+d^2\right)(ac+cd) & (ac+cd)(ab+bd)+\left(bc+d^2\right)^2 \end{pmatrix}$$

$$= \begin{pmatrix} a^2+bc & ab+bd \\ ac+cd & bc+d^2 \end{pmatrix}^2 = \left(A^2\right)^2 = A^4$$

13.4.4.3 Matrix Polynomial

Let $f(x) = a_0x^n + a_1x^{n-1} + a_2x^{n-2} + \dots + a_n$. Then $f(A) = a_0A^n + a_1A^{n-1} + \dots + a_n$ is called the matrix polynomial.

13.4.5 Elementary Transformation of a Matrix

Any one of the following operations on a matrix is called an elementary transformation.

1. Interchanging any two rows (or columns), denoted by $R_i \leftrightarrow R_j$ or $C_i \leftrightarrow C_j$
2. Multiplication of the element of any row (or column) by a non-zero quantity and denoted by $R_i \to kR_i$ or $C_i \to kC_j$
3. Addition of constant multiple of the elements of any row to the corresponding element of any other row, denoted by $R_i \to R_i + kR_j$ or $C_i \to C_i + kC_j$

13.4.6 Equivalent Matrix

- Two matrices A and B are said to be equivalent, if one can be obtained from the other by a sequence of elementary transformation.
- The symbol $\approx$ is used for equivalence.

13.4.7 Transpose of a Matrix

Let $A = [a_{ij}]_{m\times n}$, be a matrix of order m×n. Then, the n×m matrix obtained by interchanging the rows and columns of A is called the transpose of A and is denoted by A' or A^T with $A' = A^T = [a_{ji}]_{n\times m}$

13.4.7.1 Properties of Transpose

1. $(A^T)^T = A$

   This is obvious as interchange of rows and columns twice amounts to restoring the original matrix.

   $(A+B)^T = A^T + B^T$

   Proof:

   Let $A = \begin{pmatrix} a & b & c \\ d & e & f \end{pmatrix}$ and $B = \begin{pmatrix} p & q & r \\ s & t & u \end{pmatrix} \Rightarrow$

   $$A + B = \begin{pmatrix} a+p & b+q & c+r \\ d+s & e+t & f+u \end{pmatrix}$$

   $$\Rightarrow (A+B)^T = \begin{pmatrix} a+p & d+s \\ b+q & e+t \\ c+r & f+u \end{pmatrix}$$

   Now $A^T = \begin{pmatrix} a & d \\ b & e \\ c & f \end{pmatrix}$ and $B^T = \begin{pmatrix} p & s \\ q & t \\ r & u \end{pmatrix} \Rightarrow A^T + B^T = \begin{pmatrix} a+p & d+s \\ b+q & e+t \\ c+r & f+u \end{pmatrix}$

   Hence $(A + B)^T = A^T + B^T$

3. $(AB)^T = B^T A^T$

   Proof:

   Let $A_{2\times3} = \begin{pmatrix} a & b & c \\ d & e & f \end{pmatrix}$ and $B_{3\times2} = \begin{pmatrix} p & s \\ q & t \\ r & u \end{pmatrix} \Rightarrow AB_{2\times2} =$

   $$\begin{pmatrix} ap+bq+cr & as+bt+cu \\ dp+eq+fr & ds+et+fu \end{pmatrix} \Rightarrow$$

   $$(AB)^T_{2\times2} = \begin{pmatrix} ap+bq+cr & dp+eq+fr \\ as+bt+cu & ds+et+fu \end{pmatrix}$$

   Now $A^T_{3\times2} = \begin{pmatrix} a & d \\ b & e \\ c & f \end{pmatrix}$, $B_{2\times3} = \begin{pmatrix} p & q & r \\ s & t & u \end{pmatrix} \Rightarrow B^T A^T =$

   $$\begin{pmatrix} ap+bq+cr & dp+eq+fr \\ as+bt+cu & ds+et+fu \end{pmatrix}.$$

Hence $(AB)^T = B^TA^T$

Note that the product A^TB^T will be a 3×3 matrix and will not be equal to $(AB)^T$.

4. $(kA)^T = kA^T$
5. $(A^N)^T = (A^T)^N$
6. $(ABC)^T = C^TB^TA^T$

   Note: This equality can be verified by the readers with real life matrices A, B and C like it has been done in property 3 above. This will also help in practicing multiplication of matrices.

13.4.8 Trace of a Matrix

The sum of the diagonal elements of a square matrix A is called the trace of A, denoted by trace (A) or tr (A) i.e. $a_{11} + a_{22} + \dots\dots + a_{nn}$

13.4.8.1 Properties of Trace of a Matrix

1. Trace (A ± B)= Trace (A) ± Trace (B)
2. Trace (kA)= k Trace (A)
3. Trace (A^T) = Trace (A) because diagonal numbers of transpose of any matrix remains unchanged.
4. Trace (I_n) = n, as all diagonal numbers are unity and there are n diagonal numbers.in I_n
5. Trace (0) = 0
6. Trace (AB) ≠ Trace (A)×Trace (B)

   Proof:

   Let $A = \begin{pmatrix} a & b \\ c & d \end{pmatrix}$ and B = $\begin{pmatrix} e & f \\ g & h \end{pmatrix}$

   Then AB = $\begin{pmatrix} ae+bg & af+bh \\ ce+dg & cf+dh \end{pmatrix}$ leading to

   Trace (AB) = $ae+bg+cf+dh$; Trace A = $a + d$ and Trace B= $e + h$

   So Trace (A)×Trace (B) = $ae+ah+de+dh$

   We can see from the above that Trace (AB) ≠ Trace (A)×Trace (B)

7 Trace $(AA^T) \geq 0$

Let $A=\begin{pmatrix} a & b \\ c & d \end{pmatrix} \Rightarrow A^T=\begin{pmatrix} a & c \\ b & d \end{pmatrix}$

So $AA^T=\begin{pmatrix} a^2+b^2 & ac+bd \\ ac+bd & c^2+d^2 \end{pmatrix} \Rightarrow \text{Trace}(AA^T)=a^2+b^2+c^2+d^2$

So, Trace $(AA^T) \geq 0$ since sum of squares of any 4 numbers as above is positive

13.4.9 Conjugate of a Matrix

The matrix obtained from a matrix A containing complex numbers as its elements, on replacing its elements by the corresponding conjugate complex number is called conjugate of A and is denoted by $\overline{A}$

13.4.9.1 Properties of Conjugate of a Matrix

If A is a matrix of order m×n, then

1. $\overline{\left(\overline{A}\right)}=A$, because conjugate of conjugate of a complex number is the number itself.
2. For matrix B of order m×n $\overline{A+B}=\overline{A}+\overline{B}$
3. For matrix B of order n×p $\left(\overline{AB}\right)=\overline{A}\times\overline{B}$

   Let us explore with an example.

Let A $=\begin{pmatrix} 1 & 1+2i \\ 2+i & 3 \end{pmatrix}$ and B $=\begin{pmatrix} 3 & 2+3i \\ 1+i & 4 \end{pmatrix}$

This would imply that $\overline{A}=\begin{pmatrix} 1 & 1-2i \\ 2-i & 3 \end{pmatrix}$ and $\overline{B}=\begin{pmatrix} 3 & 2-3i \\ 1-i & 4 \end{pmatrix}$

Now

$$AB=\begin{pmatrix} 1 & 1+2i \\ 2+i & 3 \end{pmatrix}\times\begin{pmatrix} 3 & 2+3i \\ 1+i & 4 \end{pmatrix}=\begin{pmatrix} 2+3i & 6+11i \\ 9+6i & 13+8i \end{pmatrix}\Rightarrow$$

$$\overline{AB}=\begin{pmatrix} 2-3i & 6-11i \\ 9-6i & 13-8i \end{pmatrix}$$

$$\overline{A}\times\overline{B}=\begin{pmatrix} 1 & 1-2i \\ 2-i & 3 \end{pmatrix}\times\begin{pmatrix} 3 & 2-3i \\ 1-i & 4 \end{pmatrix}=\begin{pmatrix} 2-3i & 6-11i \\ 9-6i & 13-8i \end{pmatrix}\Rightarrow$$

$$\overline{AB}=\overline{A}\times\overline{B}$$

Note: The reader should practice multiplication of real and imaginary numbers.

If k is a scalar, then $\left(\overline{kA}\right) = k\overline{A}$

$\overline{\left(A^n\right)} = \left(\overline{A}\right)^n$

13.4.10 Transpose Conjugate of a Matrix

The transpose of the conjugate of a matrix A is called transpose conjugate of A and is denoted by A^0 or A^* i.e., $\left(\overline{A}\right)^T = A^0 = A^*$

13.4.10.1 Properties of Transpose Conjugate of a Matrix

(i) $\left(A^*\right)^* = A$

(ii) $\left(A+B\right)^* = A^* + B^*$

(iii) $(kA)^* = kA^*$

(iv) $(AB)^* = B^*A^*$

Let us explore this property with the matrices A and B considered in property (iii) of Conjugate of matrices at 13.4.9.1 above.

So A = $\begin{pmatrix} 1 & 1+2i \\ 2+i & 3 \end{pmatrix}$ and B = $\begin{pmatrix} 3 & 2+3i \\ 1+i & 4 \end{pmatrix}$ leading to

$AB = \begin{pmatrix} 2+3i & 6+11i \\ 9+6i & 13+8i \end{pmatrix}$, $A^* = \begin{pmatrix} 1 & 2-i \\ 1-2i & 3 \end{pmatrix}$ & $B^* = \begin{pmatrix} 3 & 1-i \\ 2-3i & 4 \end{pmatrix}$

Transpose of $AB = \begin{pmatrix} 2+3i & 9+6i \\ 6+11i & 13+8i \end{pmatrix}$ leading to transpose conjugate $\left(AB\right)^* = \begin{pmatrix} 2-3i & 9-6i \\ 6-11i & 13-8i \end{pmatrix}$ and

$$B^* \times A^* = \begin{pmatrix} 3 & 1-i \\ 2-3i & 4 \end{pmatrix} \times \begin{pmatrix} 1 & 2-i \\ 1-2i & 3 \end{pmatrix} = \begin{pmatrix} 2-3i & 9-6i \\ 6-11i & 13-8i \end{pmatrix}$$

Hence (AB)* $= B^*A^*$

$\left(A^n\right)^* = \left(A^*\right)^n$

13.4.11 Adjoint of a Square Matrix

Adjoint of a Square Matrix A is defined as a Matrix that when multiplied by the given matrix yields as product of a unitary matrix I and value of determinant |A| formed by elements of matrix A as shown below:

$$A \times (adj\ A) = (adj\ A) \times A = |A| \times I$$

We will learn about determinants in the next section 13.6. This is a very important property that is used for calculating Inverse of A. We will see this property in the next section 13.4.11.

Adj A is calculated by calculating different cofactors of the determinant A as depicted below:

Let $A[a_{ij}]_{n\times n}$ be a square matrix of order n and let C_{ij} be the cofactor of a_{ij} in the determinant |A|, then the adjoint of A is the transpose of the matrix, formed by the cofactors of the matrix.

As already mentioned in section 13.3, Cofactor C_{ij} of element a_{ij} of determinant $|A|_{n\times n}$ is defined as $C_{ij} = (-1)^{i+j} \times |C|$ where determinant C is formed by eliminating i^{th} row and j^{th} column of determinant|A|

Here is an example of Adjoint of a Matrix:

Let A be a 3×3 square matrix as $A = \begin{pmatrix} 3 & -6 & 5 \\ 1 & -2 & 8 \\ 4 & 7 & 3 \end{pmatrix}$

Cofactors C_{ij} of this matrix in determinant |A| will be

$C_{11} = (-1)^{1+1} \times \begin{vmatrix} -2 & 8 \\ 7 & 3 \end{vmatrix}$, $C_{12} = (-1)^{1+2} \times \begin{vmatrix} 1 & 8 \\ 4 & 3 \end{vmatrix}$, $C_{13} = (-1)^{1+3} \times \begin{vmatrix} 1 & -2 \\ 4 & 7 \end{vmatrix}$,

$C_{21} = (-1)^{2+1} \times \begin{vmatrix} -6 & 5 \\ 7 & 3 \end{vmatrix}$

$C_{22} = (-1)^{2+2} \times \begin{vmatrix} 3 & 5 \\ 4 & 3 \end{vmatrix}$, $C_{23} = (-1)^{2+3} \times \begin{vmatrix} 3 & -6 \\ 4 & 7 \end{vmatrix}$, $C_{31} = (-1)^{3+1} \times \begin{vmatrix} -6 & 5 \\ -2 & 8 \end{vmatrix}$, $C_{32} =$

$(-1)^{3+2} \times \begin{vmatrix} 3 & 5 \\ 1 & 8 \end{vmatrix}$ $C_{33} = (-1)^{3+3} \times \begin{vmatrix} 3 & -6 \\ 1 & -2 \end{vmatrix}$, leading to Matrix

$$C = \begin{pmatrix} -2\times 3 - 7\times 8 & -(1\times 3 - (4\times 8)) & 1\times 7 - 4\times(-2) \\ -(-6\times 3 - 7\times 5) & 3\times 3 - 4\times 5 & -(3\times 7 - 4\times(-6)) \\ -6\times 8 - (-2)\times 5 & -(3\times 8 - 1\times 5) & -(3\times(-2) - 1\times(-6)) \end{pmatrix} =$$

$$\begin{pmatrix} -62 & 29 & 15 \\ 53 & -11 & -45 \\ -38 & -19 & 0 \end{pmatrix}$$

Therefore ***adj(A)*** = Transpose of Matrix C=

$$\begin{pmatrix} -62 & 53 & -38 \\ 29 & -11 & -19 \\ 15 & -45 & 0 \end{pmatrix}$$

We can verify by multiplying matrix ***A*** and this ***adj(A)*** to get the following result:

$$A \times adj(A) = \begin{pmatrix} 3 & -6 & 5 \\ 1 & -2 & 8 \\ 4 & 7 & 3 \end{pmatrix} \times \begin{pmatrix} -62 & 53 & -38 \\ 29 & -11 & -19 \\ 15 & -45 & 0 \end{pmatrix} =$$

$$\begin{pmatrix} -285 & 0 & 0 \\ 0 & -285 & 0 \\ 0 & 0 & -285 \end{pmatrix} = -285 \times \begin{pmatrix} 1 & 0 & 0 \\ 0 & 1 & 0 \\ 0 & 0 & 1 \end{pmatrix}$$

As already mentioned, the result will be same even if ***adj(A)*** is multiplied by ***A*** Readers may verify this fact.

13.4.11.1 Other properties of Adjoint of a Square Matrix

If A and B are square matrices of order n, then

1. adj (A^T) = $(\text{adj } A)^T$
2. adj (AB) = (adj B) (adj A)
3. adj (kA) = k(adj A), k $\in$ R
4. adj (A^m) = $(\text{adj } A)^m$
5. adj (adj A) = |A| × A, A is a non-singular matrix.
6. |adj A| =|A|,A is a non-singular matrix.

7. |adj (adj A)| =|A| A is a non-singular matrix.

8. Adjoint of a diagonal matrix is a diagonal matrix

13.4.12 Inverse of a Square Matrix

Let A be a square matrix of order n, then a square matrix B, such that AB = BA = Identity matrix I, is called inverse of A, denoted by A^{-1}i.e.,or $AA^{-1} = A^{-1}A = I$.

There are two approaches to find Inverse of a Matrix.

Approach 1

We have seen in Property 1 in section 13.4.10.1 thatA× (adj A)= $|A| \times I$

Multiplying both sides by A^{-1}, we get $A^{-1} \times A \times$ (adj A)= $|A| \times A^{-1} \times I$ leading to (adj A)= $|A| \times A^{-1}$. This further leads to $A^{-1} = \frac{1}{|A|}(adjA)$

Taking the example of ***adj A***at 13.4.10above, let us find inverse of A.

$$A^{-1} = \frac{1}{|A|} \times \begin{pmatrix} -62 & 53 & -38 \\ 29 & -11 & -19 \\ 15 & -45 & 0 \end{pmatrix} \text{where } A = \begin{pmatrix} 3 & -6 & 5 \\ 1 & -2 & 8 \\ 4 & 7 & 3 \end{pmatrix}$$

$$\text{Now} |A| = 3 \times \begin{vmatrix} -2 & 8 \\ 7 & 3 \end{vmatrix} - (-6) \times \begin{vmatrix} 1 & 8 \\ 4 & 3 \end{vmatrix} + 5 \times \begin{vmatrix} 1 & -2 \\ 4 & 7 \end{vmatrix} \Rightarrow$$

$$|A| = 3 \times (-6 - 56) + 6 \times (3 - 32) + 5 \times (7 - (-8))$$

$$= 3 \times (-62) + 6 \times (-29) + 5 \times (15) = -186 - 174 + 75 = -285$$

$$\text{Hence } A^{-1} = \frac{1}{-285} \times \begin{pmatrix} -62 & 53 & -38 \\ 29 & -11 & -19 \\ 15 & -45 & 0 \end{pmatrix}$$

Approach 2

This is based on Reduced Echelon Form of a Matrix discussed later at section 13.5.8. Here we reduce the given matrix A into an Identity Matrix by elementary row operations while performing the same row operations on an Identity matrix.

This operation is known as ***Gauss Jordon method*** and can be akin to carrying out the following that leads to inverse of the given matrix:

$$[A \mid I] \Rightarrow \left[A^{-1} \times A \mid A^{-1} \times I\right] \Rightarrow \left[I \mid A^{-1}\right]$$

For doing this we place matrix A and Identity matrix side by side, also known as an Augmented matrix as below:

$$\left.\begin{pmatrix} 3 & -6 & 5 \\ 1 & -2 & 8 \\ 4 & 7 & 3 \end{pmatrix}\right| \left.\begin{pmatrix} 1 & 0 & 0 \\ 0 & 1 & 0 \\ 0 & 0 & 1 \end{pmatrix}\right|$$

To achieve transformation of the given matrix A to I, we need to follow the following steps:

1. Choose first non-zero column and use appropriate row operation to get 1 at the top.
2. Use multiples of row containing 1 to get zero in all remaining places in the column.
3. Repeat step 1 with sub-matrix formed by deleting row in step 2 and all rows above it mentally.
4. Repeat step 2 with entire matrix including rows deleted mentally above.

Based on the above process, let us transform the augmented matrix A discussed above (reproduced below)

$$\left.\begin{pmatrix} 3 & -6 & 5 \\ 1 & -2 & 8 \\ 4 & 7 & 3 \end{pmatrix}\right| \left.\begin{pmatrix} 1 & 0 & 0 \\ 0 & 1 & 0 \\ 0 & 0 & 1 \end{pmatrix}\right|$$

$R_1 \to R_1 - 2R_2$

$$\left.\begin{pmatrix} 1 & -2 & -11 \\ 1 & -2 & 8 \\ 4 & 7 & 3 \end{pmatrix}\right| \left.\begin{pmatrix} 1 & -2 & 0 \\ 0 & 1 & 0 \\ 0 & 0 & 1 \end{pmatrix}\right|$$

$R_2 \to R_2 - R_1$ and $R_3 \to R_3 - 4R_1$

$$\left.\begin{pmatrix} 1 & -2 & -11 \\ 0 & 0 & 19 \\ 0 & 15 & 47 \end{pmatrix}\right| \left.\begin{pmatrix} 1 & -2 & 0 \\ -1 & 3 & 0 \\ -4 & 8 & 1 \end{pmatrix}\right|$$

$R_2 \to R_2 + R_3$

$$\begin{pmatrix} 1 & -2 & -11 \\ 0 & 15 & 66 \\ 0 & 15 & 47 \end{pmatrix} \left| \begin{pmatrix} 1 & -2 & 0 \\ -5 & 11 & 1 \\ -4 & 8 & 1 \end{pmatrix} \right|$$

$R_3 \to R_3 - R_2$ leads to

$$\begin{pmatrix} 1 & -2 & -11 \\ 0 & 15 & 66 \\ 0 & 0 & -19 \end{pmatrix} \left| \begin{pmatrix} 1 & -2 & 0 \\ -5 & 11 & 1 \\ 1 & -3 & 0 \end{pmatrix} \right|$$

$R_3 \to R_3/-19$ and $R_2 \to R_2/15$

$$\begin{pmatrix} 1 & -2 & -11 \\ 0 & 1 & \frac{66}{15} \\ 0 & 0 & 1 \end{pmatrix} \left| \begin{pmatrix} 1 & -2 & 0 \\ -\frac{1}{3} & \frac{11}{15} & \frac{1}{15} \\ \frac{-1}{19} & \frac{3}{19} & 0 \end{pmatrix} \right|$$

$R_1 \to R_1 + 2R_2$

$$\begin{pmatrix} 1 & 0 & -\frac{33}{15} \\ 0 & 1 & \frac{66}{15} \\ 0 & 0 & 1 \end{pmatrix} \left| \begin{pmatrix} \frac{1}{3} & \frac{-8}{15} & \frac{2}{15} \\ -\frac{1}{3} & \frac{11}{15} & \frac{1}{15} \\ \frac{-1}{19} & \frac{3}{19} & 0 \end{pmatrix} \right|$$

$R_1 \to R_1 + \frac{33}{15} R_3,\ R_2 \to R_2 - \frac{66}{15} R_3$

$$\begin{pmatrix} 1 & 0 & 0 \\ 0 & 1 & 0 \\ 0 & 0 & 1 \end{pmatrix} \left| \begin{pmatrix} \frac{62}{285} & \frac{-53}{285} & \frac{2}{15} \\ -\frac{29}{285} & \frac{11}{285} & \frac{1}{15} \\ \frac{-1}{19} & \frac{3}{19} & 0 \end{pmatrix} \right|$$

$$\Rightarrow \left(\begin{matrix} 1 & 0 & 0 \\ 0 & 1 & 0 \\ 0 & 0 & 1 \end{matrix}\right| \left. \frac{-1}{285} \times \begin{pmatrix} -62 & 53 & -38 \\ 29 & -11 & -19 \\ 15 & -45 & 0 \end{pmatrix}\right|$$

This is the same result as obtained by Approach 1.

Let us verify the result by calculating $A \times A^{-1}$

$$A \times A^{-1} = \begin{pmatrix} 3 & -6 & 5 \\ 1 & -2 & 8 \\ 4 & 7 & 3 \end{pmatrix} \times \frac{1}{-285} \times \begin{pmatrix} -62 & 53 & -38 \\ 29 & -11 & -19 \\ 15 & -45 & 0 \end{pmatrix} \Rightarrow A \times A^{-1} =$$

$$\frac{1}{-285} \begin{pmatrix} 3\times-62+(-6\times29)+5\times15 & 3\times53+(-6\times-11)+5\times-45 & 3\times(-38)+(-6)\times-19+5\times0 \\ 1\times-62+(-2\times29)+8\times15 & 1\times53+(-2\times-11)+8\times-45 & 1\times(-38)+(-2)\times-19+8\times0 \\ 4\times-62+(7\times29)+3\times15 & 4\times53+(7\times-11)+3\times-45 & 4\times(-38)+7\times-19+3\times0 \end{pmatrix} \Rightarrow$$

$A \times A^{-1}$

$$= \frac{1}{-285} \times \begin{pmatrix} -186-174+75 & 159+66-225 & -114+114+0 \\ -62-58+120 & 53+22-360 & -38+38+0 \\ -248+203+45 & 212-77-135 & -152-133+0 \end{pmatrix}$$

$$= \frac{1}{-285} \times \begin{pmatrix} -285 & 0 & 0 \\ 0 & -285 & 0 \\ 0 & 0 & -285 \end{pmatrix} = \begin{pmatrix} 1 & 0 & 0 \\ 0 & 1 & 0 \\ 0 & 0 & 1 \end{pmatrix} = I$$ **Hence proved.**

Non-square matrices do not have **inverses**. A **square matrix** which has an **inverse** is called invertible or non-singular, and a **square matrix** without an **inverse** is called non-invertible or singular.

13.4.12.1 Properties of Inverse of a Square Matrix

1. Square matrix A is invertible if and only if $|A| \neq 0$
2. $(A^{-1})^{-1} = A$
3. $(A^{T})^{-1} = (A^{-1})^{T}$

4. $(AB)^{-1} = B^{-1}A^{-1}$. In general, $(A_1A_1A_1 \ldots A_n)^{-1} = A_n^{-1}A_{n-1}^{-1} \ldots A_3^{-1}A_2^{-1}A_1^{-1}$
5. If a non-singular square matrix A is symmetric, then A^{-1} is also symmetric.
6. $|A^{-1}| = |A|^{-1}$
7. $AA^{-1} = A^{-1}A = I$
8. $(A^k)^{-1} = (A^{-1})^k$ where k is a natural number
9. As an example, if $A = \begin{pmatrix} a & 0 & 0 \\ 0 & b & 0 \\ 0 & 0 & c \end{pmatrix}$ and $|A| = abc \neq 0$, then

$$A^{-1} = \begin{pmatrix} 1/a & 0 & 0 \\ 0 & 1/b & 0 \\ 0 & 0 & 1/c \end{pmatrix}$$

13.4.13 Rank of a Matrix

Rank of a Matrix A is defined as the order of largest square matrix of A whose determinant is NOT ZERO.

If $|A| = 0$, rank of the matrix $\rho(A) <$ size of A and equals size of the any square submatrix of the given matrix whose determinant value $\neq 0$.

In other words, Rank is a positive integer say ***r*** if there exists at least one minor (square submatrix) in A of order r which is not zero and every minor in A of order greater than r is zero.***Rank of a such a matrix A is denoted by ρ(A) = r.***

It also implies that if $|A| \neq 0$, rank of the matrix $\rho(A) =$ size of the matrix A, if A is a square matrix or lower of the two size numbers n and m if $A_{n \times m}$ is a rectangular matrix.

13.4.13.1 Methods of finding Rank of a Matrix:

There are 4 methods by which we can find rank of a matrix. These are:

Method 1: Normal form:

Process here is to

- Reduce the given matrix to Identity matrix.

- Both row and column operations are permissible
- Normal form of matrix is $\begin{pmatrix} I_n & 0 \\ 0 & 0 \end{pmatrix}$ where we have a unitary submatrix of order n at the left most position with all other elements 0.
- This yields $\rho(A) = n$

Example:

Find rank of the matrix $A = \begin{pmatrix} 1 & 2 & 3 & 2 \\ 2 & 3 & 5 & 1 \\ 1 & 3 & 4 & 5 \end{pmatrix}$

First step is to reduce the first column to 1 in position a_{11} and 0 in rest of the positions a_{21} and a_{31}. This can be achieved by

$R_2 \rightarrow R_2 - 2R_1$ and $R_3 \rightarrow R_3 - R_1$ leading to

$$A = \begin{pmatrix} 1 & 2 & 3 & 2 \\ 0 & -1 & -1 & -3 \\ 0 & 1 & 1 & 3 \end{pmatrix}$$

Next step is to reduce the first row to all 0s after 1 in position a_{11}. This can be achieved by $C_2 \rightarrow C_2 - 2C_1$, $C_3 \rightarrow C_3 - 3C_1$ and $C_4 \rightarrow C_4 - 2C_1$ leading to

$$A = \begin{pmatrix} 1 & 0 & 0 & 0 \\ 0 & -1 & -1 & -3 \\ 0 & 1 & 1 & 3 \end{pmatrix}$$

Now we need 1 in position a_{22} and 0s in rest of the positions of row 2 This can be achieved by the following 2 steps

$R_2 \rightarrow (-1) \times R_2$ yielding

$$A = \begin{pmatrix} 1 & 0 & 0 & 0 \\ 0 & 1 & 1 & 3 \\ 0 & 1 & 1 & 3 \end{pmatrix}$$ and then $C_3 \rightarrow C_3 - C_2$ and $C_4 \rightarrow C_4 - 3C_2$

leading to

$$A = \begin{pmatrix} 1 & 0 & 0 & 0 \\ 0 & 1 & 0 & 0 \\ 0 & 1 & 0 & 0 \end{pmatrix}$$

Now we need 0 in position a_{32}. This can be achieved by $R_2 \to R_3 - R_2$ yielding

$$A = \begin{pmatrix} 1 & 0 & 0 & 0 \\ 0 & 1 & 0 & 0 \\ 0 & 0 & 0 & 0 \end{pmatrix} \text{ which is same as } \begin{pmatrix} I_2 & 0 \\ 0 & 0 \end{pmatrix} \Rightarrow \rho(A) = 2$$

Note: It can be noticed after step 1 that in the resultant matrix, if $R_2 \to R_2 + R_3$ we get 0 in all positions of row 2 implying thereby, $|A| = 0$ implying thatrank is <3.

Method 2: PAQ form:

- Equate given matrix $[A_{mxn}]$ to $I_m \times A \times I_n$ where I_m and I_n are Identity matrices of size m and n respectively.
- Reduce the above combination to the normal form shown in Method 1 by the following process:
    - Both row and column operations are permissible.
    - Row operations are conducted simultaneously on the given matrix A as well as I_m while the column operations are conducted on the Given matrix and I_n

Result is $\begin{pmatrix} I_n & 0 \\ 0 & 0 \end{pmatrix} = P \times A \times Q$.

- Then $\rho(A)$ = n and P and Q are two non-singular matrices.

Example:

Find rank of the matrix $A = \begin{pmatrix} 1 & -1 & 3 & 6 \\ 1 & 3 & -3 & -4 \\ 5 & 3 & 3 & 11 \end{pmatrix}$

As per the above process

$$\begin{pmatrix} 1 & -1 & 3 & 6 \\ 1 & 3 & -3 & -4 \\ 5 & 3 & 3 & 11 \end{pmatrix} = \begin{pmatrix} 1 & 0 & 0 \\ 0 & 1 & 0 \\ 0 & 0 & 1 \end{pmatrix} \times A \times \begin{pmatrix} 1 & 0 & 0 & 0 \\ 0 & 1 & 0 & 0 \\ 0 & 0 & 1 & 0 \\ 0 & 0 & 0 & 1 \end{pmatrix}$$

$R_2 \rightarrow R_2 - R_1$ and $R_3 \rightarrow R_3 - 5R_1$ leading to

$$\begin{pmatrix} 1 & -1 & 3 & 6 \\ 0 & 4 & -6 & -10 \\ 0 & 8 & -12 & -19 \end{pmatrix} = \begin{pmatrix} 1 & 0 & 0 \\ -1 & 1 & 0 \\ -5 & 0 & 1 \end{pmatrix} \times A \times \begin{pmatrix} 1 & 0 & 0 & 0 \\ 0 & 1 & 0 & 0 \\ 0 & 0 & 1 & 0 \\ 0 & 0 & 0 & 1 \end{pmatrix}$$

$C_4 \rightarrow C_4 - 6C_1$, $C_3 \rightarrow C_3 - 3C_1$ and $C_2 \rightarrow C_2 + C_1$ leading to

$$\begin{pmatrix} 1 & 0 & 0 & 0 \\ 0 & 4 & -6 & -10 \\ 0 & 8 & -12 & -19 \end{pmatrix} = \begin{pmatrix} 1 & 0 & 0 \\ -1 & 1 & 0 \\ -5 & 0 & 1 \end{pmatrix} \times A \times \begin{pmatrix} 1 & 1 & -3 & -6 \\ 0 & 1 & 0 & 0 \\ 0 & 0 & 1 & 0 \\ 0 & 0 & 0 & 1 \end{pmatrix}$$

Now to get 0 in a_{32} and then 1 in a_{22}, $R_3 \rightarrow R_3 - 2R_2$ and $R_2 \rightarrow R_2/4$

$$\begin{pmatrix} 1 & 0 & 0 & 0 \\ 0 & 1 & -\frac{3}{2} & -\frac{5}{2} \\ 0 & 0 & 0 & 1 \end{pmatrix} = \begin{pmatrix} 1 & 0 & 0 \\ -\frac{1}{4} & \frac{1}{4} & 0 \\ -3 & -2 & 1 \end{pmatrix} \times A \times \begin{pmatrix} 1 & 1 & -3 & -6 \\ 0 & 1 & 0 & 0 \\ 0 & 0 & 1 & 0 \\ 0 & 0 & 0 & 1 \end{pmatrix}$$

Further to get 0 in position a_{23} and a_{24} we use $C_4 \rightarrow C_4 + \frac{5}{2} C_2$, $C_3 \rightarrow C_3 + \frac{3}{2} C_2$ leading to

$$\begin{pmatrix} 1 & 0 & 0 & 0 \\ 0 & 1 & 0 & 0 \\ 0 & 0 & 0 & 1 \end{pmatrix} = \begin{pmatrix} 1 & 0 & 0 \\ -\frac{1}{4} & \frac{1}{4} & 0 \\ -3 & -2 & 1 \end{pmatrix} \times A \times \begin{pmatrix} 1 & 1 & -\frac{3}{2} & -\frac{7}{2} \\ 0 & 1 & \frac{3}{2} & \frac{5}{2} \\ 0 & 0 & 1 & 0 \\ 0 & 0 & 0 & 1 \end{pmatrix}$$

Now interchanging columns 4 and 3 we get

$$\begin{pmatrix} 1 & 0 & 0 & 0 \\ 0 & 1 & 0 & 0 \\ 0 & 0 & 1 & 0 \end{pmatrix} = \begin{pmatrix} 1 & 0 & 0 \\ -\frac{1}{4} & \frac{1}{4} & 0 \\ -3 & -2 & 1 \end{pmatrix} \times A \times \begin{pmatrix} 1 & 1 & -\frac{7}{2} & -\frac{3}{2} \\ 0 & 1 & \frac{5}{2} & \frac{3}{2} \\ 0 & 0 & 0 & 1 \\ 0 & 0 & 1 & 0 \end{pmatrix}$$

$\Rightarrow \begin{pmatrix} I_3 & 0 \end{pmatrix} = P \times A \times Q$ yielding ρ(A) = 3, where

matrices $P = \begin{pmatrix} 1 & 0 & 0 \\ -\frac{1}{4} & \frac{1}{4} & 0 \\ -3 & -2 & 1 \end{pmatrix}$ and $Q = \begin{pmatrix} 1 & 1 & -\frac{7}{2} & -\frac{3}{2} \\ 0 & 1 & \frac{5}{2} & \frac{3}{2} \\ 0 & 0 & 0 & 1 \\ 0 & 0 & 1 & 0 \end{pmatrix}$ are non-singular matrices, meaning that determinant value of P and Q is non-zero.

Method 3: Echelon form

This is generally the simplest way and hence the default method of finding rank of a matrix.

- Here ONLY row operations are permissible.
- We reduce the given matrix A to upper triangular matrix form
- And ρ(A) = no. of non-zero rows.

Example:

$$\text{Let } A = \begin{pmatrix} 1 & 2 & 3 & -1 \\ -2 & -1 & -3 & -1 \\ 1 & 0 & 1 & 1 \\ 0 & 1 & 1 & -1 \end{pmatrix}$$

First step is to convert all elements of column 1 other than a_{11} to zero. This can be achieved by $R_2 \to R_2 + 2R_1$ and $R_3 \to R_3 - R_1$ leading to

$$A = \begin{pmatrix} 1 & 2 & 3 & -1 \\ 0 & 3 & 3 & -3 \\ 0 & -2 & -2 & 2 \\ 0 & 1 & 1 & -1 \end{pmatrix}$$

Now we should get zero at a_{23} and a_{24}. We can achieve this by $R_3 \to R_3 + \frac{2}{3}R_2$ and $R_4 \to R_4 - \frac{1}{3}R_2$ leading to

$$A = \begin{pmatrix} 1 & 2 & 3 & -1 \\ 0 & 3 & 3 & -3 \\ 0 & 0 & 0 & 0 \\ 0 & 0 & 0 & 0 \end{pmatrix}.$$

This is the desired upper triangular form leading to $\rho(A) = 2$

Method 4: By definition of rank

We have seen the definition of Rank of a matrix at 13.3.1 above.

Let the given matrix be $A = \begin{pmatrix} 2 & 4 & 1 \\ 3 & 6 & 2 \\ 4 & 8 & 3 \end{pmatrix}$

Here, value of determinant A

$$|A| = \begin{pmatrix} 2 & 4 & 1 \\ 3 & 6 & 2 \\ 4 & 8 & 3 \end{pmatrix} 2 \times (18 - 16) - 4(9 - 8) + 1(24 - 24) = 4 - 4 + 0 = 0$$

Hence we have to look for determinant values of submatrices of size 2×2. Here again if we see submatrices of size 2×2 of columns 1 and 2, we find determinant values of both sub-matrices to be 0 (12-12) and (24-24).

However, both submatrices of size 2×2 of columns 2 and 3 have non zero determinant values as (8–6 = 2) & (18–16 = 2).

Hence rank of the given matrix = 2.

It may be noticed that this method is fairly complicated especially if size of the matrix is larger.

13.4.1.3.2 Important properties of Rank of a Matrix

1. The rank of a null matrix is zero ie, $\rho(0) = 0$
2. If I_n is an identity matrix of order n, then $\rho(I_n) = n$.
3. If at least one minor of order r of the matrix is not equal to zero, then $\rho(A) = r$.
4. If every $(r+1)^{th}$ order minor of A is zero, then any higher order minor will also be zero.
5. If A is of order n, then for a non-singular matrix A, $\rho(A) = n$
6. Elementary operations do not change the rank of a matrix.
7. $\rho(A^T) = \rho(A)$
8. $\rho(A^*) = \rho(A)$
9. $\rho(A + B) \leq \rho(A) + \rho(B)$

10. If A and B are two matrices such that the product AB is defined, then rank (AB)cannot exceed the rank of the either matrix.
11. If A and B are square matrices of same order and ρ(A) = ρ(B) = n, then (AB)= n

13.5 Some Special Types of Matrices

13.5.1 Symmetric and Skew-Symmetric Matrices

1. A square matrix $A = [a_{ij}]_{n\times n}$, is said to be symmetric, if transpose of A, $A^T = A$ i.e., $a_{ij} = a_{ji}$, for all i and j.
2. A square matrix A is said to be skew-symmetric matrix, if $a_{ij} = - a_{ji}$. It also implies that for a skew symmetric matrix A, $A^T = -A$.

13.5.1.1 Properties of Symmetric and Skew-Symmetric Matrices

1. Elements of principal diagonals of a skew-symmetric matrix are all zero, because $a_{ii} = -a_{ii} \Rightarrow 2a_{ii} = 0$ or $a_{ii} = 0$, for all values of i.
2. If A is a square matrix, then
   (a) A + A^T is symmetric.
   (b) A - A^T is skew-symmetric matrix.

   We can use a simple 2×2 square matrix

   $$A = \begin{pmatrix} a & b \\ c & d \end{pmatrix} \Rightarrow A^T = \begin{pmatrix} a & c \\ b & d \end{pmatrix}$$

   So $A + A^T = \begin{pmatrix} 2a & b+c \\ b+c & 2d \end{pmatrix}$, ***which is a symmetric matrix.***

   Similarly $A - A^T = \begin{pmatrix} 0 & b-c \\ c-b & 0 \end{pmatrix} \Rightarrow \begin{pmatrix} 0 & (b-c) \\ -(b-c) & 0 \end{pmatrix}$, ***which is clearly a Skew symmetric matrix.***

   Note: Readers are encouraged to try and verify this property for any two 3×3 matrices.
3. If A and B are two symmetric (or skew-symmetric) matrices of the same order, then A + B is also symmetric (or skew-symmetric).

Let us see an example with two symmetric matrices

$$A = \begin{pmatrix} a & b & c \\ b & d & e \\ c & e & f \end{pmatrix} \text{ and } B = \begin{pmatrix} g & h & i \\ h & j & k \\ i & k & l \end{pmatrix}, \text{ leading to}$$

$$A + B = \begin{pmatrix} a+g & b+h & c+i \\ b+h & d+j & e+k \\ c+i & e+k & f+l \end{pmatrix}$$ ***which is clearly a symmetric matrix***.

Similarly, let us see an example with two skew symmetric matrices

$$A = \begin{pmatrix} 0 & b & c \\ -b & 0 & d \\ -c & -d & 0 \end{pmatrix} \text{ and } B = \begin{pmatrix} 0 & e & f \\ -e & 0 & g \\ -f & -g & 0 \end{pmatrix}, \text{ leading to}$$

$$A + B = \begin{pmatrix} 0 & b+e & c+f \\ -(b+e) & 0 & d+g \\ -(c+f) & -(d+g) & 0 \end{pmatrix}$$ ***which is clearly a skew symmetric matrices***.

4. If A is symmetric (or skew-symmetric), then kA (k is a scalar) is also symmetricor skew-symmetric matrix respectively.
5. If A and B are symmetric matrices of the same order, then AB+ BA is also symmetric and AB – BA is skew-symmetric.

Proof:

Since **A** and **B** are symmetric, $A=A^T$ and $B= B^T$

Considering generic matrices of example 3,

$$AB = \begin{pmatrix} a & b & c \\ b & d & e \\ c & e & d \end{pmatrix} \times \begin{pmatrix} g & h & i \\ h & j & k \\ i & k & j \end{pmatrix}$$

$$= \begin{pmatrix} ag+bh+ci & ah+bj+ck & ai+bk+cj \\ bg+dh+ei & bh+dj+ek & bi+dk+ej \\ cg+eh+di & ch+ej+dk & ci+ek+dj \end{pmatrix}$$

$$BA = \begin{pmatrix} g & h & i \\ h & j & k \\ i & k & j \end{pmatrix} \times \begin{pmatrix} a & b & c \\ b & d & e \\ c & e & d \end{pmatrix} =$$

$$\begin{pmatrix} ag+bh+ci & bg+dh+ei & cg+eh+di \\ ah+bj+ck & bh+dj+ek & ch+ej+dk \\ ai+bk+cj & bi+dk+ej & ci+ek+dj \end{pmatrix}$$

$$\text{So, } (AB)^T = \begin{pmatrix} ag+bh+ci & bg+dh+ei & cg+eh+di \\ ah+bj+ck & bh+dj+ek & ch+ej+dk \\ ai+bk+cj & bi+dk+ej & ci+ek+dj \end{pmatrix} \text{ and}$$

$$(BA)^T = \begin{pmatrix} ag+bh+ci & ah+bj+ck & ai+bk+cj \\ bg+dh+ei & bh+dj+ek & bi+dk+ej \\ cg+eh+di & ch+ej+dk & ci+ek+dj \end{pmatrix}$$

$$(AB+BA) = \begin{pmatrix} 2(ag+bh+ci) & ah+b(g+j)+ck+dh+ei & ai+bk+c(g+j)+di+eh \\ ah+b(g+j)+ck+dh+ei & 2(bh+dj+ek) & bi+ch+2dk+2ej \\ ai+bk+c(g+j)+di+eh & bi+ch+2dk+2ej & 2(ci+ek+dj) \end{pmatrix}$$

It may be noticed that (AB + BA) is symmetric. Hence (AB + BA)=(AB + BA)^T

Similarly, (AB)^T + (BA)^T

$$= \begin{pmatrix} 2(ag+bh+ci) & ah+b(g+j)+ck+dh+ei & ai+bk+c(g+j)+di+eh \\ ah+b(g+j)+ck+dh+ei & 2(bh+dj+ek) & bi+ch+2dk+2ej \\ ai+bk+c(g+j)+di+eh & bi+ch+2dk+2ej & 2(ci+ek+dj) \end{pmatrix}$$

which is also symmetric and same as (AB + BA) and (AB + BA)^T

We can see the same conclusion by the following:

(AB+BA)^T = (AB)^T + (BA)^T ……(Property 2 of Transpose matrix)

Again, we know that $(AB)^T = B^TA^T$ and $(BA)^T = A^TB^T$.

Thus, $(AB+BA)^T = B^TA^T + A^TB^T$

But we have $A = A^T$ and $B = B^T$ as A and B are symmetric.

So, $(AB+BA)^T = BA + AB = AB + BA$.

Since $(AB+BA)^T = AB + BA$, $(AB+BA)$ is a Symmetric Matrix.

Similarly, $(AB-BA)^T = (AB)^T - (BA)^T = B^TA^T - A^TB^T = BA - AB = -(AB - BA)$

Since $(AB - BA)^T = -(AB - BA)$, $(AB-BA)$ is a Skew-symmetric Matrix.

6. Every square matrix can be expressed uniquely as the sum of a symmetric and a skew-symmetric matrix.

   Let us assume that an arbitrary square matrix is A and there are two other matrices P (a symmetric matrix) and Q (a skew symmetric matrix). We need to prove that A can be expressed as P + Q.

   Let us start with the assumption $A = P + Q$

   This would mean that $A^T = (P + Q)^T = P^T + Q^T$ (from properties of transpose)

   We know that $P^T = P$ (being symmetric matrix) and $Q^T = -Q$ (being skew-symmetric matrix). Hence $A^T = P - Q$

   From the above values of A and A^T,

   we get $\frac{1}{2}\left(A + A^T\right) = \frac{1}{2}(P + Q + P - Q) = P$ ..... (1)

   Similarly, we get $\frac{1}{2}\left(A - A^T\right) = \frac{1}{2}(P + Q - P + Q) = Q$ ....(2)

   Adding (1) and (2) we get back $A = P + Q$

   Hence it is proved that any square matrix A can be expressed as sum of a symmetric matrix P and a skew-symmetric matrix Q.

7. The matrix B^TAB is symmetric or skew-symmetric according as A is symmetric or skew-symmetric matrix.

   Let us take the case of symmetric matrix A.

   A being symmetric, $A^T = A$.

   Let us consider transpose of (B^TAB). Let $AB = P$

   Then $(B^TAB)^T = (B^TP)^T = P^T(B^T)^T = P^TB$ (Since $(AB)^T = B^TA^T$)

Therefore, $(B^TAB)^T = P^TB = (AB)^TB = B^TA^TB = B^TAB$ (since $A^T=A$)

Since $(B^TAB)^T = B^TAB$, Hence B^TAB is a symmetric matrix.

Now if A is skew symmetric, $A^T = -A$.

Again let AB = P,

Then $(B^TAB)^T = (B^TP)^T = P^T(B^T)^T = P^TB$ (Since $(AB)^T = B^TA^T$)

Therefore, $(B^TAB)^T = P^TB = (AB)^TB = B^TA^TB = - B^TAB$ (since $A^T= -A$)

Since $(B^TAB)^T = - B^TAB$, Hence B^TAB is a skew symmetric matrix.

8. All positive integral powers of a symmetric matrix are symmetric.
9. All positive odd integral powers of a skew-symmetric matrix are skew-symmetric and positive even integral powers of a skew-symmetric are symmetric matrix.

   This is so because odd powers of negative numbers are negative, hence character of skew symmetric matrix remains unchanged. On the other hand, even powers of negative numbers are positive, thus converting skew symmetric matrix into a symmetric matrix
10. For a square matrix A, AA^T and A^TA are symmetric matrix.

    Let us find transpose of AA^T.

    $(AA^T)^T = (A^T)^TA^T = AA^T$.......(because $(AB)^T = B^TA^T$)

    Since transpose of AA^T is same as AA^T, AA^T is a symmetric matrix

    Similarly, $(A^TA)^T = A^T(A^T)^T = A^TA$, hence A^TA is also a symmetric matrix
11. Every skew-symmetric matrix of odd order has rank less than its order.

    Let us consider a generic skew symmetric matrix of order 3 below

$$A = \begin{pmatrix} 0 & b & c \\ -b & 0 & d \\ -c & -d & 0 \end{pmatrix}$$

We can see that $|A| = (-b)(cd) + c(bd) = -bcd + bcd = 0$

Thus every skew symmetric matrix A of odd order is singular. Hence its rank is less than its order.

13.5.2 Orthogonal Matrix

When the product of a square matrix A and its transpose A^T gives identity matrix, then the matrix is said to be an orthogonal matrix. In case of a square matrix A of order n, $AA^T = I_n = A^TA$

Let us consider a generic square matrix for clarity.

$$\text{Let } A = \begin{pmatrix} a & b \\ c & d \end{pmatrix} \Rightarrow A^T = \begin{pmatrix} a & c \\ b & d \end{pmatrix}$$

$$\therefore AA^T = \begin{pmatrix} a^2+b^2 & ac+bd \\ ac+bd & c^2+d^2 \end{pmatrix} \text{ And } I_n = \begin{pmatrix} 1 & 0 \\ 0 & 1 \end{pmatrix}$$

Leading to the conclusion that for A to be orthogonal,

$a^2+b^2 = c^2+d^2 = 1 \; and \; ac+bd = 0 \Rightarrow \frac{a}{b} = -\frac{d}{c}$ and a, b, c, d are all <1

There are many possibilities.

One simple possibility is

$$a = \cos x,\ b = \sin x,\ c = \sin x \ \&\ d = -\cos x \Rightarrow$$

$$A = \begin{pmatrix} \cos x & \sin x \\ \sin x & -\cos x \end{pmatrix} \Rightarrow A^T = \begin{pmatrix} \cos x & \sin x \\ \sin x & -\cos x \end{pmatrix} \Rightarrow$$

$$AA^T = \begin{pmatrix} \cos^2 x + \sin^2 x & \cos x \sin x - \sin x \cos x \\ \sin x \cos x - \cos x \sin x & \sin^2 x + \cos^2 x \end{pmatrix} = \begin{pmatrix} 1 & 0 \\ 0 & 1 \end{pmatrix}$$

Similarly,

$$A^T A = \begin{pmatrix} \cos^2 x + \sin^2 x & \cos x \sin x - \sin x \cos x \\ \sin x \cos x - \cos x \sin x & \sin^2 x + \cos^2 x \end{pmatrix} = \begin{pmatrix} 1 & 0 \\ 0 & 1 \end{pmatrix}$$

13.5.2.1 Properties of Orthogonal Matrix

(i) If A is orthogonal matrix, then A^T is also orthogonal matrix.

(ii) For any two orthogonal matrices A and B, AB as well as BA is also an orthogonal matrix.

(iii) If A is an orthogonal matrix, its inverse A^{-1} is also orthogonal matrix. (We will learn shortly about inverse of a matrix)

13.5.3 Idempotent Matrix

A square matrix A is said to be idempotent, if $A^2 = A$.

Considering the same example of matrix

$$A = \begin{pmatrix} a & b \\ c & d \end{pmatrix} \Rightarrow A^2 = \begin{pmatrix} a^2 + bc & ab + bd \\ ac + cd & bc + d^2 \end{pmatrix}$$

Since $A^2 = A$, we get $a = a^2 + bc$,

$b = ab + bd \Rightarrow b(1 - a - d) = 0$ yielding either $b = 0$ or $d = 1 - a$ ....(1)

$c = ac + cd \Rightarrow c(1 - a - d) = 0$ yielding either $c = 0$ or $d = 1 - a$ ...(2)

And $d = bc + d^2 \Rightarrow bc = d(1 - d)$ ...(3)

Let us assume that b and c are not zero. From (1) and (2) it will imply that $d = 1 - a$.

Substituting this value of d in eq. (3) above, we get $bc = d\left(1 - \left(1 - a\right)\right) = ad$

Hence for A to be Idempotent, we get bc = ad and b ≠ 0 and c ≠ 0.

Example: Let us see an example with real numbers.

Let $a = 3 \Rightarrow d = 1 - 3 = -2$, leading further to $bc = 3 \times -2 = -6$

Any possible combinations of b & c like (-3, 2), (-6. 1) etc. satisfying the above equality will work

$b = -6$ & $c = 1$ gives us $A = \begin{pmatrix} 3 & -6 \\ 1 & -2 \end{pmatrix}$

Let us calculate A^2

$$\Rightarrow A^2 = \begin{pmatrix} 3 \times 3 + 1 \times -6 & 3 \times -6 + -6 \times -2 \\ 1 \times 3 + -2 \times 1 & 1 \times -6 + -2 \times -2 \end{pmatrix} = \begin{pmatrix} 3 & -6 \\ 1 & -2 \end{pmatrix} = A$$

Hence A is Idempotent.

Similarly $A = \begin{pmatrix} 3 & -3 \\ 2 & -2 \end{pmatrix}$ *with other set of values for b and c* will be Idempotent.

As already learnt that determinant value of matrix A denoted as

$$|A| = \begin{vmatrix} a & b \\ c & d \end{vmatrix} = ad - bc$$

(diagonal multiplication of top left to tight bottom - diagonal multiplication of top right and left bottom)

We have seen above that for a generic 2×2 idempotent matrix $ad = bc$

This implies that for a generic 2×2 idempotent matrix $|A| = 0$. However, there is an exception to this property listed at 1 below.

13.5.3.1 Other properties of Idempotent Matrix

1. Except for Identity matrix, every Idempotent matrix is singular i.e. it is a squarematrix of any size whose determinant is 0.

   We have already seen this property for generic 2×2 idempotent matrix.

   We can now check whether a 2×2 identity matrix $A = \begin{pmatrix} 1 & 0 \\ 0 & 1 \end{pmatrix}$ is idempotent and whether $|A| = 0$.

   For $A = \begin{pmatrix} 1 & 0 \\ 0 & 1 \end{pmatrix}$, $A^2 = \begin{pmatrix} 1\times1+0\times0 & 1\times0+0\times1 \\ 0\times1+1\times0 & 0\times0+1\times1 \end{pmatrix} = \begin{pmatrix} 1 & 0 \\ 0 & 1 \end{pmatrix} = A$

   Hence Identity matrix A is idempotent.

   However, in this case $|A| = 1\times1 - 0\times0 = 1 \neq 0$

   Hence the above property is proved

2. If A and B are two idempotent matrices, let C be a linear combination of the form $C = k_1A + k_2B$, with non-zero scalars k_1 and k_2, then

   I. C is idempotent and AB = BA if

   i. C = A + B and AB = 0.

   ii. C = B - A and AB = A

   iii. C = A - B and AB = B And

   II. C is idempotent and AB ≠ BA if

   $k_1 \in \backslash\{0,\ 1 \mid k_1 \neq 1\}$, $k_2 = 1 - k_1$ & $(A - B)^2 = 0$

Proof for set I:

For C to be idempotent, C^2 should be equal to C. This leads to

$$k_1A + k_2B = (k_1A + k_2B)(k_1A + k_2B) \Rightarrow$$

$$k_1A + k_2B = k_1^2A^2 + k_1A \times k_2B + k_2B \times k_1A + k_2^2B^2 \Rightarrow$$

$$k_1A + k_2B = k_1^2A + k_1k_2AB + k_2k_1BA + k_2^2B \Rightarrow \quad (\because A^2 = A \,\&\, B^2 = B)$$

$$k_1(1-k_1)A - k_1k_2AB - k_2k_1BA + k_2(1-k_2)B = 0 \;........(1)$$

Now we have two possibilities as stated above i.e. ***AB = BA or AB ≠ BA***

For set I, AB = BA and we need to prove 3 cases as mentioned above:

Let us multiply eq (1) above, first by A and then by B to get the following:

$$k_1(1-k_1)A + k_2(1-2k_1-k_2)AB = 0......(2)\; (\because A^2 = A)$$

$$k_2(1-k_2)B + k_1(1-2k_2-k_1)AB = 0......(3)\; (\because B^2 = B)$$

Post multiplying (2) by B or pre-multiplying (3) by A yields

$$\left[k_1 + k_2 - (k_1 + k_2)^2\right]AB = 0......(4)$$

Case 1:

With $\boldsymbol{AB=0}$, and knowing that $A \neq 0,\ B \neq 0,\ k_1 \neq 0\ \&\ k_2 \neq 0$ eq (2) and (3) yield $k_1 = k_2 = 1$ implying that $\boldsymbol{C = A + B}$. ***This is our case 1***

Case 2 and 3:

For ***AB ≠ 0, equation (4) yields***

$(k_1 + k_2)\left[1 - (k_1 + k_2)\right] = 0$ leading to

$$k_1 + k_2 = 1 \;........(5)\quad or\quad k_1 + k_2 = 0.........(6)$$

When substituting values at (5), eq (2) and (3) yield respectively

$$k_1k_2A + k_2(-k_1)AB = 0 \;\; \& \;\; k_1k_2B + k_1(-k_2)AB = 0 \Rightarrow$$

$$k_1k_2(A - AB) = 0 \text{ and } k_1k_2(B - AB) = 0 \;..........(7)$$

When substituting values at (6), eq (2) and (3) yield

$$k_1(1-k_1)(A - AB) = 0 \text{ and } k_1(1+k_1)(B - AB) = 0 \;....(8)$$

In view of the assumption that $A \neq 0$ & $B \neq 0$, the equalities AB = A & AB = B cannot hold simultaneously. Consequently, since $k_1 \neq 0$ & $k_2 \neq 0$, it follows that the pair of conditions (7) can never be fulfilled, whereas the pair of conditions (8) for $k_1 + k_2 = 0$ is fulfilled if and only if either ***$k_1 = -1$ (implying C = B-A)*** and ***AB=A***, which is the ***Case 2*** or ***$k_1 = 1$ (implying C = A-B)*** and ***AB=B, which is the case 3***

Proof for set II

The given conditions are ***AB ≠ BA** with*

$$k_1 \in \backslash\{0,\ 1 \mid k_1 \neq 1\},\ k_2 = 1 - k_1$$

We again refer to the equation (1) reproduced below:

$$k_1(1-k_1)A - k_1k_2AB - k_2k_1BA + k_2(1-k_2)B = 0 \;........(1)$$

Pre-multiplying and post-multiplying it by A leads, respectively, to

$$k_1(1-k_1)A + k_2(1-k_2-k_1)AB - k_1k_2ABA = 0 \;.......(9)$$

$$\left(\because A^2 = A \;\&\; A^2B = AB\right)$$

$$k_1(1-k_1)A + k_2(1-k_2-k_1)BA - k_1k_2ABA = 0 \;.......(10)$$

$$\left(\because A^2 = A \;\&\; BA^2 = BA\right)$$

Subtracting (10) from (9), $k_2(1-k_2-k_1)(AB - BA) = 0$ and since ***AB ≠ BA and*** $k_2 \neq 0$, it leads to the same condition as (5) above

i.e $k_1 + k_2 = 1$

Substituting (5) in (1) simplifies the latter to

$$k_1k_2\left(A-AB-BA+B\right)=0$$

Which in view of $k_1k_2 \neq 0$ yields $\left(A-AB-BA+B\right)=0$

Since A and B are idempotent, $A=A^2$ & $B=B^2$ that changes the above condition to

$$\left(A^2-AB-BA+B^2\right)=0 \Rightarrow \left\{A\left(A-B\right)-B(A-B\right\} \Rightarrow \left(A-B\right)^2=0$$

Hence proved that ***C is idempotent when AB ≠ BA if*** $k_1 \in \backslash\{0,\ 1 \mid k_1 \neq 1\}$, $k_2=1-k_1$ $\&\left(A-B\right)^2=0$

3. ***Diagonal matrix (1, 1, 1, ...,1) is an idempotent matrix.***

   Diagonal matrix is nothing but an Identity Matrix I.

   Let us verify I^2 for a 2×2 matrix.

$$I=\begin{pmatrix}1 & 0\\ 0 & 1\end{pmatrix} \Rightarrow I^2=\begin{pmatrix}1\times1+0\times0 & 1\times0+0\times1\\ 0\times1+0\times1 & 0\times0+1\times1\end{pmatrix}=\begin{pmatrix}1 & 0\\ 0 & 1\end{pmatrix}=I$$

   Thus I is idempotent. This can be verified for larger I matrices as well.

4. If A is an idempotent matrix and A + B = I, then B is also idempotent and

   AB = BA= 0

   Proof:

   We know that an Identity matrix I is idempotent.

   A + B = I leads to B = I – A

   Squaring this equation, we get

   B^2 = (I-A) × (I-A) = I – A –A +A^2.

   Since A^2 = A,

   B^2 = $(I-A)^2$ = I – A –A +A = I – A = B

   Hence B is Idempotent.

   Since A, B and I are idempotent, it follows from case I that AB = BA= 0.

5. ***If*** l_1, l_2 and l_3 are direction cosines of a 3-d vector, then

$$\Delta = \begin{pmatrix} l_1^2 & l_1 l_2 & l_1 l_3 \\ l_1 l_2 & l_2^2 & l_2 l_3 \\ l_3 l_1 & l_3 l_2 & l_3^2 \end{pmatrix}$$ ***is an idempotent matrix and*** $|\Delta| = 0$

(Readers should refer to chapter 14 before going through this property)

Proof:

If l_1, l_2 and l_3 are direction cosines of a 3-d vector, then we know that

$$l_1^2 + l_2^2 + l_3^2 = 1$$

Now if we multiply matrix Δ by Δ, we get

$\Delta^2 =$

$$\begin{pmatrix} l_1^4 + l_1^2 l_2^2 + l_1^2 l_3^2 & l_1^3 l_2 + l_1 l_2^3 + l_1 l_2 l_3^2 & l_1^3 l_3 + l_1 l_2^2 l_3 + l_1 l_3^3 \\ l_1^3 l_2 + l_1 l_2^3 + l_1 l_2 l_3^2 & l_1^2 l_2^2 + l_2^4 + l_2^2 l_3^2 & l_1^2 l_2 l_3 + l_2^3 l_3 + l_2 l_3^3 \\ l_1^3 l_3 + l_1 l_2^2 l_3 + l_1 l_3^3 & l_1^2 l_2 l_3 + l_2^3 l_3 + l_2 l_3^3 & l_1^2 l_3^2 + l_2^2 l_3^2 + l_3^4 \end{pmatrix} \Rightarrow$$

$\Delta^2 =$

$$\begin{pmatrix} l_1^2 \left(l_1^2 + l_2^2 + l_3^2\right) & l_1 l_2 \left(l_1^2 + l_2^2 + l_3^2\right) & l_1 l_3 \left(l_1^2 + l_2^2 + l_3^2\right) \\ l_1 l_2 \left(l_1^2 + l_2^2 + l_3^2\right) & l_2^2 \left(l_1^2 + l_2^2 + l_3^2\right) & l_2 l_3 \left(l_1^2 + l_2^2 + l_3^2\right) \\ l_1 l_3 \left(l_1^2 + l_2^2 + l_3^2\right) & l_2 l_3 \left(l_1^2 + l_2^2 + l_3^2\right) & l_3^2 \left(l_1^2 + l_2^2 + l_3^2\right) \end{pmatrix} =$$

$$= \begin{pmatrix} l_1^2 & l_1 l_2 & l_1 l_3 \\ l_1 l_2 & l_2^2 & l_2 l_3 \\ l_3 l_1 & l_3 l_2 & l_3^2 \end{pmatrix} = \Delta \text{ because } l_1^2 + l_2^2 + l_3^2 = 1$$

Hence it is proved that $\Delta^2 = \Delta$ proving thereby that Δ is idempotent matrix leading to a further conclusion that determinant of Δ i.e. $|\Delta| = 0$ (reference generic property of Idempotent matrices listed above at 13.5.3.1. this can also be verified by calculating determinant value of the matrix with parameters given above.

6. A square matrix A is said to be involutory, if $A^2 = I$

13.5.4 Nilpotent Matrix

A square matrix $A_{n\times n}$ which vanishes when raised to some positive integral power m is called Nilpotent matrix, where $m \le n$ i.e. $A^m = 0$. If m is the least positive integer such that $A^m = 0$, then m is called the index of the Nilpotent matrix A.

Let us see a couple of real life examples for clarity:

Example 1: Square of a 2×2 matrix = 0

$$A = \begin{pmatrix} ab & b^2 \\ -a^2 & -ab \end{pmatrix} \Rightarrow A^2 = \begin{pmatrix} a^2b^2 - a^2b^2 & ab^3 - ab^3 \\ -a^3b + a^3b & -a^2b^2 + a^2b^2 \end{pmatrix} = \begin{pmatrix} 0 & 0 \\ 0 & 0 \end{pmatrix} = 0$$

Hence A is a Nilpotent matrix

Example 2: Cube of a 3×3 matrix = 0.

$$\text{Let } B = \begin{pmatrix} 0 & 1 & 0 \\ 0 & 0 & 1 \\ 0 & 0 & 0 \end{pmatrix}$$

$$\Rightarrow B^2 = \begin{pmatrix} 0 & 1 & 0 \\ 0 & 0 & 1 \\ 0 & 0 & 0 \end{pmatrix} \times \begin{pmatrix} 0 & 1 & 0 \\ 0 & 0 & 1 \\ 0 & 0 & 0 \end{pmatrix} =$$

$$\begin{pmatrix} 0+0+0 & 0+0+0 & 0+1+0 \\ 0+0+0 & 0+0+0 & 0+0+0 \\ 0+0+0 & 0+0+0 & 0+0+0 \end{pmatrix} = \begin{pmatrix} 0 & 0 & 1 \\ 0 & 0 & 0 \\ 0 & 0 & 0 \end{pmatrix} \Rightarrow$$

$$B^3 = \begin{pmatrix} 0 & 1 & 0 \\ 0 & 0 & 1 \\ 0 & 0 & 0 \end{pmatrix} \times \begin{pmatrix} 0 & 0 & 1 \\ 0 & 0 & 0 \\ 0 & 0 & 0 \end{pmatrix} = \begin{pmatrix} 0+0+0 & 0+0+0 & 0+0+0 \\ 0+0+0 & 0+0+0 & 0+0+0 \\ 0+0+0 & 0+0+0 & 0+0+0 \end{pmatrix} = 0$$

Hence B is a Nilpotent matrix.

13.5.5 Unitary Matrix

A square matrix A is said to be unitary, if its Inverse is equal to its Conjugate Transpose i.e. $A^{-1} = A^T$ and their product $A^T \times A^{-1} = I$ Unitary matrix is same as Orthogonal Matrix discussed at 13.4.2.1 above.

13.5.6 Hermitian Matrix

A Hermitian matrix (or **self-adjoint matrix**) is a complex **square matrix** that is equal to its own conjugate transpose i.e. the element in the i[th] row and j[th] column is equal to the complex conjugate of the element in the j[th] row and i[th]column, for all indices i and j. i.e. $a_{ij} = \overline{a_{ji}}$.

It implies that the diagonal elements of a Hermitian matrix must be real because, they have to be their own conjugate.

Examples of Hermitian matrices are as below:

$$\begin{pmatrix} -1 & 1-2i & 0 \\ 1+2i & 0 & -i \\ 0 & i & 1 \end{pmatrix} \text{ and } \begin{pmatrix} -1 & 1+i & 2i \\ 1-i & 5 & -3 \\ -2i & -3 & 0 \end{pmatrix}$$

13.5.6.1 Properties of Hermitian Matrix

1. If A is Hermitian matrix, then kA is also Hermitian matrix for any non-zero real number k.
2. If A and B are Hermitian matrices of same order, then $\lambda_1 A + \lambda B$ is also Hermitian for any non-zero real numbers λ_1, and λ.
3. If A is any square matrix, then AA* and A*A are also Hermitian. This is so because by the very definition of Hermitian Matrix, A = A*. Hence AA* = A*A=A^2
4. If A and B are Hermitian, then AB is also Hermitian, if AB = BA
5. If A is a Hermitian matrix, then A* is also Hermitian.
6. If A and B are Hermitian matrix of same order, then AB+BA is also Hermitian.
7. If A is a square matrix, then A + A* is also Hermitian,
8. Any square matrix can be uniquely expressed as A + iB, where A and B are Hermitian matrices.

13.5.7 Skew-Hermitian Matrix

A square matrix A is said to be Skew-Hermitian if it is equal to negative of its own Conjugate Transpose i.e. A= – A′ or $a_{ij} = -\overline{a_{ji}}$ for every i & j.

Here is an example of skew Hermitian Matrix $A=\begin{pmatrix} 2i & 3 & i \\ -3 & 0 & -2-i \\ i & 2-i & i \end{pmatrix}$

So, Conjugate of A,

$$\overline{A}=\begin{pmatrix} -2i & 3 & -i \\ -3 & 0 & -2+i \\ -i & 2+i & -i \end{pmatrix} \Rightarrow A^{*}=\left(\overline{A}\right)^{T}=\begin{pmatrix} -2i & -3 & -i \\ 3 & 0 & 2+i \\ -i & -2+i & -i \end{pmatrix}$$

We can thus see that $-A^{*}=\begin{pmatrix} 2i & 3 & i \\ -3 & 0 & -2-i \\ i & 2-i & i \end{pmatrix}=A$.

Hence A is a Skew Hermitian Matrix.

13.5.7.1 Properties of Skew-Hermitian Matrix

1. If A is Skew-Hermitian matrix, then kA is Skew-Hermitian matrix, where k is anynon-zero real number.
2. If A and B are Skew-Hermitian matrix of same order, then $\lambda_1 A + \lambda_2 B$ is also Skew-Hermitian for any real number λ_1 and λ_2.
3. If A and B are Hermitian matrices of same order, then AB–BA is Skew-Hermitian.
4. If A is any square matrix, then A - A^* is a Skew-Hermitian matrix.
5. Every square matrix can be uniquely expressed as the sum of a Hermitian and A Skew-Hermitian matrices.
6. If A is a Skew-Hermitian matrix, then A is a Hermitian matrix.

13.5.8 Echelon Form of a Matrix

A non-zero matrix A is said to be in Echelon form, if A satisfies the following conditions

1. All the non-zero rows of A, if any precede the zero rows.
2. The number of zeros preceding the first non-zero element in a row is less than the number of such zeros in the successive row.
3. The first non-zero element in a row is unity.

4. The number of non-zero rows of a matrix given in the Echelon form is its rank

   Note: ***The Echelon form of matrix also has a Reduced Echelon form where all diagonal elements are 1 and all other elements are zero. In other words, reduced Echelon form is nothing but an identity matrix of given order. This form is used in solution for simultaneous linear equations.***

13.6 Properties of determinants

Even though these properties are valid for determinants of matrices of order n, we will look at their proofs for determinants of order 3.

1. ***The value of a determinant is equal to sum of the product of each element of ANY ROW OR COLUMN and its respective cofactor***

   Cofactor is defined at section 13.3.11, as noted there, value of determinant with elements of row 1 is

$$(-1)^{1+1} \times a_{11} \times \begin{vmatrix} a_{22} & a_{23} \\ a_{32} & a_{33} \end{vmatrix} + (-1)^{1+2} \times a_{12} \times \begin{vmatrix} a_{21} & a_{23} \\ a_{31} & a_{33} \end{vmatrix} +$$

$$(-1)^{1+3} \times a_{13} \times \begin{vmatrix} a_{21} & a_{22} \\ a_{31} & a_{32} \end{vmatrix}$$

$$= a_{11}(a_{22}a_{33} - a_{32}a_{23}) - a_{12}(a_{21}a_{33} - a_{31}a_{23}) + a_{13}(a_{21}a_{32} - a_{31}a_{22})$$

Or $= a_{11}a_{22}a_{33} - a_{11}a_{32}a_{23} - a_{12}a_{21}a_{33} + a_{12}a_{31}a_{23} + a_{13}a_{21}a_{32} - a_{13}a_{31}a_{22}$

Now considering elements of column 2, we get value of det as

$$(-1)^{1+2} \times a_{12} \times \begin{vmatrix} a_{21} & a_{23} \\ a_{31} & a_{33} \end{vmatrix} + (-1)^{2+2} \times a_{22} \times \begin{vmatrix} a_{11} & a_{13} \\ a_{31} & a_{33} \end{vmatrix} +$$

$$(-1)^{2+3} \times a_{32} \times \begin{vmatrix} a_{11} & a_{13} \\ a_{21} & a_{23} \end{vmatrix}$$

$$= -a_{12}(a_{21}a_{33} - a_{31}a_{23}) + a_{22}(a_{11}a_{33} - a_{31}a_{13}) - a_{32}(a_{11}a_{23} - a_{21}a_{13})$$

By rearranging, it equals

$$= a_{11}a_{22}a_{33} - a_{11}a_{32}a_{23} - a_{12}a_{21}a_{33} + a_{12}a_{31}a_{23} + a_{13}a_{21}a_{32} - a_{13}a_{31}a_{22}$$

This value is exactly same as the value obtained by considering elements of row 1.

Readers can try and work out value of the det considering elements of row 2 and 3 as well as column 1 and 3. Result will be the same.

This property is key to finding value of determinants of matrices of order n.

2. The value of a determinant remains unchanged if corresponding rows and columns are interchanged; i.e. det A = det A^T

$$\text{Let } \det A = \begin{vmatrix} a_{11} & a_{12} & a_{13} \\ a_{21} & a_{22} & a_{23} \\ a_{31} & a_{32} & a_{33} \end{vmatrix} \text{ and so } \det A^T = \begin{vmatrix} a_{11} & a_{21} & a_{31} \\ a_{12} & a_{22} & a_{32} \\ a_{13} & a_{23} & a_{33} \end{vmatrix}$$

We have already seen that

$$\det A = a_{11}a_{22}a_{33} + a_{12}a_{23}a_{31} + a_{13}a_{21}a_{32} - a_{31}a_{22}a_{13} - a_{32}a_{23}a_{11} - a_{33}a_{21}a_{12}$$

$$\text{Similarly } \det A^T = a_{11}a_{22}a_{33} + a_{21}a_{32}a_{13} + a_{31}a_{12}a_{23} - a_{13}a_{22}a_{31} - a_{23}a_{23}a_{11} - a_{33}a_{12}a_{21}$$

If we note carefully, the two determinant values are the same with some difference in placement of sub-items of each item.

Hence det A = det A^T

There is special type of determinant called ***skew symmetric determinant***. A determinant D is said to be skew-symmetric if $d_{ij} = -d_{ji}$. It implies that the transpose of a skew symmetric determinant D i.e. $D^T = -D$.

But we have seen above that for any determinant D, value of determinant $D^T = D$.

This leads us to the conclusion that for a skew symmetric determinant D = -D leading to $|D| = 0$

3. ***If any two rows of a determinant are interchanged, then sign of the value of determinant changes.***

$$\text{Let } \det A = \begin{vmatrix} a_{11} & a_{12} & a_{13} \\ a_{21} & a_{22} & a_{23} \\ a_{31} & a_{32} & a_{33} \end{vmatrix} \text{ and } \det B = \begin{vmatrix} a_{21} & a_{22} & a_{23} \\ a_{11} & a_{12} & a_{13} \\ a_{31} & a_{32} & a_{33} \end{vmatrix}$$

with row 1 and row 2 interchanged.

Value of det A noted in Property 1 above =

$$a_{11}a_{22}a_{33} + a_{12}a_{23}a_{31} + a_{13}a_{21}a_{32} - a_{31}a_{22}a_{13} - a_{32}a_{23}a_{11} - a_{33}a_{21}a_{12}$$

By diagonal method, we can see that value of det B =

$$a_{21}a_{12}a_{33} + a_{22}a_{13}a_{31} + a_{23}a_{11}a_{32} - a_{31}a_{12}a_{23} - a_{32}a_{13}a_{21} - a_{33}a_{11}a_{22}$$

On careful comparison, it can be seen that value of **det A = - value of det B**

Hence proved.

4. ***If every element of a row of a determinant is 0, then value of the determinant is zero.***

   We can notice from value of det A and B that the value is sum/difference of six components, each of which is product of three elements of the matrix that contains one element from each row. Hence if one of the rows has all zero elements that each component of the value of the det will be zero.

   Hence proved.

5. ***If every element of a row of a determinant is multiplied by a scalar factor k, then value of the determinant gets multiplied by k.***

   Here again by the same logic discussed in Property 4, since each component of value of the det has one element from each row, and elements of the row which has been multiplied by k will result in value of each component going k times the earlier value. As a result, value of the determinant will be k times its previous value.

6. ***If a determinant has any two rows (or columns) identical, then its value is zero.***

   Let $\det A = \begin{vmatrix} a_{11} & a_{12} & a_{13} \\ a_{11} & a_{12} & a_{13} \\ a_{31} & a_{32} & a_{33} \end{vmatrix}$ with rows 1 and 2 being identical.

   We have seen in Property 3 above that if any two rows of a determinant are interchanged, then sign of the value of determinant changes. Now in this case interchange of row 1 and 2 does not change the determinant as the two rows are identical. Since Value of det A $|A|$ should have gone to -$|A|$, it implies that Value of such a det A = 0

7. ***Value of a determinant remains unchanged if each element of a row (or a column) is increased by a scalar multiple of corresponding element of another row.***

$$\det A = \begin{vmatrix} a_{11} & a_{12} & a_{13} \\ a_{21} & a_{22} & a_{23} \\ a_{31} & a_{32} & a_{33} \end{vmatrix} \text{ and let}$$

$$\det A' = \begin{vmatrix} a_{11} & a_{12} & a_{13} \\ a_{21} + ka_{31} & a_{22} + ka_{32} & a_{23} + ka_{33} \\ a_{31} & a_{32} & a_{33} \end{vmatrix}$$

with row 2 increased by k times the corresponding elements of row 3.

Value of det A′ will have three additional positive elements as $ka_{11}a_{32}a_{33} + ka_{12}a_{33}a_{31} + ka_{13}a_{31}a_{32}$ and three additional negative elements as $-(ka_{31}a_{32}a_{13} + ka_{32}a_{33}a_{11} + ka_{33}a_{31}a_{12})$. On comparison, it will be seen that these two set of elements cancel each other. Hence value of the determinant remains unchanged.

Note: While applying this property atleast one row (or column) must remain unchanged

13.6.1 Solved examples of determinants.

1. Evaluate $A = \begin{vmatrix} 2 & -3 & 1 \\ 0 & 5 & 2 \\ -1 & -2 & 3 \end{vmatrix}$

Here it will be easiest to calculate value of the determinant by using column 1 as one of the elements is 0.

Value of the given det $|A| =$

$$(-1)^{1+1} \times 2 \times \begin{vmatrix} 5 & 2 \\ -2 & 3 \end{vmatrix} + (-1)^{3+1} \times (-1) \times \begin{vmatrix} -3 & 1 \\ 5 & 2 \end{vmatrix}$$

$$= 2 \times (5 \times 3 - (-2) \times 2) + (-1) \times ((-3) \times 2 - (5 \times 1)) =$$

$$2 \times (15 + 4) - (-6 - 5) = 38 + 11 = 49$$

2. Prove that value of the determinant of a matrix of order n is zero if two rows are identical.

We have seen in the Property 7 above that "*Value of a determinant remains unchanged if each element of a row is increased by a scalar multiple of corresponding element of another row*".

If the scalar multiple k is -1 and if each element of one of the two identical rows is multiplied by -1 and added to the other identical row, ***each element of that row will turn into zero.***

We have also seen in the Property 4 above that *If every element of a row of a determinant is 0, then value of the determinant is zero.* Hence it is proved that value of the determinant of a matrix of order n is zero if two rows are identical.

3. Prove that for any determinant of a matrix of order 3, the sum of the products of elements of any row and the cofactors of the corresponding elements of another row is zero.

$$\text{Let } \det A = \begin{vmatrix} a_{11} & a_{12} & a_{13} \\ a_{21} & a_{22} & a_{23} \\ a_{31} & a_{32} & a_{33} \end{vmatrix}$$

Cofactors C_{ij} of row 2 are

$$C_{21} = (-1)^{2+1} \times \begin{vmatrix} a_{12} & a_{13} \\ a_{32} & a_{33} \end{vmatrix},\ C_{22} = (-1)^{2+2} \times \begin{vmatrix} a_{11} & a_{13} \\ a_{31} & a_{33} \end{vmatrix},$$

$$C_{23} = (-1)^{2+3} \times \begin{vmatrix} a_{11} & a_{12} \\ a_{31} & a_{32} \end{vmatrix}$$

Multiplying elements of row 1 with respective cofactors of row 2, we get sum of these products say B as

$$B = a_{11} \times (-1) \times (a_{12}a_{33} - a_{32}a_{13}) + a_{12} \times (a_{11}a_{33} - a_{31}a_{13}) + a_{13} \times (-1) \times (a_{11}a_{32} - a_{31}a_{12}) \Rightarrow$$

$$B = a_{11}a_{32}a_{13} - a_{11}a_{12}a_{33} + a_{12}a_{11}a_{33} - a_{12}a_{31}a_{13} - a_{13}a_{11}a_{32} + a_{13}a_{31}a_{12}$$

We can see that items 1&5, 2&3 and 4&6 cancel out giving value of B = 0.

Note: This property is true for any determinant of a matrix of order n.

4. ***Prove that det AB = det A × det B for any two square matrices A and B of order two.***

Let $A = \begin{vmatrix} a & b \\ c & d \end{vmatrix}$ & $B = \begin{vmatrix} e & f \\ g & h \end{vmatrix} \Rightarrow AB = \begin{vmatrix} ae+bg & af+bh \\ ce+dg & cf+dh \end{vmatrix}$

So det A = $ad - bc$,

det B = $eh - gf$ and

det AB = $(ae+bg)(cf+dh)-(ce+dg)(af+bh)$

$= acef + bcfg + adeh + bdgh - acef - adfg - bceh - bdgh$

Items 1,5 and 4,8 cancel each other leaving det AB = $bcfg + adeh - adfg - bceh$

Now det A× det B =

$$(ad-bc)(eh-fg) = adeh - bceh - adfg + bcfg$$
$$= bcfg + adeh - bceh - adfg$$

Hence det AB = det A × det B

Note: This result is true for square matrices of any order n.

5. Prove that value of det

$$A = \begin{vmatrix} c+d & c & c \\ c & c+d & c \\ c & c & c+d \end{vmatrix} \quad \textit{is equal to} \quad d^2(3c+d)$$

Straight approach

Taking elements of row one with their respective cofactors, value of the given determinant =

$$(c+d)\left[(c+d)^2 - c^2\right] - c\left[c(c+d) - c^2\right] + c\left[c^2 - c(c+d)\right]$$

$$= (c+d)\left[d^2 + 2cd\right] - c^2(c+d) + c^3 + c^3 - c^2(c+d)$$
$$= (c+d)\left[d^2 + 2cd\right] - c^3 - c^2d + c^3 + c^3 - c^3 - c^2d$$
$$= (c+d)\left[d^2 + 2cd\right] - 2c^2d = cd^2 + 2c^2d + d^3 + 2cd^2 - 2c^2d$$
$$= d^3 + 3cd^2 = d^2(3c+d) \text{.........Hence proved}$$

Smart approach:

Using property 7, let us add (-1) times row 2 to each of rows 1 and 3 leading to

$$A = \begin{vmatrix} d & -d & 0 \\ c & c+d & c \\ 0 & -d & d \end{vmatrix}$$

Taking elements of row one with their respective cofactors, value of the given determinant =

$$d \times (-1)^{1+1} \left[(c+d) \times d - (-d) \times c \right] + (-d) \times (-1)^{1+2} \left[c \times d - 0 \times c \right]$$

$$= d\left[d(c+d) + cd + cd \right] = d^2 \left[(c+d) + c + c \right] = d^2 (3c + d)$$

It will be seen that with a little thinking, many problems can be solved more easily.

Applications of Matrices:

Matrices have many applications for real life problems. Let us explore a few.

13.6.2 Solution of System of Linear Equations

As we know, values of the variables satisfying all the linear equations in a system of linear equations is called solution of system of linear equations. There are two kinds of systems of linear equations- **Homogeneous and Non-Homogeneous System of Linear Equations.**

Consider a system of m linear equations in n variables $x_1, x_2, .., x_n$;

$$\left\{ \begin{array}{l} a_{11}x_1 + a_{12}x_2 + \dots\dots\dots\dots a_{1n}x_n = c_1 \\ a_{21}x_1 + a_{22}x_2 + \dots\dots\dots\dots a_{2n}x_n = c_2 \\ \dots\dots\dots \\ a_{m1}x_1 + a_{m2}x_2 + \dots\dots\dots\dots a_{mn}x_n = c_m \end{array} \right\} \dots\dots(1)$$

The above system (1) of linear equations can be written in matrix form as below:

$$\begin{pmatrix} a_{11} & a_{12} & \dots\dots\dots\dots & a_{1n} \\ a_{21} & a_{22} & \dots\dots\dots\dots & a_{2n} \\ \dots\dots\dots & & & \\ a_{m1} & a_{m2} & \dots\dots\dots & a_{mn} \end{pmatrix} \begin{pmatrix} x_1 \\ x_2 \\ \dots \\ x_n \end{pmatrix} = \begin{pmatrix} c_1 \\ c_2 \\ \dots \\ c_m \end{pmatrix} \dots\dots(2)$$

If we denote ***Matrix of Coefficients*** of order m×n as **A, *Matrix of variables*** as **X** and ***Matrix of constants*** as **B**, then Matrix representation of the above system of equations can be written as ***AX = B***

This matrix A of ***mn*** elements plus an additional column representing Matrix B is called the ***Augmented Matrix*** of the system i.e.

$$\begin{pmatrix} a_{11} & a_{12} & \cdots\cdots\cdots & a_{1n} & c_1 \\ a_{21} & a_{22} & \cdots\cdots\cdots & a_{2n} & c_2 \\ \cdots\cdots & & & & \\ a_{m1} & a_{m2} & \cdots\cdots & a_{mn} & c_m \end{pmatrix}$$

is the Augmented matrix of system (2).

A solution to the above system, if it exists can be determined by transforming the augmented matrix by row transformations until the matrix is in Echelon form. The system of linear equations represented by the transformed augmented matrix will have the same solution set as the original augmented matrix.

If the system of linear equations at (1) above is such that the equations are all satisfied by at least one set of values of variables, then it is said to be ***Consistent***. And it is said to be ***Inconsistent,*** if the equations are not satisfied simultaneously by any set of values of variables.

The system ***AX = B*** is called a ***Homogeneous systemif B = 0*** and a ***non-homogeneous system*** of equations ***if B ≠ 0.***

(i) Non-Homogeneous System of Equations Let AX = B be a system of n linear equations in n variables.

- If |A| ≠ 0, then the system of equations is consistent and has a unique solution given by X = A^{-1}B.
- If |A| = 0 and (adj A)B = 0, then the system of equations is consistent and has infinitely many solutions.
- If |A| = 0 and (adj A) B ≠ 0, then the system of equations is inconsistent i.e., having no solution

(ii) Homogeneous System of Equations Let AX = 0 is a system of n linear equations in n variables.

- If I |A| ≠ 0, then its only solution X = 0, is called the trivial solution.

- If I $|A| = 0$, then the system has infinitely many solutions, called non-trivial solution.

13.6.3 Solution of System of Equations by Augmented Matrix/ Rank Method

(i) Non-Homogeneous System of Equations Let AX = B, be a system of n linear quations in n variables, then

- **Step I:** Write the augmented matrix [A:B]
- **Step II{** Reduce the augmented matrix to Echelon form using elementary row-transformation.
- **Step III:** Determine the rank of coefficient matrix A and augmented matrix [A:B] by counting the number of non-zero rows in A and [A:B].

Important Results

1. If $\rho(A) \neq \rho(AB)$, then the system of equations is inconsistent.
2. If $\rho(A) = \rho(AB)$ = the number of unknowns, then the system of equations is consistent and has a unique solution.
3. If $\rho(A) = \rho(AB) <$ the number of unknowns, then the system of equationsis consistent and has infinitely many solutions.

(ii) Homogeneous System of Equations

Let AX = 0, be a homogeneous system of linear equations then,

- If $\rho(A)$ = number ofunknown, then AX = 0, have a non-trivial solution, i.e., X = 0.
- If $\rho(A) <$ number of unknowns, then AX = 0, have a non-trivial solution, withinfinitely many solutions.

Here are a couple of examples:

1. $2x - 2y + z = 3$
   $3x + y - z = 7$
   $x - 3y + 2z = 0$

   Augmented matrix for this system is as below:

   $$\begin{pmatrix} 2 & -2 & 1 & 3 \\ 3 & 1 & -1 & 7 \\ 1 & -3 & 2 & 0 \end{pmatrix}$$

   $R_1 \to R_1 - R_3$ and $R_2 \to R_2 - 3R_3$

$$\begin{pmatrix} 1 & 1 & -1 & 3 \\ 0 & 10 & -7 & 7 \\ 1 & -3 & 2 & 0 \end{pmatrix}$$

$R_3 \rightarrow R_3 - R_1$ and $R_2 \rightarrow R_2/10$

$$\begin{pmatrix} 1 & 1 & -1 & 3 \\ 0 & 1 & -0.7 & 0.7 \\ 0 & -4 & 3 & -3 \end{pmatrix}$$

$R_1 \rightarrow R_1 - R_2$

$$\begin{pmatrix} 1 & 0 & -0.3 & 2.3 \\ 0 & 1 & -0.7 & 0.7 \\ 0 & -4 & 3 & -3 \end{pmatrix}$$

$R_3 \rightarrow R_3 + 4R_2$

$$\begin{pmatrix} 1 & 0 & -0.3 & 2.3 \\ 0 & 1 & -0.7 & 0.7 \\ 0 & 0 & 0.2 & -0.2 \end{pmatrix}$$

$R_1 \rightarrow R_1 + 1.5\ R_2$ & $R_3 \rightarrow 5R_3$

$$\begin{pmatrix} 1 & 0 & 0 & 2 \\ 0 & 1 & -0.7 & 0.7 \\ 0 & 0 & 1 & -1 \end{pmatrix}$$

$R_2 \rightarrow R_2 + 0.7\ R_3$

$$\begin{pmatrix} 1 & 0 & 0 & 2 \\ 0 & 1 & 0 & 0 \\ 0 & 0 & 1 & -1 \end{pmatrix}$$

The augmented matrix is now in Echelon form and represents the following:

$x = 2,\ \ y = 0\ \ \&\ z = -1$

2. x + 3y + z = 6
   3x – 2y – 8z = 7
   4x + 5y – 3z = 17

Augmented matrix for this system is as below:

$$\begin{pmatrix} 1 & 3 & 1 & 6 \\ 3 & -2 & -8 & 7 \\ 4 & 5 & -3 & 17 \end{pmatrix}$$

$R_3 \to R_3 - 4R_1$ and $R_2 \to R_2 - 3R_1$

$$\begin{pmatrix} 1 & 3 & 1 & 6 \\ 0 & -11 & -11 & -11 \\ 0 & -7 & -7 & -7 \end{pmatrix}$$

$R_2 \to -R_2/11$ & $R_3 \to -R_3/7$

$$\begin{pmatrix} 1 & 3 & 1 & 6 \\ 0 & 1 & 1 & 1 \\ 0 & 1 & 1 & 1 \end{pmatrix}$$

$R_3 \to R_3 - R_2$

$$\begin{pmatrix} 1 & 3 & 1 & 6 \\ 0 & 1 & 1 & 1 \\ 0 & 0 & 0 & 0 \end{pmatrix}$$

The augmented matrix is now in Echelon form and represents the following system of equations:

$x + 3y + z = 6$

$y + z = 1$

$0 = 0$

This implies that the system is consistent and there can be infinite number of solutions with any value of z yielding various values of x and y from $y = 1 - z$ and $x = 3 + 2z$.

13.7 Applications of determinants:

There are many applications of the determinants. Let us see one of the very simple applications.

13.7.1 Test for co-linearity of three distinct points:

Recall the following formula for area of a triangle given coordinates of the 3 vertices $(x_1, y_1), (x_2, y_2)$ & (x_3, y_3) of the triangle given in section 7.2.7.3

$$\text{Area} = \frac{1}{2}\left[x_1(y_2 - y_3) + x_2(y_3 - y_1) + x_3(y_1 - y_2)\right]$$

The above expression for area can be expressed in determinant form as below:

$$\Delta Area = \frac{1}{2} \times \begin{vmatrix} x_1 & x_2 & x_3 \\ y_1 & y_2 & y_3 \\ 1 & 1 & 1 \end{vmatrix}$$

Now if value of the determinant is zero, it would imply that the three given points are co-linear i.e. they lie on a straight line.

13.7.2 Solution of System of Linear Equations by Cramer's Rule

Cramer's Rule can be used to solve a system of 3 equations in 3 variables say x, y and z. Cramer's Rule involves the use of determinants to find the solution and like any other technique it has its advantages and disadvantages. So let's take a look at Cramer's Rule.

Cramer's Rules makes use of 4 determinants say D, D_x, D_y and D_z, Determinant D is based on coefficients of x, y and z, D_x is same as D but column 1 is replaced by the constants. Similarly, D_y and D_z are same as D but columns 2 & 3 are replaced respectively by the constants. Value of different variables then is

$$x = \frac{D_x}{D},\ y = \frac{D_y}{D}\ \&\ z = \frac{D_z}{D}$$

Let us see solution based on Cramer's Rule for the example 1 considered above for clarity.

2x – 2y + z = 3

3x + y – z = 7

x – 3y + 2z = 0

$$\text{So } D = \begin{vmatrix} 2 & -2 & 1 \\ 3 & 1 & -1 \\ 1 & -3 & 2 \end{vmatrix},\ D_x = \begin{vmatrix} 3 & -2 & 1 \\ 7 & 1 & -1 \\ 0 & -3 & 2 \end{vmatrix},\ D_y = \begin{vmatrix} 2 & 3 & 1 \\ 3 & 7 & -1 \\ 1 & 0 & 2 \end{vmatrix},$$

$$D_z = \begin{vmatrix} 2 & -2 & 3 \\ 3 & 1 & 7 \\ 1 & -3 & 0 \end{vmatrix}$$

Calculating the values of these determinants, we get

$$|D| = 2\times(2-3)+2\times(6+1)+1\times(-9-1) = -2+14-10 = 2$$

$$|D_x| = 3\times(2-3)+2\times(14-0)+1\times(-21-0) = -3+28-21 = 4$$

$$|D_y| = 2\times(14-0)-3\times(6+1)+1\times(0-7) = 28-21-7 = 0$$

$$|D_z| = 2\times(0+21)+2\times(0-7)+3\times(-9-1) = 42-14-30 = -2$$

Hence $x=\frac{4}{2}=2,\ y=\frac{0}{2}=0\ \&\ z=\frac{-2}{2}=-1$, the same result as obtained above.

Let us see another example as below:

2x + 3y − z = 1

4x + y − 3z = 11

3x − 2y + 5z = 21

In this case, $D=\begin{vmatrix}2 & 3 & -1\\ 4 & 1 & -3\\ 3 & -2 & 5\end{vmatrix},\ D_x=\begin{vmatrix}1 & 3 & -1\\ 11 & 1 & -3\\ 21 & -2 & 5\end{vmatrix},$

$$D_y=\begin{vmatrix}2 & 1 & -1\\ 4 & 11 & -3\\ 3 & 21 & 5\end{vmatrix},\ D_z=\begin{vmatrix}2 & 3 & 1\\ 4 & 1 & 11\\ 3 & -2 & 21\end{vmatrix}$$

Values of these determinants are

$$|D| = 2\times(5-6)-3\times(20+9)-1\times(-8-3) = -2-87+11 = -78$$

$$|D_x| = 1\times(5-6)-3\times(55+63)-1\times(-22-21) = -1-354+43 = -312$$

$$|D_y| = 2\times(55+63)-1\times(20+9)-1\times(84-33) = 236-29-51 = 156$$

$$|D_z| = 2\times(21+22)-3\times(84-33)+1\times(-8-3) = 86-153-11 = -78$$

Hence $x=\frac{-312}{-78}=4,\ y=\frac{156}{-78}=-2\ \&\ z=\frac{-78}{-78}=1$

13.7.2.1 Advantages and Disadvantages of Cramer's Rule

Advantages:

1. We can find the value of x, y, or z without having to know any of the other values of x, y, or z.
2. If any of the values of x, y, or z is/are fractions, we do not have to plug in a fraction to find the other values. Each value can be found independently.

Disadvantages

One of the only disadvantages to using Cramer's rule is if the value of D is zero then Cramer's Rule will not work. However, if the value of D is zero then we know that the solution is either "No Solution" or "Infinite Solutions".

13.8 Unsolved problems:

1. If a+b+c = 0 and $\begin{vmatrix} a-x & c & b \\ c & b-x & a \\ b & a & c-x \end{vmatrix} = 0$, then prove that either x= 0 or $x = \pm\sqrt{\frac{3}{2}\left(a^2+b^2+c^2\right)}$
2. If $A = \begin{bmatrix} \cos\theta & -\sin\theta \\ \sin\theta & \cos\theta \end{bmatrix}$ and A + A^T= I, Find possible values of θ
3. Prove that if $\begin{vmatrix} x-3 & x-4 & x-a \\ x-2 & x-3 & x-b \\ x-1 & x-2 & x-c \end{vmatrix} = 0$, a, b, c are in AP
4. Using properties of Determinants, prove that

$$\begin{vmatrix} a^2 & bc & ac+c^2 \\ a^2+ab & b^2 & ac \\ ab & b^2+bc & c^2 \end{vmatrix} = 4a^2b^2c^2$$

5. Find the matrix A such that $\begin{bmatrix} 2 & -1 \\ 0 & 1 \\ -2 & 4 \end{bmatrix} \times A = \begin{bmatrix} -1 & -8 & -10 \\ 3 & 4 & 0 \\ 10 & 20 & 10 \end{bmatrix}$

6. Given that x = -9 is a root of $\begin{vmatrix} x & 3 & 7 \\ 2 & x & 2 \\ 7 & 6 & x \end{vmatrix} = 0$, find the other roots.

7. If A and B are two matrices such that AB = B and BA = A, then find A^2+B^2

8. Let A and B be 3×3 matrices such that $A^T = -A$ and $B^T = B$, then find that the matrix λ(AB + 3BA) is skew symmetric for what value of λ

9. If $\begin{vmatrix} x & b & c \\ a & y & c \\ a & b & z \end{vmatrix} = 0$, then find the value of $\frac{x}{x-a} + \frac{y}{y-b} + \frac{z}{z-c}$

10. If $A = \begin{vmatrix} 1 & -2 & 1 \\ -2 & 3 & 1 \\ 1 & 1 & 5 \end{vmatrix}$, verify that $(\text{Adj}A)^{-1} = \text{Adj}(A^{-1})$

11. If $P(\alpha) = \begin{bmatrix} \cos\alpha & \sin\alpha \\ -\sin\alpha & \cos\alpha \end{bmatrix}$, show that $P(\alpha) \times P(\beta) = P(\alpha + \beta)$

12. If $A + B + C = \pi$, show that $\begin{vmatrix} \sin(A+B+C) & \sin B & \cos C \\ -\sin B & 0 & \tan A \\ \cos(A+B)) & -\tan A & 0 \end{vmatrix} = 0$

13. If x, y, z, all positive, are p^{th}, q^{th} and r^{th} terms of a GP, prove that $\begin{vmatrix} \log x & p & 1 \\ \log y & q & 1 \\ \log z & r & 1 \end{vmatrix} = 0$

14. Find x if $\begin{bmatrix} x & -5 & -1 \end{bmatrix} \begin{bmatrix} 1 & 0 & 2 \\ 0 & 2 & 1 \\ 2 & 0 & 3 \end{bmatrix} \begin{bmatrix} x \\ 4 \\ 1 \end{bmatrix} = 0$

15. If points (2, 0), (0, 5) and (x, y) are collinear, find relation between x and y.

16. If $A=\begin{bmatrix} a & b \\ 0 & 1 \end{bmatrix}$, prove that $A^n=\begin{bmatrix} a^n & b\dfrac{a^n-1}{a-1} \\ 0 & 1 \end{bmatrix}$

17. For the matrix $A=\begin{bmatrix} 3 & 1 \\ -1 & 2 \end{bmatrix}$, $A^2-5A+7I=0$, then find A^{-1}

18. Given $A=\begin{bmatrix} 2 & 2 & -4 \\ -4 & 2 & -4 \\ 2 & -1 & 5 \end{bmatrix}$, $B=\begin{bmatrix} 1 & -1 & 0 \\ 2 & 3 & 4 \\ 0 & 1 & 2 \end{bmatrix}$, Find BA and use it to solve the system of equations $y + 2z = 7$, $x - y = 3$ and $2x + 3y + 4z = 17$

19. Using properties of Determinants, show that if $\begin{vmatrix} 3 & -2 & \sin 3\theta \\ -7 & 8 & \cos 2\theta \\ -11 & 14 & 2 \end{vmatrix}=0$, then $\sin\theta = 0$ or ½

20. Express the matrix A as the sum of a symmetric and a skew symmetric matrix where $A=\begin{bmatrix} 3 & -2 & -4 \\ 3 & -2 & -5 \\ -1 & 1 & 2 \end{bmatrix}$

Answers to Unsolved Problems

2. $\theta = \pm\frac{\pi}{3}$

5. $A = \begin{bmatrix} 1 & -2 & -5 \\ 3 & 4 & 0 \end{bmatrix}$

6. 2 and 7

7. $A^2 + B^2 = \begin{bmatrix} 2 & 0 \\ 0 & 2 \end{bmatrix}$

8. $\lambda = 3$

9. 2

14. $x = \pm 4\sqrt{3}$

15. $2y + 5x = 10$

17. $A^{-1} = \frac{1}{7} \times \begin{bmatrix} 2 & -1 \\ 1 & 3 \end{bmatrix}$

18. $x = 2,\ y = -1$ and $z = 4$

14

VECTOR ALGEBRA AND 3-D GEOMETRY

14.1 Introduction

In our day to day life, we come across two types of quantities like speed of a car, wind velocity, hitting of ball by cricket batsman etc. Observe that speed of a car is stated by only a magnitude (kms/hour) without any reference to direction in which car is travelling whereas wind velocity involves both - its magnitude as well as direction of wind flow. Similarly, when a cricketer hits the ball, it involves force with which the ball is hit as well as direction of the hit.

Quantities like speed of a car that have only magnitude are called scalar quantities whereas the quantities like wind velocity that have both magnitude and direction, are called vectors. Other examples of scalar quantities are mass, volume, temperature etc. and of vector quantities like displacement, acceleration, momentum etc. Even weight of a substance which equals force exerted by the gravitational pull {m(mass)×g(acceleration due to gravity) is a vector quantity as it has both magnitude and direction.

In mathematics, physics and engineering, we frequently come across both types of Quantities.

In this Chapter, we will deal with basic concepts of ***vectors***, their algebraic and geometrical properties and operations as well as concepts of 3-D Geometry.

14.2 Basic Properties of Vectors

Since vector has both magnitude and direction, it can be expressed as a straight line with a direction arrow in 2-dimensional or 3-dimensionalspace (refer figure 14.1 below). Such a line segment is also called a directed line. Direction could be in any of the two directions i.e. A to B or B to A. Length of the line AB represents its magnitude. Vector AB represented as $\overrightarrow{AB}$ can be symbolically represented as $\vec{a}$ and its magnitude is expressed as $|\vec{a}|$.

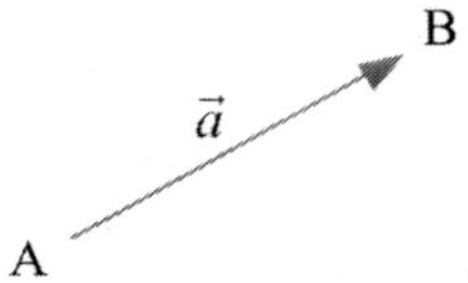

14.2.1 Types of vectors:

Zero vector: A vector whose initial and terminal points coincide, is called a zerovector (or null vector), and denoted as $\vec{0}$. Zero vector cannot be assigned a definitedirection as it has zero magnitude. Or, alternatively, it may be regarded as having any direction. The vectors with same start and end point say A and B ($\overrightarrow{AA}$, $\overrightarrow{BB}$) represent **Zero Vectors**.

Unit Vector:A vector whose magnitude is unity (i.e., 1 unit) is called a unit vector. Theunit vector in the direction of a given vector $\vec{a}$ is denoted by $\hat{a}$.

Co-initial Vectors: Two or more vectors having the same initial point are called co-initial vectors.

Collinear Vectors: Two or more vectors are said to be collinear if they are parallel to the same line, irrespective of their magnitudes and directions.

Equal Vectors: Two vectors $\vec{a}$ and $\vec{b}$ are said to be equal, if they have the same magnitude and direction regardless of the positions of their initial points, and written as $\vec{a} = \vec{b}$

Negative of a Vector: A vector whose magnitude is the same as that of a given vector (say $\overrightarrow{AB}$), but direction is opposite to that of it, is called negative of the given vector. For example, vector $\overrightarrow{BA}$ is negative of the vector $\overrightarrow{AB}$ and written as $\overrightarrow{BA} = -\overrightarrow{AB}$

The vectors defined above are such that with initial point not fixed, any of them may be subject to its parallel displacement without changing its magnitude and direction. Such vectors are called free vectors. We will be dealing with free vectors only.

14.2.2 Representation of vectors in 2-dimensional and 3-dimensional space

Vectors can be expressed on two dimensional (x-y) or on 3-dimensional (x-y-z) coordinate system, (refer adjoining figure for 2-dimentional representation)

We can see that O being center of the coordinate system, $\overrightarrow{OP}$ (denoted as vector $\vec{r}$) is a vector from O to P and is also known as position vector of point P. Since point P has coordinates (x,y) and if line OP makes an angle θ to the x-axis, these co-ordinates can be expressed as (rcosθ, rsinθ) which are also known as its polar coordinates.

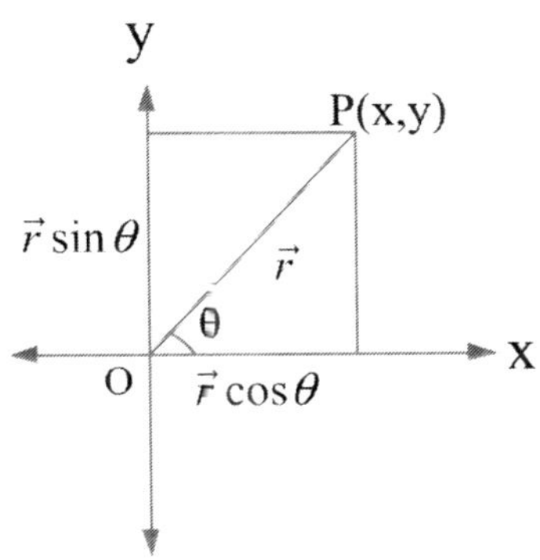

2-dimentional vector

In 2-dimensional system, coordinates of point P can be represented as (rcosθ, rsinθ) with magnitude $|\vec{r}|$ of $\overrightarrow{OP} = \sqrt{x^2 + y^2}$

It is a little tricky to represent 3-dimensional system on a plain paper. Assume that a point P in space has coordinates (x,y,z) with vector $\overrightarrow{OP}$ $(\vec{r})$ of magnitude $|\vec{r}|$ making angles of α, β, γ with positive direction of x, yand z coordinate axis respectively as shown in figure 14.1 below.

For better understanding, one can assume that the point P is in space in a room and rectangular floor with one corner of the room presumed to be center of the coordinate system O. Two lines (joining the floor and respective walls) originating from this corner O are axes x and y whereas vertical line from the given corner to the roof is z-axis.

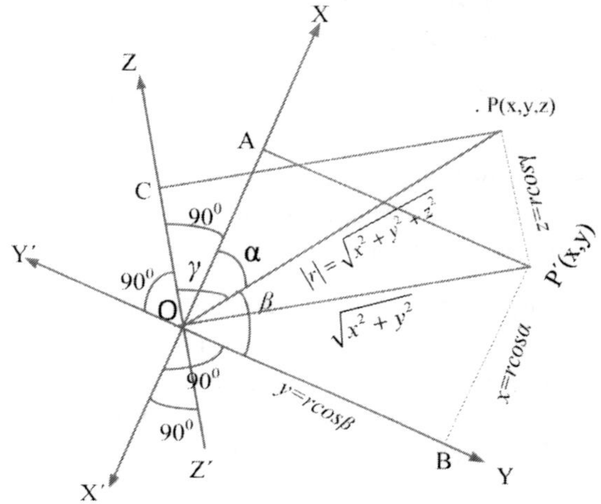

Figure 14.1

In order to find relationship between magnitude $|\vec{r}|$ and x, y, z parameters, let a perpendicular be dropped from point P to the floor (x-y plane) meeting it at point P'. Vector $\overrightarrow{OP'}$ can also be seen as projection of vector $\overrightarrow{OP}$ on the x-y plane and is denoted as P_x. Coordinates of P' will be (x, y) implying that if perpendicular lines are drawn from P' to x-axis and y-axis meeting these axes at A and B respectively. Length of line $P'B = OA = r\cos\alpha$, $P'A = OB = r\cos\beta$ and $P'P = r\cos\gamma$ will equal ***x***, ***y*** and ***z*** respectively. Just like in 2-dimensional system, magnitude of vector $|\overrightarrow{OP'}| = |\overrightarrow{P_x}|$ will be $\sqrt{x^2 + y^2}$. Similarly, P_y and P_z are projections of vector $\overrightarrow{OP}$ on the y-z and z-x planes and equal $\sqrt{y^2 + z^2}$ and $\sqrt{z^2 + x^2}$ respectively.

Knowing that $PP' =$ z and $OP'(P_x)$ being on the floor, is perpendicular to PP', we can easily see that $\Delta PP'O$ is a right angled triangle such that magnitude of $\overrightarrow{OP} = |\vec{r}| = \sqrt{OP'^2 + PP'^2} = \sqrt{x^2 + y^2 + z^2}$

The tricky part is to appreciate that in the figure 14.1 above, angles between z-axis and x and y axes ($\angle COA$ and $\angle COB$) are right angles though they appear acute/obtuse because these are in space and not in the plane of the paper.

In the 3-dimentional coordinate system, let us name the unit vectors along x, y and z axis as $\hat{i}, \hat{j}$ & $\hat{k}$ respectively. Therefore, the 3 components of vector $\vec{r}$ will be written as $x\hat{i}, y\hat{j}$ & $z\hat{k}$ respectively. From $\Delta P'OA$ we notice that vector $\overrightarrow{OA} = x\hat{i}$ and $\overrightarrow{OP'} = \overrightarrow{OA} + \overrightarrow{AP'} = \overrightarrow{OA} + \overrightarrow{OB} = x\hat{i} + y\hat{j}$. Similarly, from $\Delta P'OP$, we notice that vector $\overrightarrow{PP'} = z\hat{k}$ and $\overrightarrow{OP} = \overrightarrow{OP'} + \overrightarrow{PP'} = x\hat{i} + y\hat{j} + z\hat{k}$ leading to position vector $\vec{r} = x\hat{i} + y\hat{j} + z\hat{k}$

This is a very important concept. This form of any vector is known as its component form. Here x, y, z are the scalar components of vector $\vec{r}$ along x, y and z axis and $x\hat{i}$, $y\hat{j}$ & $z\hat{k}$ are called vector components along the three axes. Sometimes x, y & z are also termed as rectangular components of vector $\vec{r}$.

14.2.3 Direction cosines and Direction ratios of a vector:

As already mentioned, point P when joined to the origin, makes angles α , β and γ respectively to the x, y and z axis (of course in three different planes) as shown in figure 14.1 above. We have already seen that x, y and z can be represented as $r\cos\alpha$, $r\cos\beta$ and $r\cos\gamma$ respectively. α, β ***and*** γ ***are known asdirection angles*** and **cos** α, $\cos\beta$ ***and*** $\cos\gamma$ ***are known as direction cosines of direction vector*** $\vec{r}$.

Since a given vector in space can be extended in two opposite directions, it has two sets of direction cosines. In order to have a unique set of direction cosines, we must take into account direction of the given vector. These unique direction cosines are usually denoted by ***l, m and n.***

Any three numbers which are proportional to the direction cosines of a vector are called ***direction ratios*** of the vector. If ***l, m, n*** are direction cosines and ***a, b, c*** are direction ratios of a vector, then $\frac{a}{l} = \frac{b}{m} = \frac{c}{n} = p$, where p is any non zero constant which is a real number. leading to $l = \frac{a}{p}, m = \frac{b}{p}$ & $n = \frac{c}{p}$ .....(1)

If we assume p to be equal to distance (say $|\vec{r}|$) of the given point from the origin, which is a fair assumption as 'p' can be any real number. co-ordinates of the given point P can be represented as

$$x = |\vec{r}|l = |\vec{r}| \times \frac{a}{|\vec{r}|} = a, y = |\vec{r}|m = |\vec{r}| \times \frac{b}{|\vec{r}|} = b \,\&\, z = |\vec{r}|n = |\vec{r}| \times \frac{c}{|\vec{r}|} = c$$

Thus Direction vector $\vec{r}$ can be represented as

$$\vec{r} = |\vec{r}|l\hat{i} + |\vec{r}|m\hat{j} + |\vec{r}|n\hat{k} = \left(a\hat{i} + b\hat{j} + c\hat{k}\right) \quad ...(2)$$

Note: It is important to remember that we have assumed that 'p'=$|\vec{r}|$

We know that if $\vec{r}$ is direction vector of an arbitrary point with coordinates (x, y, z), vector $\vec{r}$ can be represented as $\vec{r} = x\hat{i} + y\hat{j} + z\hat{k}$.(3)

The above expression is an extension of 2-dimesional representation.

From equations (2) and (3), we reach a very important conclusion that co-ordinates of any point in space are same as its direction ratios with reference to the origin.

We can also see from eq. (2) that $\frac{\vec{r}}{|\vec{r}|} = \left(l\hat{i} + m\hat{j} + n\hat{k}\right) \Rightarrow$ $\hat{r} = \left(l\hat{i} + m\hat{j} + n\hat{k}\right)$, where $\hat{r}$ is the unit vector along the direction vector $\overrightarrow{OP}$.

If a given vector in space does not pass through the origin, then, in order to find its direction cosines, we draw a line through the origin and parallel to the given vector. Now we find direction cosines of the parallel vector passing through the origin as two parallel lines have the same set of direction cosines.

Direction ratios are sometime called as direction numbers.

14.2.3.1 Relation between Direction cosines and direction ratios of a vector

We have seen that magnitude of direction vector $\vec{r}$ is $|\vec{r}| = \sqrt{x^2 + y^2 + z^2}$

By substituting values of x, y and z in terms of direction cosines, we get $|r| = \sqrt{(rl)^2 + (rm)^2 + (rn)^2} \Rightarrow \sqrt{l^2 + m^2 + n^2} = 1 \Rightarrow l^2 + m^2 + n^2 = 1$.

If 'p' is the ratio between direction ratios ***a, b, c*** and direction cosines l, ***m, n*** of a line, we can see that $(l, m, n = ap, bp, cp.....(1))$ and so

$$(ap)^2 + (bp)^2 + (cp)^2 = 1 \Rightarrow p^2\left(a^2 + b^2 + c^2\right) = 1 \Rightarrow$$

$$p = \pm\frac{1}{\sqrt{a^2 + b^2 + c^2}}$$

Substituting these values in equation (1) above, we get values of direction cosines as

$$l = \pm\frac{a}{\sqrt{a^2 + b^2 + c^2}},\ m = \pm\frac{b}{\sqrt{a^2 + b^2 + c^2}} \text{ and } n = \pm\frac{c}{\sqrt{a^2 + b^2 + c^2}}$$

.....Equation 14.1

All these are very essential concepts for understanding 3-dimensional geometry.

Example 1:

If a line makes angles of 90⁰, 60⁰ and 30⁰ with the positive directions of x, y and z-axis respectively, find its direction cosines.

Solution:

Since direction cosines are cosines of respective angles, we get direction cosines *l, m, n* as

$l = \cos 90^0 = 1$, $m = \cos 60^0 = \frac{1}{2}$ and $n = \cos 30^0 = \frac{\sqrt{3}}{2}$

Example 2:

If a line has direction ratios -2, 1, 2, find its direction cosines.

Solution:

Given the direction ratios, we use the following formulae for determining direction cosines,

$$l = \pm\frac{a}{\sqrt{a^2 + b^2 + c^2}},\ m = \pm\frac{b}{\sqrt{a^2 + b^2 + c^2}} \text{ and } n = \pm\frac{c}{\sqrt{a^2 + b^2 + c^2}}$$

So $l = \pm\frac{-2}{\sqrt{4+1+4}} = \mp\frac{2}{3}$, $m = \pm\frac{1}{\sqrt{4+1+4}} = \pm\frac{1}{3}$, and

$$n = \pm\frac{2}{\sqrt{4+1+4}} = \pm\frac{2}{3},$$

Hence the direction cosines are $\left(-\frac{2}{3},\frac{1}{3},\frac{2}{3}\right) or \left(\frac{2}{3},-\frac{1}{3},-\frac{2}{3}\right)$

14.2.3.2 Direction Cosines of a line passing through two points

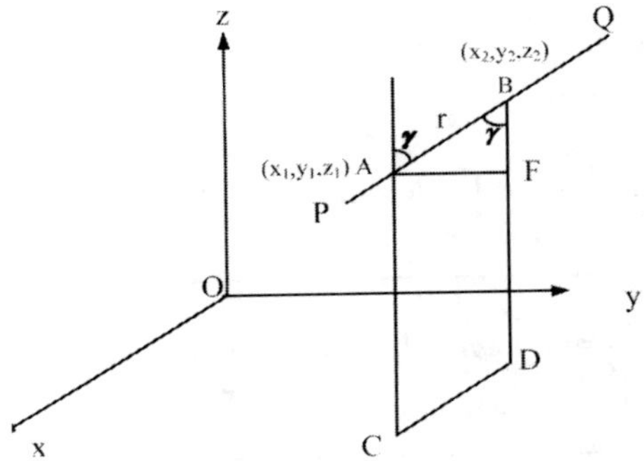

Figure 14.2

Let l, m, n be the direction cosines of line PQ passing through points A (x_1, y_1, z_1) and B (x_2, y_2, z_2) as shown in figure 14.4 above and let it make angles α,β,γ with positive direction of x, y and z axis respectively.

Perpendiculars are drawn from A and B on x-y plane meeting it at C and D respectively. Now we draw a perpendicular from A to BD meeting it at F. In right angled triangle AFB, $\angle ABF = \gamma$ leading to

$$\cos\gamma = \frac{z_2 - z_1}{AB}$$

Similarly $\cos\alpha = \frac{x_2 - x_1}{AB}$ and $\cos\beta = \frac{y_2 - y_1}{AB}$

Hence direction cosines of line segment joining points $A(x_1, y_1, z_1)$ and $B(x_2, y_2, z_2)$ are

$\frac{x_2 - x_1}{AB}$, $\frac{y_2 - y_1}{AB}$ and $\frac{z_2 - z_1}{AB}$ where

$$AB = \sqrt{(x_2 - x_1)^2 + (y_2 - y_1)^2 + (z_2 - z_1)^2}.$$

Since AB is a constant, $(x_2 - x_1), (y_2 - y_1)$ & $(z_2 - z_1)$ ***are also Direction ratios of the line segment AB or vector*** $\overrightarrow{AB}$

...Equation set 14.2

Example 3:

Find direction cosines of a line passing through two points (-2, 4, -5) and (1, 2, 3)

Solution:

Using the formulae derived above, we get length of the line as

$$\sqrt{(x_2 - x_1)^2 + (y_2 - y_1)^2 + (z_2 - z_1)^2} = \sqrt{(1-(-2))^2 + (2-4)^2 + (3-(-5))^2}$$

$$= \sqrt{3^2 + (-2)^2 + 8^2} = \sqrt{9 + 4 + 64} = \sqrt{77}$$

Hence the direction cosines of the line segment are

$$\frac{(1-(-2))}{\sqrt{77}}, \frac{(2-4)}{\sqrt{77}}, \frac{(3-(-5))}{\sqrt{77}} = \frac{3}{\sqrt{77}}, \frac{-2}{\sqrt{77}}, \frac{8}{\sqrt{77}}$$

Example 4:

Using directionratios show that the points A (2, 3, -4), B (1, -2, 3) and C (3, 8, -11) are collinear.

Solution:

Direction ratios of a line joining points (x_1, y_1, z_1) and (x_2, y_2, z_2) are $(x_2 - x_1)$, $(y_2 - y_1)$ and $(z_2 - z_1)$

Hence direction ratios of line segment AB are

$$(1-2), (-2-3), (3-(-4)) = (-1, -5, 7)$$

Similarly direction ratios of line segment BC are

$$(3-1),(8-(-2)),(-11-(3))=(2,10,-14)=-2\times(-1,-5,7))$$

We see that the direction ratios of the line segments AB and BC are proportional. Hence these two lines should be parallel. However since point B lies on both the lines, it implies that points A, B and C lie on the same line and are hence collinear.

14.2.4 Arithmetic operations on Vectors

12.2.4.1 Addition of vectors:

A vector $\overrightarrow{PQ}$ simply means the displacement from a point P to the point Q. As an example, consider a situation that a person moves from point P to point Q and then from point Q to point R in a different direction as depicted in figure 14.3 below:

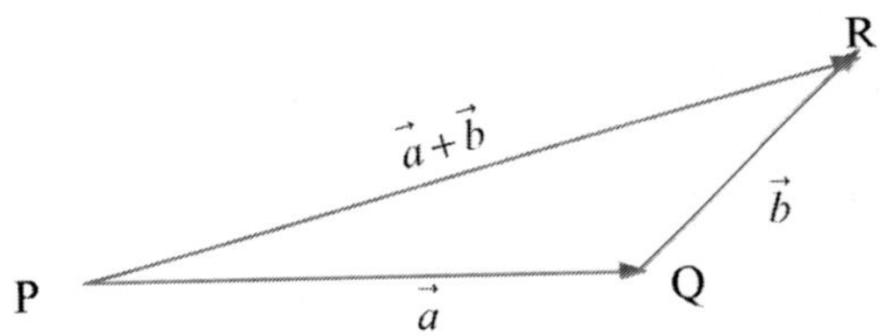

Figure 14.3

The net displacement made by the person from point P to the point R is given by the vector $\overrightarrow{PR}$ and is expressed as $\overrightarrow{PR} = \overrightarrow{PQ} + \overrightarrow{QR}$ ..(1)

This is known as the ***triangle law of vector addition.***

Similarly, we can visualize subtraction of two vectors. In Figure 14.5, if direction of vector QR is reversed we will get vector QR′ as depicted in figure 14.4 below:

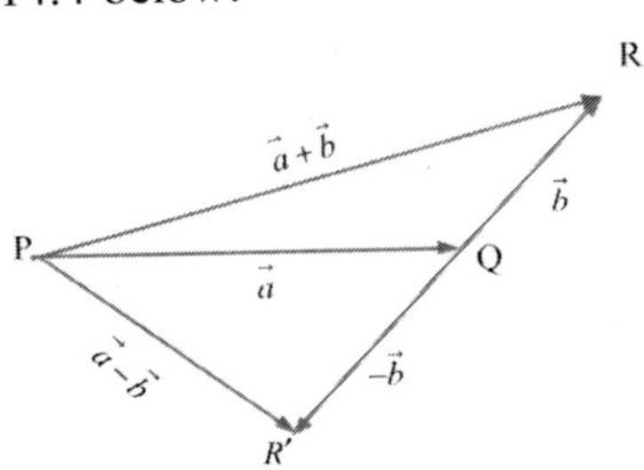

Figure 14.4

Now consider that a person moves from point P to point Q and then from point Q to point R′ where QR′ is equal and opposite to QR. Resultant of $\overrightarrow{PQ} + \overrightarrow{QR'} = \overrightarrow{PR'}$. We can now see that PR' equals difference of vectors PQ and QR.

Direction of different vectors is to be noted carefully. If we consider vector $\overrightarrow{PR}$ in the above diagram, it would equal to - $\overrightarrow{RP}$, Substituting in the above equation (1), we get $\overrightarrow{PQ} + \overrightarrow{QR} = - \overrightarrow{RP}$ leading to $\overrightarrow{PQ} + \overrightarrow{QR} + \overrightarrow{RP} = 0$. Here the person moves from point P and comes back to the same point P after moving to point Q and then point R.

This is again a very important result implying that when the sides of a triangle are taken in order, it leads to zero resultant as the initial and terminal points coincide.

A related result is that if two vectors are represented by two adjacent sides of a parallelogram, as depicted in figure 14.5 below, its diagonal gives the resultant vector. This is known as ***parallelogram law of vectors***.

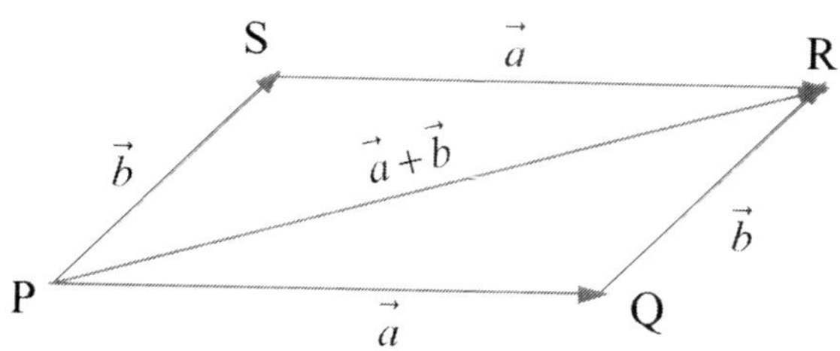

Figure 14.5

14.2.4.1.1 Properties of Vector addition:

1. For any two vectors $\vec{a}$ and $\vec{b}$,

   $\vec{a} + \vec{b} = \vec{b} + \vec{a}$ (Commutative property) 14.4

   This is evident from the vector diagram shown in figure 14.5. Diagonal PR represents both (PQ + QR) i.e. $\left(\vec{a} + \vec{b}\right)$ as well as (PS + SR) i.e. $\left(\vec{b} + \vec{a}\right)$

2. For any three vectors $\vec{a}$, $\vec{b}$ and $\vec{c}$

$\left(\vec{a}+\vec{b}\right)+\vec{c}=\vec{a}+\left(\vec{b}+\vec{c}\right)$ (Associative property) .. Equation 14.5

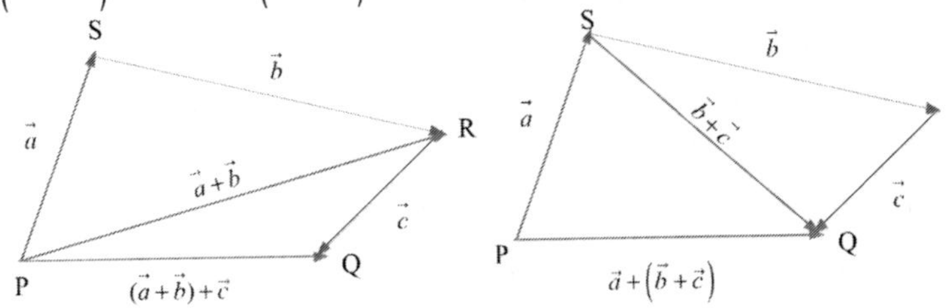

Figure 14.6

Reference diagram 14.6 above. The two vector diagrams are exactly the same. It will be noticed that vector PQ represents vector $(\vec{a}+\vec{b})+\vec{c}$ in the left side diagram and vector $\vec{a}+\left(\vec{b}+\vec{c}\right)$ in the right side diagram respectively. Hence Associative property is also applicable to the vector addition. This also implies that we can write vector addition of three vectors $\vec{a}$, $\vec{b}$ and $\vec{c}$ without brackets as $\vec{a}+\vec{b}+\vec{c}$

It may also be realized that $\vec{a}+\vec{0}=\vec{0}+\vec{a}$. Hence zero vector $\vec{0}$ is known as ***Additive identity*** of vector addition.

14.2.4.2 Multiplication of a vector by a scalar

It would be recalled that a vector has both magnitude and direction. Magnitude is a scalar quantity and magnitude together with direction data makes it a vector. Hence if a vector say $\vec{a}$ is multiplied by a scalar quantity say ***n***, magnitude of the vector $|\vec{a}|$ changes to $|n\vec{a}|=n|\vec{a}|$. ***If the scalar is positive, direction of the vector remains unchanged whereas if the scalar is negative, direction of the vector reverses.***

If the scalar n = -1, the vector $|\vec{a}|$ changes to vector -$|\vec{a}|$. If we add these two vectors, we get $\vec{a}+(-\vec{a})=\vec{a}-(\vec{a})=\vec{0}$. ***Hence vector $-\vec{a}$ is called the negative (or additive inverse) of vector $\vec{a}$***

And if the scalar n $=\frac{1}{|\vec{a}|}$, provided the vector $\vec{a}$ is not a Null vector, vector $|n\vec{a}|=\frac{1}{|\vec{a}|}\times|\vec{a}|=1$. ***So, vector $n\vec{a}$ here represents the unit vector in the direction of vector $\vec{a}$. We write it as $\hat{a}=\frac{1}{|\vec{a}|}\times\vec{a}$***

From the above discussion, it will be realized that for any scalar k, $k\vec{0} = \vec{0}$

14.2.4.3 Product of two vectors:

Multiplication of two vectors is defined in two ways, namely, scalar (or dot)product, where the result is a scalar, and vector (or cross) product, where theresult is a vector. Based upon these two types of products for vectors, various applications in geometry, mechanics and engineering haveevolved.

14.2.4.3.1 Scalar or dot Product of two vectors:

The scalar product of two nonzero vectors $\vec{x}$ and $\vec{y}$ is denoted as $\vec{x}.\vec{y}$ and is defined as $\vec{x}.\vec{y} = |\vec{x}|.|\vec{y}|\cos\theta$ where θ is the angle between $\vec{x}$ and $\vec{y}$ with $0 \le \theta \le \pi$ as shown in figure 14.7 below:

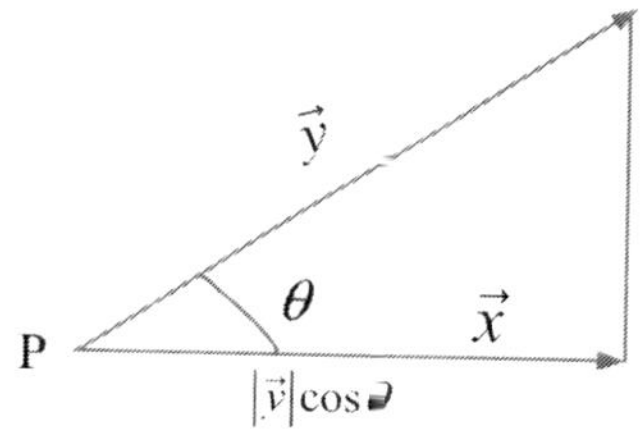

Figure 14.7

In practical terms, a dot product of two vectors also defines the product of one of the vectors with component of the other vector in the direction of the first vector. Considering vector $\vec{y}$, we can see that it has two components- the component in the direction of vector $\vec{x}$ is $|\vec{y}|\cos\theta$ and the component perpendicular to the direction of vector $\vec{x}$ is $|\vec{y}|\sin\theta$. The perpendicular component has no effect on vector $\vec{x}$. Hence the dot product is product of vector $\vec{x}$ and $|\vec{y}|\cos\theta$.

Hence $\vec{x}.\vec{y} = |\vec{x}||\vec{y}|\cos\theta$ or $\cos\theta = \dfrac{\vec{x}.\vec{y}}{|\vec{x}||\vec{y}|}$ ..... Equation 14.6

From definition of the scalar product, we can observe the following:

i. $\vec{x}.\vec{y}$ is a real number.

ii. $\vec{x}.\vec{y} = \vec{y}.\vec{x}$ as the scalar product value by interchanging the vectors remains unchanged.

iii. If the two vectors are in the same direction, $\theta = 0^0$ implying that $\cos\theta = 1$, leading to $\vec{x}.\vec{y} = |\vec{x}||\vec{y}|$, i.e. product of magnitude of the two vectors.

iv. If the two vectors are in the opposite direction, $\theta = \pi$ implying that $\cos\theta = -1$, leading to $\vec{x}.\vec{y} = -|\vec{x}|.|\vec{y}|$, i.e. negative of the product of magnitude of the two vectors. In particular, if the vectors are of the same magnitude say $|\vec{x}|$, $\vec{x}.\vec{y} = |\vec{x}| \times (-|\vec{x}|) = -|\vec{x}|^2$

v. If the two vectors are perpendicular to each other, $\theta = \frac{\pi}{2}$ implying that $\cos\theta = 0$, leading to $\vec{x}.\vec{y} = 0$. Since unit vectors $\hat{i}, \hat{j}$ & $\hat{k}$ along x, y and z axis respectively are perpendicular to each other, we can see that $\hat{i}.\hat{j} = \hat{j}.\hat{k} = \hat{k}.\hat{i} = 0$ Similarly, $\hat{i}.\hat{i} = \hat{j}.\hat{j} = \hat{k}.\hat{k} = 1$ since angle between them is 0^0.

vi. If either $\vec{x}$ or $\vec{y}$ is 0, θ is not defined and hence $\vec{x}.\vec{y} = 0$

vii. The angle θ between two non –zero vectors $\vec{x}$ & $\vec{y}$ is given by

$$\cos\theta = \frac{\vec{x}.\vec{y}}{|\vec{x}||\vec{y}|} \text{ or } \theta = \cos^{-1}\left(\frac{\vec{x}.\vec{y}}{|\vec{x}||\vec{y}|}\right)$$

Dot product in Cartesian coordinate form:

Let there be two vectors $\vec{a}$ & $\vec{b}$ with Cartesian form respectively as $\vec{a} = a_1\hat{i} + a_2\hat{j} + a_3\hat{k}$ and $\vec{b} = b_1\hat{i} + b_2\hat{j} + b_3\hat{k}$

So $\vec{a}.\vec{b} = \left(a_1\hat{i} + a_2\hat{j} + a_3\hat{k}\right).\left(b_1\hat{i} + b_2\hat{j} + b_3\hat{k}\right) \Rightarrow$

$$\vec{a}.\vec{b} = (a_1b_1)\hat{i}.\hat{i} + (a_1b_2)\hat{i}.\hat{j} + (a_1b_3)\hat{i}.\hat{k} + (a_2b_1)\hat{j}.\hat{i} + (a_2b_2)\hat{j}.\hat{j}$$
$$+ (a_2b_3)\hat{j}.\hat{k} + (a_3b_1)\hat{k}.\hat{i} + (a_3b_2)\hat{k}.\hat{j} + (a_3b_3)\hat{k}.\hat{k}$$

Since vectors $\hat{i}, \hat{j}$ *and* $\hat{k}$ are mutually perpendicular, we can see that

$\vec{a}.\vec{b} = a_1b_1 + a_2b_2 + a_3b_3$, because terms with $\hat{i}.\hat{j}$, $\hat{j}.\hat{k}$ & $\hat{k}.\hat{i}$ disappear and $\hat{i}.\hat{i}$, $\hat{j}.\hat{j}$ & $\hat{k}.\hat{k}$ are replaced by 1

An interesting corollary relating to dot product of sum of two vectors:

$\left(\vec{a} + \vec{b}\right).\left(\vec{a} + \vec{b}\right) = (\vec{a}.\vec{a}) + \vec{b}.\vec{b} + 2\vec{a}.\vec{b} = |\vec{a}|^2 + \left|\vec{b}\right|^2 + 2|\vec{a}|\left|\vec{b}\right|\cos\theta$ (As

$\vec{a}.\vec{b} = |\vec{a}||\vec{b}|\cos\theta$)

Since all terms are scalar, this leads us to

$$\left|\left(\vec{a}+\vec{b}\right)^2\right| = |\vec{a}|^2 + |\vec{b}|^2 + 2|\vec{a}||\vec{b}|\cos\theta \Rightarrow |\vec{a}+\vec{b}| =$$

$$\sqrt{|\vec{a}|^2 + |\vec{b}|^2 + 2|\vec{a}||\vec{b}|\cos\theta} \quad \ldots 14.7$$

Proof:

Let $\vec{a} = a_1\hat{i} + a_2\hat{j} + a_3\hat{k}$ and $\vec{b} = b_1\hat{i} + b_2\hat{j} + b_3\hat{k}$

Then $\left(\vec{a}+\vec{b}\right) = (a_1+b_1)\hat{i} + (a_2+b_2)\hat{j} + (a_3+b_3)\hat{k}$

$$\left(\vec{a}+\vec{b}\right).\left(\vec{a}+\vec{b}\right) = \left|\left(\vec{a}+\vec{b}\right)\right|^2 = (a_1+b_1)^2 + (a_2+b_2)^2 + (a_3+b_3)^2$$

Magnitude of the two vectors are $|\vec{a}| = \sqrt{a_1^{\,2} + a_2^{\,2} + a_3^{\,2}}$ and $|\vec{b}| = \sqrt{b_1^{\,2} + b_2^{\,2} + b_3^{\,2}}$

$$\left|\left(\vec{a}+\vec{b}\right)\right| = \sqrt{(a_1+b_1)^2 + (a_2+b_2)^2 + (a_3+b_3)^2}$$

$$= \sqrt{\left(a_1^{\,2} + a_2^{\,2} + a_3^{\,2}\right) + \left(b_1^{\,2} + b_2^{\,2} + b_3^{\,2}\right) + 2(a_1b_1 + a_2b_2 + a_3b_3)}$$

Since $\vec{a}.\vec{b} = a_1b_1 + a_2b_2 + a_3b_3$, we get

$\left|\left(\vec{a}+\vec{b}\right)\right| = \sqrt{|\vec{a}|^2 + |\vec{b}|^2 + 2\vec{a}.\vec{b}} = \sqrt{|\vec{a}|^2 + |\vec{b}|^2 + 2|\vec{a}||\vec{b}|\cos\theta}$, where θ is the angle between the two vectors.

Similarly we can prove that

$$\left|\left(\vec{a}-\vec{b}\right)\right|^2 = |\vec{a}|^2 + |\vec{b}|^2 - 2\vec{a}.\vec{b} \Rightarrow |\vec{a}-\vec{b}| = \sqrt{|\vec{a}|^2 + |\vec{b}|^2 - 2|\vec{a}||\vec{b}|\cos\theta} \quad \ldots 14.8$$

Example 5:

If $\hat{a}$ & $\hat{b}$ are unit vectors inclined at an angle θ, prove that $\cos\frac{\theta}{2} = \frac{1}{2}\left|\hat{a}+\hat{b}\right|$

Solution:

Magnitude of sum of two vectors $\vec{a}\,\&\,\vec{b}$ is

$$\left|\vec{a}+\vec{b}\right|=\sqrt{\left|\vec{a}\right|^2+\left|\vec{b}\right|^2+2\left|\vec{a}\right|\left|\vec{b}\right|\cos\theta}$$

Since $\left|\hat{a}\right|=1$ and $\left|\hat{b}\right|=1$, by substituting in the above equation, we get

$$\left|\hat{a}+\hat{b}\right|=\sqrt{1+1+2\cos\theta}=\sqrt{2\left(1+\cos\theta\right)}=\sqrt{2\times2\cos^2\frac{\theta}{2}}=2\cos\frac{\theta}{2}\Rightarrow$$

$$\cos\frac{\theta}{2}=\frac{1}{2}\left|\hat{a}+\hat{b}\right|$$

Example 6:

If $\hat{a}\,\&\,\hat{b}$ are unit vectors inclined at an angle θ, prove that

$$\tan\frac{\theta}{2}=\frac{\left|\hat{a}-\hat{b}\right|}{\left|\hat{a}+\hat{b}\right|}$$

Solution:

We have seen in example 5 that $\left|\vec{a}+\vec{b}\right|=\sqrt{\left|\vec{a}\right|^2+\left|\vec{b}\right|^2+2\left|\vec{a}\right|\left|\vec{b}\right|\cos\theta}$

leading to $\left|\hat{a}+\hat{b}\right|=\sqrt{1+1+2\cos\theta}=\sqrt{2\left(1+\cos\theta\right)}=2\cos\frac{\theta}{2}$

Similarly, we will get

$$\left|\hat{a}-\hat{b}\right|=\sqrt{1+1-2\cos\theta}=\sqrt{2\left(1-\cos\theta\right)}=2\sin\frac{\theta}{2}$$

Dividing the above equations, we get $\tan\frac{\theta}{2}=\frac{\left|\hat{a}-\hat{b}\right|}{\left|\hat{a}+\hat{b}\right|}$

Example 7:

If $\left|\vec{a}+\vec{b}\right|=60$, $\left|\vec{a}-\vec{b}\right|=40$ & $\left|\vec{b}\right|=46$, find $\left|\vec{a}\right|$

Solution:

Magnitude of sum and difference of two vectors $\vec{a}\,\&\,\vec{b}$ are

$$\left|\vec{a}+\vec{b}\right|=\sqrt{\left|\vec{a}\right|^2+\left|\vec{b}\right|^2+2\left|\vec{a}\right|\left|\vec{b}\right|\cos\theta} \text{ and}$$

$$\left|\vec{a}-\vec{b}\right|=\sqrt{\left|\vec{a}\right|^2+\left|\vec{b}\right|^2-2\left|\vec{a}\right|\left|\vec{b}\right|\cos\theta}$$

Substituting the given values we get

$$60=\sqrt{\left|\vec{a}\right|^2+46^2+2\times 46\left|\vec{a}\right|\cos\theta} \text{ and}$$

$$40=\sqrt{\left|\vec{a}\right|^2+46^2-2\times 46\left|\vec{a}\right|\cos\theta}$$

Squaring and adding,

$$60^2+40^2=2\left|\vec{a}\right|^2+2\times 46^2 \Rightarrow 2\left|\vec{a}\right|^2=\left\{\left(60^2-46^2\right)+\left(40^2-46^2\right)\right\}$$

$$2\left|\vec{a}\right|^2=\left\{(60+46)(60-46)+(40+46)(40-46)\right\}=$$

$$106\times 14+(86\times -6)=1484-516=968$$

Or $\left|\vec{a}\right|^2=\dfrac{968}{2}=484 \Rightarrow \left|\vec{a}\right|=22$

Example 8:

Show that the unit vectors

$$\hat{a}=\frac{1}{7}\left(2\hat{i}+3\hat{j}+6\hat{k}\right),\ \hat{b}=\frac{1}{7}\left(3\hat{i}-6\hat{j}+2\hat{k}\right) \ \&\ \hat{c}=\frac{1}{7}\left(6\hat{i}+2\hat{j}-3\hat{k}\right)$$

are mutually perpendicular vectors

Solution:

For unit vectors $\hat{a}\ \&\ \hat{b}$, $\left|\hat{a}+\hat{b}\right|=2\cos\dfrac{\theta}{2}$

Substituting the given values we get,

$$\left|\frac{1}{7}\left(2\hat{i}+3\hat{j}+6\hat{k}\right)+\frac{1}{7}\left(3\hat{i}-6\hat{j}+2\hat{k}\right)\right|=\left|\frac{1}{7}\left(5\hat{i}-3\hat{j}+8\hat{k}\right)\right|=2\cos\frac{\theta}{2}$$

Solving LHS, we get

$$\frac{1}{7}\sqrt{5^2+(-3)^2+8^2}=\frac{1}{7}\sqrt{98}=\frac{7\sqrt{2}}{7}=\sqrt{2}\Rightarrow 2\cos\frac{\theta}{2}=\sqrt{2}$$

So, $\cos\dfrac{\theta}{2}=\dfrac{\sqrt{2}}{2}=\dfrac{1}{\sqrt{2}}\Rightarrow\dfrac{\theta}{2}=45^0\Rightarrow\theta=90^0$

Similarly, we can find angle between unit vectors $\hat{b}$ & $\hat{c}$. Let the angle be α

So $\left|\hat{b}+\hat{c}\right| = 2\cos\frac{\alpha}{2} \Rightarrow 2\cos\frac{\alpha}{2} = \left|\frac{1}{7}\left(3\hat{i}-6\hat{j}+2\hat{k}\right)+\frac{1}{7}\left(6\hat{i}+2\hat{j}-3\hat{k}\right)\right| = \left|\frac{1}{7}\left(9\hat{i}-4\hat{j}-1\hat{k}\right)\right| \Rightarrow$

$$2\cos\frac{\alpha}{2} = \frac{1}{7}\sqrt{9^2+(-4)^2+(-1)^2} \Rightarrow 2\cos\frac{\alpha}{2} = \frac{1}{7}\sqrt{98} = \frac{7\sqrt{2}}{7} = \sqrt{2}$$

So $\cos\frac{\alpha}{2} = \frac{\sqrt{2}}{2} = \frac{1}{\sqrt{2}} \Rightarrow \frac{\alpha}{2} = 45^0 \Rightarrow \alpha = 90^0$

Thus the given vectors $\vec{a},\vec{b}$ & $\vec{c}$ are mutually perpendicular.

Example 9

Show that for any two vectors $\vec{a}$ & $\vec{b}$ which are not mutually perpendicular, if $\left(\vec{a}+\vec{b}\right).\left(\vec{a}-\vec{b}\right)=0$, $|\vec{a}| = |\vec{b}|$

Solution:

For vectors $\vec{a}$ & $\vec{b}$, $\left(\vec{a}+\vec{b}\right).\left(\vec{a}-\vec{b}\right) = \left|\vec{a}+\vec{b}\right| \times \left|\vec{a}-\vec{b}\right| \cos\alpha$ ....(1), where α is the angle between $\left(\vec{a}+\vec{b}\right)$ & $\left(\vec{a}-\vec{b}\right)$

Since $\left(\vec{a}+\vec{b}\right).\left(\vec{a}-\vec{b}\right)=0$ (given), we get $\left|\vec{a}+\vec{b}\right| \times \left|\vec{a}-\vec{b}\right| \cos\alpha = 0$

Since $\cos\alpha \neq 0$, hence either $\left|\vec{a}+\vec{b}\right|$ *or* $\left|\vec{a}-\vec{b}\right| = 0$, leading to the conclusion that $\vec{a} = -\vec{b}$ *or* $\vec{a} = \vec{b}$

This leads us to the result that $|\vec{a}| = |\vec{b}|$

14.2.4.3.1.1 *Distributive property of dot product:*

$$\vec{a}.(\vec{b}+\vec{c}) = (\vec{a}.\vec{b}) + (\vec{a}.\vec{c})$$

Proof:

Reference figure 14.8. As we know, dot product of two vectors is nothing but product of magnitude of one vector and magnitude of projection of the other vector in the direction of the first vector.

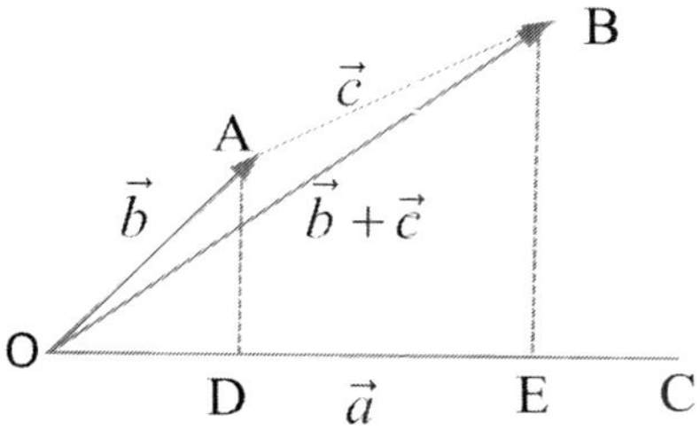

Figure 14.8

Magnitude of vector $\vec{a}$ is OC and that of vectors $\vec{b}$, $\vec{c}$ and $(\vec{b}+\vec{c})$ are OA, AB and OB respectively.

OD is projection of vector $\vec{b}$ on vector $\vec{a}$ while DE and OE are projections of vectors $\vec{c}$ and $(\vec{b}+\vec{c})$ respectively.

Hence $\vec{a}.\vec{b}$ = OC×OD, $\vec{a}.\vec{c}$ = OC×DE and $\vec{a}.\left(\vec{b}+\vec{c}\right)$ = OC×OE

Since OE = OD + DE, we can conclude that OC×OE = OC× OD + OC × DE ⇒

$$\vec{a}.\left(\vec{b}+\vec{c}\right)=\left(\vec{a}.\vec{b}\right)+\left(\vec{a}.\vec{c}\right) \qquad ....14.9$$

Hence proved

Example 10

For what value of λ are the vectors $\vec{a}$ & $\vec{b}$ perpendicular to each other, if $\vec{a}=\lambda\hat{i}+2\hat{j}+\hat{k}$ *and* $\vec{b}=4\hat{i}-9\hat{j}+2\hat{k}$

Solution:

For the two vectors $\vec{a}$ & $\vec{b}$ to be perpendicular, $\vec{a}.\vec{b}=0$

Hence $\left(\lambda\hat{i}+2\hat{j}+\hat{k}\right).\left(4\hat{i}-9\hat{j}+2\hat{k}\right)=0 \Rightarrow 4\lambda-18+2=0$

Or $4\lambda=16 \Rightarrow \lambda=4$

Example 12

Find the angles which the vector $\vec{a}=\hat{i}-\hat{j}+\sqrt{2}\hat{k}$ makes with the coordinate axis.

Solution:

Let α,β,γ be the angles made by the given vector with x, y and z axis respectively.

From the given vector equation, we get value of $|\vec{a}| = \sqrt{1+1+2} = \sqrt{4} = 2$

If we consider dot product of the given vector $\vec{a}$ with x-axis $(\hat{i})$, we get

$\vec{a} \,.\, \hat{i} = |\vec{a}| \times 1 \times \cos\alpha = (\hat{i} - \hat{j} + \sqrt{2}\hat{k}).\hat{i} = 1 \Rightarrow \cos\alpha = \frac{1}{|\vec{a}|} = \frac{1}{2} \Rightarrow \alpha = 60^0$

Similarly for the y-axis, we get

$\vec{a} \,.\, \hat{j} = |\vec{a}| \times 1 \times \cos\beta = |\vec{a}|\cos\beta = (\hat{i} - \hat{j} + \sqrt{2}\hat{k}).\hat{j} =$

$$-1 \Rightarrow \cos\beta = \frac{1}{|\vec{a}|} = -\frac{1}{2} \Rightarrow \beta = 120^0$$

And for the z-axis, we get

$\vec{a} \,.\, \hat{k} = |\vec{a}| \times 1 \times \cos\gamma = |\vec{a}|\cos\gamma = (\hat{i} - \hat{j} + \sqrt{2}\hat{k}).\hat{k} =$

$$\sqrt{2} \Rightarrow \cos\gamma = \frac{\sqrt{2}}{|\vec{a}|} = \frac{1}{\sqrt{2}} \Rightarrow \gamma = 45^0$$

Example 13

Dot product of a vector $\vec{a}$ with $\hat{i} + \hat{j} - 3\hat{k}$, $\hat{i} + 3\hat{j} - 2\hat{k}$ *and* $2\hat{i} + \hat{j} + 4\hat{k}$ are 0, 5 and 8 respectively. Find the vector $\vec{a}$

Solution:

Let the vector $\vec{a}$ be $x\hat{i} + y\hat{j} + z\hat{k}$

Then $\vec{a}.(\hat{i} + \hat{j} - 3\hat{k}) = 0 \Rightarrow x + y - 3z = 0$ ... (1)

Similarly $\vec{a}.(\hat{i} + 3\hat{j} - 2\hat{k}) = 5 \Rightarrow x + 3y - 2z = 5$ ... (2)

And $\vec{a}.(2\hat{i} + \hat{j} + 4\hat{k}) = 8 \Rightarrow 2x + y + 4z = 8$ ...(3)

Subtracting eq. (1) from eq. (2), we get $2y + z = 5$ ...(4)

Similarly by subtracting 2 times eq. (2) from eq. (3), we get $-5y + 8z = -2$ ...(5)

Now adding 5 times eq. (4) to 2 times eq. (5), we get $21z = 21 \Rightarrow \boldsymbol{z = 1}$

Substituting value of z in eq. (4), we get $2y + 1 = 5 \Rightarrow \boldsymbol{y = 2}$

Substituting value of y and z in eq. (2), we get $x + 3 \times 2 - 2 = 5 \Rightarrow \boldsymbol{x = 1}$

Hence vector $\vec{a}$ is $\hat{i} + 2\hat{j} + \hat{k}$

14.2.4.3.2 Vector or Cross Product of two vectors:

The Vector product or cross product of two vectors $\vec{x}$ and $\vec{y}$ is a third vector that is perpendicular to the two original vectors and is denoted as $\vec{x} \times \vec{y}$.

Let us consider a plane containing the two vectors $\vec{x}$ and $\vec{y}$. Then vector representing cross productis perpendicular to this plane implying that the cross product vector is always perpendicular to both component vectors and its direction is determined by Right Hand Screw Rule according to which if the screw rotates in the direction $\vec{x}$ *to* $\vec{y}$ (anticlockwise in this case), the screw will move up.

Geometrically, it can be represented as in figure 14.9.

Mathematically, magnitude of the cross product $\vec{x} \times \vec{y} = |\vec{x}||\vec{y}|\sin\theta$,,,,,Eq. 14.10.

Now we can see from the figure that if we form a parallelogram with vectors $\vec{x}$ & $\vec{y}$ as its sides with angle θ between them, height of the parallelogram so formed is $|y|\sin\theta$. And area of this parallelogram works out to be $|\vec{x}||\vec{y}|\sin\theta$

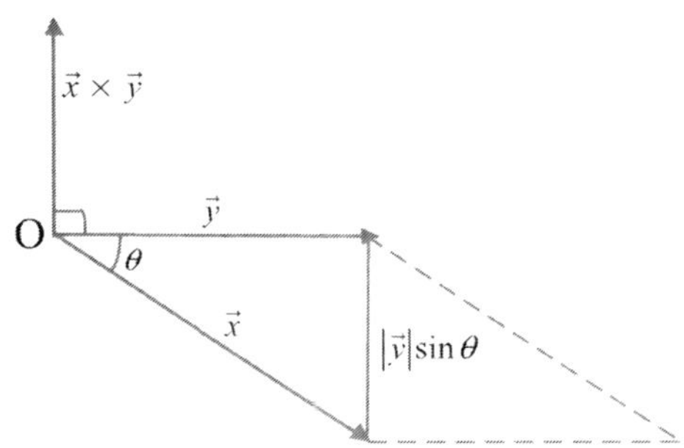

Figure 14.9

Thus magnitude of cross product of two vectors can be understood as area of the parallelogram formed by the two vectors as its arms.

Magnitude *of cross product is maximum if the component vectors are perpendicular to each other and zero if the component vectors overlap or are parallel/anti-parallel.*

Direction of the cross product is determined by Right Hand Thumb Rule which states that if index finger of the Right Hand points in the direction of vector $\vec{x}$ and middle finger in the direction

of vector $\vec{y}$, then thumb points in the direction of vector product.

Another common method of finding direction of the vector product is by the corkscrew right-hand rule. As per this Rule, place a corkscrew in the direction perpendicular to the plane that contains vectors $\vec{x}$ and $\vec{y}$ & turn it in the direction from $\vec{x}$ to $\vec{y}$. The cork screw moves in the direction of vector product. If the direction $\vec{x}$ to $\vec{y}$ is anti-clockwise, $x \times y$ will point upwards, else it points downwards. That is these two vector products are anti-parallel. This implies that $\vec{x} \times \vec{y} = -\vec{y} \times \vec{x}$.

This leads us to an important conclusion regarding Cross product of unit vectors $\hat{i}, \hat{j}$ & $\hat{k}$ along x, y, z axis respectively.

Since $\sin 0^0 = 0$, $\hat{i} \times \hat{i} = \hat{j} \times \hat{j} = \hat{k} \times \hat{k} = 0$

Similarly, since $\sin 90^0 = 1$,

$\hat{i} \times \hat{j} = \hat{k},\ \hat{j} \times \hat{k} = \hat{i}$ & $\hat{k} \times \hat{i} = \hat{j}$ ($\because \hat{i}, \hat{j}$ & $\hat{k}$ are anticlockwise to each other)

$\hat{j} \times \hat{i} = -\hat{k},\ \hat{k} \times \hat{j} = -\hat{i}$ & $\hat{i} \times \hat{k} = -\hat{j}$ ($\because \hat{k}, \hat{j}$ & $\hat{i}$ are clockwise to each other)

Cross product of two vectors in Cartesian form:

Let vectors $\vec{a}$ & $\vec{b}$ in Cartesian coordinates be represented as

$\vec{a} = a_1\hat{i} + a_2\hat{j} + a_3\hat{k}$ & $\vec{b} = b_1\hat{i} + b_2\hat{j} + b_3\hat{k}$

$$\left(\vec{a} \times \vec{b}\right) = \left(a_1\hat{i} + a_2\hat{j} + a_3\hat{k}\right) \times \left(b_1\hat{i} + b_2\hat{j} + b_3\hat{k}\right) \Rightarrow$$

$$\left(\vec{a} \times \vec{b}\right) = \begin{bmatrix} (a_1b_1)\hat{i} \times \hat{i} + (a_1b_2)\hat{i} \times \hat{j} + (a_1b_3)\hat{i} \times \hat{k} + (a_2b_1)\hat{j} \times \hat{i} + (a_2b_2)\hat{j} \times \hat{j} \\ +(a_2b_3)\hat{j} \times \hat{k} + (a_3b_1)\hat{k} \times \hat{i} + (a_3b_2)\hat{k} \times \hat{j} + (a_3b_3)\hat{k} \times \hat{k} \end{bmatrix}$$

Using the property of cross product of unit vectors ($\hat{i}, \hat{j}$ & $\hat{k}$)listed above, the expression changes to

$$\left(\vec{a} \times \vec{b}\right) = \left[(a_2b_3 - a_3b_2)\hat{i} - (a_1b_3 - a_3b_1)\hat{j} + (a_1b_2 - a_2b_1)\hat{k}\right]$$

This can be represented by expansion of the determinant

$$\begin{vmatrix} \hat{i} & \hat{j} & \hat{k} \\ a_1 & a_2 & a_3 \\ b_1 & b_2 & b_3 \end{vmatrix}$$

Hence $\left(\vec{a}\times\vec{b}\right)=\begin{vmatrix}\hat{i} & \hat{j} & \hat{k}\\ a_1 & a_2 & a_3\\ b_1 & b_2 & b_3\end{vmatrix}$ ... Equation 14.11

Cross product is a very important concept that has many applications in Physics.

Interesting corollary related to dot and cross product of two vectors.

We have seen that $\vec{x}.\vec{y}=|\vec{x}||\vec{y}|\cos\theta$ and $\vec{x}\times\vec{y}=|\vec{x}||\vec{y}|\sin\theta$

Squaring and adding, we get

$$(\vec{x}.\vec{y})^2+(\vec{x}\times\vec{y})^2=|\vec{x}|^2|\vec{y}|^2\cos^2\theta+|\vec{x}|^2|\vec{y}|^2\sin^2\theta\Rightarrow$$

Or $(\vec{x}.\vec{y})^2+(\vec{x}\times\vec{y})^2=|\vec{x}|^2|\vec{y}|^2\left(\cos^2\theta+\sin^2\theta\right)=|\vec{x}|^2|\vec{y}|^2$

14.2.4.3.3 Other properties of Cross Product of two vectors:

1. Cross product with a zero vector is zero.

$\vec{a}\times\vec{0}=\vec{0}$ ...Eq.14.12

This is quite evident because with magnitude of one of the vectors being zero, there will be no parallelogram. Hence cross product of any vector with a null vector is zero.

2. Multiplication by a scalar quantity

$c(\vec{a}\times\vec{b})=(c\vec{a}\times\vec{b})=(\vec{a}\times c\vec{b})$ ... Eq. 14.13

Cross product of two vectors ($\vec{a}$ & $\vec{b}$ here) multiplied with a scalar quantity ***c***, implies that area of the parallelogram formed by the two vectors is multiplied by the scalar quantity. Now area of a parallelogram is ***c*** times its original value if either of the arms of the parallelogram are multiplied by the scalar ***c***.

Hence it is clear that $c(\vec{a}\times\vec{b})=(c\vec{a}\times\vec{b})=(\vec{a}\times c\vec{b})$

3. Distributive property:

$\vec{a}\times(\vec{b}+\vec{c})=(\vec{a}\times\vec{b})+(\vec{a}\times\vec{c})$

We have already seen that in terms ofCartesian coordinates, cross product of two vectors $\left(\vec{a}\times\vec{b}\right)=\begin{vmatrix}i & j & k\\ a_1 & a_2 & a_3\\ b_1 & b_2 & b_3\end{vmatrix}$

$$\text{Hence } \vec{a}\times(\vec{b}+\vec{c}) = \begin{vmatrix} i & j & k \\ a_1 & a_2 & a_3 \\ b_1+c_1 & b_2+c_2 & b_3+c_3 \end{vmatrix}$$

Using properties of determinants, we can split Row 3 of the above determinant to get sum of following two determinants:

$$\vec{a}\times(\vec{b}+\vec{c}) = \begin{vmatrix} i & j & k \\ a_1 & a_2 & a_3 \\ b_1 & b_2 & b_3 \end{vmatrix} + \begin{vmatrix} i & j & k \\ a_1 & a_2 & a_3 \\ c_1 & c_2 & c_3 \end{vmatrix} \Rightarrow$$

$$\vec{a}\times(\vec{b}+\vec{c}) = (\vec{a}\times\vec{b}) + (\vec{a}\times\vec{c}) \quad \text{... Eq. 14.14}$$

Geometric proof:

Let us consider figure 14.10 below.

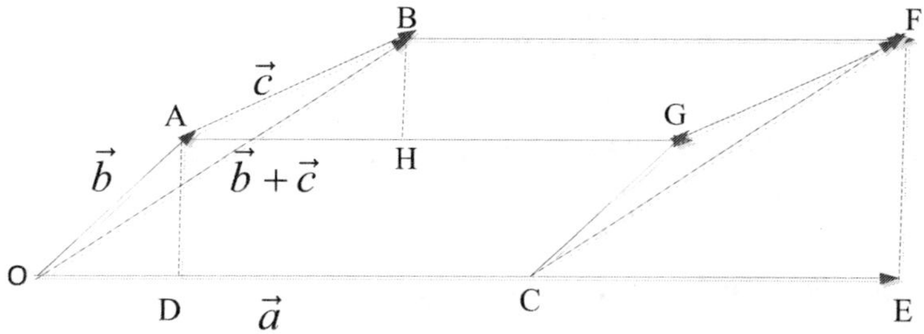

Figure 14.10

In this figure, $\vec{a}$ = OC, $\vec{b}$ = OA, $\vec{c}$ = AB and $(\vec{b}+\vec{c})$ = OB

Knowing that magnitude of cross product of two vectors equals area of the parallelogram formed by them, we can see from figure 14.12 that

$|\vec{a}\times\vec{b}|$ = Area OAGC, $|\vec{a}\times\vec{c}|$ = Area ABFG &

$|\vec{a}\times(\vec{b}+\vec{c})|$ = Area OBFC

Area OAGC = OC×AD, Area ABFG = AG×BH & Area OBFC = OC×FE

We can also observe that AG = OC and FE = AD + BH........ (1)

Hence $(\vec{a}\times\vec{b})$ = OC×AD, $(\vec{a}\times\vec{c})$ = OC×BH and $\vec{a}\times(\vec{b}+\vec{c})$ = OC×FE

Substituting value of FE as at eq. (1) above,

$\vec{a}\times(\vec{b}+\vec{c})=$ OC× (AD + BH)= OC×AD + OC×BH = Area OAGC + Area ABFG

Hence $\vec{a}\times(\vec{b}+\vec{c})=(\vec{a}\times\vec{b})+(\vec{a}\times\vec{c})$

An interesting corollary related to cross product of two vectors:

We can find area of a parallelogram even if we know its diagonal vectors (and not its side vectors).

Proof:

Refer to the figure 14.11. We can see that diagonals d_1 and d_2 can be expressed as

$$\vec{d}_1=\vec{a}+\vec{b} \ \ \& \ \ \vec{d}_2=\vec{b}-\vec{a}$$

$$d_1\times d_2=\left(\vec{a}+\vec{b}\right)\times\left(\vec{b}-\vec{a}\right)\Rightarrow d_1\times d_2=\vec{a}\times\vec{b}-\vec{a}\times\vec{a}+\vec{b}\times\vec{b}-\vec{b}\times\vec{a}$$

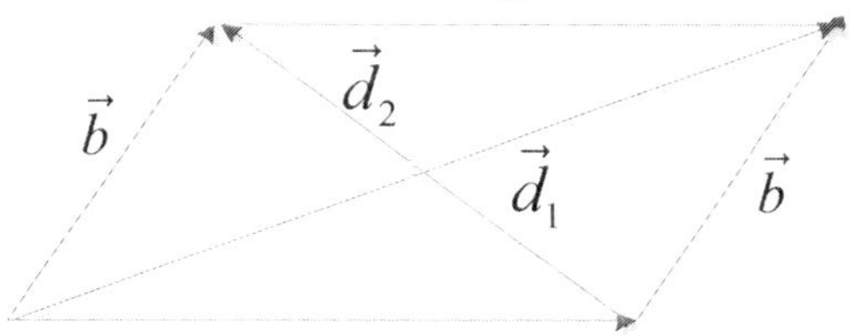

Figure 14.11

Since $\vec{a}\times\vec{a}$ and $\vec{b}\times\vec{b}$ are zero, we get

$$d_1\times d_2=\vec{a}\times\vec{b}+\vec{a}\times\vec{b}\left\{\because -\left(\vec{b}\times\vec{a}\right)=\vec{a}\times\vec{b}\right\} \text{ Or } d_1\times d_2=2\left(\vec{a}\times\vec{b}\right)\Rightarrow$$

Area of the parallelogram $\left(\vec{a}\times\vec{b}\right)=\frac{1}{2}(d_1\times d_2)$

Example: 14

If $\vec{a}=3\hat{i}+4\hat{j}$ and $\vec{b}=\hat{i}+\hat{j}+\hat{k}$, find value of $\left|\vec{a}\times\vec{b}\right|$

Solution:

$$\left(\vec{a}\times\vec{b}\right)=\begin{vmatrix}\hat{i} & \hat{j} & \hat{k}\\ a_1 & a_2 & a_3\\ b_1 & b_2 & b_3\end{vmatrix}$$

Substituting the values of Cartesian coordinates of $\vec{a}$ & $\vec{b}$, we get

$$\left(\vec{a}\times\vec{b}\right)=\begin{vmatrix}\hat{i} & \hat{j} & \hat{k}\\ 3 & 4 & 0\\ 1 & 1 & 1\end{vmatrix}=\hat{i}\times(4\times1-1\times0)-\hat{j}\times(3\times1-1\times0)+\hat{k}\times(3\times1-1\times4)\Rightarrow$$

$$\left(\vec{a}\times\vec{b}\right)=4\hat{i}-3\hat{j}+\hat{k}\times(-1)\Rightarrow\left|\vec{a}\times\vec{b}\right|=\sqrt{4^2+(-3)^2+(-1)^2}=\sqrt{26}$$

Example: 15

If $\vec{a}=3\hat{i}-\hat{j}-2\hat{k}$ and $\vec{b}=2\hat{i}+3\hat{j}+\hat{k}$, find $\left(\vec{a}+2\vec{b}\right)\times\left(2\vec{a}-\vec{b}\right)$

Solution:

$\left(\vec{a}+2\vec{b}\right)=7\hat{i}+5\hat{j}$ & $\left(2\vec{a}-\vec{b}\right)=4\hat{i}-5\hat{j}-5\hat{k}$ Hence

$$\left(\vec{a}+2\vec{b}\right)\times\left(2\vec{a}-\vec{b}\right)=\begin{vmatrix}\hat{i} & \hat{j} & \hat{k}\\ 7 & 5 & 0\\ 4 & -5 & -5\end{vmatrix}=$$

$$\hat{i}\times(5\times(-5))-\hat{j}\times(7\times(-5))+\hat{k}\times(7\times(-5)-4\times5)\Rightarrow$$

$$\left(\vec{a}+2\vec{b}\right)\times\left(2\vec{a}-\vec{b}\right)=-25\hat{i}+35\hat{j}-55\hat{k}$$

Example: 16

If $\vec{a}=\hat{i}+\hat{j}-\hat{k}$, $\vec{b}=\hat{i}+\hat{j}$, $\vec{c}=\hat{i}-2\hat{j}-2\hat{k}$, $\vec{r}\times\vec{a}=\vec{c}\times\vec{a}$ & $\vec{r}.\vec{b}=0$, then find $\vec{r}.\vec{a}$

Solution:

$$\vec{r}\times\vec{a}=\vec{c}\times\vec{a}\Rightarrow(\vec{r}-\vec{c})\times\vec{a}=0$$

This means that vector $(\vec{r}-\vec{c})$ is parallel to vector $\vec{a}$.

Hence

$$(\vec{r}-\vec{c})=\lambda\vec{a}\Rightarrow\vec{r}=\vec{c}+\lambda\vec{a}\Rightarrow\vec{r}=(1+\lambda)\hat{i}+(-2+\lambda)\hat{j}+(-1-\lambda)\hat{k}$$

$$\vec{r}.\vec{b}=0\Rightarrow(1+\lambda)+(-2+\lambda)=0\Rightarrow2\lambda-1=0\Rightarrow\lambda=\frac{1}{2}$$

Hence $\vec{r}.\vec{a}=(1+\lambda)+(-2+\lambda)-(-1-\lambda)=3\lambda=\frac{3}{2}$

Example: 17

Find area of the parallelogram whose diagonals are $4\hat{i}-\hat{j}-3\hat{k}$ and $-2\hat{i}+\hat{j}-2\hat{k}$

Solution:

We have seen that if the two diagonals of a parallelogram are $\vec{d}_1$ & $\vec{d}_2$, its area $=\frac{1}{2}\left|\vec{d}_1\times\vec{d}_2\right|$

Here $\vec{d}_1=4\hat{i}-\hat{j}-3\hat{k}$ & $\vec{d}_2=-2\hat{i}+\hat{j}-2\hat{k}$

Hence Area of the parallelogram $=\frac{1}{2}\left|d_1\times d_2\right|=\begin{vmatrix}\hat{i} & \hat{j} & \hat{k}\\ 4 & -1 & -3\\ -2 & 1 & -2\end{vmatrix}$

$$=\left|\hat{i}\times(-1\times(-2)-1\times(-3))-\hat{j}\times(4\times(-2)-(-2)\times(-3))+\hat{k}\times(4\times1-(-2)\times(-1))\right|$$

Or $\left|d_1\times d_2\right|=2\times\left|\hat{i}\times(2+3)-\hat{j}\times(-8-6)+\hat{k}\times(4-2)\right|=2\times\left|5\hat{i}+14\hat{j}+2\hat{k}\right|$

So Area of the parallelogram =

$$\frac{1}{2}\times\sqrt{5^2+14^2+2^2}=\frac{1}{2}\times\sqrt{25+196+4}=\frac{15}{2}$$

14.2.5 Product of three or more vectors:

Three or more vectors may have product as mix of dot or cross product. If $\vec{a}$, $\vec{b}$, $\vec{c}$ & $\vec{d}$ are any four vectors, properties of their dot/ cross products are as follows:

14.2.5.1 Dot product:

$$\left(\vec{a}.\vec{b}\right)\vec{c}\neq\vec{a}\left(\vec{b}.\vec{c}\right)$$

Proof:

We know that dot product of two vectors is a scalar quantity determined by magnitude of the two vectors and angle between them. Since $\vec{a}$ & $\vec{c}$ are different vectors with different magnitude and varying angles to vector $\vec{b}$, scalar product $\left(\vec{a}.\vec{b}\right)\neq\left(\vec{b}.\vec{c}\right)$ and also scalar multiple of vectors $\vec{a}$ & $\vec{c}$ with different multipliers will also be unequal.

Hence $\left(\vec{a}.\vec{b}\right)\vec{c} \neq \vec{a}\left(\vec{b}.\vec{c}\right)$ ...Eq. 14.15

14.2.5.2 *Mix of Dot and Cross product*:

1. $\vec{a}.\left(\vec{b}\times\vec{c}\right)=\begin{vmatrix} a_1 & a_2 & a_3 \\ b_1 & b_2 & b_3 \\ c_1 & c_2 & c_3 \end{vmatrix}$

Proof:

Let vectors $\vec{a}$, $\vec{b}$ & $\vec{c}$ in Cartesian coordinates be represented as $\vec{a}=a_1\hat{i}+a_2\hat{j}+a_3\hat{k}$, $\vec{b}=b_1\hat{i}+b_2\hat{j}+b_3\hat{k}$ & $\vec{c}=c_1\hat{i}+c_2\hat{j}+c_3\hat{k}$

$$\left(\vec{b}\times\vec{c}\right)=\left(b_1\hat{i}+b_2\hat{j}+b_3\hat{k}\right)\times\left(c_1\hat{i}+c_2\hat{j}+c_3\hat{k}\right)\Rightarrow$$

$$\left(\vec{b}\times\vec{c}\right)=\begin{bmatrix}(b_1c_1)\hat{i}\times\hat{i}+(b_1c_2)\hat{i}\times\hat{j}+(b_1c_3)\hat{i}\times\hat{k}+(b_2c_1)\hat{j}\times\hat{i}+(b_2c_2)\hat{j}\times\hat{j} \\ +(b_2c_3)\hat{j}\times\hat{k}+(b_3c_1)\hat{k}\times\hat{i}+(b_3c_2)\hat{k}\times\hat{j}+(b_3c_3)\hat{k}\times\hat{k}\end{bmatrix}$$

Using the property of cross product of unit vectors $\hat{i},\hat{j},\hat{k}$, the expression changes to

$$\left(\vec{b}\times\vec{c}\right)=\Big[(b_1c_2)\hat{k}+(b_1c_3)\left(-\hat{j}\right)+(b_2c_1)\left(-\hat{k}\right)+(b_2c_3)\hat{i}+$$

$$(b_3c_1)\hat{j}+(b_3c_2)\left(-\hat{i}\right)\Big]\Rightarrow$$

$$\left(\vec{b}\times\vec{c}\right)=\left[(b_2c_3-b_3c_2)\hat{i}+(b_3c_1-b_1c_3)\hat{j}+(b_1c_2-b_2c_1)\hat{k}\right]\Rightarrow$$

Now taking dot product of $\left(a_1\hat{i}+a_2\hat{j}+a_3\hat{k}\right)$ with the above expression we get

$$\vec{a}.\left(\vec{b}\times\vec{c}\right)=\left(a_1\hat{i}+a_2\hat{j}+a_3\hat{k}\right).\left[(b_2c_3-b_3c_2)\hat{i}+(b_3c_1-b_1c_3)\hat{j}+(b_1c_2-b_2c_1)\hat{k}\right]\Rightarrow$$

$$\vec{a}.\left(\vec{b}\times\vec{c}\right)=\begin{bmatrix}(a_1b_2c_3-a_1b_3c_2)\hat{i}.\hat{i}+(a_1b_3c_1-a_1b_1c_3)\hat{i}.\hat{j}+(a_1b_1c_2-a_1b_2c_1)\hat{i}.\hat{k}+ \\ (a_2b_2c_3-a_2b_3c_2)\hat{j}.\hat{i}+(a_2b_3c_1-a_2b_1c_3)\hat{j}.\hat{j}+(a_2b_1c_2-a_2b_2c_1)\hat{j}.\hat{k}+ \\ (a_3b_2c_3-a_3b_3c_2)\hat{k}.\hat{i}+(a_3b_3c_1-a_3b_1c_3)\hat{k}.\hat{j}+(a_3b_1c_2-a_3b_2c_1)\hat{k}.\hat{k}\end{bmatrix}\Rightarrow$$

Using the property of dot product of unit vectors $\hat{i},\hat{j},\hat{k}$, the expression changes to

$$\vec{a}.\left(\vec{b}\times\vec{c}\right)=\left[\left(a_1b_2c_3-a_1b_3c_2\right)+\left(a_2b_3c_1-a_2b_1c_3\right)+\left(a_3b_1c_2-a_3b_2c_1\right)\right]$$

$$=\left[a_1\left(b_2c_3-b_3c_2\right)-a_2\left(b_1c_3-b_3c_1\right)+a_3\left(b_1c_2-b_2c_1\right)\right]$$

This is nothing but expansion of determinant $\begin{vmatrix} a_1 & a_2 & a_3 \\ b_1 & b_2 & b_3 \\ c_1 & c_2 & c_3 \end{vmatrix}$

Hence proved that $\vec{a}.\left(\vec{b}\times\vec{c}\right)=\begin{vmatrix} a_1 & a_2 & a_3 \\ b_1 & b_2 & b_3 \\ c_1 & c_2 & c_3 \end{vmatrix}$ ...Eq. 14.16

An easy way to remember this equality is to recall that

$\vec{b}\times\vec{c}=\begin{vmatrix} \hat{i} & \hat{j} & \hat{k} \\ b_1 & b_2 & b_3 \\ c_1 & c_2 & c_3 \end{vmatrix}$ and to get, $\vec{a}\left(\vec{b}\times\vec{c}\right)$, we replace $\hat{i}$, $\hat{j}$ & $\hat{k}$ by components of vector $\vec{a}\left(a_1, a_2 \;\&\; a_3\right)$ respectively.

2. $\vec{a}.\left(\vec{b}\times\vec{c}\right)=\vec{b}.\left(\vec{c}\times\vec{a}\right)=\vec{c}.\left(\vec{c}\times\vec{b}\right)$ Eq.14.17

Proof:

By Cartesian coordinates:

From properties of determinants we know that if any two rows of a determinant are interchanged, then sign of the value of determinant changes. So if rows are interchanged twice, value of the determinant remains unchanged.

Hence

$$\begin{vmatrix} a_1 & a_2 & a_3 \\ b_1 & b_2 & b_3 \\ c_1 & c_2 & c_3 \end{vmatrix}=\begin{vmatrix} b_1 & b_2 & b_3 \\ c_1 & c_2 & c_3 \\ a_1 & a_2 & a_3 \end{vmatrix}=\begin{vmatrix} c_1 & c_2 & c_3 \\ a_1 & a_2 & a_3 \\ b_1 & b_2 & b_3 \end{vmatrix}\Rightarrow\vec{a}.\left(\vec{b}\times\vec{c}\right)=\vec{b}.\left(\vec{c}\times\vec{a}\right)=\vec{c}.\left(\vec{a}\times\vec{b}\right)$$

An easy way to recall this identity of scalar triple products is that we can interchange dot and cross products as depicted below:

$\vec{a}.\left(\vec{b}\times\vec{c}\right)=\left(\vec{a}\times\vec{b}\right).\vec{c}=\vec{c}.\left(\vec{a}\times\vec{b}\right)$ $\because$ ***value of scalar product remains unchanged by interchanging the vectors.***

Similarly $\vec{c}.\left(\vec{a}\times\vec{b}\right)=\left(\vec{c}\times\vec{a}\right).\vec{b}=\vec{b}.\left(\vec{c}\times\vec{a}\right)$

Geometric proof:

Reference figure 14.12 below.

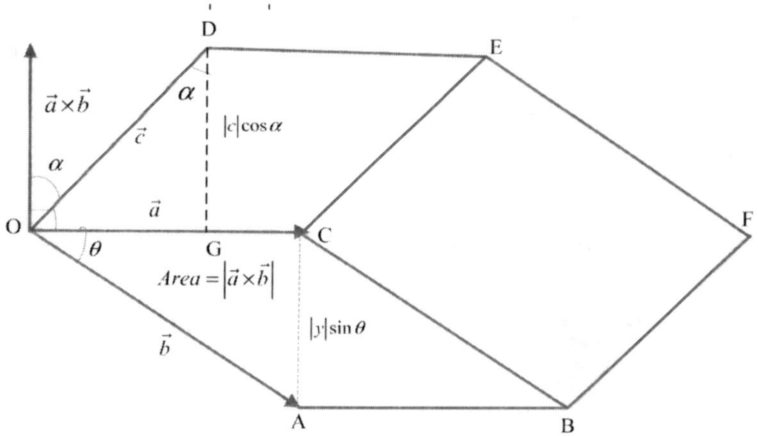

Figure 14.12

We have already seen in section 14.2.4.3.2 that magnitude of $\vec{a}\times\vec{b}$ is equal to area of the parallelogram formed by the two vectors as its sides. Hence in figure 14.12, $\vec{a}\times\vec{b}$ = area of the parallelogram OABC. Vector $\vec{c}$ in the figure forms the third side of the parallelepiped formed by vectors $\vec{a},\vec{b}$ & $\vec{c}$. DG is perpendicular to the side OC of the parallelogram. If DG makes an angle α to side OD (vector $\vec{c}$), height of the parallelepiped = $|c|\cos\alpha$ yielding

Volume of the parallelepiped = $\left|\vec{a}\times\vec{b}\right|\times|c|\cos\alpha$

Since angle between vector $\vec{a}\times\vec{b}$ and vector $\vec{c}$ is α (being alternate angle as vector $\vec{a}\times\vec{b}$ is also perpendicular to the base parallelogram), $\left(\vec{a}\times\vec{b}\right).\vec{c} = \left|\vec{a}\times\vec{b}\right|\times|c|\cos\alpha$ ***Therefore*** $\left(\vec{a}\times\vec{b}\right).\vec{c}$ ***being a scalar quantity equals volume of the parallelepiped formed by the three vectors.***

There is an interesting implication of this result i.e. if the three vectors are co-planar, volume of parallelepiped formed by them will be zero. In other words, if vectors $\vec{a},\vec{b}$ & $\vec{c}$ ***are coplanar,*** $\left(\vec{a}\times\vec{b}\right).\vec{c} = 0$ ***This is also so because in this case,*** $\left(\vec{a}\times\vec{b}\right)$ ***is perpendicular to*** $\vec{c}$ ***i.e.*** α = ***90° leading*** $\cos\alpha = 0$.

14.2.5.3 Triple Cross product:

3. $\vec{a}\times(\vec{b}\times\vec{c})=(\vec{a}.\vec{c})\vec{b}-(\vec{a}.\vec{b})\vec{c}$ ...14.18

Proof:

The cross product of vector $\vec{a}$ with the cross product of the other two vectors $\vec{b}$ & $\vec{c}$ i.e. $\vec{a}\times(\vec{b}\times\vec{c})$ is the triple cross product of the given vectors. ***The result of the triple cross product is a vector that lies in the plane*** of the first two vectors. This is so because vector $(\vec{b}\times\vec{c})$ ***is perpendicular to the plane containing vectors*** $\vec{b}$ & $\vec{c}$ ***and similarly vectors*** $\vec{a}$ ***and*** $(\vec{b}\times\vec{c})$ ***have to be coplanar and their cross product will be perpendicular to the vectors*** $\vec{a}$ ***as well as*** $(\vec{b}\times\vec{c})$ ***implying thereby that vector*** $\vec{a}\times(\vec{b}\times\vec{c})$ ***will be in the same plane as*** $\vec{b}$ & $\vec{c}$.

Refer to figure 14.12.

Algebraically, let vectors $\vec{a}$, $\vec{b}$ & $\vec{c}$ in Cartesian coordinates be represented as $\vec{a}=a_1\hat{i}+a_2\hat{j}+a_3\hat{k}$; $\vec{b}=b_1\hat{i}+b_2\hat{j}+b_3\hat{k}$ and $\vec{c}=c_1\hat{i}+c_2\hat{j}+c_3\hat{k}$

$$\vec{a}\times\left(\vec{b}\times\vec{c}\right)=\left(a_1\hat{i}+a_2\hat{j}+a_3\hat{k}\right)\times\left[\left(b_1\hat{i}+b_2\hat{j}+b_3\hat{k}\right)\times\left(c_1\hat{i}+c_2\hat{j}+c_3\hat{k}\right)\right]$$

Expanding $\left(\vec{b}\times\vec{c}\right)=\left[\left(b_1\hat{i}+b_2\hat{j}+b_3\hat{k}\right)\times\left(c_1\hat{i}+c_2\hat{j}+c_3\hat{k}\right)\right]$ we get

$$\left(\vec{b}\times\vec{c}\right)=\begin{bmatrix}(b_1c_1)\hat{i}\times\hat{i}+(b_1c_2)\hat{i}\times\hat{j}+(b_1c_3)\hat{i}\times\hat{k}+(b_2c_1)\hat{j}\times\hat{i}+(b_2c_2)\hat{j}\times\hat{j}\\+(b_2c_3)\hat{j}\times\hat{k}+(b_3c_1)\hat{k}\times\hat{i}+(b_3c_2)\hat{k}\times\hat{j}+(b_3c_3)\hat{k}\times\hat{k}\end{bmatrix}$$

Using the property of cross product of unit vectors $\hat{i},\hat{j}$ & $\hat{k}$, the expression changes to

$$\left(\vec{b}\times\vec{c}\right)=\left[(b_1c_2)\hat{k}+(b_1c_3)\left(-\hat{j}\right)+(b_2c_1)\left(-\hat{k}\right)+(b_2c_3)\hat{i}+(b_3c_1)\hat{j}+(b_3c_2)\left(-\hat{i}\right)\right]\Rightarrow$$

$$\left(\vec{b}\times\vec{c}\right)=\left[(b_2c_3-b_3c_2)\hat{i}+(b_3c_1-b_1c_3)\hat{j}+(b_1c_2-b_2c_1)\hat{k}\right]\Rightarrow$$

$$\vec{a}\times\left(\vec{b}\times\vec{c}\right)=\left(a_1\hat{i}+a_2\hat{j}+a_3\hat{k}\right)\times\left[(b_2c_3-b_3c_2)\hat{i}+(b_3c_1-b_1c_3)\hat{j}+(b_1c_2-b_2c_1)\hat{k}\right]\Rightarrow$$

$$\vec{a}\times\left(\vec{b}\times\vec{c}\right)=$$

$$\begin{bmatrix}\left(a_1b_2c_3-a_1b_3c_2\right)\hat{i}\times\hat{i}+\left(a_1b_3c_1-a_1b_1c_3\right)\hat{i}\times\hat{j}+\left(a_1b_1c_2-a_1b_2c_1\right)\hat{i}\times\hat{k}+\\ \left(a_2b_2c_3-a_2b_3c_2\right)\hat{j}\times\hat{i}+\left(a_2b_3c_1-a_2b_1c_3\right)\hat{j}\times\hat{j}+\left(a_2b_1c_2-a_2b_2c_1\right)\hat{j}\times\hat{k}+\\ \left(a_3b_2c_3-a_3b_3c_2\right)\hat{k}\times\hat{i}+\left(a_3b_3c_1-a_3b_1c_3\right)\hat{k}\times\hat{j}+\left(a_3b_1c_2-a_3b_2c_1\right)\hat{k}\times\hat{k}\end{bmatrix}$$

Once again using the property of cross product of unit vectors $\hat{i}, \hat{j}$ & $\hat{k}$, the expression changes to

$$\vec{a}\times\left(\vec{b}\times\vec{c}\right)=$$

$$\begin{bmatrix}\left(a_1b_3c_1-a_1b_1c_3\right)\hat{k}+\left(a_1b_1c_2-a_1b_2c_1\right)\left(-\hat{j}\right)+\left(a_2b_2c_3-a_2b_3c_2\right)\left(-\hat{k}\right)+\\ \left(a_2b_1c_2-a_2b_2c_1\right)\hat{i}+\left(a_3b_2c_3-a_3b_3c_2\right)\hat{j}+\left(a_3b_3c_1-a_3b_1c_3\right)\left(-\hat{i}\right)\end{bmatrix}\Rightarrow$$

$$\vec{a}\times\left(\vec{b}\times\vec{c}\right)=$$

$$\left.\begin{matrix}\left(a_2b_1c_2-a_2b_2c_1-a_3b_3c_1+a_3b_1c_3\right)\hat{i}+\left(a_3b_2c_3-a_3b_3c_2-a_1b_1c_2+a_1b_2c_1\right)\hat{j}+\\ \left(a_1b_3c_1-a_1b_1c_3-a_2b_2c_3+a_2b_3c_2\right)\hat{k}\end{matrix}\right]\Rightarrow$$

By adding and subtracting similar terms $a_1b_1c_1$, $a_2b_2c_2$ and $a_3b_3c_3$ in $\hat{i}$, $\hat{j}$ and $\hat{k}$ part of the above expression respectively, we get

$$\vec{a}\times\left(\vec{b}\times\vec{c}\right)=\begin{bmatrix}\left(a_2b_1c_2-a_2b_2c_1-a_3b_3c_1+a_3b_1c_3+a_1b_1c_1-a_1b_1c_1\right)\hat{i}+\\ \left(a_3b_2c_3-a_3b_3c_2-a_1b_1c_2+a_1b_2c_1+a_2b_2c_2-a_2b_2c_2\right)\hat{j}+\\ \left(a_1b_3c_1-a_1b_1c_3-a_2b_2c_3+a_2b_3c_2+a_3b_3c_3-a_3b_3c_3\right)\hat{k}\end{bmatrix}\Rightarrow$$

$$\vec{a}\times\left(\vec{b}\times\vec{c}\right)=\begin{bmatrix}b_1\left(a_1c_1+a_2c_2+a_3c_3\right)\hat{i}-c_1\left(a_1b_1+a_2b_2+a_3b_3\right)\hat{i}+\\ b_2\left(a_1c_1+a_2c_2+a_3c_3\right)\hat{j}-c_2\left(a_1b_1+a_2b_2+a_3b_3\right)\hat{j}+\\ b_3\left(a_1b_3c_1+a_2b_3c_2+a_3b_3c_3\right)\hat{k}-c_3\left(a_1b_1+a_2b_2+a_3b_3\right)\hat{k}\end{bmatrix}$$

Substituting $\vec{a}.\vec{b}=a_1b_1+a_2b_2+a_3b_3$ and $\vec{a}.\vec{c}=a_1c_1+a_2c_2+a_3c_3$ (as derived in section 14.2.4.3.1) in the above expression, we get

$$\vec{a}\times\left(\vec{b}\times\vec{c}\right)=\left[b_1\left(\vec{a}.\vec{c}\right)\hat{i}-c_1\left(\vec{a}.\vec{b}\right)\hat{i}+b_2\left(\vec{a}.\vec{c}\right)\hat{j}-c_2\left(\vec{a}.\vec{b}\right)\hat{j}+b_3\left(\vec{a}.\vec{c}\right)\hat{k}-c_3\left(\vec{a}.\vec{b}\right)\hat{k}\right]\Rightarrow$$

$$\vec{a}\times\left(\vec{b}\times\vec{c}\right)=\left[\left(\vec{a}.\vec{c}\right)\left(b_1\hat{i}+b_2\hat{j}+b_3\hat{k}\right)-\left(\vec{a}.\vec{b}\right)\left(c_1\hat{i}+c_2\hat{j}+c_3\hat{k}\right)\right]=\left(\vec{a}.\vec{c}\right)\vec{b}-\left(\vec{a}.\vec{b}\right)\vec{c}$$

Hence proved.

The above triple cross product identity can also be written as $\vec{a}\times\left(\vec{b}\times\vec{c}\right)=\vec{b}\left(\vec{a}.\vec{c}\right)-\vec{c}\left(\vec{a}.\vec{b}\right)$

We can notice that the dot product being a scalar, triple cross product $\vec{a}\times(\vec{b}\times\vec{c})$ ***ends up being a vector in the plane of vectors*** $\vec{b}$ & $\vec{c}$***, as mentioned already.***

An easy way to remember this triple cross product identity is that "the first vector ($\vec{a}$ here) attracts the farther vector first ($\vec{c}$) and the nearer vector ($\vec{b}$) later and changes cross to dot product. The resulting dot products, simply multiply with the third vector in each case and we subtract second product from the first".

Another easy way to recall is that triple cross product of ABC is $\left(\vec{a}\times\vec{b}\right)\times\vec{c}=BAC-CAB$

Example 17:

$\vec{c}$ is coplanar with $\vec{a}=-\hat{i}+\hat{j}+\hat{k}$ and $\vec{b}=2\hat{i}+\hat{k}$, $\vec{a}.\vec{c}=7$ and $\vec{c}\perp\vec{b}$ find the value of $2\left|\vec{a}+\vec{b}+\vec{c}\right|^2$

Solution:

We have to find value of $\left|\vec{a}+\vec{b}+\vec{c}\right|^2$ which equals $\left(\vec{a}+\vec{b}+\vec{c}\right).\left(\vec{a}+\vec{b}+\vec{c}\right)$. Vectors $\vec{a}$ and $\vec{b}$ are known. So we need to find vector $\vec{c}$

Let $\vec{c}=c_1\hat{i}+c_2\hat{j}+c_3\hat{k}$

Since vectors $\vec{a},\vec{b}$ and $\vec{c}$ are coplanar, $\vec{c}.\left(\vec{a}\times\vec{b}\right)=0$

Using property of triple dot product, we get

$$\vec{c}.\left(\vec{a}\times\vec{b}\right)=\begin{vmatrix}c_1 & c_2 & c_3\\ a_1 & a_2 & a_3\\ b_1 & b_2 & b_3\end{vmatrix}=\begin{vmatrix}c_1 & c_2 & c_3\\ -1 & 1 & 1\\ 2 & 0 & 1\end{vmatrix}=c_1-c_2\left(-1-2\right)+c_3\left(-2\right)=0\Rightarrow$$

$$c_1+3c_2-2c_3=0 \qquad ...(1)$$

Since $\vec{a}.\vec{c} = 7 \Rightarrow c_1 \times (-1) + c_2 \times 1 + c_3 \times 1 = -c_1 + c_2 + c_3 = 7$ ...(2)

Since $\vec{c} \perp \vec{b} \Rightarrow \vec{c}.\vec{b} = 0 \Rightarrow c_1 \times 2 + c_3 \times 1 = 2c_1 + c_3 = 0 \Rightarrow c_3 = -2c_1$ ...(3)

Substituting value of c_3 from equation (3) in equations (1) and (2), we get

$$c_1 + 3c_2 - 2\times(-2c_1) = 0 \Rightarrow 5c_1 + 3c_2 = 0 \qquad ...(4)$$

And $-c_1 + c_2 - 2c_1 = 7 \Rightarrow -3c_1 + c_2 = 7$ ...(5)

Multiplying (5) by 3 and subtracting from (4) we get,

$$-14c_1 = 21 \Rightarrow c_1 = -\frac{3}{2}$$

Hence from (4), $-\frac{3}{2}\times 5 + 3c_2 = 0 \Rightarrow c_2 = \frac{5}{2}$ and

from (2) we get $\frac{3}{2} + \frac{5}{2} + c_3 = 7 \Rightarrow c_3 = 3$

Hence $\vec{c} = -\frac{3}{2}\hat{i} + \frac{5}{2}\hat{j} + 3\hat{k}$

So

$$\left(\vec{a} + \vec{b} + \vec{c}\right) = \left(-1 + 2 - \frac{3}{2}\right)\hat{i} + \left(1 + \frac{5}{2}\right)\hat{j} + (1 + 1 + 3)\hat{k} = -\frac{1}{2}\hat{i} + \frac{7}{2}\hat{j} + 5\hat{k}$$

$$2\left|\vec{a} + \vec{b} + \vec{c}\right|^2 = 2\times\left[\left(-\frac{1}{2}\right)^2 + \left(\frac{7}{2}\right)^2 + 5^2\right]$$

$$= 2\times\left(\frac{1}{4} + \frac{49}{4} + 25\right) = 2\times\left(\frac{25}{2} + 25\right) = 75$$

Alternate (a smarter) solution:

The above solution is a normal way of solving this problem. If we consider the given information carefully, we can find the following alternate solution.

It is given that $\vec{c}$ is coplanar with $\vec{a}$ & $\vec{b}$ and $\vec{c} \perp \vec{b}$.

We know that $\vec{a} \times \vec{b}$ is perpendicular to the plane containing $\vec{a}$ & $\vec{b}$. Now if we consider cross product of $\vec{b}$ & $\left(\vec{a} \times \vec{b}\right)$, it will be perpendicular to both these vectors implying thereby that it will be coplanar with $\vec{a} \times \vec{b}$ and perpendicular to $\vec{b}$.

Since $\vec{c}$ is also coplanar with $\vec{a}$ & $\vec{b}$ and perpendicular to $\vec{b}$, it is clear that $\vec{c}$ and $\vec{b}\times\left(\vec{a}\times\vec{b}\right)$ are parallel.

So vector $\vec{c}$ can be expressed as

$$\vec{c}=\lambda\left\{\vec{b}\times\left(\vec{a}\times\vec{b}\right)\right\}=\lambda\left\{\left(\vec{b}.\vec{b}\right)\vec{a}-\left(\vec{b}.\vec{a}\right)\vec{b}\right\}=\lambda\left(\left|\vec{b}\right|^2\vec{a}-\left(\vec{b}.\vec{a}\right)\vec{b}\right)$$

From the given vector equations of vectors $\vec{a}$ & $\vec{b}$,

$\left|\vec{b}\right|^2=2^2+1^2=5$ and $\left(\vec{b}.\vec{a}\right)=(-1)\times2+1\times0+1\times1=-1$

Hence

$$\vec{c}=\lambda\left(\left|\vec{b}\right|^2\vec{a}-\left(\vec{b}.\vec{a}\right)\vec{b}\right)=\lambda\left\{5\left(-\hat{i}+\hat{j}+\hat{k}\right)-(-1)\left(\hat{i}+2\hat{k}\right)\right\}=$$

$$\lambda\left(-6\hat{i}+5\hat{j}+3\hat{k}\right)$$

Now since it is given that $\vec{a}\,\vec{c}=7$,

$$\left(-\hat{i}+\hat{j}+\hat{k}\right).\left\{\lambda\left(-6\hat{i}+5\hat{j}+3\hat{k}\right)\right\}=7\Rightarrow\lambda(6+5+3)=7\Rightarrow\lambda=\frac{1}{2}$$

Hence $\vec{c}=\frac{1}{2}\times\left(-6\hat{i}+5\hat{j}+3\hat{k}\right)=\left(-3\hat{i}+\frac{5}{2}\hat{j}+\frac{3}{2}\hat{k}\right)$

This is the same equation of vector $\vec{c}$ derived in the earlier solution, giving us the final result as $2\left|\vec{a}+\vec{b}+\vec{c}\right|^2=75$

4. $\left(\vec{a}\times\vec{b}\right)\times\vec{c}=\left[\left(\vec{b}.\vec{c}\right)\vec{a}-\left(\vec{a}.\vec{c}\right)\vec{b}\right]$ ...Eq. 14.19

Proof:

Using cross product of $\left(\vec{b}\times\vec{c}\right)$ derived in property 3 above, we can see that

$$\left(\vec{a}\times\vec{b}\right)=\left[\left(a_2b_3-a_3b_2\right)\hat{i}+\left(a_3b_1-a_1b_3\right)\hat{j}+\left(a_1b_2-a_2b_1\right)\hat{k}\right]\Rightarrow$$

Now cross multiplying the above expression with $\left(c_1\hat{i}+c_2\hat{j}+c_3\hat{k}\right)$ we get

$$\left(\vec{a}\times\vec{b}\right)\times\vec{c}=\left[(a_2b_3-a_3b_2)\hat{i}+(a_3b_1-a_1b_3)\hat{j}+(a_1b_2-a_2b_1)\hat{k}\right]\times\left(c_1\hat{i}+c_2\hat{j}+c_3\hat{k}\right)$$

Knowing that

$\hat{i}\times\hat{i}=\hat{j}\times\hat{j}=\hat{k}\times\hat{k}=0$, $\hat{i}\times\hat{j}=\hat{k}$, $\hat{j}\times\hat{k}=\hat{i}$ & $\hat{k}\times\hat{i}=\hat{j}$ and $\hat{j}\times\hat{i}=-\hat{k}$, $\hat{k}\times\hat{j}=-\hat{i}$ & $\hat{i}\times\hat{k}=-\hat{j}$, we get

$$\left(\vec{a}\times\vec{b}\right)\times\vec{c}=\left[\begin{matrix}\{c_2(a_1b_2-a_2b_1)-c_3(a_3b_1-a_1b_3)\}\hat{i}+\{c_3(a_2b_3-a_3b_2)-c_1(a_1b_2-a_2b_1)\}\hat{j}\\+\{c_1(a_3b_1-a_1b_3)-c_2(a_2b_3-a_3b_2)\}\hat{k}\end{matrix}\right]$$

$$\left(\vec{a}\times\vec{b}\right)\times\vec{c}=\left[\left\{\begin{matrix}(a_1b_2c_2+a_1b_3c_3-a_3b_1c_3-a_2b_1c_2)\hat{i}+(a_2b_3c_3+a_2b_1c_1-a_3b_2c_3-a_1b_2c_1)\hat{j}\\+(a_3b_1c_1+a_3b_2c_2-a_1b_3c_1-a_2b_3c_2)\hat{k}\end{matrix}\right\}\right]$$

By adding and subtracting similar terms- $a_1b_1c_1$, $a_2b_2c_2$ and $a_3b_3c_3$ in $\hat{i}$, $\hat{j}$ and $\hat{k}$ part of the above expression respectively, we get

$$\left(\vec{a}\times\vec{b}\right)\times\vec{c}=\left[\left\{\begin{matrix}(a_1b_2c_2+a_1b_3c_3+a_1b_1c_1-a_1b_1c_1-a_3b_1c_3-a_2b_1c_2)\hat{i}\\+(a_2b_3c_3+a_2b_1c_1+a_2b_2c_2-a_2b_2c_2-a_3b_2c_3-a_1b_2c_1)\hat{j}\\+(a_3b_1c_1+a_3b_2c_2+a_3b_3c_3-a_3b_3c_3-a_1b_3c_1-a_2b_3c_2)\hat{k}\end{matrix}\right\}\right]$$

$$\left(\vec{a}\times\vec{b}\right)\times\vec{c}=\left[\left\{\begin{matrix}a_1(b_1c_1+b_2c_2+b_3c_3)-b_1(a_1c_1+a_2c_2+a_3c_3)\hat{i}\\+a_2(b_1c_1+b_2c_2+b_3c_3)-b_2(a_1c_1+a_2c_2+a_3c_3)\hat{j}\\+a_3(b_1c_1+b_2c_2+b_3c_3)-b_3(a_1c_1+a_2c_2+a_3c)\hat{k}\end{matrix}\right\}\right]$$

Since $b_1c_1+b_2c_2+b_3c_3=\vec{b}.\vec{c}$ and $a_1c_1+a_2c_2+a_3c_3=\vec{a}.\vec{c}$, we get

$$\left(\vec{a}\times\vec{b}\right)\times\vec{c}=\left[\left(\vec{b}.\vec{c}\right)\left(a_1\hat{i}+a_2\hat{j}+a_3\hat{k}\right)-\left(\vec{a}.\vec{c}\right)\left(b_1\hat{i}+b_2\hat{j}+b_3\hat{k}\right)\right]\Rightarrow$$

$\left(\vec{a}\times\vec{b}\right)\times\vec{c}=\left[\left(\vec{b}.\vec{c}\right)\vec{a}-\left(\vec{a}.\vec{c}\right)\vec{b}\right]$ .... Hence proved.

From property (3) and (4) above, we conclude that

$\vec{a}\times(\vec{b}\times\vec{c})\neq\left(\vec{a}\times\vec{b}\right)\times\vec{c}$

5. $\vec{a}\times\left(\vec{b}\times\vec{c}\right)+\vec{b}\times\left(\vec{c}\times\vec{a}\right)+\vec{c}\times\left(\vec{a}\times\vec{b}\right)=0$ ...Eq.14.20

Generalizing property (3) above, we can say that

$\vec{a}\times(\vec{b}\times\vec{c})=(\vec{a}.\vec{c})\vec{b}-(\vec{a}.\vec{b})\vec{c}$

$\vec{b}\times(\vec{c}\times\vec{a})=(\vec{b}.\vec{a})\vec{c}-(\vec{b}.\vec{c})\vec{a}$ and

$\vec{c}\times(\vec{a}\times\vec{b})=(\vec{c}.\vec{b})\vec{a}-(\vec{c}.\vec{a})\vec{b}$

Adding these three equalities, we get

$$\vec{a}\times\left(\vec{b}\times\vec{c}\right)+\vec{b}\times\left(\vec{c}\times\vec{a}\right)+\vec{c}\times\left(\vec{a}\times\vec{b}\right)=$$

$$(\vec{a}.\vec{c})\vec{b}-(\vec{a}.\vec{b})\vec{c}+(\vec{b}.\vec{a})\vec{c}-(\vec{b}.\vec{c})\vec{a}+(\vec{c}.\vec{b})\vec{a}-(\vec{c}.\vec{a})\vec{b}$$

Or

$$\vec{a}\times\left(\vec{b}\times\vec{c}\right)+\vec{b}\times\left(\vec{c}\times\vec{a}\right)+\vec{c}\times\left(\vec{a}\times\vec{b}\right)=$$

$$\vec{b}(\vec{a}.\vec{c}-\vec{c}.\vec{a})+\vec{c}(\vec{b}.\vec{a}-\vec{a}.\vec{b})+\vec{a}(\vec{c}.\vec{b}-\vec{b}.\vec{c})$$

Since $\vec{a}.\vec{c}=\vec{c}.\vec{a}$; $\vec{a}.\vec{b}=\vec{b}.\vec{a}$; $\vec{c}.\vec{b}=\vec{b}.\vec{c}$, we get

$$\vec{a}\times\left(\vec{b}\times\vec{c}\right)+\vec{b}\times\left(\vec{c}\times\vec{a}\right)+\vec{c}.\times\left(\vec{a}\times\vec{b}\right)=0$$

6. $\left(\vec{a}\times\vec{b}\right).\left(\vec{c}\times\vec{d}\right)=(\vec{a}.\vec{c})(\vec{b}.\vec{d})-(\vec{a}.\vec{d})(\vec{b}.\vec{c})$ ...Eq.14.21

This is a scalar product identity known as Lagrange's Identity.

To prove this we can use the property of interchanging dot and cross as discussed at property 2 of scalar triple product above.

Thus $\left(\vec{a}\times\vec{b}\right).\left(\vec{c}\times\vec{d}\right)=\vec{a}.\left\{\vec{b}\times\left(\vec{c}\times\vec{d}\right)\right\}$

Using expansion of triple cross product we get

$$\left(\vec{a}\times\vec{b}\right).\left(\vec{c}\times\vec{d}\right)=\vec{a}.\left\{\left(\vec{b}.\vec{d}\right).\vec{c}-\left(\vec{b}.\vec{c}\right)\vec{d}\right\}$$

$\left(\vec{a}\times\vec{b}\right).\left(\vec{c}\times\vec{d}\right)=(\vec{a}.\vec{c})(\vec{b}.\vec{d})-(\vec{a}.\vec{d})(\vec{b}.\vec{c})$ ..... **Hence proved**.

This scalar identity can also be expressed in determinant form as

$$\left(\vec{a}\times\vec{b}\right).\left(\vec{c}\times\vec{d}\right)=\begin{vmatrix}(\vec{a}.\vec{c}) & (\vec{a}.\vec{d})\\(\vec{b}.\vec{c}) & (\vec{b}.\vec{d})\end{vmatrix}$$ ...Eq. 14.22, for easy recall.

Example 18

If $\vec{a}=2\hat{i}+5\hat{j}-7\hat{k}$, $\vec{b}=-3\hat{i}+4\hat{j}+\hat{k}$ and $\vec{c}=\hat{i}-2\hat{j}-3\hat{k}$, compute $\vec{a}\times\left(\vec{b}\times\vec{c}\right)$ & $\left(\vec{a}\times\vec{b}\right)\times\vec{c}$ and verify that these are not equal

Solution:

Reference the identity $\vec{a}\times(\vec{b}\times\vec{c})=(\vec{a}.\vec{c})\vec{b}-(\vec{a}.\vec{b})\vec{c}$

Substituting the values of $\vec{a}$, $\vec{b}$ & $\vec{c}$ we get

$$\vec{a}\times(\vec{b}\times\vec{c})=$$

$$\left\{\left(2\hat{i}+5\hat{j}-7\hat{k}\right).\left(\hat{i}-2\hat{j}-3\hat{k}\right)\right\}\left(-3\hat{i}+4\hat{j}+\hat{k}\right)-\left\{\left(2\hat{i}+5\hat{j}-7\hat{k}\right).\left(-3\hat{i}+4\hat{j}+\hat{k}\right)\right\}\left(\hat{i}-2\hat{j}-3\hat{k}\right)$$

$$=\left\{(2-10+21)\right\}\left(-3\hat{i}+4\hat{j}+\hat{k}\right)-\left\{(-6+20-7)\right\}\left(\hat{i}-2\hat{j}-3\hat{k}\right)$$

$$=13\left(-3\hat{i}+4\hat{j}+\hat{k}\right)-7\left(\hat{i}-2\hat{j}-3\hat{k}\right)=-46\hat{i}+66\hat{j}+34\hat{k}$$

And $\left(\vec{a}\times\vec{b}\right)\times\vec{c}=(\vec{a}.\vec{c})\vec{b}-(\vec{b}.\vec{c})\vec{a}$

Solving it we get $\left(\vec{a}\times\vec{b}\right)\times\vec{c}=$

$$\left\{\left(2\hat{i}+5\hat{j}-7\hat{k}\right).\left(\hat{i}-2\hat{j}-3\hat{k}\right)\right\}\left(-3\hat{i}+4\hat{j}+\hat{k}\right)-\left\{\left(-3\hat{i}+4\hat{j}+\hat{k}\right).\left(\hat{i}-2\hat{j}-3\hat{k}\right)\right\}\left(2\hat{i}+5\hat{j}-7\hat{k}\right)$$

$$=\left\{(2-10+21)\right\}\left(-3\hat{i}+4\hat{j}+\hat{k}\right)-\left\{(-3-8-3)\right\}\left(2\hat{i}+5\hat{j}-7\hat{k}\right)$$

$$=13\left(-3\hat{i}+4\hat{j}+\hat{k}\right)+14\left(2\hat{i}+5\hat{j}-7\hat{k}\right)=-11\hat{i}+122\hat{j}-85\hat{k}$$

Hence proved that $\vec{a}\times\left(\vec{b}\times\vec{c}\right)\neq\left(\vec{a}\times\vec{b}\right)\times\vec{c}$.

Alternative solution:

We can prove the two triple cross products directly as well. In that case

$$\left(\vec{a}\times\vec{b}\right)=\begin{vmatrix}\hat{i} & \hat{j} & \hat{k}\\ 2 & 5 & -7\\ -3 & 4 & 1\end{vmatrix}=\hat{i}\left(5-4(-7)\right)-\hat{j}\left(2-(-3)(-7)\right)+\hat{k}\left(2\times4-(-3)(-5)\right)\Rightarrow$$

$$=33\hat{i}+19\hat{j}+23\hat{k}$$

Cross multiplying further with $\vec{c}$, we get

$$\left(\vec{a}\times\vec{b}\right)\times\vec{c}=\begin{vmatrix}\hat{i} & \hat{j} & \hat{k}\\ 33 & 19 & 23\\ 1 & -2 & -3\end{vmatrix}=$$

$$19(-3)-(-2)\times23)-\hat{j}\left(33(-3)-23\right)+\hat{k}\left(33\times(-2)-19\right)\Rightarrow$$

$$=-11\hat{i}+122\hat{j}-85\hat{k}$$

Similarly

$$\left(\vec{b}\times\vec{c}\right)=\begin{vmatrix}\hat{i} & \hat{j} & \hat{k}\\ -3 & 4 & 1\\ 1 & -2 & -3\end{vmatrix}=$$

$$\hat{i}\left(4(-3)-(-2)\right)-\hat{j}\left((-3)(-3)-1\right)+\hat{k}\left((-3)(-2)-4\right)\Rightarrow$$

$$=-10\hat{i}-8\hat{j}+2\hat{k}$$

Cross multiplying $\vec{a}$ with the above product, we get

$$\vec{a}\times\left(\vec{b}\times\vec{c}\right)\times=\begin{vmatrix}\hat{i} & \hat{j} & \hat{k}\\ 2 & 5 & -7\\ -10 & -8 & 2\end{vmatrix}=$$

$$=\hat{i}\left(10-56\right)-\hat{j}\left(4-(-10)(-7)\right)+\hat{k}\left(2\times(-8)-(-10)\times5\right)$$

$$=-46\hat{i}+66\hat{j}+34\hat{k}$$

We arrive at the same result as in the previous method and see that the previous method is a bit easier. Students can use either of these methods to prove that $\vec{a}\times\left(\vec{b}\times\vec{c}\right)\neq\left(\vec{a}\times\vec{b}\right)\times\vec{c}$

Example 19

If $|\vec{a}|=13$, $\left|\vec{b}\right|=5$ and $\vec{a}.\vec{b}=60$, find $\left|\vec{a}\times\vec{b}\right|$

Solution:

$\vec{a}.\vec{b}=|\vec{a}|\left|\vec{b}\right|\cos\theta$, θ being angle between them.

Substituting the given values, $60=(13\times5)\cos\theta\Rightarrow\cos\theta=\frac{60}{65}=\frac{12}{13}$

$$\left|\vec{a}\times\vec{b}\right|=\left||\vec{a}|\left|\vec{b}\right|\sin\theta\right|=(13\times5)\sin\theta=65\sqrt{1-\cos^2\theta}=$$

$$65\times\sqrt{1-\left(\frac{12}{13}\right)^2}=65\times\frac{5}{13}=25$$

Example 20

If $\vec{a}\times\vec{b}=\vec{b}\times\vec{c}\neq0$, then show that $\vec{a}+\vec{c}=\lambda\vec{b}$ where λ is any scalar.

Solution:

Given that $\vec{a}\times\vec{b}=\vec{b}\times\vec{c}$, we get

$$\left(\vec{a}\times\vec{b}\right)=-\left(\vec{c}\times\vec{b}\right)\quad\because\left\{\left(\vec{b}\times\vec{c}\right)=-\left(\vec{c}\times\vec{b}\right)\right\}$$

Hence $\left(\vec{a}\times\vec{b}\right)+\left(\vec{c}\times\vec{b}\right)=0\Rightarrow(\vec{a}+\vec{c})\times\vec{b}=0$

It implies that vector $(\vec{a}+\vec{c})$ is parallel to vector $\vec{b}$. Hence the two vectors can be expressed as $\vec{a}+\vec{c}=\lambda\vec{b}$ for a scalar λ.

Example 21

If $\vec{a},\vec{b}$ & $\vec{c}$, are lengths of the sides BC, CA and AB respectively of a triangle, prove that $\overrightarrow{AB}+\overrightarrow{BC}+\overrightarrow{CA}=0$ and deduce that $\frac{a}{\sin A}=\frac{b}{\sin B}=\frac{c}{\sin C}$.

Solution:

Refer to the triangle in the adjoining figure. $AB = |\vec{c}| = c$, $BC = |\vec{a}| = a$ and $CA = |\vec{b}| = b$

We can see that by triangle law of addition

$\vec{a} + \vec{b} = -\vec{c} \Rightarrow \vec{a} + \vec{b} + \vec{c} = 0$.

Hence $\overrightarrow{BC} + \overrightarrow{CA} + \overrightarrow{AB} = 0$

Now by vector cross multiplication, we get

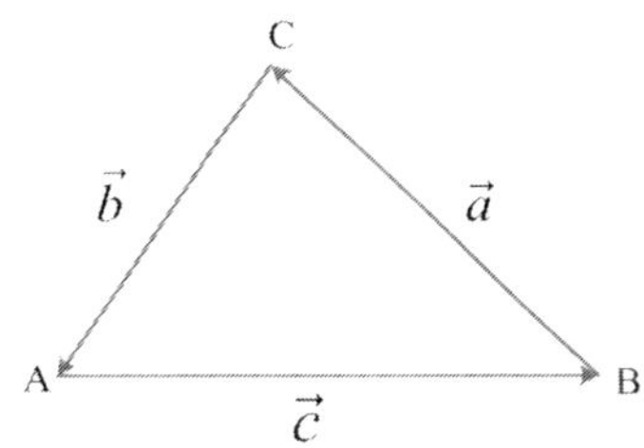

$$\vec{a} \times \left(\vec{a} + \vec{b} + \vec{c}\right) = \vec{a} \times 0 = 0 \Rightarrow \vec{a} \times \vec{a} + \vec{a} \times \vec{b} + \vec{a} \times \vec{c} = 0 \Rightarrow$$

$$0 + \vec{a} \times \vec{b} - \vec{c} \times \vec{a} = 0 \left(\because \vec{a} \times \vec{c} = -\vec{c} \times \vec{a}\right) \Rightarrow \vec{a} \times \vec{b} = \vec{c} \times \vec{a} \Rightarrow$$

$$ab \sin C = ac \sin B \Rightarrow b \sin C = c \sin B \Rightarrow \frac{b}{\sin B} = \frac{c}{\sin C}$$

Similarly we can prove that $\vec{c} \times \left(\vec{a} + \vec{b} + \vec{c}\right) = 0 \Rightarrow \vec{c} \times \vec{a} = \vec{b} \times \vec{c} \Rightarrow$

$$ac \sin B = bc \sin A \Rightarrow a \sin B = b \sin A \Rightarrow \frac{b}{\sin B} = \frac{a}{\sin A}$$

Thus $\dfrac{a}{\sin A} = \dfrac{b}{\sin B} = \dfrac{c}{\sin C}$

Example 22

If $\vec{a} = \hat{i} + 4\hat{j} + 2\hat{k}$, $\vec{b} = 3\hat{i} - 2\hat{j} + 7\hat{k}$ and $\vec{c} = 2\hat{i} - \hat{j} + 4\hat{k}$, find a vector $\vec{d}$ which is perpendicular to both $\vec{a}$ & $\vec{b}$ with $\vec{c}.\vec{d} = 15$

Solution:

Vector perpendicular to both $\vec{a}$ & $\vec{b}$ will be parallel to $\vec{a} \times \vec{b}$.

Hence $\vec{d} = \lambda\left(\vec{a} \times \vec{b}\right)$ where λ is a scalar.

Therefore, $\vec{d} = \lambda \begin{vmatrix} \hat{i} & \hat{j} & \hat{k} \\ 1 & 4 & 2 \\ 3 & -2 & 7 \end{vmatrix} = \hat{i}\left(28-(-4)\right) - \hat{j}(7-6) + \hat{k}\left(1\times(-2) - 3\times 4\right) \Rightarrow$

$\vec{d} = \lambda\left\{32\hat{i} - \hat{j} - 14\hat{k}\right\}$

Now it is given that $\vec{c}.\vec{d} = 15$

Hence $\vec{c}.\vec{d} = \left(2\hat{i} - \hat{j} + 4\hat{k}\right).\lambda\left\{32\hat{i} - \hat{j} - 14\hat{k}\right\} = 15 \Rightarrow$

$\lambda(64+1-56) = 15 \Rightarrow \lambda = \frac{15}{9} = \frac{5}{3}$

So $\vec{d} = \frac{5}{3}\left\{32\hat{i} - \hat{j} - 14\hat{k}\right\} = \frac{1}{3}\left\{160\hat{i} - 5\hat{j} - 70\hat{k}\right\}$

Example 23

If $\vec{a} = a_1\hat{i} + a_2\hat{j} + a_3\hat{k}$, $\vec{b} = b_1\hat{i} + b_2\hat{j} + b_3\hat{k}$ & $\vec{c} = c_1\hat{i} + c_2\hat{j} + c_3\hat{k}$,

prove that $\vec{a}\times\left(\vec{b}+\vec{c}\right) = \vec{a}\times\vec{b} + \vec{a}\times\vec{c}$

Solution:

We know that

$$\vec{a}\times\vec{b} = \begin{vmatrix} \hat{i} & \hat{j} & \hat{k} \\ a_1 & a_2 & a_3 \\ b_1 & b_2 & b_3 \end{vmatrix} == \hat{i}\left\{a_2b_3 - b_2a_3\right\} - \hat{j}\left\{a_1b_3 - b_1a_3\right\} + \hat{k}\left\{a_1b_2 - a_2b_1\right\}$$

$$\vec{a}\times\vec{c} = \begin{vmatrix} \hat{i} & \hat{j} & \hat{k} \\ a_1 & a_2 & a_3 \\ c_1 & c_2 & c_3 \end{vmatrix} == \hat{i}\left\{a_2c_3 - c_2a_3\right\} - \hat{j}\left\{a_1c_3 - c_1a_3\right\} + \hat{k}\left\{a_1c_2 - a_2c_1\right\}$$

Now $\vec{b}+\vec{c} = (b_1+c_1)\hat{i} + (b_2+c_2)\hat{j} + (b_3+c_3)\hat{k}$

$$\vec{a}\times\left(\vec{b}+\vec{c}\right) = \begin{vmatrix} \hat{i} & \hat{j} & \hat{k} \\ a_1 & a_2 & a_3 \\ (b_1+c_1) & (b_2+c_2) & (b_3+c_3) \end{vmatrix} = \begin{vmatrix} \hat{i} & \hat{j} & \hat{k} \\ a_1 & a_2 & a_3 \\ b_1 & b_2 & b_3 \end{vmatrix} + \begin{vmatrix} \hat{i} & \hat{j} & \hat{k} \\ a_1 & a_2 & a_3 \\ c_1 & c_2 & c_3 \end{vmatrix} \Rightarrow$$

$\vec{a}\times\left(\vec{b}+\vec{c}\right) = \left(\vec{a}\times\vec{b}\right) + \left(\vec{a}\times\vec{c}\right)$. Hence proved.

Example 24

Find area of the triangle formed by O, A, B where $\overrightarrow{OA} = \hat{i} + 2\hat{j} + 3\hat{k}$ $\overrightarrow{OB} = -3\hat{i} - 2\hat{j} + \hat{k}$

Solution:

Area of a triangle $= \frac{1}{2}\left|\vec{a} \times \vec{b}\right|$ where $\vec{a}$ & $\vec{b}$ are vectors representing two adjoining sides of a triangle.

Hence area of the given triangle $\frac{1}{2}\left|\vec{a} \times \vec{b}\right| = \frac{1}{2}\begin{vmatrix} \hat{i} & \hat{j} & \hat{k} \\ 1 & 2 & 3 \\ -3 & -2 & 1 \end{vmatrix}$

$$= \frac{1}{2}\left|\hat{i}\left(2-(-2\times 3)\right) - \hat{j}\left(1-(-9)\right) + \hat{k}\left(1\times(-2)-(-3)\times 2\right)\right| \Rightarrow$$

$$= \frac{1}{2}\left|8\hat{i} - 10\hat{j} - 4\hat{k}\right| = \frac{1}{2}\sqrt{64+100+16} = \frac{1}{2}\sqrt{180} = \frac{1}{2}\times 6\sqrt{5} = 3\sqrt{5}$$

14.3 Different forms of equation of a line in space

A line in space is unique if it passes through two points or if it passes through a known point and has known direction. Just like equation of a line in different formats in x-y plane, we shall now study Vector and Cartesian equations of a line in 3-dimensional space.

14.3.1 *Equation of a line through a given point and parallel to a given vector:*

Vector form:

Let ***l*** be the line which passes through the given point A and is parallel to a vector $\vec{b}$ as shown in figure 14.13 below. Let $\vec{a}$ be the position vector of point A and let $\vec{r}$ be the position vector of an arbitrary point P on the line.

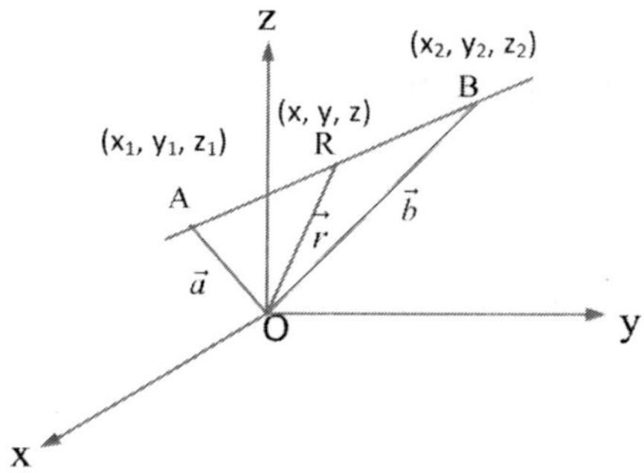

Figure 14.13

Since vector $\overrightarrow{AP}$ is parallel to vector $\vec{b}$, we can represent vector $\overrightarrow{AP}$ as $\overrightarrow{AP} = \lambda\vec{b}$.

We can also see that
$\overrightarrow{OA} + \overrightarrow{AP} = \overrightarrow{OP} \Rightarrow \vec{a} + \lambda\vec{b} = \vec{r} \Rightarrow \vec{r} = \vec{a} + \lambda\vec{b}$.....*Eq.* 14.23

The eq. 14.23 above is vector equation of the line passing through point A and parallel to vector $\vec{b}$.

Cartesian form:

Let a. b, c are direction ratios of the line $\overrightarrow{AP}$. We will see in section 14.4 that direction ratios of two parallel vectors are equal. Hence vector $\vec{b}$ being parallel to the line $\overrightarrow{AP}$ will also have the same direction ratios and it can be expressed as $\vec{b} = a\hat{i} + b\hat{j} + c\hat{k}$

Let coordinates of point A be (x_1, y_1, z_1) and of any point on the line say P be (x, y, z), then we can write direction vector $\vec{r} = x\hat{i} + y\hat{j} + z\hat{k}$; $\vec{a} = x_1\hat{i} + y_1\hat{j} + z_1\hat{k}$

Substituting the above Cartesian values in vector equation 14.23 derived above and equating the three coordinate values, we get:

$x = x_1 + \lambda a; \quad y = y_1 + \lambda b \quad \& \quad z = z_1 + \lambda c$

Eliminating value of λ, we get

$$\frac{x - x_1}{a} = \frac{y - y_1}{b} = \frac{z - z_1}{c} \; \ldots \text{Eq. } 14.24$$

This is Cartesian equation of the line where ***a, b, c are direction ratios of the line.***

Since direction cosines l, m, n of the line $\overrightarrow{AP}$, are proportional to the respective direction ratios(a, ***b, c),*** it leads to the conclusion that equation of the line $\overrightarrow{AP}$ is also $\dfrac{x-x_1}{l}=\dfrac{y-y_1}{m}=\dfrac{z-z_1}{n}$ ....Eq. 14.25

Example 25

Let (a, b) be real numbers $(a, b \in R)$.

If mirror image of the point $(a,5,9)$ with respect to the line $\dfrac{x-3}{7}=\dfrac{y-2}{5}=\dfrac{z-1}{-9}$ is $\{20, b, (-a-9)\}$, find $|a+b|$

Solution:

Since mirror image is at a the same distance as the object in front of the mirror, the point on the line joining the two given points will be midpoint of the two given points i.e. point on the line will be

$$\left\{\frac{a+20}{2}, \frac{6+b}{2}, \frac{9-a-9}{2}\right\} \Rightarrow \left\{\left(\frac{a}{2}+10\right), \left(3+\frac{b}{2}\right), \frac{-a}{2}\right\}$$

Substituting in the equation of the line, we get

$$\frac{a/2+10-3}{7}=\frac{3+b/2-2}{5}=\frac{-a/2-1}{-9} \Rightarrow$$

$$\frac{a+14}{14}=\frac{b+2}{10}=\frac{a+2}{18} \Rightarrow 18a+14\times 18=14a+28 \Rightarrow$$

$$4a=14(2-18) \Rightarrow a=-\frac{14\times 16}{4}=-56$$

And

$$14b+28=10a+14\times 10 \Rightarrow 14b=-560+140-28 \Rightarrow$$

$$14(-40+10-2) \Rightarrow b=-32$$

Hence $|a+b|=|-56-32|=88$

14.3.2 *Equation of a vector passing through two points in 3-dimensional space:*

Vector form: Let $\vec{a}$ and $\vec{b}$ be position vectors of the two points

A(x_1, y_1, z_1) and B (x_2, y_2, z_2) respectively lying on the given vector as shown in figure 14.14 below. Let (x, y, z) be the coordinates of an arbitrary point R. Then point R will lie on the givenvector if and only if vectors $\overrightarrow{AR} = \vec{r} - \vec{a}$ and $\overrightarrow{AB} = \vec{b} - \vec{a}$ are collinear.

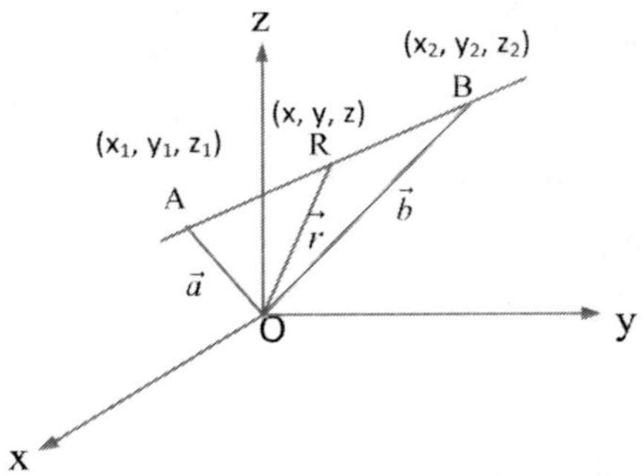

Figure 14.14

This leads us to the conclusion that point R lies on the vector AB if and only if $\vec{r} - \vec{a} = \lambda\left(\vec{b} - \vec{a}\right)$, where λ is an arbitrary constant.

Thus $\vec{r} = \vec{a} + \lambda\left(\vec{b} - \vec{a}\right)$***...Eq. 14.26 is vector form of equation of the line.***

Cartesian form:

Now we can easily derive **Cartesian form** of the above equation.

We know that $\vec{r} = x\hat{i} + y\hat{j} + z\hat{k}$; $\vec{a} = x_1\hat{i} + y_1\hat{j} + z_1\hat{k}$ and $\vec{b} = x_2\hat{i} + y_2\hat{j} + z_2\hat{k}$

Substituting these values in the vector equation derived above, we get

$$\vec{r} = x\hat{i} + y\hat{j} + z\hat{k} = \left(x_1\hat{i} + y_1\hat{j} + z_1\hat{k}\right) + \lambda\left[\left(x_2\hat{i} + y_2\hat{j} + z_2\hat{k}\right) - \left(x_1\hat{i} + y_1\hat{j} + z_1\hat{k}\right)\right]$$

or

$$x\hat{i} + y\hat{j} + z\hat{k} = \left(x_1 + \lambda\left(x_2 - x_1\right)\right)\hat{i} + \left(y_1 + \lambda\left(y_2 - y_1\right)\right)\hat{j} + \left(z_1 + \lambda\left(z_2 - z_1\right)\right)\hat{k} \Rightarrow$$

$$x = \left\{x_1 + \lambda\left(x_2 - x_1\right)\right\}; y = \left\{y_1 + \lambda\left(y_2 - y_1\right)\right\} \& z = \left\{z_1 + \lambda\left(z_2 - z_1\right)\right\}$$

Eliminating λ, we get

$\frac{x-x_1}{x_2-x_1}=\frac{y-y_1}{y_2-y_1}=\frac{z-z_1}{z_2-z_1}$ .....Eq. 14.27 ***which is the Cartesian form of the equation of the vector.***

Here we can infer an important conclusion that that (x_2-x_1), (y_2-y_1) ***and*** (z_2-z_1) ***are direction ratios of the vector with points*** (x_1, y_1, z_1) ***and*** (x_2, y_2, z_2) ***lying on it.***

Example 26

Find the equation of the line passing through point A (0, 1, 2) and perpendicular to the line $\frac{x-1}{2}=\frac{y+1}{2}=\frac{z-1}{3}$

We have seen that equation of a line passing through two known points with coordinates $(x_1, y_1, z_1)\&(x_2, y_2, z_2)$ respectively is $\frac{x-x_1}{x_2-x_1}=\frac{y-y_1}{y_2-y_1}=\frac{z-z_1}{z_2-z_1}$

Here one point has coordinates (0, 1, 2) yielding $x_1=0,\ y_1=1,\ z_1=2$

We need to find coordinates of the other point B on the given line such that AB is perpendicular to the given line.

From equation of the given line, we see that Direction ratios of this line are 2, 2, 3. Hence its vector equation is $\vec{r}_1=a\hat{i}+b\hat{j}+c\hat{k}\Rightarrow\vec{r}_1=2\hat{i}+2\hat{j}+3\hat{k}$

Let us assume that $\frac{x-1}{2}=\frac{y+1}{2}=\frac{z-1}{3}=\lambda$. This implies that coordinates of a point on the given line, say B, are $(2\lambda+1,\ 2\lambda-1,\ 3\lambda+1)$.

If this point is assumed as (x_2, y_2, z_2), direction ratios of line AB will be $\{(2\lambda+1-0),(2\lambda-1-1),(3\lambda+1-2)\}\Rightarrow\{(2\lambda+1),(2\lambda-2),(3\lambda-1)\}$. This yields equation of line AB in terms of direction ratios as $\vec{r}_2=(2\lambda+1)\hat{i}+(2\lambda-2)\hat{j}+(3\lambda-1)\hat{k}$

Since line AB is perpendicular to the given line, their dot product equals zero. Hence

$$\{(2\lambda+1)\hat{i}+(2\lambda-2)\hat{j}+(3\lambda-1)\hat{k}\}.(2\hat{i}+2\hat{j}+3\hat{k})=$$

$$4\lambda+2+4\lambda-4+3\times(3\lambda-1)=0\Rightarrow$$

$$17\lambda-5=0\Rightarrow\lambda=\frac{5}{17}.$$

Hence direction ratios of line AB $\{(x_2-x_1),(y_2-y_1),(z_2-z_1)\}$ are

$$\left(\left(2\times\frac{5}{17}+1\right),\left(2\times\frac{5}{17}-2\right),\left(3\times\frac{5}{17}-1\right)\right)\Rightarrow\left(\frac{27}{17},\frac{-24}{17},\frac{-2}{17}\right)$$

Therefore, equation of the required line is

$$\frac{x-0}{27/17}=\frac{y-1}{-24/17}=\frac{z-2}{-2/17}\Rightarrow\frac{x}{27}=\frac{y-1}{-24}=\frac{z-2}{-2}$$

14.4 Angle between two vectors in 3-D space:

Let us consider two vectors P and Q passing through the origin with direction cosines (l_1, m_1, n_1) and (l_2, m_2, n_2) and magnitude r_1 and r_2 respectively.

Therefore, $\vec{P}=r_1(l_1\hat{i}+m_1\hat{j}+n_1\hat{k})$ and $\vec{Q}=r_2(l_2\hat{i}+m_2\hat{j}+n_2\hat{k})$

Dot product $\vec{P}.\vec{Q}=|P||Q|\cos\theta$

$$\text{or }\cos\theta=\frac{\vec{P}.\vec{Q}}{|P||Q|}=\frac{r_1(l_1\hat{i}+m_1\hat{j}+n_1\hat{k}).r_2(l_2\hat{i}+m_2\hat{j}+n_2\hat{k})}{r_1\sqrt{(l_1^2+m_1^2+n_1^2)}\times r_2\sqrt{(l_2^2+m_2^2+n_2^2)}}$$

$$\text{or }\cos\theta=\frac{r_1r_2\times(l_1\hat{i}+m_1\hat{j}+n_1\hat{k}).(l_2\hat{i}+m_2\hat{j}+n_2\hat{k}).}{r_1\sqrt{(l_1^2+m_1^2+n_1^2)}\times r_2\sqrt{(l_2^2+m_2^2+n_2^2)}}$$

We know that $i.i=1, j.j=1\ \&\ k.k=1$ where as $i.j=0, j.k=0\ \&\ i.k=0$

Hence $(l_1\hat{i}+m_1\hat{j}+n_1\hat{k}).(l_2\hat{i}+m_2\hat{j}+n_2\hat{k})=l_1l_2+m_1m_2+n_1n_2$

$$\cos\theta=\frac{(l_1l_2+m_1m_2+n_1n_2)}{\sqrt{(l_1^2+m_1^2+n_1^2)}\times\sqrt{(l_2^2+m_2^2+n_2^2)}}\quad\text{......Eq. 14.28}$$

We also know that $\sin\theta=\sqrt{1-\cos^2\theta}\Rightarrow$

$$\sin\theta = \sqrt{1 - \frac{(l_1 l_2 + m_1 m_2 + n_1 n_2)^2}{(l_1^2 + m_1^2 + n_1^2) \times (l_2^2 + m_2^2 + n_2^2)}} \Rightarrow$$

$$\sin\theta = \sqrt{\frac{(l_1^2 + m_1^2 + n_1^2) \times (l_2^2 + m_2^2 + n_2^2) - (l_1 l_2 + m_1 m_2 + n_1 n_2)^2}{(l_1^2 + m_1^2 + n_1^2) \times (l_2^2 + m_2^2 + n_2^2)}}$$

Expanding and cancelling out like terms, we get

$$\sin\theta = \sqrt{\frac{(l_1^2 m_2^2 + l_1^2 n_2^2 + m_1^2 l_2^2 + m_1^2 n_2^2 + n_1^2 l_2^2 + n_1^2 m_2^2) - (2l_1 l_2 m_1 m_2 + 2m_1 m_2 n_1 n_2 + 2n_1 n_2 l_1 l_2)}{(l_1^2 + m_1^2 + n_1^2) \times (l_2^2 + m_2^2 + n_2^2)}}$$

$$\therefore \sin\theta = \sqrt{\frac{(l_1 m_2 - m_1 l_2)^2 + (l_1 n_2 - n_1 l_2)^2 + (m_1 n_2 - n_1 m_2)^2}{(l_1^2 + m_1^2 + n_1^2) \times (l_2^2 + m_2^2 + n_2^2)}}$$

.......Eq. 14.29

Since $l^2 + m^2 + n^2 = 1$ for any vector,

we get $l_1^2 + m_1^2 + n_1^2 = 1$ and $l_2^2 + m_2^2 + n_2^2 = 1$

Hence $\cos\theta = (l_1 l_2 + m_1 m_2 + n_1 n_2)$

And $\sin\theta = \sqrt{(l_1 m_2 - m_1 l_2)^2 + (l_1 n_2 - n_1 l_2)^2 + (m_1 n_2 - n_1 m_2)^2}$

Expressions for $\cos\theta$ & $\sin\theta$ can also be expressed in terms of direction ratios (a_1, b_1, c_1) and (a_2, b_2, c_2) of the two vectors respectively.

Substituting values of direction cosines in terms of direction ratios in the above equation 14.28 where k_1 and k_2 are two arbitrary constants, we get

$$\cos\theta = \frac{(a_1 k_1 a_2 k_2 + b_1 k_1 b_2 k_2 + c_1 k_1 c_2 k_2)}{\sqrt{((a_1 k_1)^2 + (b_1 k_1)^2 + (c_1 k_1)^2)} \times \sqrt{((a_2 k_2)^2 + (b_2 k_2)^2 + (c_2 k_2)^2)}} \Rightarrow$$

$$\cos\theta = \frac{k_1 k_2 (a_1 a_2 + b_1 b_2 + c_1 c_2)}{k_1 \sqrt{a_1^2 + b_1^2 + c_1^2} \times k_2 \sqrt{a_2^2 + b_2^2 + c_2^2}} = \frac{(a_1 a_2 + b_1 b_2 + c_1 c_2)}{\sqrt{a_1^2 + b_1^2 + c_1^2} \times \sqrt{a_2^2 + b_2^2 + c_2^2}}$$

..... Eq. 14.30

which is exactly same as the expression with direction cosines but with direction ratios substituted respectively in place of direction cosines.

Similarly, expression for $\sin\theta$ using direction ratios is

$$\sin\theta = \frac{\sqrt{(a_1b_2 - a_2b_1)^2 + (b_1c_2 - b_2c_1)^2 + (c_1a_2 - c_2a_1)^2}}{\sqrt{(a_1^2 + b_1^2 + c_1^2)} \times \sqrt{(a_2^2 + b_2^2 + c_2^2)}} \quad \text{..Eq.14.31}$$

In case the vector lines are not passing through the origin, we can take two other vectors P' and Q' parallel to the vectors P and Q and passing through the origin. Direction cosines of a vector parallel to a given vector remain unchanged. Hence expressions for $\cos\theta \,\&\, \sin\theta$ derived above remain valid for all vectors.

Special cases:

1. If two lines are perpendicular,

   $\cos\theta = \cos 90^0 = 0 \Rightarrow a_1a_2 + b_1b_2 + c_1c_2 = 0$ or
   $l_1l_2 + m_1m_2 + n_1n_2 = 0$

2. If the lines are parallel, $\theta = 0^0 \Rightarrow \sin\theta = 0$

   So $\sqrt{(a_1b_2 - a_2b_1)^2 + (b_1c_2 - b_2c_1)^2 + (c_1a_2 - c_2a_1)^2} = 0$

   Since all three components of the equation on LHS are positive (being square of a number), it implies that each of these components = 0

   So we get $\frac{a_1}{a_2} = \frac{b_1}{b_2} = \frac{c_1}{c_2}$ or $\frac{l_1}{l_2} = \frac{m_1}{m_2} = \frac{n_1}{n_2}$

Example 27

Find the angle between the vectors with direction ratios proportional to (1, -2, 1) and (4, 3, 2)

Solution:

If θ is the angle between two lines with direction ratios (a_1, b_1, c_1) and (a_2, b_2, c_2), $\cos\theta = \frac{(a_1a_2 + b_1b_2 + c_1c_2)}{\sqrt{a_1^2 + b_1^2 + c_1^2} \times \sqrt{a_2^2 + b_2^2 + c_2^2}}$

Hence in the given example,

$$\cos\theta = \frac{\left(1\times 4+(-2)\times 3+1\times 2\right)}{\sqrt{\left(1^2+(-2)^2+1^2\right)}\times\sqrt{\left(4^2+3^2+2^2\right)}}$$

$$= \frac{4-6+2}{\sqrt{6}\times\sqrt{29}} = 0$$

Hence the angle between the lines is 90^0.

Example 28

Find angle between the lines whose direction ratios are proportional to a, b, c and b –c, c –a, a –b.

Solution:

If θ is the angle between two lines with direction ratios (a_1, b_1, c_1) and (a_2, b_2, c_2), $\cos\theta = \dfrac{\left(a_1a_2+b_1b_2+c_1c_2\right)}{\sqrt{a_1^{\,2}+b_1^{\,2}+c_1^{\,2}}\times\sqrt{a_2^{\,2}+b_2^{\,2}+c_2^{\,2}}}$

Hence in the given example,

$$\cos\theta = \frac{a\times(b-c)+b(c-a)+c(a-b)}{\sqrt{\left(a^2+b^2+c^2\right)}\times\sqrt{(b-c)^2+(c-a)^2+(a-b)^2}}$$

$$= \frac{ab-ac+bc-ab+ac-bc}{\sqrt{\left(a^2+b^2+c^2\right)}\times\sqrt{(b-c)^2+(c-a)^2+(a-b)^2}} = 0$$

Hence the angle between the lines is 90^0.

Example 29

If coordinates of points A, B, C, D are (1, 2, 3), (4, 5, 7), (-4, 3, -6) and (2, 9, 2) respectively. Find the angle between AB and CD

Solution:

We know that direction ratios of a line joining points (x_1, y_1, z_1) and (x_2, y_2, z_2) are $\left(x_2-x_1\right)$, $\left(y_2-y_1\right)$ and $\left(z_2-z_1\right)$

Hence direction ratios of line segment AB $\left(a_1,b_1,c_1\right)$ are

$$(1-4)+(2-5)+(3-7)=(-3,-3,-4)$$

Similarly direction ratios of line segment CD $\left(a_2,b_2,c_2\right)$ are $(-4-2),(3-9),(-6-2)=(-6,-6,-8)=2(-3,-3,-4)$

Since the direction ratios are proportional to each other, the lines are parallel. We can verify this result by finding angle θ between the lines, given by the following formula.

$$\cos\theta = \frac{(a_1a_2 + b_1b_2 + c_1c_2)}{\sqrt{a_1^2 + b_1^2 + c_1^2} \times \sqrt{a_2^2 + b_2^2 + c_2^2}}$$

$$= \frac{((-3\times-6)+(-3\times-6)+(4\times8))}{\sqrt{(-3)^2+(-3)^2+4^2} \times \sqrt{(-6)^2+(-6)^2+8^2}} = \frac{68}{\sqrt{34}\times\sqrt{136}} = \frac{68}{68} = 1$$

Thus angle between the lines is 0^0 meaning thereby that the lines are parallel.

Example 30

Find the angle between the lines whose direction cosines are given by the equations $(i)\, l+m+n=0$ & $(ii)\; l^2+m^2-n^2=0$

Solution:

First we try to solve the given two equations.

From eq. (i), we get $l+m=-n$ .....(iii) and from eq. (ii),

we get $l^2+m^2=n^2$ ...(iv)

Squaring equation (iii) and subtracting eq. (iv) from it, we get

$(l+m)^2-(l^2+m^2)=0 \Rightarrow 2lm=0$

This implies that either $l=0$ *or* $m=0$.

If *l* = *0*, *m* = *-n* implying thereby that direction cosines of one line are *(0, -n, n)*

And if *m* = *0*, *l* = *-n* implying thereby that direction cosines of the other line are

(-n, 0, n)

If angle between the lines is θ, $\cos\theta = (l_1l_2 + m_1m_2 + n_1n_2)$

Substituting the values of direction cosines for the two lines derived above,

$\cos\theta = (0\times-n+(-n\times0)+n\times n) = n^2$ .... (1).

We also know that $l^2+m^2+n^2=1 \Rightarrow l^2+m^2=1-n^2$

Substituting in the given relationship (ii), $l^2 + m^2 = n^2$, we get

$$n^2 = 1 - n^2 \Rightarrow n^2 = \frac{1}{2}$$

Hence from equation (1) above, $\cos\theta = n^2 = \frac{1}{2} \Rightarrow \theta = 60^0$

14.5 Shortest distance between two vectors in space:

If two lines in space intersect at a point, then the shortest distance between them is zero. And, if two lines in space are parallel, there is a plane in space that contains both these parallel lines. The shortest distance between them is constant and is equal to the perpendicular distance between them i.e. equal to length of the perpendicular drawn from a point on one line onto the other line. Further, in space, there are lines which are neither intersecting nor parallel. In fact, such pair of lines are *non-coplanar* and are called *skew lines*.

For clarity, let us consider a room with floor along x and y axis with one vertical joint of walls along z axis as shown in figure 14.17.

The line SQ is drawn diagonally across the ceiling and line PC is drawn from one corner of the ceiling (P) just above corner A diagonally across the room to corner C on the floor. *These two lines are neither parallel nor meet at any point. Hence these lines are skew lines.*

By shortest distance between such skew lines, we mean length of a join between a point on one line to a point on the other line such that length of the segment so obtained is the shortest. ***Such a segment is perpendicular to both the lines.***

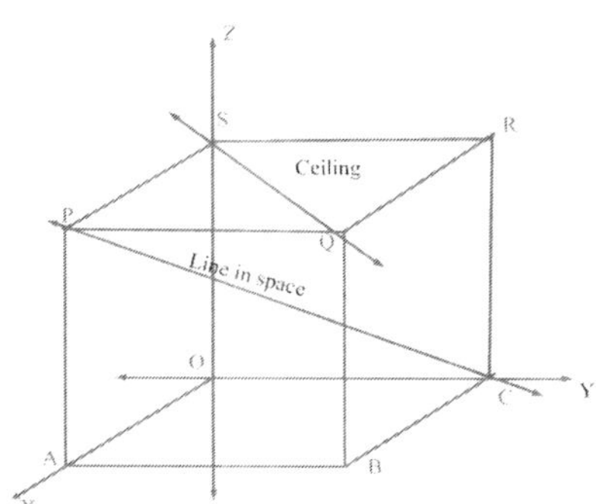

Figure 14.15

14.5.1 Distance between two skew lines:

Let us now find shortest distance between two skew lines, l_1 *and* l_2 as in figure 14.16.

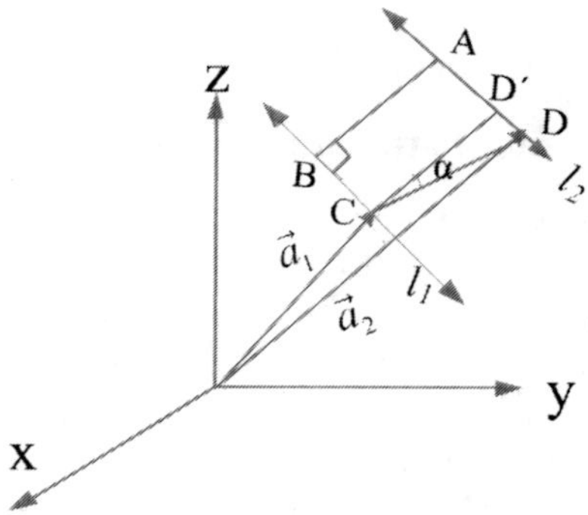

Figure 14.16

Polar form:

Let $\vec{a}_1$ be the position vector of point C on line l_1 and $\vec{a}_2$ be the position vector of point D on line l_2. Let there be two co-planar lines $\vec{b}_1$ and $\vec{b}_2$ parallel respectively to the lines l_1 *and* l_2. Then equation of skew lines l_1 *and* l_2 will be $\vec{r}_1 = \vec{a}_1 + \lambda\vec{b}_1$ and $\vec{r}_2 = \vec{a}_2 + \mu\vec{b}_2$

Let there be two vectors $\overrightarrow{AB}$ and $\overrightarrow{CD}$ such that Vector $\overrightarrow{AB}$ is perpendicular to both the lines $\vec{b}_1$ and $\vec{b}_2$ and Vector $\overrightarrow{CD}$ is any vector joining two arbitrary points on the two lines. Letα be the angle between them. Shortest distance between these vectors will be projection of vector $\overrightarrow{CD}$ in the direction of vector $\overrightarrow{AB}$ (shown as $\overline{CD}$)
$\Rightarrow \left|\overrightarrow{CD'}\right| = \left|\overrightarrow{CD}\right| \cos\alpha = \left|\overrightarrow{AB}\right|$.

Since vector is perpendicular to both the lines $\vec{b}_1$ and $\vec{b}_2$, unit vector $\hat{n}$ in the direction of vector $\overrightarrow{AB}$ will be

$$\hat{n} = \frac{\vec{b}_1 \times \vec{b}_2}{\left|\vec{b}_1 \times \vec{b}_2\right|} \qquad \text{...(1)}$$

If d is the shortest distance along AB, $\overrightarrow{AB} = \mathrm{d}\,\hat{n}$ or $\hat{n} = \dfrac{\overrightarrow{AB}}{\left|\overrightarrow{AB}\right|}$ ... (2)

$\vec{a}_1$ & $\vec{a}_2$ being position vectors of points C and D, $\overrightarrow{CD} = \vec{a}_1 - \vec{a}_2$. (3)

We know from section 14.4 above that $\cos\alpha = \dfrac{\overrightarrow{CD}.\overrightarrow{AB}}{\left|\overrightarrow{CD}\right|\left|\overrightarrow{AB}\right|}$ .... (4)

Substituting values of $\overrightarrow{AB}$ and $\overrightarrow{CD}$ from the above equations 2 and 3 ineq. (4),

$$\cos\alpha = \frac{\overrightarrow{CD}.\overrightarrow{AB}}{\left|\overrightarrow{CD}\right|\left|\overrightarrow{AB}\right|} = \frac{\left(\overrightarrow{a_1} - \overrightarrow{a_2}\right).\hat{n}d}{d.\left|\overrightarrow{CD}\right|} = \frac{\left(\overrightarrow{a_1} - \overrightarrow{a_2}\right).\left(\overline{b_1} \times \overline{b_2}\right)}{\left|\overrightarrow{CD}\right|\left|\overline{b_1} \times \overline{b_2}\right|} \Rightarrow$$

$$\left|\overrightarrow{CD}\right|\cos\alpha = d = \left|\frac{\left(\overrightarrow{a_1} - \overrightarrow{a_2}\right).\left(\overline{b_1} \times \overline{b_2}\right)}{\left|\overline{b_1} \times \overline{b_2}\right|}\right| \text{...... Eq. 14.32}$$

Cartesian form:

Let a_1, b_1, c_1 and a_2, b_2, $c_2$be the direction ratios of vectors $\overrightarrow{b_1}$ and $\overrightarrow{b_2}$ respectively and coordinates of points C and D on lines $\vec{l}_1$ & $\vec{l}_2$ be x_1, y_1, z_1 and x_2, y_2, $z_2$respectively. Equations of vectors $\overrightarrow{b_1}$, $\overrightarrow{b_2}$ will therefore be

$$\vec{b}_1 = a_1\hat{i} + b_1\hat{j} + c_1\hat{k}\ ,\ \vec{b}_2 = a_2\hat{i} + b_2\hat{j} + c_2\hat{k}$$

And equation of vector $\overrightarrow{CD} = \left(\overrightarrow{a_1} - \overrightarrow{a_2}\right)$ will be $(x_1 - x_2)\hat{i} + (y_1 - y_2)\hat{j} + (z_1 - z_2)\hat{k}$

Now the shortest distance d can be found by substituting these values in equation 14.32 derived in vector form above. Hence

$$d = \left|\frac{\left\{(x_1 - x_2)\hat{i} + (y_1 - y_2)\hat{j} + (z_1 - z_2)\hat{k}\right\}.\left\{\left(a_1\hat{i} + b_1\hat{j} + c_1\hat{k}\right) \times \left(a_2\hat{i} + b_2\hat{j} + c_2\hat{k}\right)\right\}}{\left|\left(a_1\hat{i} + b_1\hat{j} + c_1\hat{k}\right) \times \left(a_2\hat{i} + b_2\hat{j} + c_2\hat{k}\right)\right|}\right| \Rightarrow$$

$$d = \left|\frac{\left\{(x_1 - x_2)\hat{i} + (y_1 - y_2)\hat{j} + (z_1 - z_2)\hat{k}\right\}.\left\{(b_1c_2 - b_2c_1)\hat{i} + (a_1c_2 - a_2c_1)\hat{j} + (a_1b_2 - a_2b_1)\hat{k}\right\}}{\left|\left\{(b_1c_2 - b_2c_1)\hat{i} + (a_1c_2 - a_2c_1)\hat{j} + (a_1b_2 - a_2b_1)\hat{k}\right\}\right|}\right| \Rightarrow$$

$$d = \frac{\left\{(x_1 - x_2)(b_1c_2 - b_2c_1) + (y_1 - y_2)(a_1c_2 - a_2c_1) + (z_1 - z_2)(a_1b_2 - a_2b_1)\right\}}{\left|\sqrt{(b_1c_2 - b_2c_1)^2 + (a_1c_2 - a_2c_1)^2 + (a_1b_2 - a_2b_1)^2}\right|} \Rightarrow$$

$$\text{Or } d = \frac{\begin{vmatrix} x_1 - x_2 & y_1 - y_2 & z_1 - z_2 \\ a_1 & b_1 & c_1 \\ a_2 & b_2 & c_2 \end{vmatrix}}{\left\| \begin{matrix} \hat{i} & \hat{j} & \hat{k} \\ a_1 & b_1 & c_1 \\ a_2 & b_2 & c_2 \end{matrix} \right\|} \text{ in determinant form} \ldots \text{Eq. 14.33}$$

Example 31

Let λ be an integer. If the shortest distance between the lines $x - \lambda = 2y - 1 = -2z$ and $x = y + 2\lambda = z - \lambda$ is $\frac{\sqrt{7}}{2\sqrt{2}}$, then find value of λ.

Solution:

Equation of the given lines can be written in standard form as below:

$$\frac{x-\lambda}{1} = \frac{y - \frac{1}{2}}{\frac{1}{2}} = \frac{z}{-\frac{1}{2}} \text{ and } \frac{x}{1} = \frac{y+2\lambda}{1} = \frac{z-\lambda}{1}$$

The respective direction ratios are $\left(1, \frac{1}{2}, -\frac{1}{2}\right)$ and $(1,1,1)$

Since their Direction ratios (denominators in the above equations) are not equal, these are skew line. And coordinates of a point on the two lines are $\left(\lambda, \frac{1}{2}, 0\right)$ and $(0, -2\lambda, \lambda)$ respectively. We know that the shortest distance between two skew lines is

$$d = \frac{\begin{vmatrix} x_1 - x_2 & y_1 - y_2 & z_1 - z_2 \\ a_1 & b_1 & c_1 \\ a_2 & b_2 & c_2 \end{vmatrix}}{\left\| \begin{matrix} \hat{i} & \hat{j} & \hat{k} \\ a_1 & b_1 & c_1 \\ a_2 & b_2 & c_2 \end{matrix} \right\|}$$

Substituting respective values from the given example

$$d=\frac{\begin{vmatrix}\lambda-0 & \frac{1}{2}-(-2)\lambda & 0-\lambda\\ 1 & \frac{1}{2} & -\frac{1}{2}\\ 1 & 1 & 1\end{vmatrix}}{\left\|\begin{matrix}\hat{i} & \hat{j} & \hat{k}\\ 1 & \frac{1}{2} & -\frac{1}{2}\\ 1 & 1 & 1\end{matrix}\right\|}=\frac{\begin{vmatrix}\lambda & \frac{1}{2}+2\lambda & -\lambda\\ 2 & 1 & -1\\ 1 & 1 & 1\end{vmatrix}}{\left\|\begin{matrix}\hat{i} & \hat{j} & \hat{k}\\ 2 & 1 & -1\\ 1 & 1 & 1\end{matrix}\right\|}\Rightarrow$$

$$d=\frac{\lambda(1-(-1))-\left(\frac{1}{2}+2\lambda\right)(2-(-1))+(-\lambda)(2-1)}{\left|\hat{i}(1-(-1))-\hat{j}(2-(-1))+\hat{k}(2-1)\right|}\Rightarrow$$

$$d=\frac{2\lambda-3\left(\frac{1}{2}+2\lambda\right)-\lambda}{\left|2\hat{i}-3\hat{j}-\hat{k}\right|}=\frac{-5\lambda-\frac{3}{2}}{\sqrt{2^2+(-3)^2+(-1)^2}}=\frac{-5\lambda-\frac{3}{2}}{\sqrt{14}}=\frac{\sqrt{7}}{2\sqrt{2}}\text{ (given)}$$

Hence $-5\lambda-\frac{3}{2}=\frac{\sqrt{7}\times\sqrt{14}}{2\sqrt{2}}=\frac{7}{2}\Rightarrow -5\lambda=\frac{7}{2}+\frac{3}{2}\Rightarrow\lambda=-1$ or λ

being an integer, $|\lambda|=1$

14.5.2 Distance between two parallel vectors:

Reference figure 14.17

Since lines l_1 and l_2 are parallel, they are coplanar and let line l_2 be denoted as vector $\vec{b}$ Let $\overrightarrow{a_1}$ and $\overrightarrow{a_2}$ be the position vectors of point A on l_1 and point C on l_2. Hence vector $\overrightarrow{AC}=\vec{a}_1-\vec{a}_2$

$\overrightarrow{AB}$ being perpendicular to both the lines l_1 and l_2, its length equals distance between the two parallel lines. We, therefore, need to find length of line AB.

We can see that magnitude of vector product $\left(\overrightarrow{AC}\times\overrightarrow{BC}\right)$ equal $\left|\overrightarrow{BC}\right|\left|\overrightarrow{AC}\right|\sin\theta=\left|\overrightarrow{BC}\right|\left|\overrightarrow{AB}\right|\Rightarrow$

$$\left|\overrightarrow{AB}\right|=\frac{\left|\overrightarrow{AC}\times\overrightarrow{BC}\right|}{\left|\overrightarrow{BC}\right|}=\left|\frac{(\vec{a}_1-\vec{a}_2)\times\vec{b}}{\left|\vec{b}\right|}\right| \quad \text{Eq 14.35}$$

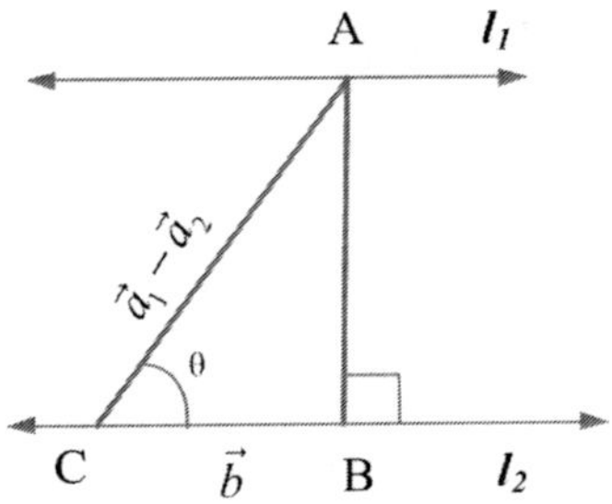

Figure 14.17

14.6 Plane in 3-dimentional space

A plane can be determined uniquely by one of the following known parameters:

(i). If a point on the plane is known and the plane is perpendicular to a known direction, equation of the plane based on these parameters is known as "Equation in the Normal form" Coordinates of the foot of the normal to a plane from any known point and distance of normal to the plane can be found using this equation.

(ii) Three non collinear points are known that lie on the plane.

We shall now see vector and Cartesian form of equations for these type of planes.

14.6.1 Equation of a Plane in Normal form

Polar form:

In figure 14.18, we can see that ABC is a plane and a point Q with position vector $\vec{a}$ and coordinates (x_0,y_0,z_0) lies on the plane. Vector $\vec{N}$ is normal to the plane. Let point P (x,y,z) be an arbitrary point on the plane with position vector $\vec{r}$

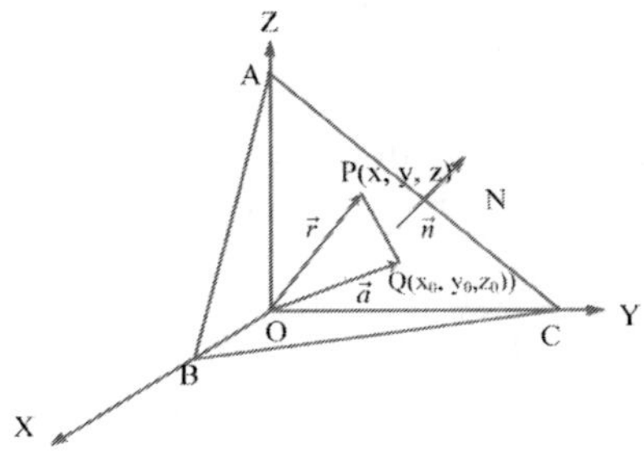

Figure 14.18

Since both Q and P lie on the plane, vector $\overrightarrow{QP}$ will be perpendicular to the normal vector $\vec{n}$ of the plane. Hence their dot product $\overrightarrow{QP}.\vec{n} = 0$ ….(1)

$\overrightarrow{OQ} + \overrightarrow{QP} = \overrightarrow{OP} \Rightarrow \vec{a} + \overrightarrow{QP} = \vec{r} \Rightarrow$

$\overrightarrow{QP} = \vec{r} - \vec{a}$. Substituting this value in eq. (1) above,

$(\vec{r} - \vec{a}).\vec{n} = 0 \Rightarrow \vec{r}.\vec{n} - \vec{a}.\vec{n} = 0 \Rightarrow \vec{r}.\vec{n} = \vec{a}.\vec{n}$ …. (2)

$(\vec{a}.\vec{n})$ ***being a scalar quantity, it can be denoted as D. Hence*** $\vec{r}.\vec{n} = D$.. Eq. 14.36

This is the vector form of the equation of a plane where D is a scalar quantity.

Cartesian Form:

We have seen that coordinates of a point in space are same as direction ratios of its position vector. Since point P *(x,y,z)* lies on the plane, equation of position vector $\vec{r}$ of P ($\overrightarrow{OP}$) is $\vec{r} = x\hat{i} + y\hat{j} + z\hat{k}$. Similarly equation of a known point Q is $\vec{a} = x_0\hat{i} + y_0\hat{j} + z_0\hat{k}$.

Hence $(\vec{r} - \vec{a}) = (x - x_0)\hat{i} + (y - y_0)\hat{j} + (z - z_0)\hat{k}$

Let a, b, c be the direction ratios of the normal vector, then $\vec{n} = a\hat{i} + b\hat{j} + c\hat{k}$

Substituting in eq. (2) above,

$$\{(x - x_0)\hat{i} + (y - y_0)\hat{j} + (z - z_0)\hat{k}\}.(a\hat{i} + b\hat{j} + c\hat{k}) = 0 \Rightarrow$$

$$\{a(x - x_0) + b(y - y_0) + c(z - z_0)\} = 0 \Rightarrow (ax + by + cz) - (ax_0 + by_0 + cz_0) = 0 \Rightarrow$$

$(ax + by + cz) - D = 0$ ***or*** $(ax + by + cz) = D$ **…..** Eq. 14.37

This is Cartesian equation of the plane in normal form where (a, b, c) are direction ratios of the normal vector and D is scalar constant.It is pertinent to note that indices of (x, y, z) in equation of a plane are direction ratios of normal to the plane. This is a very important concept.

Let us recall that equations of a generic line in normal form in Cartesian coordinates is $\frac{x - x_0}{a} = \frac{y - y_0}{b} = \frac{z - z_0}{c} = \lambda$ where a, b, c are its direction ratios and it passes through a known point (x_0, y_0, z_0). If such a line is normal to the plane and the generic point (x, y, z) is foot of the normal on the plane, the above equalities yield coordinates of the generic point as $x = a\lambda + x_0$, $y = b\lambda + y_0$ & $z = c\lambda + z_0$

Substituting these values in the equation of the plane we get

$$a(a\lambda + x_0) + b(b\lambda + y_0) + c(c\lambda + z_0) - D = 0 \Rightarrow,$$

$$\lambda(a^2 + b^2 + c^2) + (ax_0 + by_0 + cz_0 - D) = 0 \Rightarrow$$

$$\lambda = -\frac{ax_0 + by_0 + cz_0 - D}{a^2 + b^2 + c^2} \Rightarrow$$

Equation of the normal from point (x_0, y_0, z_0) to the plane $ax + by + cz - D = 0$ as

$$\frac{x - x_0}{a} = \frac{y - y_0}{b} = \frac{z - z_0}{c} = -\frac{ax_0 + by_0 + cz_0 - D}{a^2 + b^2 + c^2} \quad \text{.......Eq. 14.38}$$

From this equation, knowing direction ratios of the normal (a, b, c) from equation of the plane, coordinates of the foot of the normal (x, y, z) ***from a known external point*** (x_0, y_0, z_0) ***can be derived. This is a very useful equation for solving problems related to normal of a plane from an external point.***

As a special case, if we want to find coordinates of foot of the normal (point N) from origin, we can substitute value of (x_0, y_0, z_0) as (0, 0, 0). Let the coordinates of foot of normal be (α, β, γ). Then distance of the plane from origin = $\sqrt{\alpha^2 + \beta^2 + \gamma^2}$

And equation of the normal $\overrightarrow{ON}$ from eq. 14.38 is

$$\frac{\alpha - 0}{a} = \frac{\beta - 0}{b} = \frac{\gamma - 0}{c} = -\frac{a \times 0 + b \times 0 + c \times 0 - D}{a^2 + b^2 + c^2} \Rightarrow \frac{\alpha}{a} = \frac{\beta}{b} = \frac{\gamma}{c} = \frac{D}{a^2 + b^2 + c^2}$$

Hence $\alpha = \dfrac{aD}{a^2 + b^2 + c^2}, \beta = \dfrac{bD}{a^2 + b^2 + c^2}, \gamma = \dfrac{cD}{a^2 + b^2 + c^2}$ ...(3)

Since $d^2 = \alpha^2 + \beta^2 + \gamma^2$, substituting values of (α, β, γ) from eq. (3), we get

$$d^2 = \left(\frac{aD}{a^2 + b^2 + c^2}\right)^2 + \left(\frac{bD}{a^2 + b^2 + c^2}\right)^2 + \left(\frac{cD}{a^2 + b^2 + c^2}\right)^2 = D^2\left(\frac{a^2 + b^2 + c^2}{(a^2 + b^2 + c^2)^2}\right) \Rightarrow$$

$$d = \frac{D}{\sqrt{a^2 + b^2 + c^2}} \quad \text{........Eq. 14.39}$$

14.6.2 Equation of a plane passing through three non collinear points

Polar form:

Let R, S, T be three non collinear points in a plane ABCD with position vectors $\vec{a}, \vec{b}$ & $\vec{c}$ respectively as shown in figure 14.19. Since vectors $\overrightarrow{RS}$ & $\overrightarrow{RT}$ lie on the given plane, vector $\overrightarrow{RS} \times \overrightarrow{RT}$ will be perpendicular of the plane. Let $\vec{r}$ be the position vector of an arbitrary point P in the plane. Therefore equation of the plane passing through P and perpendicular to the vector $\overrightarrow{RS} \times \overrightarrow{RT}$ will be $(\vec{r} - \vec{a}).\left(\overrightarrow{RS} \times \overrightarrow{RT}\right) = 0$

Now vector $\overrightarrow{RS} = \vec{b} - \vec{a}$ & $\overrightarrow{RT} = \vec{c} - \vec{a}$.

$$\therefore (\vec{r} - \vec{a}).\left[\left(\vec{b} - \vec{a}\right) \times (\vec{c} - \vec{a})\right] = 0 \quad \text{.... Eq. 14.40}$$

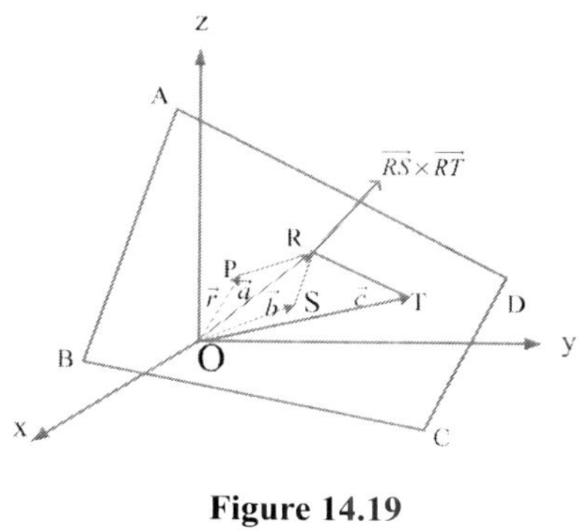

Figure 14.19

This is the equation of the plane passing through three non-collinear points.

It is important to note that there is only one plane that passes through three non-collinear points. And there can be any number of planes passing through three collinear points like pages of a book hinged on one linear edge.

Cartesian form:

Let (x_1, y_1 and z_1), (x_2, y_2 and z_2) and (x_3, y_3 and z_3) be coordinates of points R, S and T respectively. And let (x, y and z) be coordinates of point P. Then

$$\overrightarrow{RP}=(\vec{r}-\vec{a})=(x-x_1)\hat{i}+(y-y_1)\hat{j}+(z-z_1)\hat{k}$$

$$\overrightarrow{RS}=(\vec{b}-\vec{a})=(x_2-x_1)\hat{i}+(y_2-y_1)\hat{j}+(z_2-z_1)\hat{k}$$

$$\overrightarrow{RT}=(\vec{c}-\vec{a})=(x_3-x_1)\hat{i}+(y_3-y_1)\hat{j}+(z_3-z_1)\hat{k}$$

Substituting in the vector equation and using the triple scalar product value, we get

$$\overrightarrow{RP}.\left(\overrightarrow{RS}\times\overrightarrow{RT}\right)=(\vec{r}-\vec{a}).\left[\left(\vec{b}-\vec{a}\right)\times(\vec{c}-\vec{a})\right]=\begin{vmatrix}x-x_1 & y-y_1 & z-z_1\\ x_2-x_1 & y_2-y_1 & z_2-z_1\\ x_3-x_1 & y_3-y_1 & z_3-z_1\end{vmatrix}=0 \quad \text{Eq.14.41}$$

This is the Cartesian equation of the plane passing through points R, S and T.

14.6.3 Intercept form of equation of a plane

We will now find equation of a plane in terms of intercepts made by the plane on the three axes. Let equation of the plane be Ax + By + Cz + D = 0 $(D \neq 0)$..... (1)

As shown in figure 14.20, let e, f and g be the intercepts made by the plane on x, y and z axis respectively. Hence the plane meets x, y and z axis at (*e*, 0, 0), (0, *f*, 0), (0, 0, *g*).

This leads us to $Ae+D=0 \Rightarrow A=\dfrac{-D}{e} \Rightarrow$

$$Bf+D=0 \Rightarrow B=\frac{-D}{f}$$

$$Cg+D=0 \Rightarrow C=\frac{-D}{g}$$

Substituting in eq. (1) above, we get

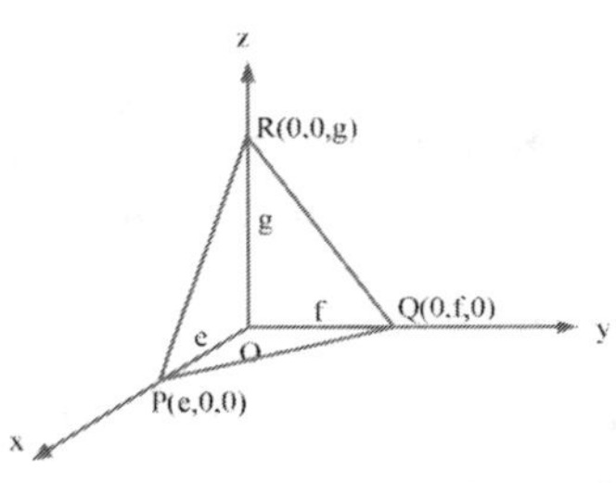

Figure 14.20

$$\frac{-D}{e}x+\frac{-D}{f}y+\frac{-D}{g}z+D=0 \Rightarrow$$

$$(-D)\left\{\frac{x}{e}+\frac{y}{f}+\frac{z}{g}\right\}=-D \Rightarrow \left\{\frac{x}{e}+\frac{y}{f}+\frac{z}{g}\right\}=1 \ldots \text{Eq. } 14.42$$

This is the required equation in the Intercept form and is 3-dimensional extension of 2-dimensional line representation.

14.6.4 Plane passing through intersection of two given planes

Polar form:

Refer figure 14.21

Let equations of the planes A_1 and $A_2$with position vector $\vec{r}$ and normal to vector $\vec{n}$ be $\vec{r}.\hat{n}_1 = d_1$ & $\vec{r}.\hat{n}_2 = d_2$. The position vector of any point on the intersection of the two planes must satisfy both the equations. If $\vec{t}$ is the position vector of an arbitrary point on line of intersection, $\vec{t}.\hat{n}_1 = d_1$ & $\vec{t}.\hat{n}_2 = d_2$

Therefore for all real values of α, $\vec{t}.(\hat{n}_1+\alpha\hat{n}_2)=d_1+\alpha d_2$ .Eq. 14.43

Hence the equation $\vec{t}.(\hat{n}_1+\alpha\hat{n}_2)=d_1+\alpha d_2$ represents the plane A_3 such that if any arbitrary vector $\vec{t}$ satisfies equations for both the planes A_1 and A_2, it also satisfies the equation for plane A_3, implying thereby that ***any plane passing through the intersection of planes A_1 and A_2 with equations,*** $\vec{t}.\hat{n}_1 = d_1$ & $\vec{t}.\hat{n}_2 = d_2$

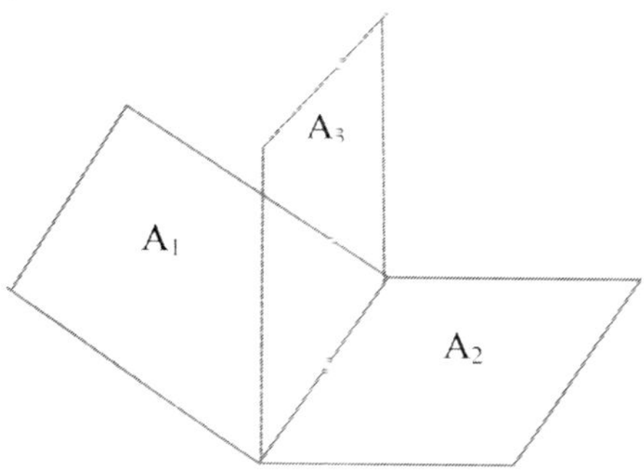

Figure 14.21

Hence the equation $\vec{r}.(\hat{n}_1 + \alpha\hat{n}_2) = d_1 + \alpha d_2$ represents the plane A_3 such that if any arbitrary vector $\vec{r}$ satisfies equations for both the planes A_1 and A_2, it also satisfies the equation for plane A_3, implying thereby that ***any plane passing through the intersection of planes A_1 and A_2 with equations, $\vec{r}.\hat{n}_1 = d_1$ & $\vec{r}.\hat{n}_2 = d_2$ has the equation $\vec{r}.(\hat{n}_1 + \alpha\hat{n}_2) = d_1 + \alpha d_2$, where α is a non-zero real number***

Cartesian form: .

Let Cartesian equations of vectors $\hat{n}_1$, $\hat{n}_2$ & $\vec{r}$ be

$\hat{n}_1 = a_1\hat{i} + b_1\hat{j} + c_1\hat{k}$, $\hat{n}_2 = a_2\hat{i} + b_2\hat{j} + c_2\hat{k}$ and $\vec{r} = x\hat{i} + y\hat{j} + z\hat{k}$

Substituting in the polar equation of the plane, we get

$$\left(x\hat{i} + y\hat{j} + z\hat{k}\right).\left\{\left(a_1\hat{i} + b_1\hat{j} + c_1\hat{k}\right) + \alpha\left(a_2\hat{i} + b_2\hat{j} + c_2\hat{k}\right)\right\} = d_1 + \alpha d_2 \Rightarrow$$

$$\left(x\hat{i} + y\hat{j} + z\hat{k}\right).\left\{(a_1 + \alpha a_2)\hat{i} + (b_1 + \alpha b_2)\hat{j} + (c_1 + \alpha c_2)\hat{k}\right\} = d_1 + \alpha d_2 \Rightarrow$$

$$\left\{x(a_1 + \alpha a_2) + y(b_1 + \alpha b_2) + z(c_1 + \alpha c_2)\right\} = d_1 + \alpha d_2 \Rightarrow$$

$$\left\{(a_1x + b_1y + c_1z - d_1) + \alpha(a_2x + b_2y + c_2z - d_2)\right\} = 0 \text{ ....Eq. 14.44}$$

This is the required equation of the plane in Cartesian form.

Example 32

Find equation of the plane perpendicular to the planes 3x+y–2z+1 = 0 and 2x-5y-z+3 = 0 such that it passes through point (1, 2,-3).

Solution:

Equation of Direction vectors of normal $\vec{N}$ perpendicular to both the planes can be found by cross product of the normal vectors of the two planes viz

$$\begin{vmatrix} \hat{i} & \hat{j} & \hat{k} \\ 3 & 1 & -2 \\ 2 & -5 & -1 \end{vmatrix} = \hat{i}(-1-10) - \hat{j}(-3+4) + \hat{k}(-15-2) = -11\hat{i} - \hat{j} - 17\hat{k}$$

Thus (-11,-1,-17) are the Direction ratios of the normal to the plane.

Since the plane passes through point (1,2,-3). Equation of the plane will be $a(x - x_1) + b(y - y_1) + c(z - z_1) = 0$, where a, b and c are the direction ratios of its normal.

So the equation is of the plane is

$$-11(x-1)+(-1)(y-2)+(-17)(z-(-3))=-11x+11-y+2-17z-51=0\Rightarrow$$

The equation as

$$11x-11+y-2+17z+51=0\Rightarrow 11x+y+17z+38=0$$

Example 33

Find equation of the plane passing through point (1,0,2) and intersection of planes $\vec{r}.(\hat{i}+\hat{j}+\hat{k})=1$ and $\vec{r}.(\hat{i}-2\hat{j})=-2$

Solution:

In Cartesian form, equation of the two given planes are $(x\hat{i}+y\hat{j}+z\hat{k}).(\hat{i}+\hat{j}+\hat{k})=1\Rightarrow x+y+z-1=0$ and

$$(x\hat{i}+y\hat{j}+z\hat{k}).(\hat{i}-2\hat{j})=-2\Rightarrow x-2y+2=0$$

Equation of the plane passing through intersection of these two planes will be $(x+y+z-1)+\alpha(x-2y+2)=0$

Since this plane passes through the point (1, 0, 2),

$$(1+2-1)+\alpha(1+2)=0\Rightarrow\alpha=-\frac{2}{3}$$

Hence equation of the required plane is

$$(x+y+z-1)-\frac{2}{3}(x-2y+2)=0\Rightarrow 3x+3y+3z-3-2x+4y-4=0\Rightarrow$$

$$x+7y+3z-7=0$$

In polar form, the equation can be written as $\vec{r}.(\hat{i}+7\hat{j}+3\hat{k})=7$

14.6.5 Co-planarity of two lines

Let there be two lines, with line-1 passing through point A with position vector $\vec{a}_1$ and parallel to vector $\vec{b}_1$ and line-2 passing through point B with position vector $\vec{a}_2$ and parallel to vector $\vec{b}_2$

Vector $\overrightarrow{AB}=\vec{a}_1-\vec{a}_2$

The given lines will be co-planar only if vector $\overrightarrow{AB}$ is perpendicular to the cross product of vectors $\vec{b}_1$ & $\vec{b}_2$

Hence equation $(\vec{a}_1 - \vec{a}_2).(\vec{b}_1 \times \vec{b}_2) = 0$ represents the condition for co-planarity of the two lines.

Cartesian form:

Let (x_1, y_1, z_1) and (x_2, y_2, z_2) be coordinates of points A and B and let (a_1, b_1, c_1) and (a_2, b_2, c_2) be the direction ratios of the lines $\vec{b}_1$ & $\vec{b}_2$ respectively. Hence

$$\overrightarrow{AB} = (x_2 - x_1)\hat{i} + (y_2 - y_1)\hat{j} + (z_2 - z_1)\hat{k}$$

$$\vec{b}_1 = a_1\hat{i} + b_1\hat{j} + c_1\hat{k} \text{ and } \vec{b}_2 = a_2\hat{i} + b_2\hat{j} + c_2\hat{k}$$

The given lines are co-planar if and only if $\overrightarrow{AB}.(\vec{b}_1 \times \vec{b}_2) = 0$

Then Cartesian form of this condition is

$$\overrightarrow{AB}.(\vec{b}_1 \times \vec{b}_2) = \begin{vmatrix} x_2 - x_1 & y_2 - y_1 & z_2 - z_1 \\ a_1 & b_1 & c_1 \\ a_2 & b_2 & c_2 \end{vmatrix} = 0 \quad \text{..........Eq 14.45}$$

14.6.6 Angle between two planes:

Vector form:

Angle between two planes is defined as angle between their normals. Angle could be acute or obtuse. Normally the acute angle is considered,

Reference figure 14.22. Here $\vec{n}_1$ and $\vec{n}_2$ be the normals to the two planes with angle θ between them. Let $\vec{r}$ be an arbitrary direction vector of a point on intersection of the planes and d_1 and d_2 be the distance of the planes from centre of the coordinate system. From the given data we get $\vec{r}.\hat{n}_1 = d_1$ and $\vec{r}.\hat{n}_2 = d_2$

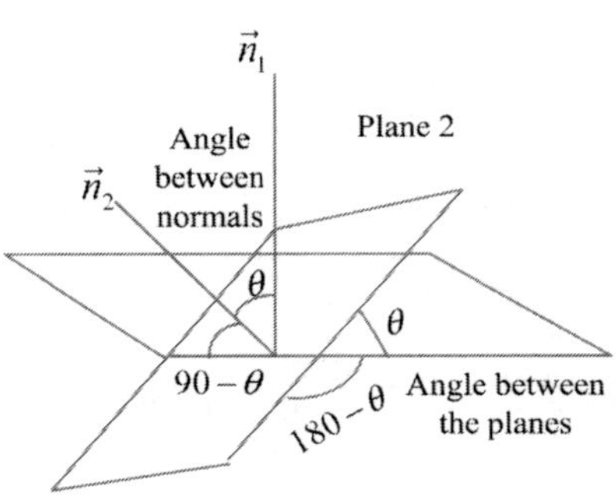

Figure 14.22

From these two equations, we get $\cos\theta = \left|\frac{\vec{n}_1.\vec{n}_2}{|\vec{n}_1||\vec{n}_2|}\right|$

Thus the planes are perpendicular to each other if $\vec{n}_1.\vec{n}_2 = 0$ and parallel if $\vec{n}_1$ is parallel to $\vec{n}_2$ ......Eq. 14.46

Cartesian form:

Let equation of the planes be $a_1x + b_1y + c_1z + d_1 = 0$ and $a_2x + b_2y + c_2z + d_2 = 0$

Thus direction ratios of the normals are a_1, b_1, c_1 and a_2, b_2, c_2 respectively yielding $\vec{n}_1 = a_1\hat{i} + b_1\hat{j} + c_1\hat{k}$ & $\vec{n}_2 = a_2\hat{i} + b_2\hat{j} + c_2\hat{k}$

Therefore, $\cos\theta = \left|\frac{(a_1\hat{i} + b_1\hat{j} + c_1\hat{k}).(a_2\hat{i} + b_2\hat{j} + c_2\hat{k})}{|a_1\hat{i} + b_1\hat{j} + c_1\hat{k}||a_2\hat{i} + b_2\hat{j} + c_2\hat{k}|}\right|$ or

$$\cos\theta = \left|\frac{a_1a_2 + b_1b_2 + c_1c_2}{\sqrt{a_1^2 + b_1^2 + c_1^2}\sqrt{a_2^2 + b_2^2 + c_2^2}}\right|$$

This is exactly the same formula we derived in section 14.4 above for angle between two vectors.Hence we can conclude that

- If planes are at right angle $a_1a_2 + b_1b_2 + c_1c_2 = 0$ ....Eq. 14.47
- If planes are parallel $\frac{a_1}{a_2} = \frac{b_1}{b_2} = \frac{c_1}{c_2}$ ....Eq. 14.48

14.6.7 Distance of a point from a plane:

Vector form:

Refer Figure 14.23 below.

Consider a plane A_1 with point A on it with position vector $\vec{r}$ and vector ON $(\vec{n})$ of length d as perpendicular to the plane from the origin. Equation of plane A_1 will be $\vec{r}.\hat{n} = d$.

Let P be an arbitrary point in space with position vector $\vec{a}$ and let PN′ be perpendicular from P to the plane A_1. Consider another plane A_2 containing point P and parallel to the plane A_1.Let Vector ON meet plane A_2 at M such that $OM = d'$, being distance of plane A_2 from the origin. Hence equation of plane A_2 is $\vec{a}.\hat{n} = d'$.

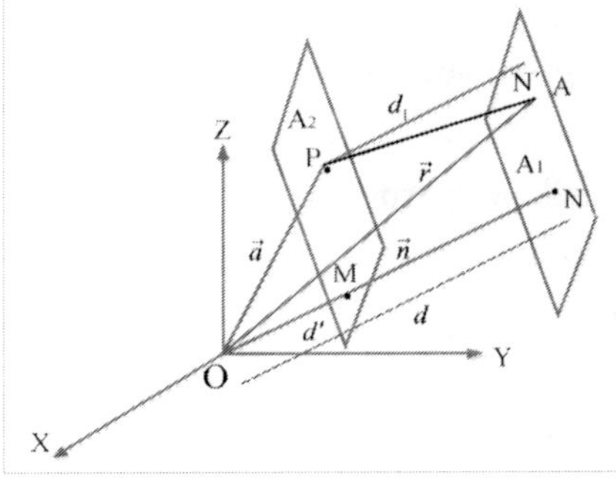

Figure 14.23

Let distance of point P from plane $A_1 = PN' = d_1$. We need to find d_1.

Since both Planes A_1 and A_2 are parallel to each other, ON and PN' will also beparallel as they are perpendicular to Planes A_1 and A_2, respectively and $PN' = MN$, both being perpendicular distance between two parallel planes.

Hence $PN' = d - d'$, leading to the equation of distance to the given plane A_1 from point P $= \vec{r}.\hat{n} - \vec{a}.\hat{n} = \left|\hat{n}.(\vec{r} - \vec{a})\right|$

Alternatively, we can visualize that vector $\overrightarrow{PA} = \vec{r} - \vec{a}$. Since PN' is normal to plane A_1, distance of point P from the Plane $A_1 = \left|\overrightarrow{PN'}\right| = \left|\overrightarrow{PA}.\hat{n}\right| = \left|(\vec{r} - \vec{a}).\hat{n}\right|$ ..Eq. 14.49

Cartesian form

Let $P(x_1, y_1, z_1)$ be the given point with position vector $\vec{a}$ in plane $A_2$and $ax + by + cz - D = 0$ be the Cartesian equation of the given plane A_1. Hence a, b and c are direction ratios of the normal $\vec{n}$.

Then $\vec{a} = \left(x_1\hat{i} + y_1\hat{j} + z_1\hat{k}\right)$ and $\vec{n} = a\hat{i} + b\hat{j} + c\hat{k}$ leading to unit vector $\hat{n} = \dfrac{a\hat{i} + b\hat{j} + c\hat{k}}{\left|\sqrt{a^2 + b^2 + c^2}\right|}$

Further $d = \dfrac{D}{|\vec{n}|} = \dfrac{D}{\left|\sqrt{a^2 + b^2 + c^2}\right|}$

Hence length of perpendicular to the plane A_2 from origin is

$$\vec{a}.\hat{n} = d' \Rightarrow \left(x_1\hat{i} + y_1\hat{j} + z_1\hat{k}\right).\frac{a\hat{i} + b\hat{j} + c\hat{k}}{\left|\sqrt{a^2 + b^2 + c^2}\right|} = \frac{ax_1 + by_1 + cz_1}{\left|\sqrt{a^2 + b^2 + c^2}\right|} = d'$$

Perpendicular from P to the plane $A_1 = d - d'$

Hence distance of plane A_1 from point

$$P = \frac{D}{\left|\sqrt{a^2 + b^2 + c^2}\right|} - \frac{ax_1 + by_1 + cz_1}{\left|\sqrt{a^2 + b^2 + c^2}\right|}$$

$$= \left|\frac{ax_1 + by_1 + cz_1 - D}{\left|\sqrt{a^2 + b^2 + c^2}\right|}\right| \text{..... Eq. 14.50}$$

Example 34

Find distance of a point P (1,1,9) from point of intersection of plane x + y + z = 17 and the line $\frac{x-3}{1} = \frac{y-4}{2} = \frac{z-5}{2}$

Solution:

Let $\frac{x-3}{1} = \frac{y-4}{2} = \frac{z-5}{2} = \lambda \Rightarrow x = \lambda + 3,\ y = 2\lambda + 4,\ z = 2\lambda + 5$

Since this point lies on the plane, we get

$\lambda + 3 + 2\lambda + 4 + 2\lambda + 5 = 17 \Rightarrow 5\lambda + 12 = 17 \Rightarrow \lambda = 1$

So the point of intersection is

$x = 1 + 3 = 4,\ y = 2 + 4 = 6,\ z = 2 + 5 = 7,\ (4,6,7)$

Hence the distance of point P

$= \sqrt{(4-1)^2 + (6-1)^2 + (7-9)^2} = \sqrt{9 + 25 + 4} = \sqrt{38}$

Example 35

Find all vectors of magnitude $10\sqrt{3}$ that are perpendicular to the plane of vectors $\hat{i} + 2\hat{j} + \hat{k}$ and $-\hat{i} + 3\hat{j} + 4\hat{k}$

Solution:

Let the two given vectors be identified as $\vec{a} = \hat{i} + 2\hat{j} + \hat{k}$ and $\vec{b} = -\hat{i} + 3\hat{j} + 4\hat{k}$

Equation of the vector perpendicular to both these vectors will be given by $\vec{a}\times\vec{b}$

$$\vec{a}\times\vec{b}=\begin{vmatrix}\hat{i} & \hat{j} & \hat{k}\\ 1 & 2 & 1\\ -1 & 3 & 4\end{vmatrix}=\hat{i}\times(2\times4-3\times1)-\hat{j}\times(1\times4-(-1)\times1)+\hat{k}\times(1\times3-(-1)\times2)\Rightarrow$$

Or $\vec{a}\times\vec{b}=5\hat{i}-5\hat{j}+5\hat{k}\Rightarrow\left|\vec{a}\times\vec{b}\right|=\sqrt{5^2+(-5)^2+5^2}=\sqrt{75}=5\sqrt{3}$

Unit vectors perpendicular to both vectors $\vec{a}\,\&\,\vec{b}$ will be $\pm\dfrac{5\hat{i}-5\hat{j}+5\hat{k}}{5\sqrt{3}}=\pm\dfrac{\hat{i}-\hat{j}+\hat{k}}{\sqrt{3}}$

Hence the required vectors at a distance of $10\sqrt{3}$ or of length $10\sqrt{3}$ will be $\pm10\sqrt{3}\times\dfrac{\hat{i}-\hat{j}+\hat{k}}{\sqrt{3}}=\pm10\left(\hat{i}-\hat{j}+\hat{k}\right)$

Example 36

A plane passes through three points A (1, 2, 3), B (2, 3, 1) and C (2, 4, 2). If O is the origin and P is (2, -1, 1), find length of projection of $\overrightarrow{OP}$ on the plane.

Solution:

We know that equation of the plane passing through three known points with coordinates (x_1,y_1,z_1), $(x_2,y_2,z_2)\,\&\,(x_3,y_3,z_3)$ is

$$\begin{vmatrix}x-x_1 & y-y_1 & z-z_1\\ x_2-x_1 & y_2-y_1 & z_2-z_1\\ x_3-x_1 & y_3-y_1 & z_3-z_1\end{vmatrix}=0\Rightarrow\begin{vmatrix}x-1 & y-2 & z-3\\ 2-1 & 3-2 & 1-3\\ 2-1 & 4-2 & 2-3\end{vmatrix}=0\Rightarrow$$

$$(x-1)(3)-(y-2)(1)+(z-3)(1)=0\Rightarrow 3x-y+z-4=0$$

Since $\overrightarrow{OP}$ is a vector in space, its projection on the given plane will be $\overrightarrow{NP'}$ where N is foot of the normal from the origin and P′ is the foot of the perpendicular from point P. Hence ON and PP′ are parallel.

We have learnt in section 14.6.1that if $ax+by+cz=D$ be the Cartesian equation of a plane, a, b and c are direction ratios of the normal $\vec{N}$ to the plane.

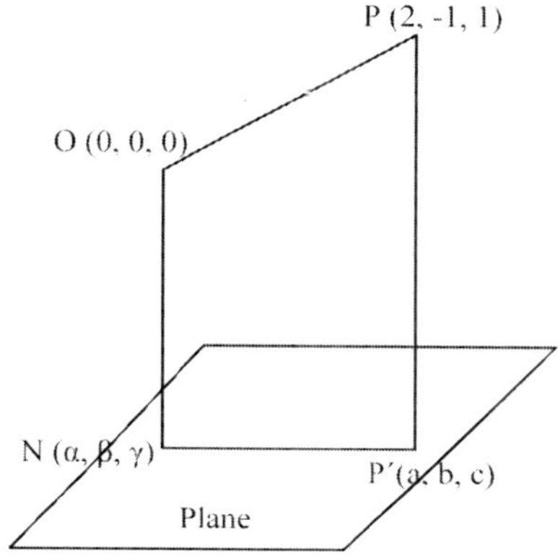

Comparing equation of the given plane derived above with standard equation of the plane in normal form, we can see that direction ratios of foot of the normal to the plane (N)are (3, -1, 1)

We can now use the following equation of a line in normal form derived in section 14.6.1

$$\frac{x-x_1}{a}=\frac{y-y_1}{b}=\frac{z-z_1}{c}=-\frac{\left(ax_1+by_1+cz_1-d\right)}{\left(a^2+b^2+c^2\right)}$$

where a, b, c are direction ratios and (x_1, y_1, z_1) is the external point.

If α, β and γ are the coordinates of point N, we get equation of normal vector $\overrightarrow{ON}$ as

$$\frac{\alpha-0}{3}=\frac{\beta-0}{-1}=\frac{\gamma-0}{1}=-\frac{\left(3\times 0+(-1)\times 0+1\times 0-4\right)}{3^2+(-1)^2+1^2}=\frac{4}{11}\Rightarrow \alpha=\frac{12}{11},\beta=\frac{-4}{11},\gamma=\frac{4}{11}$$

Similarly if a, b and c are coordinates of foot of perpendicular P′ from point P, we get PP' as

$$\frac{a-2}{3}=\frac{b+1}{-1}=\frac{c-1}{1}=-\frac{\left(3\times 2+(-1)(-1)+1\times 1-4\right)}{11}=-\frac{4}{11}\Rightarrow a=\frac{10}{11},\ b=\frac{-7}{11},\ c=\frac{7}{11}$$

Knowing coordinates of point N and P′, projection of OP on the plane, NP′ equals

$$\sqrt{\left(\frac{12}{11}-\frac{10}{11}\right)^2+\left(\frac{-4}{11}-\frac{-7}{11}\right)^2+\left(\frac{4}{11}-\frac{7}{11}\right)^2}=\sqrt{\left(\frac{2}{11}\right)^2+\left(\frac{3}{11}\right)^2+\left(\frac{-3}{11}\right)^2}=\sqrt{\frac{22}{121}}=\sqrt{\frac{2}{11}}$$

14.6.8 Angle between a line and a plane

Refer to figure 14.24. It will be noticed that angle between a line and a plane is the complement of the angle between the line and normal to the plane, shown as θ.

Hence angle between line and the plane is 90 – θ.

If equation of the line is $\vec{r} = \vec{a} + \lambda\vec{b}$ and the equation of the plane is $\vec{r}.\hat{n} = d$. Then the angle θbetween the line and normal to the plane is given as $\cos\theta = \left|\frac{\vec{b}\,.\,\hat{n}}{|\vec{b}||\hat{n}|}\right|$

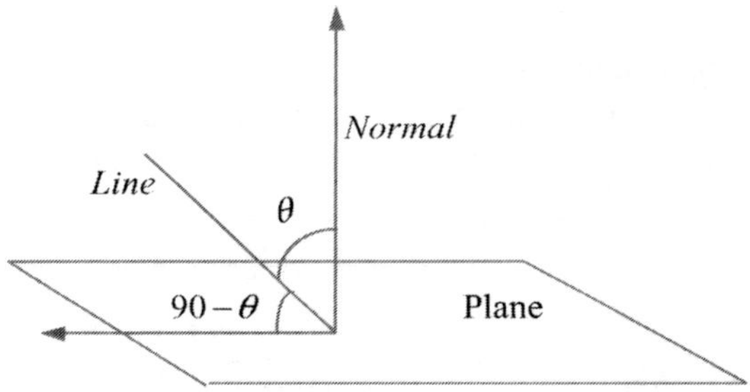

Figure 14.24

If angle between the line and the plane is denoted as ϕ , then $\phi = 90 - \theta$. Hence $\text{Sin}\,\phi = \cos\theta = \left|\frac{\vec{b}\,.\,\hat{n}}{|\vec{b}||\hat{n}|}\right|$ Or $\phi = \sin^{-1}\left|\frac{\vec{b}\,.\,\hat{n}}{|\vec{b}||\hat{n}|}\right|$. 14.51

14.7 Practice problems:

1. If $\vec{a}, \vec{b}$ & $\vec{c}$ are three mutually perpendicular unit vectors, prove that $\left|\vec{a} + \vec{b} + \vec{c}\right| = \sqrt{3}$
2. Let α be the angle between two lines, say AB and BC, whose direction cosines satisfy the equations $l + m - n = 0$ and $l^2 + m^2 - n^2 = 0$. Find value of $\sin^4\alpha + \cos^4\alpha$
3. Find a unit vector perpendicular to both the vectors $\vec{a} = 4\hat{i} - \hat{j} + 3\hat{k}$ and $\vec{b} = -2\hat{i} + \hat{j} - 2\hat{k}$.

4. If $\vec{a}$ & $\vec{b}$ are the two vectors such that $|\vec{a}| = 4$ and $|\vec{b}| = 3$ and $\vec{a}.\vec{b} = 6$, find the angle between them

5. Let $\vec{a} = \hat{i} + \alpha\hat{j} + 3\hat{k}$ and $\vec{b} = 3\hat{i} - \alpha\hat{j} + \hat{k}$. If area of parallelogram whose adjacent sides are represented by vectors $\vec{a}$ & $\vec{b}$ is $8\sqrt{3}$, then find value of $|\vec{a}.\vec{b}|$

6. If $\vec{a} \perp \vec{b}$ then find the value of $\vec{a} \times \left(\vec{a} \times \left(\vec{a} \times \left(\vec{a} \times \vec{b}\right)\right)\right)$

7. If A (1, 5, 35), B (7, 5, 5), C (1, λ, 7) and D (2λ, 1, 2) are coplanar. Find sum of all possible values of λ

8. Let (λ, 2, 1) be a point on the plane which passes through (4, -2, 2). If the plane is perpendicular to the line joining (2, -21, 29) and (-1, -16, 23), then find the value of $\left(\left(\frac{\lambda}{11}\right)^2 - \frac{4\lambda}{11} - 4\right)$

9. If point (α, β, γ) is the image of point (1, 3, 5) to the plane 4x – 5y + 2 z = 8, find value of $5(\alpha + \beta + \gamma)$

10. Find the coordinates of the foot of the perpendicular from the points $(1, -2, 1)$ on the plane containing the lines, $\frac{x+1}{6} = \frac{y-1}{7} = \frac{z-3}{8}$ & $\frac{x-1}{3} = \frac{y-2}{5} = \frac{z-3}{7}$

11. Vectors $\hat{i} + y\hat{j} + \hat{k}$ & $x\hat{i} - \hat{j} + z\hat{k}$ are parallel. Find unit vector parallel to $x\hat{i} - y\hat{j} + z\hat{k}$

12. Let L be a line obtained from the intersection of two planes $x + 2y + z = 6$ & . $y + 2z = 4$. If (α, β, γ) is foot of the perpendicular from $(3, 2, 1)$ on L, then find value of $21(\alpha + \beta + \gamma)$

13. Let $\vec{a} = 2\hat{i} + \lambda_1\hat{j} + 3\hat{k}$, $\vec{b} = 4\hat{i} + (3 - \lambda_2)\hat{j} + 6\hat{k}$ & $\vec{c} = 3\hat{i} + 6\hat{j} + (\lambda_3 - 1)\hat{k}$ be three vectors such that $\vec{b} = 2\vec{a}$ & $\vec{a} \perp \vec{c}$, then find a possible value of $(\lambda_1, \lambda_2, \lambda_3)$

14. Direction cosines of two lines satisfy the equations $(i)\, 2l + 2m - n = 0$ & $(ii)\, lm + mn + nl = 0$. Find angle between the lines

15. Let $\hat{a}, \hat{b}, \hat{c}$ be three unit vectors such that $\left|\left(\hat{a}-\hat{b}\right)^2\right| + \left|\left(\hat{a}-\hat{c}\right)^2\right| = 8$. Find the value of $\left|\hat{a}+2\hat{b}\right|^2 + \left|\hat{a}+2\hat{c}\right|^2$

16. Find area of the parallelogram formed by the vectors $\vec{a} = 3\hat{i} + \hat{j} - 2\hat{k}$ and $-2\hat{i} + \hat{j} + 2\hat{k}$

17. If $|\vec{a}| = \sqrt{3}, \left|\vec{b}\right| = 5, \vec{b}.\vec{c} = 10$ angle between $\vec{b}$ & $\vec{c}$ is $\frac{\pi}{3}$. If $\vec{a} \perp \left(\vec{b} \times \vec{c}\right)$, find value of $\left|\vec{a} \times \left(\vec{b} \times \vec{c}\right)\right|$

18. Let $\vec{a} = \hat{i} - 2\hat{j} + \hat{k}, \vec{b} = \hat{i} - \hat{j} + \hat{k}$ If $\vec{c}$ is a vector such that $\vec{b} \times \vec{c} = \vec{b} \times \vec{a}$ and $\vec{a}.\vec{c} = 0$, find value of $\vec{c}.\vec{b}$?

19. Let $\vec{u}$ be a vector coplanar with vectors $\vec{a} = 2\hat{i} + 3\hat{j} - \hat{k}$ & $\vec{b} = \hat{j} + \hat{k}$ If vector $\vec{u}$ is perpendicular vector $\vec{a}$ and $\vec{u}.\vec{b} = 24$, find the value of $|\vec{u}|^2$

20. Let $\hat{a}$ & $\hat{b}$ be unit vectors such that $\left|\hat{a}+\hat{b}\right| = \frac{3}{2}$, then find the value of $\left(2\hat{a}+7\hat{b}\right).\left(4\hat{a}+3\hat{b}+2020\left(\hat{a}\times\hat{b}\right)\right)$

21. Let $\vec{q}$ & $\vec{r}$ be non-collinear vectors. If $\vec{p}.(\vec{q}+\vec{r}) = 4$ and $\vec{p} \times (\vec{q} \times \vec{r}) = \left(x^2 - 2x + 6\right)\vec{q} + (\sin y)\vec{r}$, find the line on which point (x, y) lies?

22. For three vectors $\vec{a}, \vec{b}, \vec{c}$, if $|\vec{a}| = 2$, $\left|\vec{b}\right| = 1$, $\left(\vec{a} \times \vec{b}\right) = \vec{c}$ and $\left(\hat{b} \times \vec{c}\right) = \vec{a}$, find value of $\left[\vec{a}+\vec{b} \quad \vec{b}+\vec{c} \quad \vec{c}+\vec{a}\right]$

23. Let L be the line passing through intersection of planes $3x - y + 2z + 1 = 0$ and $3x - 2y + z = 3$. Then find equation of the plane passing through point (2, 1, 4) and perpendicular to the line L

24. Find the interval in which lies the shortest distance between the lines $\frac{x-2}{2} = \frac{y-3}{2} = \frac{z-0}{1}$ and $\frac{x+4}{-1} = \frac{y-7}{8} = \frac{z-5}{4}$

25. Find distance of point $(1, -2, 3)$ from the plane $x - y + z = 5$ measured parallel to the line $\frac{x}{2} = \frac{y}{3} = \frac{z}{-6}$

26. If foot of the perpendicular drawn from $(1, 0, 3)$ on a line passing through $(\alpha, 7, 1)$ is $\left(\frac{5}{3}, \frac{7}{3}, \frac{17}{3}\right)$, find α ?

27. If a point $R(4, y, z)$ lies on the line segment joining the points $P\ (2,-3,4)$ and $Q\ (8,0,10)$, find distance of point R from the origin?

28. Let ABC be a Δ, position vectors of whose vertices A, B, C are respectively, $\hat{i}+2\hat{j}+4\hat{k}$, $-2\hat{i}+2\hat{j}+\hat{k}$ & $2\hat{i}+4\hat{j}-3\hat{k}$. Find the type of the triangle.

29. For non-zero vectors $\vec{a}$, $\vec{b}$, $\vec{c}$, if $\left|\left(\vec{a}\times\vec{b}\right).\vec{c}\right| = |\vec{a}||\vec{b}||\vec{c}|$, find which of the following identities is true:

    (i) $\vec{a}.\vec{b}=0,\ \vec{b}.\vec{c}=0$

    (ii) $\vec{c}.\vec{a}=0,\ \vec{b}.\vec{c}=0$

    (iii) $\vec{c}.\vec{a}=0,\ \vec{a}.\vec{b}=0$

    (iv) $\vec{a}.\vec{b}=\vec{b}.\vec{c}=\vec{c}.\vec{a}=0$

30. Find equation of a plane passing through points (2, 2, 1), (9, 3, 6) and Perpendicular to the plane $2x+6y+6z-1=0$

Answers to Practice Problems:

1. $\sqrt{3}$
2. $\dfrac{5}{8}$
3. $\dfrac{-\hat{i}+2\hat{j}+2\hat{k}}{3}$
4. 60^0
5. 2
6. $|\vec{a}|^4\,\vec{b}$
7. $\dfrac{44}{5}$
8. 8
9. 47
10. (0, 0, 0)
11. $\dfrac{1}{\sqrt{3}}\left(\hat{i}+\hat{j}+\hat{k}\right)$
12. 102
13. $\left(-\dfrac{1}{2},4,0\right)$
14. $\dfrac{\pi}{2}$
15. 2
16. $10\sqrt{3}$ sq. units
17. 30
18. $-\dfrac{1}{2}$
19. 336
20. $\dfrac{133}{4}$
21. $x=1$
22. 8
23. $x+y-z+1=0$
24. $(2,3]$
25. 1
26. 4
27. $2\sqrt{14}$
28. Right Angled triangle.
29. (iv) $\vec{a}.\vec{b}=\vec{b}.\vec{c}=\vec{c}.\vec{a}=0$
30. $3x+4y-5z-9=0$